# Introduction to Process Control

**CHEMICAL INDUSTRIES**

A Series of Reference Books and Textbooks

*Consulting Editor*

**HEINZ HEINEMANN**
*Berkeley, California*

1. *Fluid Catalytic Cracking with Zeolite Catalysts,* Paul B. Venuto and E. Thomas Habib, Jr.
2. *Ethylene: Keystone to the Petrochemical Industry,* Ludwig Kniel, Olaf Winter, and Karl Stork
3. *The Chemistry and Technology of Petroleum,* James G. Speight
4. *The Desulfurization of Heavy Oils and Residua,* James G. Speight
5. *Catalysis of Organic Reactions,* edited by William R. Moser
6. *Acetylene-Based Chemicals from Coal and Other Natural Resources,* Robert J. Tedeschi
7. *Chemically Resistant Masonry,* Walter Lee Sheppard, Jr.
8. *Compressors and Expanders: Selection and Application for the Process Industry,* Heinz P. Bloch, Joseph A. Cameron, Frank M. Danowski, Jr., Ralph James, Jr., Judson S. Swearingen, and Marilyn E. Weightman
9. *Metering Pumps: Selection and Application,* James P. Poynton
10. *Hydrocarbons from Methanol,* Clarence D. Chang

11. *Form Flotation: Theory and Applications,* Ann N. Clarke and David J. Wilson
12. *The Chemistry and Technology of Coal,* James G. Speight
13. *Pneumatic and Hydraulic Conveying of Solids,* O. A. Williams
14. *Catalyst Manufacture: Laboratory and Commercial Preparations,* Alvin B. Stiles
15. *Characterization of Heterogeneous Catalysts,* edited by Francis Delannay
16. *BASIC Programs for Chemical Engineering Design,* James H. Weber
17. *Catalyst Poisoning,* L. Louis Hegedus and Robert W. McCabe
18. *Catalysis of Organic Reactions,* edited by John R. Kosak
19. *Adsorption Technology: A Step-by-Step Approach to Process Evaluation and Application,* edited by Frank L. Slejko
20. *Deactivation and Poisoning of Catalysts,* edited by Jacques Oudar and Henry Wise
21. *Catalysis and Surface Science: Developments in Chemicals from Methanol, Hydrotreating of Hydrocarbons, Catalyst Preparation, Monomers and Polymers, Photocatalysis and Photovoltaics,* edited by Heinz Heinemann and Gabor A. Somorjai
22. *Catalysis of Organic Reactions,* edited by Robert L. Augustine
23. *Modern Control Techniques for the Processing Industries,* T. H. Tsai, J. W. Lane, and C. S. Lin
24. *Temperature-Programmed Reduction for Solid Materials Characterization,* Alan Jones and Brian McNichol
25. *Catalytic Cracking: Catalysts, Chemistry, and Kinetics,* Bohdan W. Wojciechowski and Avelino Corma
26. *Chemical Reaction and Reactor Engineering,* edited by J. J. Carberry and A. Varma
27. *Filtration: Principles and Practices: Second Edition,* edited by Michael J. Matteson and Clyde Orr
28. *Corrosion Mechanisms,* edited by Florian Mansfeld
29. *Catalysis and Surface Properties of Liquid Metals and Alloys,* Yoshisada Ogino

30. *Catalyst Deactivation,* edited by Eugene E. Petersen and Alexis T. Bell
31. *Hydrogen Effects in Catalysis: Fundamentals and Practical Applications,* edited by Zoltán Paál and P. G. Menon
32. *Flow Management for Engineers and Scientists,* Nicholas P. Cheremisinoff and Paul N. Cheremisinoff
33. *Catalysis of Organic Reactions,* edited by Paul N. Rylander, Harold Greenfield, and Robert L. Augustine
34. *Powder and Bulk Solids Handling Processes: Instrumentation and Control,* Koichi Iinoya, Hiroaki Masuda, and Kinnosuke Watanabe
35. *Reverse Osmosis Technology: Applications for High-Purity-Water Production,* edited by Bipin S. Parekh
36. *Shape Selective Catalysis in Industrial Applications,* N. Y. Chen, William E. Garwood, and Frank G. Dwyer
37. *Alpha Olefins Applications Handbook,* edited by George R. Lappin and Joseph L. Sauer
38. *Process Modeling and Control in Chemical Industries,* edited by Kaddour Najim
39. *Clathrate Hydrates of Natural Gases,* E. Dendy Sloan, Jr.
40. *Catalysis of Organic Reactions,* edited by Dale W. Blackburn
41. *Fuel Science and Technology Handbook,* edited by James G. Speight
42. *Octane-Enhancing Zeolitic FCC Catalysts,* Julius Scherzer
43. Oxygen in Catalysis, Adam Bielanski and Jerzy Haber
44. *The Chemistry and Technology of Petroleum: Second Edition, Revised and Expanded,* James G. Speight
45. *Industrial Drying Equipment: Selection and Application,* C. M. van't Land
46. *Novel Production Methods for Ethylene, Light Hydrocarbons, and Aromatics,* edited by Lyle F. Albright, Billy L. Crynes, and Siegfried Nowak
47. *Catalysis of Organic Reactions,* edited by William E. Pascoe

48. *Synthetic Lubricants and High-Performance Functional Fluids*, edited by Ronald L. Shubkin
49. *Acetic Acid and Its Derivatives*, edited by Victor H. Agreda and Joseph R. Zoeller
50. *Properties and Applications of Perovskite-Type Oxides*, edited by L. G. Tejuca and J. L. G. Fierro
51. *Computer-Aided Design of Catalysts*, edited by E. Robert Becker and Carmo J. Pereira
52. *Models for Thermodynamic and Phase Equilibria Calculations*, edited by Stanley I. Sandler
53. *Catalysis of Organic Reactions*, edited by John R. Kosak and Thomas A. Johnson
54. *Composition and Analysis of Heavy Petroleum Fractions*, Klaus H. Altgelt and Mieczyslaw M. Boduszynski
55. *NMR Techniques in Catalysis*, edited by Alexis T. Bell and Alexander Pines
56. *Upgrading Petroleum Residues and Heavy Oils*, Murray R. Gray
57. *Methanol Production and Use*, edited by Wu-Hsun Cheng and Harold H. Kung
58. *Catalytic Hydroprocessing of Petroleum and Distillates*, edited by Michael C. Oballah and Stuart S. Shih
59. *The Chemistry and Technology of Coal: Second Edition, Revised and Expanded*, James G. Speight
60. *Lubricant Base Oil and Wax Processing*, Avilino Sequeira, Jr.
61. *Catalytic Naphtha Reforming: Science and Technology*, edited by George J. Antos, Abdullah M. Aitani, and José M. Parera
62. *Catalysis of Organic Reactions*, edited by Mike G. Scaros and Michael L. Prunier
63. *Catalyst Manufacture*, Alvin B. Stiles and Theodore A. Koch
64. *Handbook of Grignard Reagents*, edited by Gary S. Silverman and Philip E. Rakita
65. *Shape Selective Catalysis in Industrial Applications: Second Edition, Revised and Expanded*, N. Y. Chen, William E. Garwood, and Francis G. Dwyer

66. *Hydrocracking Science and Technology,* Julius Scherzer and A. J. Gruia
67. *Hydrotreating Technology for Pollution Control: Catalysts, Catalysis, and Processes,* edited by Mario L. Occelli and Russell Chianelli
68. *Catalysis of Organic Reactions,* edited by Russell E. Malz, Jr.
69. *Synthesis of Porous Materials: Zeolites, Clays, and Nanostructures,* edited by Mario L. Occelli and Henri Kessler
70. *Methane and Its Derivatives,* Sunggyu Lee
71. *Structured Catalysts and Reactors,* edited by Andrzej Cybulski and Jacob A. Moulijn
72. *Industrial Gases in Petrochemical Processing,* Harold Gunardson
73. *Clathrate Hydrates of Natural Gases: Second Edition, Revised and Expanded,* E. Dendy Sloan, Jr.
74. *Fluid Cracking Catalysts,* edited by Mario L. Occelli and Paul O'Connor
75. *Catalysis of Organic Reactions,* edited by Frank E. Herkes
76. *The Chemistry and Technology of Petroleum: Third Edition, Revised and Expanded,* James G. Speight
77. *Synthetic Lubricants and High-Performance Functional Fluids: Second Edition, Revised and Expanded,* Leslie R. Rudnick and Ronald L. Shubkin
78. *The Desulfurization of Heavy Oils and Residua, Second Edition, Revised and Expanded,* James G. Speight
79. *Reaction Kinetics and Reactor Design: Second Edition, Revised and Expanded,* John B. Butt
80. *Regulatory Chemicals Handbook,* Jennifer M. Spero, Bella Devito, and Louis Theodore
81. *Applied Parameter Estimation for Chemical Engineers,* Peter Englezos and Nicolas Kalogerakis
82. *Catalysis of Organic Reactions,* edited by Michael E. Ford
83. *The Chemical Process Industries Infrastructure: Function and Economics,* James R. Couper, O. Thomas Beasley, and W. Roy Penney

84. *Transport Phenomena Fundamentals,* Joel L. Plawsky
85. *Petroleum Refining Processes,* James G. Speight and Baki Özüm
86. *Health, Safety, and Accident Management in the Chemical Process Industries,* Ann Marie Flynn and Louis Theodore
87. *Plantwide Dynamic Simulators in Chemical Processing and Control,* William L. Luyben
88. *Chemicial Reactor Design,* Peter Harriott
89. *Catalysis of Organic Reactions,* edited by Dennis G. Morrell
90. *Lubricant Additives: Chemistry and Applications,* edited by Leslie R. Rudnick
91. *Handbook of Fluidization and Fluid-Particle Systems,* edited by Wen-Ching Yang
92. *Conservation Equations and Modeling of Chemical and Biochemical Processes,* Said S. E. H. Elnashaie and Parag Garhyan
93. *Batch Fermentation: Modeling, Monitoring, and Control,* Ali Çinar, Gülnur Birol, Satish J. Parulekar, and Cenk Ündey
94. *Industrial Solvents Handbook, Second Edition,* Nicholas P. Cheremisinoff
95. *Petroleum and Gas Field Processing,* H. K. Abdel-Aal, Mohamed Aggour, and M. Fahim
96. *Chemical Process Engineering: Design and Economics,* Harry Silla
97. *Process Engineering Economics,* James R. Couper
98. *Re-Engineering the Chemical Processing Plant: Process Intensification,* edited by Andrzej Stankiewicz and Jacob A. Moulijn
99. *Thermodynamic Cycles: Computer-Aided Design and Optimization,* Chih Wu
100. *Catalytic Naptha Reforming: Second Edition, Revised and Expanded,* edited by George T. Antos and Abdullah M. Aitani
101. *Handbook of MTBE and Other Gasoline Oxygenates,* edited by S. Halim Hamid and Mohammad Ashraf Ali
102. *Industrial Chemical Cresols and Downstream Derivatives,* Asim Kumar Mukhopadhyay

103. *Polymer Processing Instabilities: Control and Understanding*, edited by Savvas Hatzikiriakos and Kalman B . Migler
104. *Catalysis of Organic Reactions*, John Sowa
105. *Gasification Technologies: A Primer for Engineers and Scientists*, edited by John Rezaiyan and Nicholas P. Cheremisinoff
106. *Batch Processes*, edited by Ekaterini Korovessi and Andreas A. Linninger
107. *Introduction to Process Control*, Jose A. Romagnoli and Ahmet Palazoglu

# Introduction to Process Control


**Jose A. Romagnoli**

*University of Sydney*
*Sydney, Australia*

**Ahmet Palazoglu**

*University of California*
*Davis, California*

Boca Raton London New York Singapore

A CRC title, part of the Taylor & Francis imprint, a member of the Taylor & Francis Group, the academic division of T&F Informa plc.

Published in 2006 by
CRC Press
Taylor & Francis Group
6000 Broken Sound Parkway NW, Suite 300
Boca Raton, FL 33487-2742


Printed in the United States of America on acid-free paper
10 9 8 7 6 5 4 3 2 1

International Standard Book Number-10: 0-8493-3496-9 (Hardcover)
International Standard Book Number-13: 978-0-8493-3496-2 (Hardcover)
Library of Congress Card Number 2005041371



**Library of Congress Cataloging-in-Publication Data**

Romagnoli, José A. (José Alberto)
Introduction to process control / José a. Romagnoli, Ahmet Palazoglu.
p. cm.
Includes bibliographical references and index.
ISBN 0-8493-3496-9 (alk. paper)

TP155.75.R655 2005
660'.2815—dc22 2005041371

Taylor & Francis Group
is the Academic Division of T&F Informa plc.

**Visit the Taylor & Francis Web site at
http://www.taylorandfrancis.com**

**and the CRC Press Web site at
http://www.crcpress.com**

To Liliana, Jose, Juliana and the memory of my parents (Jose A. Romagnoli)

and

To Mine, Aycan, Ömer and my parents (Ahmet Palazoglu)

With our everlasting love and gratitude

"Everyone has his or her own way of learning things," he said to himself. "His way isn't the same as mine, nor mine as his. But we're both in search of our Personal Legends, and I respect him for that."

Paulo Coelho, *The Alchemist*

# Preface

In recent years, changing industrial needs and advances in computer technology had a profound impact on the education, research, and practice of process control. From the industrial perspective, improved productivity, efficiency, and product quality goals generated a demand for more effective operational strategies to be incorporated in the production line, while the developments in digital computers and communications have revolutionized the practice of process control and allowed more advanced tools to be implemented. As a consequence, we witness a vast broadening of the domain of what is technologically and economically achievable in the application of computers to control industrial processes. This domain now includes process information and data gathering, control, and online optimization, and even production scheduling and maintenance planning functions. Modern process control should then be viewed as an efficient integration of real-time information management with the traditional concept of control. As such, the process control education needs to conform to the current practice and follow the tendencies in the field.

With this book, our goal is to provide a bridge between the traditional view (role) of process control with the current and expanded view (role). This is accomplished by blending the traditional topics with a broader perspective of more integrated process operation, control, and information systems.

We believe that this book contains a number of innovative features both in terms of teaching and learning principles as well as in terms of its content that offers some unique perspectives to the education of process control. These features are:

- The traditional and expanded roles of process control in modern manufacturing are explicitly introduced and formally defined, setting the stage for new concepts to be incorporated and blended within a typical process control textbook.
- In discussing the traditional topics of process control, the strategy of "concept followed by an example" is adopted throughout the book, thus, allowing the reader to grasp the theoretical concepts in a practical manner.
- A Continuing Problem is introduced in the very first chapter and new concepts and strategies are subsequently applied to this example, culminating, at the conclusion of the book, with a complete control design strategy. Furthermore, the implementation of these strategies, for the Continuing Problem, within a typical distributed control system environment is provided through a virtual application.

- The plantwide control problem is fully addressed, and a practical application to a complete industrial example is provided. This is further expanded by incorporating the ideas of environmentally conscious manufacturing practices.
- An introduction to modern architectures of industrial computer control systems is incorporated with real case studies and applications to pilot-scale operations. This provides the basis for a full discussion of the expanded role of process control in modern manufacturing.
- Data processing and reconciliation and process monitoring are for the first time incorporated and blended as integral components of the overall control system architecture.
- A complete, user-friendly software environment accompanying the book allows the reader to interactively study the examples provided in each chapter. The website (GIVE URL) contains three MATLAB® toolboxes (APC_Tool, SFB_Tool and MPC_Tool). MATLAB is a registered trademark of The MathWorks, Inc. for product information, please contact:

  The MathWorks, Inc.
  Apple Hill Drive
  Natick, MA 01760-2098 USA
  Tel: 508 847-700
  Fax: 508 647-7001
  E-mail: info@mathworks.com
  Web: www.mathworks.com

We also provide a number of simulations in HYSYS that can be used to study the control design and tuning concepts discussed in the book. HYSYS is a registered trademark of Aspen Technology, Inc. for product information, please contact:

Aspen Technology, Inc.
Ten Canal Park
Cambridge, Massachusetts 02141-2201 USA
Tel: 617 949-1000
Fax: 617 949-1030
E-mail: info@aspentech.com
Web: www.aspentech.com

The book is primarily aimed at students of chemical engineering, at different curricular levels, as well as industrial practitioners to familiarize them with the key concepts of process control and their implementation. It also intends to engage them in the appreciation of the evolution of traditional process control into an integrated operational environment, typically used to run modern manufacturing facilities. Accordingly, the book is divided into seven sections. One can envision a first course on process control focusing on the first four or five sections, while the remaining sections may be better suited for an advanced undergraduate course or to kick off a graduate course.

Section I (Chapters 1 and 2) provides a general introduction to the issues of controlling and operating chemical process. It discusses the role (or roles) of process control and identifies why process control is a fundamental discipline within chemical engineering education and practice. Furthermore, some key concepts and definitions as well as the general terminology are presented.

Section II (Chapters 3, 4, 5, and 6) introduces the reader to the principles of process modeling, and discusses some of the key aspects of modeling based on conservation laws. This sets the stage for developing input-output models for control design and analysis purposes. Recognizing the importance of process simulations in modeling and analysis, we also discuss the current trend toward open systems architecture in simulations and its implications for process control design.

Section III (Chapters 7, 8, and 9) is devoted to various analytical techniques that can be used to study process performance characteristics such as stability and transient response. A number of model structures are investigated and the frequency response analysis is introduced. Here, the mathematical rigor of stability concepts is blended with a practical insight to convey some of the fundamental ideas behind the proper behavior of systems.

Section IV (Chapters 10, 11, and 12) covers the analysis and design of feedback control systems. The effect of various control modes is fully investigated and a number of design strategies are presented based on the knowledge acquired in previous chapters. Full use of the software environment provides the reader with an interactive environment for practical implementation of the concepts discussed though the chapters.

Section V (Chapters 13, 14, and 15) covers the analysis and design of more complex control configurations with the possibility of having more than one input or output. In particular, Chapters 14 and 15 deal with the analysis and control of multivariable systems. Again, the strategy of "concept-example" combined with the software environment facilitates the illustration of concepts involved in multivariable systems theory.

Section VI (Chapters 16, 17, 18, and 19) deals with the fundamental and practical aspects of model-based process control, in which the model of the process plays an important role in designing the control system. Model Predictive Control (MPC) strategy, the most popular model-based approach among industrial practitioners, is covered along with an introduction to the important concepts of robustness and uncertainty characterization.

Section VII (Chapters 20, 21, 22, 23, and 24) first focuses on the analysis and control design of complete processing plants. A heuristic approach to design the plant control system is fully discussed and applied to an industrial case study. Special consideration is given to the ideas of environmental-friendly process design and its implications on the control system structure. Next, we describe the current technology and the main components of a control system that supplies the control functions often encountered in an industrial application. These (and the discussion on the supporting hardware and software) provide the basis for the analysis of the expanded role of control in modern manufacturing in subsequent chapters. These capabilities are fully demonstrated through practical examples implemented within industrial distributed control systems. Furthermore, as a unique feature of this book, we show how the concepts of data processing and reconciliation and process monitoring are incorporated into the control system environment.

We have more than 15 years of teaching experience in process control and related subjects at The University of Sydney and the University of California, Davis. This book represents the collection of many experiences in educating third- and fourth-year students as well as industrial practitioners in an exciting field of chemical engineering, namely, process control.

A number of people contributed to the conceptualization and preparation of this textbook. Firstly, we acknowledge our graduate students for their valuable assistance throughout the different steps in developing the material associated with the book. Dr. A. Bakthazad and Dr. D. Wang were invaluable throughout the development of the software environment used so many times as a teaching tool. Dr. A. Abbas and P. Rolandi were instrumental in providing new material and exercises, which helped to improve the quality of the book through the years. We acknowledge Dr. J. Zeaiter, Dr. H. Alhammadi, R. Willis, B. Alhamad, Dr. F. Azimzadeh, Dr. O. Galan, and R. Chew for providing, through their research projects, some of the novel applications of control technology, which are discussed in detail throughout many chapters. We also acknowledge J. Orellana, the Laboratory Manager at The University of Sydney for his enthusiasm and support in implementing most of the ideas discussed on pilot-scale experiments. Karen Gould and Scott Beaver offered very valuable suggestions both in terms of content and presentation after reviewing the draft version of the book. Secondly, we would like to thank our colleagues and general staff both at the University of Sydney and the University of California, Davis for providing the ideal environment to reach the successful conclusion of the book. Thirdly, we would like to thank all our undergraduate students, who, for so many years, have sustained our enthusiasm for teaching process control and, of course, for their continuous feedback, which helped tremendously in shaping the material covered in the book. The last but not the least, we would like to thank our families for their continued support throughout the many steps of the preparation of the manuscript and for their understanding of many days and nights expended in this pursuit.

Jose A. Romagnoli<br>
Ahmet Palazoglu<br>
Davis, CA

# Author Bios

**Jose A. Romagnoli** currently holds the Chair in Process Systems Engineering at the Department of Chemical Engineering at The University of Sydney, Australia, where he is also the Director of the Laboratory for Process Systems Engineering. He received his B.Sc. in Chemical Engineering at the National University of the South (Argentina) in 1974, and his Ph.D. in Chemical Engineering at University of Minnesota in 1980. He has been a visiting professor in many Universities, including University of Minnesota, University of California, Davis, National University of Buenos Aires (Argentina), and Nanyang Technological University (Singapore). He is the author of more than 250 international publications and a coauthor of the book *Data Processing and Reconciliation for Chemical Process Operations* published by Academic Press International. He is a member of the Australian Academy of Technological Sciences and Engineering and consultant for several international companies. He has been awarded the Centenary Medal by the Prime Minister of Australia for his contributions to Chemical Engineering. His research interests cover all aspects of Process Systems Engineering, specifically, data processing and reconciliation, modeling of complex systems, advanced-model-based control and intelligent process monitoring and supervision.

**Ahmet Palazoglu** has received his B.S. degree in chemical engineering from the Middle East Technical University (Turkey) in 1978 and his M.S. degree in Chemical Engineering from Bogazici University (Turkey) in 1980. He received his Ph.D. degree in chemical engineering from Rensselaer Polytechnic Institute in 1984 and immediately joined University of California, Davis, where he is currently a Professor of Chemical Engineering and Materials Science. He had visiting appointments at the National University of the South (Argentina), Bogazici University (Turkey), University of Stuttgart (Germany),and Koc University (Turkey). He has more than 110 publications and offered several short courses to academic and industrial audiences on process monitoring applications. His research interests are in process control, nonlinear dynamics, process monitoring, and statistical modeling.

# Abstract / Introduction to Process Control

Changing industrial needs and increased competition in the global marketplace as well as innovations in computer and communication technologies have a profound impact on the practice, research and education of process control. Today, process control is viewed as an efficient integration of real-time information management systems with the common regulatory functions to help operate a process safely and efficiently. As such, the process control education needs to be consistent with the current practice and reveal the tendencies in the field. In this book, we offer the readers a bridge between the traditional role of process control and the current expanded role. This is accomplished by blending the basic concepts with a broader perspective of more integrated process operation, control and information systems. The book is primarily aimed at students in chemical engineering as well as industrial practitioners to initiate them into the fundamentals and current advances in process control and their implementation.

# Contents

**Chapter 17**



**Chapter 18**



**Chapter 19**

# *Section I*

## *Introduction*

# 1 Why Process Control?

The word *control* is encountered often in everyday conversation. We could be looking for the missing *remote control* for the TV, expressing our opinion on the issue of *gun control*, or try to *control our anger* when faced with a frustrating situation. The Merriam-Webster dictionary (http://www.m-w.com/) offers the following abbreviated definitions for this word, which is most relevant to our topic.

---

con·trol *transitive verb*, (1) to check, test, or verify by evidence or experiments, (2) to exercise restraining or directing influence over.

*noun*, (1) an act or instance of controlling; also: power or authority to guide or manage, (2) a device or mechanism used to regulate or guide the operation of a machine, apparatus, or system.

---

By controlling process systems, we basically imply influencing their operation. This influence can take the form of guiding or regulating the operation in such a way as to ensure a desired outcome. Perhaps you may have already witnessed such instances in the laboratory where control was part of the process operation: a heated water bath included a control unit to maintain the desired temperature in the bath; or a pressure regulator used to deliver a constant air supply to the laboratory. In all these instances, the so-called control unit is present to ensure steady operation for an otherwise dynamic process that would exhibit significant (and sustained) deviations from the desired target. In all these and other practical situations, it is hard not to notice a simple fact: all industrial processes are dynamic. In other words, the process behavior changes with time. Thus, the control engineer is called upon first and foremost to understand fundamentally the dynamic nature of processes and then, based on this knowledge, devise appropriate mechanisms to influence their behavior over time.

Control engineering is no longer regarded as a narrow specialty, but as an essential topic for all chemical engineers. For example, design engineers must carefully consider the dynamic operation of all equipment and its consequences, as the plant will never operate at steady-state, which is often the assumption that one makes to simplify analysis and design tasks. Engineers entrusted with the operation of processing plants must find the appropriate response to the ever-present upsets (e.g., feed quality or quantity fluctuations, variations in the availability of various utilities), so that the plant operates smoothly. Finally, engineers performing experiments must control their equipment and the environment meticulously to obtain the conditions prescribed by their experimental designs. In fact, without process control, it would not be possible to operate most modern processing facilities safely, reliably, and profitably, while satisfying quality standards.

## 1.1 HISTORICAL BACKGROUND

The history of control, especially the concept of feedback, can be traced back to 2000 years when ancient water clocks were being operated successfully using float regulators in Alexandria (Egypt) and Baghdad (Iraq). In an issue commemorating the *13th World Congress* of the *International Federation of Automatic Control* (IFAC), *Control Systems* journal of the *International Electronics and Electrical Engineers* (IEEE) Society, published a series of articles detailing the evolution of automatic control (*Control Systems*, Vol. 16, No. 3, 1996), among which the paper by Stuart Bennett is an excellent reference for scientific history enthusiasts.

Prior to the 1940s, most chemical processing plants were essentially run manually, as the plant operators adjusted material and energy flows by manipulating rather large valves by hand. Only the most elementary types of controllers were used to regulate a limited number of process equipment. Many instrument companies existed, producing a variety of devices for sensing, recording, and control. Due to the large-scale nature of the plants, many operators were required to keep watch on hundreds of process variables, and they needed a strong process knowledge and expertise to respond to unwanted trends in the best possible way. Moreover, large tanks were employed to act as buffers or surge capacities between various units in the plant. These tanks, although sometimes are quite costly capital investments, served the function of absorbing the majority of upsets by isolating parts of the process from incidents occurring upstream.

With increasing labor and equipment costs, and with the development of higher capacity, high-performance equipment and processes in the 1940s and early 1950s, it became uneconomical and often impossible to run plants without automatic control devices. At this stage, feedback controllers were added to the plants with little real consideration of, or appreciation for, the underlying process dynamics. Rule-of-thumb guidelines and process experience were the only design techniques, which were employed on a case-by-case basis.

By the 1960s, chemical engineers were experimenting with the new developments in dynamic analysis and control theory, which during that decade focused on the concept of optimal control. Most of the techniques were adopted from the work in the aerospace and electrical engineering fields, responding to the escalation of the arms race by dealing with missile guidance control and radar tracking systems. Foss[1] and Lee and Weekman[2] harshly critiqued these practices as being unresponsive to the actual needs of chemical processes and argued the uniqueness of the process control problem in chemical industries as distinct from others.

The rapid rise in energy costs in the 1970s highlighted additional needs for effective control applications, which are referred to as "Advanced Control Systems." The design and redesign of many plants to reduce energy consumption resulted in more complex, integrated plants that were much more interacting than before. Consequently, the challenges facing the process control engineer have continued to grow steadily over the years. The textbook by Stephanopoulos[3] represents a milestone in the conceptualization of process control as a plantwide effort, underscoring its fundamental link to process design.

This set the scene for the trends in the 1980s and 1990s. On the one hand, theoretical studies helped to establish the credibility of the field as a rigorous discipline, although many of the developed algorithms did not find immediate application in practice, a notable exception being the model predictive control technology, marketed today in various forms by vendors like Aspen Technology and Honeywell. On the other, control hardware went through a substantial transition from being pneumatic to analog or electric and to being microprocessor-based. The control hardware that has always been designed to implement single-loop controllers and their variations, started enjoying the benefits of distributed control systems (DCS) that come equipped with various advanced control options.

Exciting challenges still lay ahead as control engineers exploit the advances in computer and networking technologies to fulfill their mission more effectively.

## 1.2 ROLE OF CONTROL IN PROCESS INDUSTRIES

As discussed above, the theory and application of control has been shaped by a number of technological developments over the years. Accordingly, the role of process control is being continuously reshaped following these trends. Today, we can clearly observe two types of roles:

1. Traditional role of process control
2. Expanded role of process control

### 1.2.1 TRADITIONAL ROLE OF PROCESS CONTROL

In industrial operations, control contributes to safety, environmental impact minimization, and process optimization, by maintaining process variables near their desired values. The chemical reactor with highly exothermic reaction in Figure 1.1 demonstrates a clear example of safety through control. Many input variables, such as feed composition, feed temperature, and cooling temperature, can vary, which could lead to dangerous overflow of the liquid and large temperature excursions (runaway). The control system sketched in Figure 1.1 shows that the level is maintained near its desired value by adjusting the outlet flow rate, and the temperature is kept near its desired value by adjusting the coolant flow rate. At the same time, by controlling the temperature at some optimal value, the selectivity may be maximized, thus improving process economics. This tight temperature control may also minimize the amount of by-products produced in the reaction, thereby minimizing the environmental burden through the reduction of potential waste products.

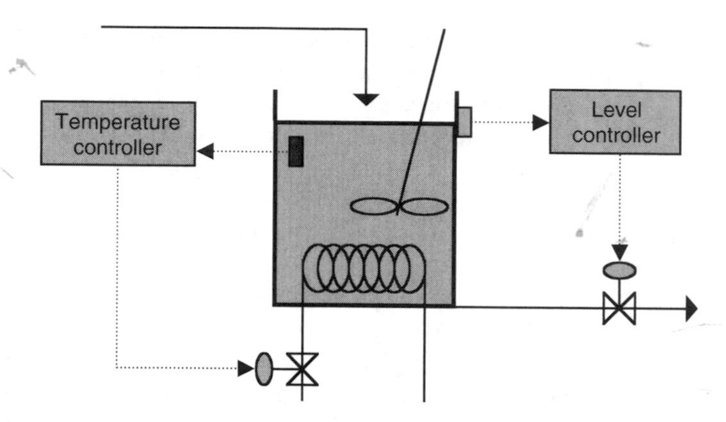


**FIGURE 1.1** A chemical reactor and its control system.

As demonstrated by the above example, during its operation, a chemical plant must satisfy a variety of requirements imposed by its designers along with the general technical, economical, and social constraints, in the presence of ever-changing external influences such as variability in supply and demand. Among such requirements are the following:

1. *Economics* The operation of a plant (or a process unit) must conform to the market conditions, recognizing the availability and the quality of raw materials as well as the demand for the final products. Furthermore, it should be as economical as possible in its utilization of raw materials, energy, capital, and human labor. Thus, the operating conditions must be controlled effectively to minimize operating costs, and maximize profits.
2. *Production specifications* A processing plant should maintain or surpass the production capacity that it is designed for, while delivering the final products with consistent quality.
3. *Operational constraints* The various types of equipments used in a chemical plant have limitations as a result of their particular design, and inherent to their operation. A centrifugal pump, for instance, can deliver only a certain flow rate as determined by its impeller size and the available pressure drop in the line. Control systems need to recognize and satisfy all such operational constraints.
4. *Safety* The safe operation of a chemical process is a primary requirement for the well being of plant personnel, the community at large, and the economic viability of the company. Thus, the operating pressures, temperatures, and concentration of chemicals should always be maintained within allowable limits.
5. *Environmental regulations* Many countries as well as local governments have enacted laws mandating that any air, water, and ground emissions from chemical plants must conform to specified limits in chemical composition and flow rate.

All requirements listed above dictate the need for continuous monitoring of the operation of a chemical plant, with intervention when necessary, to guarantee the satisfaction of each operational objective. This is accomplished through a rational arrangement of equipment and instrumentation (e.g., measuring devices, valves, controllers, computers) along with human expertise (e.g., plant designers, plant operators, and engineers), which altogether help deploy the control system.

### 1.2.2 Expanded Role of Process Control

An industrial plant is a complex combination of various unit operations that are designed to produce a desired product. An integral part of this complex combination is the control room, where plant operators keep a watchful eye on the production (Figure 1.2). They use computer-based control systems for collecting process data and adjusting the appropriate variables to maintain desired operations.

During normal operation, the plant operators should be able to rely on automatic control systems to maintain desired process behavior. The basic unit in this effort is the DCS that communicates with field instruments through Ethernet links. Figure 1.3 illustrates a computer control configuration combining a number of control systems.

Such a unit allows the plant operators to visualize key process variables, check for alarms, and interfere with the operation, when necessary. Such interference can take many forms ranging from adjusting the performance of a unit to shutting down the plant during an emergency.

**FIGURE 1.2** Control room of a pulp and paper mill (courtesy of Visy Pulp and Paper, Australia).

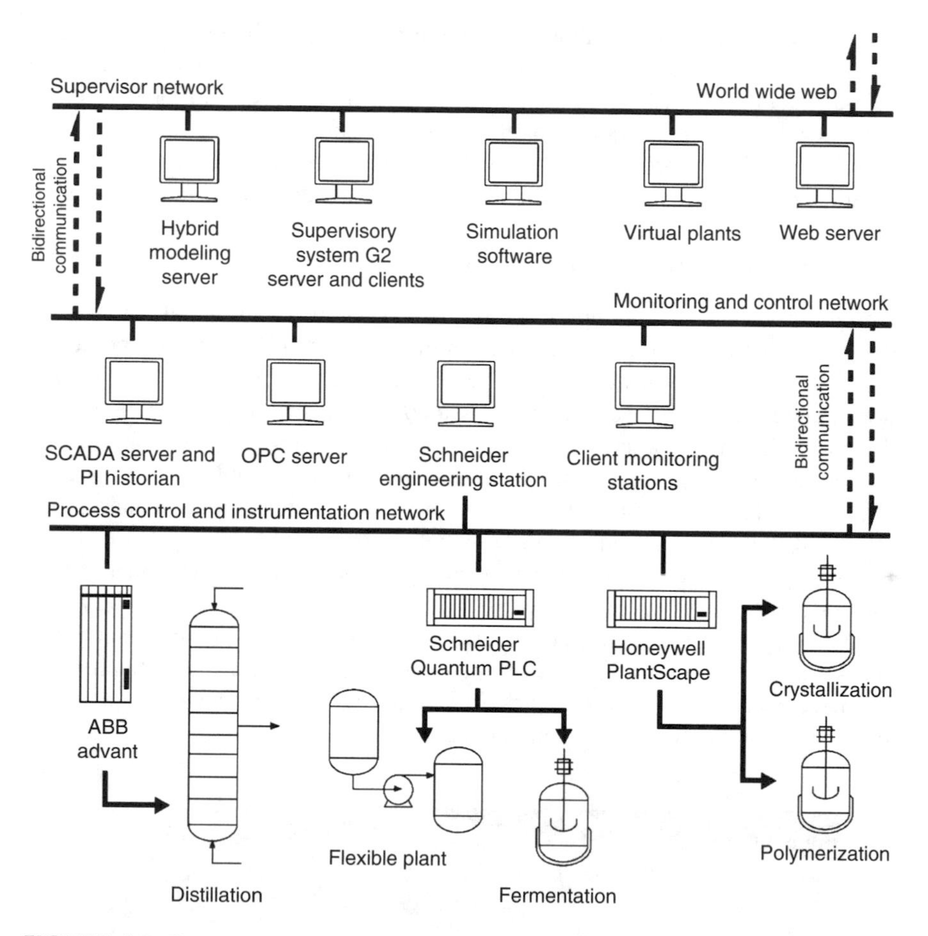


**FIGURE 1.3** Computer control architecture and its links (implemented at the Process Systems Engineering Laboratory at the University of Sydney).

The application of computers to control the process had a profound effect on the direction of current efforts in industrial systems control. Specifically, they have opened up new opportunities for system configuration based on- (1) distributed data acquisition and control and (2) hierarchical computer control where each computer performs selected and coordinated tasks.

A consequence of these developments has been a vast broadening of the domain of what is technologically and economically feasible to achieve in the application of computers to control industrial systems. Today, all aspects of information processing, data gathering, process control, online optimization, even scheduling, production, and maintenance planning functions, may be included in the range of tasks to be carried out by the computer control system. This has made possible the realization of integrated systems control (ISC) in which all factors influencing plant performance are taken into account.

---

Control systems today have an expanded role, replacing manual manufacturing activities with full automation. Modern process control is the functional integration of real-time information management with the traditional concept of control.[4]

---

In the next section and the following chapters, we will focus on exploring several issues within the traditional view of process control to establish a series of fundamental concepts. Later in the book, we will return to the expanded role of control in modern manufacturing and build on this foundation.

## 1.3 OBJECTIVES OF CONTROL

A control system is generally asked to perform one or both of the following tasks:

1. *Maintaining the process at the desired operating point* A process is expected to operate at the steady-state conditions prescribed by its design. This operating point is generally the most attractive one, satisfying among many others, economic, safety, and quality objectives. Thus, if the process strays away from this condition, substantial losses may be incurred. There may be two chief reasons why a process may not stay at this operating point (regime).

   First, the process may be *unstable*, implying that process variables may not remain within their physical bounds for perturbations of limited magnitude. Imagine a tank without any drainage being constantly filled with water. The water level in the tank will keep rising until the tank overflows. Such an uncontrolled behavior cannot be tolerated in a chemical process. Many chemical processes happen to be stable or *self-regulating* and need no external intervention for their stabilization. On the other hand, some are inherently *unstable* and require external manipulation for the stabilization of their behavior. A chemical reactor with an exothermic reaction is a good example in which the reactor temperature may stray away from desired and acceptable levels if sufficient external cooling is not employed. Such systems can only be operated with the aid of a control system.

   Secondly, a process operating at steady-state experiences frequent upsets due to various changes in its operating environment. A typical example is when the feed-stream properties (such as composition, temperature, flow rate, etc.) are fluctuating, thus continually pushing the process away from its steady-state. Depending on the magnitude and period of these fluctuations, process operation may be significantly compromised. Again, a control system would be required to provide compensation for such disturbances and maintain the process as close as possible to the desired operating point for as long as possible.

2. *Moving the process from one operating point to another* In many process operations, the plant personnel may want to change the operating point that a process is at, for a variety of reasons. One example is a petroleum refinery that processes different types of crude oil. The desired operating point for a light crude would not be so desirable when a heavier crude is processed. Or, in a polyethylene plant, quality of the product may change from time to time depending on the customer specifications. In both these examples, the plant needs to adjust the operating point so that it corresponds to the conditions best suited to the changed environment. A control system can perform this, often complex, task in an automated manner.

Riding a bicycle (see Figure 1.4) is an attempt to stabilize an unstable system, and we accomplish this by pedaling, steering, and leaning our body right or left. Of course, all this is done to keep us on our chosen trajectory towards a desired destination. The traffic or the pedestrians on the road constitute typical disturbances for our ride. We act as the control system and may even decide to change our destination, or stop safely for a brief rest.

### 1.3.1 What About Performance?

However, merely surviving a fall and staying on course are not the only objectives in riding a bicycle. One should enjoy the trip as well. Therefore, an additional objective of the control system is maintaining some level of desired *performance*. For process systems, performance expectations are set by plant engineers and typically reflect how some key process variables evolve over time.

In control system design, good performance typically requires aggressive control actions and fast responses, while stability is generally preserved with conservative control actions and slow responses. This leads to the perennial *design trade-off* between performance and stability of process control systems, as the primary goal in advanced control design is to maximize performance while ensuring stability.

### 1.3.2 Economic Benefits of Control

The *optimization* of process operating conditions can be performed in such a way as to maximize an economic objective, such as the annual profit. Key factors in

**FIGURE 1.4** A bicycle is an inherently unstable system.

good plant management are the determination of the most desirable operating targets, and the deployment of an effective automatic control strategy to maintain these targets within tolerable limits. Therefore, setting the control objectives requires a clear understanding of how the plant operates. A chemical engineer is best suited for this task.

One can determine the most desirable plant operating conditions by first defining the region of possible operation (feasible region) in the plane of free variables (Figure 1.5). The feasible region is bounded by the physical limits of the variables. Secondly, the plant economics are evaluated by superimposing the contours of increasing profits. While the optimum lies at the corner of the feasible region,[5] in practice, the plant will be operated sufficiently far from this point to ensure feasible operation in the presence of ever-present plant disturbances (upsets). The control strategy aims to keep the operating condition variations at a minimum, and to allow the operating target to stay as close as possible to the true (optimal) maximum profit. The control system is expected to minimize the variations around the operating target (performance objective) while, in turn, shrinking the tolerable operating limits. The more sophisticated (or advanced) the control system is, the better the chances are that the plant will operate even closer to the optimum target. This gain, quantified by the move toward a more profitable regime, helps establishing the financial benefits of the control system.

In summary, one can conclude that good control performance has the potential to yield substantial benefits for safe and profitable plant operation. By applying the fundamental process control principles, the engineer will be able to design plants and implement control strategies that can achieve the control objectives set forth by plant operations.

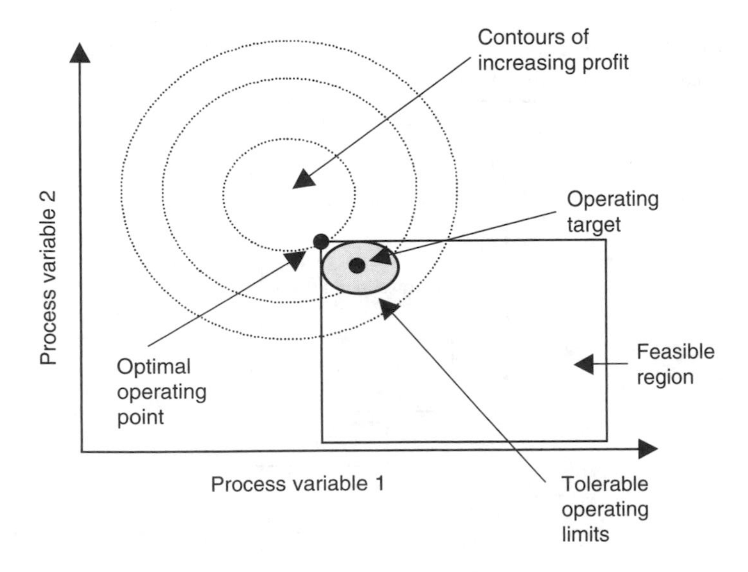


**FIGURE 1.5** Optimal plant operation.

## 1.4 SUMMARY

Good control design addresses a hierarchy of control objectives, ranging from safety to product quality and plant profitability, which depend on the operating objectives for the plant. These objectives are determined by both steady-state and dynamic analysis of the plant performance. Process control aims to reduce the plant variations and help deliver consistently high product quality and maintain operation close to the maximum profit target.

As we learned, good performance demands "tight" control of key variables. Clearly, understanding the dynamic behavior of the process is essential in designing control strategies. Only with a thorough knowledge of the process dynamics, we can design control systems that can satisfy conflicting objectives and yield lasting benefits.

## CONTINUING PROBLEM

Let us consider a simple blending process depicted in Figure 1.6, with two feed streams and one product stream. The feed streams contain aqueous mixture of a component with different compositions (mass fractions), hence requiring a blending operation to deliver a product stream with the desired composition. We will denote the volumetric flow rate ($m^3$/min) of a stream with $F$ and $x$, which will represent the component mass fraction. The subscripts clearly point to the streams.

While the flow rates of the feed streams can be adjusted by the valves, the stream compositions will vary depending on upstream processing conditions. Moreover, the product stream is likely to have a specification on its composition

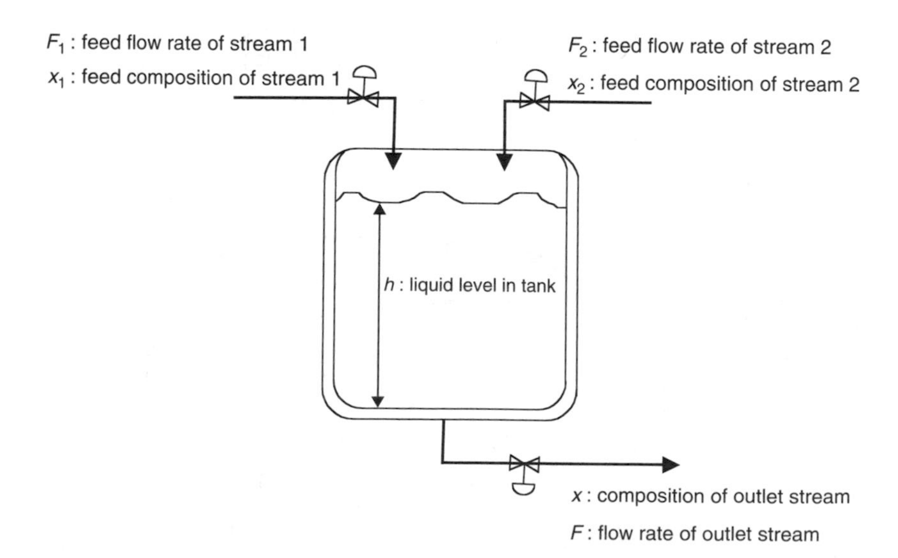


**FIGURE 1.6** A blending process.

(quality), and a composition sensor may be placed on this stream to monitor its variation:

1. Analyze this process and cite the specific reasons why control may be necessary for its operation.
2. What would be the specific benefits of a control system?

### SOLUTION

In the blending process, the main operational concern is the quality (composition) of the final blend stream, $x$. The target composition can be achieved by operating the valves on the feed flow rates. However, by doing so, even if the target is reached, the liquid level in the tank may be changing. Thus, a mechanism (a control system) becomes necessary to manage the combination of feed flow rates not only to achieve the desired target composition of the blend but also to maintain an acceptable level of liquid in the tank.

Due to consumer demands, the target product composition may vary and the operation needs to respond swiftly to such demands. By maintaining the product composition at the desired target, we minimize the production of off-specification product that may have to be discarded or reprocessed. This strategy has clear economic benefits. Furthermore, by maintaining a desired liquid level in the tank, we can effectively manage the inventories, and this would have positive economic consequences. We also have the added benefit of ensuring a safe and reliable operation (i.e., the tank does not overflow or run dry).

## REFERENCES


1. Foss, A.S., Critique of chemical process control theory, *AIChE J.*, 19, 209, 1976.
2. Lee, W. and V.W. Weekman, Advanced control practice in the chemical industry, *AIChE J.*, 22, 27, 1976.
3. Stephanopoulos, G., *Chemical Process Control – An Introduction to Theory and Practice*, Prentice-Hall, New York, 1984.
4. Erickson, K.T. and J.L. Hedrick, *Plantwide Process Control*, Wiley, New York, 1999.
5. Edgar, T.F. and D.M. Himmelblau, *Optimization of Chemical Processes*, McGraw-Hill, New York, 1988.

# 2 Definitions and Terminology

Before tackling the main features of a control design problem, it is necessary to have a clear understanding of some key concepts and definitions as well as the general terminology. In this chapter, we will start with the classification of process variables from the control viewpoint. These variables will then be used to analyze the characteristics of a control system and to formulate the problems that must be solved during design. Finally, we will discuss the elements of a control design project.

## 2.1 CONCEPTS AND DEFINITIONS

Let us consider a system that can comprise of a processing unit, a set of units, or a section of the plant. Figure 2.1 depicts such a process system that interacts with its surroundings through its boundary. From the control viewpoint, the variables that characterize the dynamic behavior of this process can be classified into two groups:

1. *Input variables* ($u$) represent the effect of the surroundings on the process.
2. *Output variables* ($y$) represent the effect of the process on the surroundings.

The input variables may or may not be available for a deliberate action on the process, resulting in a further classification as follows:

1. *Manipulated inputs* ($m$) can be adjusted freely by a human operator or a control system.
2. *Disturbances* ($d$) are not the result of an adjustment by an operator or a control system.

According to their direct measurability, disturbances themselves can be further classified as (1) measured disturbances ($d_m$) and (2) unmeasured disturbances ($d_u$). As examples of typical process disturbances, feed flow rates, feed temperatures, and pressures can be easily measured using available sensor technology. Feed composition, on the other hand, is difficult to measure, as on-line composition sensors are often expensive, unreliable, or just unavailable. This classification is displayed in Figure 2.2.

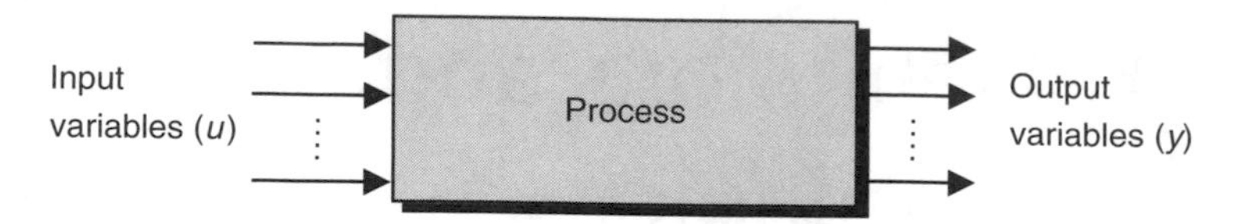


**FIGURE 2.1** Schematic representation of a process and its interactions with the surroundings.

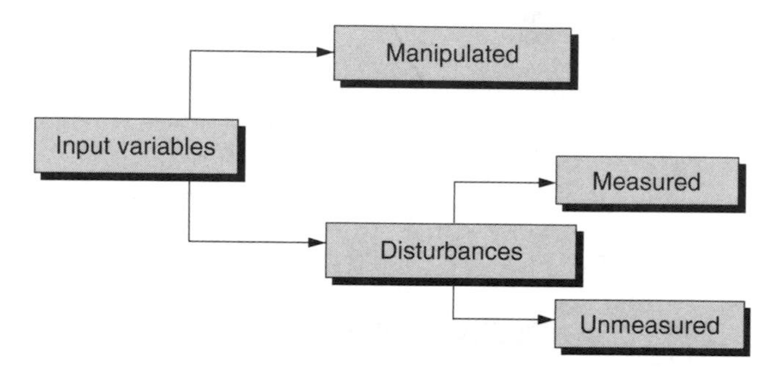


**FIGURE 2.2** Classification of input variables.

---

We shall see later that the unmeasured disturbances generate more difficult control problems, since information is lacking on their impact on the process.

---

The output variables, also referred to as *control variables*, are generally associated with the control objectives, and are related to the process variables that indicate product quality, process safety, and economics. They are further classified into the following categories:

1. *Measured outputs* ($y_m$) are those whose values are known on direct measurement.
2. *Unmeasured outputs* ($y_u$) are not or cannot be measured directly.

This classification is displayed in Figure 2.3. In summary, a process can be represented in a more detailed schematic as shown in Figure 2.4.

Depending on how many output and input variables are considered for the control problem, we can distinguish two major control structures:

1. *Single input–single output (SISO)* There is a single output variable (control objective) and a single input (manipulated) variable is used to affect the process.
2. *Multiple input–multiple output (MIMO)* There is more than one output variable (control objective), and more than one input (manipulated) variable is used to affect the process.

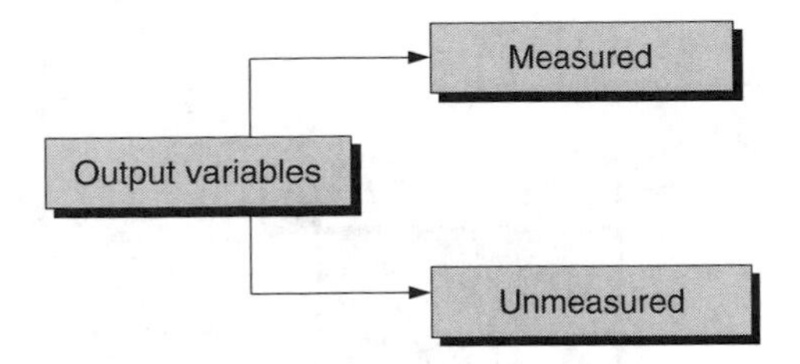


**FIGURE 2.3** Classification of output variables.

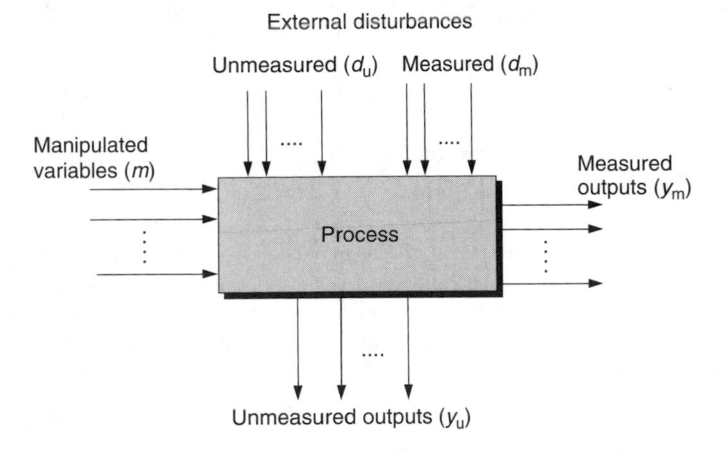


**FIGURE 2.4** Schematic representation of a process.

## Example 2.1

Consider the stirred-tank heater shown in Figure 2.5. The liquid is heated by a coil mechanism through which a low-pressure steam is passed. Here, $F$ and $T$ represent a stream flow rate and a stream temperature, respectively, and the subscripts in, out and st refer to the inlet stream, outlet stream, and the steam coil, respectively. For this process, we have:

- Input variables: $F_{in}$, $T_{in}$, $F_{st}$, and $F_{out}$
- Manipulated: $F_{st}$ and $F_{out}$
- Disturbances: $F_{in}$ and $T_{in}$ (both can be easily measured)
- Output variables: $V$ and $T_{out}$ (both can be easily measured)

---

$F_{out}$ can be considered either as input or output. Why?

---

- If there is a control valve in the effluent stream, so that its flow can be manipulated, then $F_{out}$ may be an input variable.
- Otherwise, $F_{out}$ is an output variable, reflecting the influence of input variables on the process.

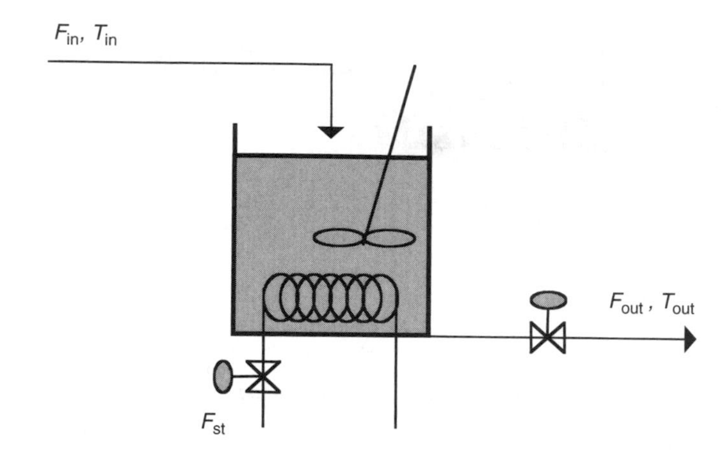


**FIGURE 2.5** A tank heater process for Example 2.1.

An operational goal for this heater process may be to maintain a constant tank level to avoid spills and also to maintain a desired liquid temperature. If we attempt to measure the liquid holdup in the tank in addition to the effluent stream temperature and use, for instance, the effluent flow rate and the steam flow rate as manipulated variables, we have a MIMO control problem.

## 2.2 CONTROL DESIGN PROBLEM

The fundamental problem in control theory is to bring about a desired behavior, expressed by a reference target ($y_r$), of an output (control) variable ($y$) through the manipulation of an input (manipulated) variable ($m$). The reference target is also referred to as the set-point ($y_{sp}$) and this will be the common use in later chapters.

A plant operator, being familiar with the operation of a process, can view the control variable through a display and make adjustments to an input variable (typically a valve on a process stream) that is expected to produce the desired outcome (Figure 2.6). This is referred to as *manual control*. One can imagine many drawbacks with this effort: one operator is tied up with a specific process (or a process unit) all the time; the operator needs to be an expert in the operation of the process, etc.

To solve this problem more effectively, we can, in principle, visualize that there is a special mechanism, called *the control mechanism*, yet to be determined. The control mechanism will adjust the manipulated variable in such a way as to cause the output variable to follow the specified reference value as dictated by the control objectives. Conceptually, this is depicted in Figure 2.7 and is referred to as *open-loop control*.

Let us assume that a *perfect* mathematical model of the process is available, which *exactly* describes how $y(t)$ changes in response to $m(t)$. Then, one can ask: Can

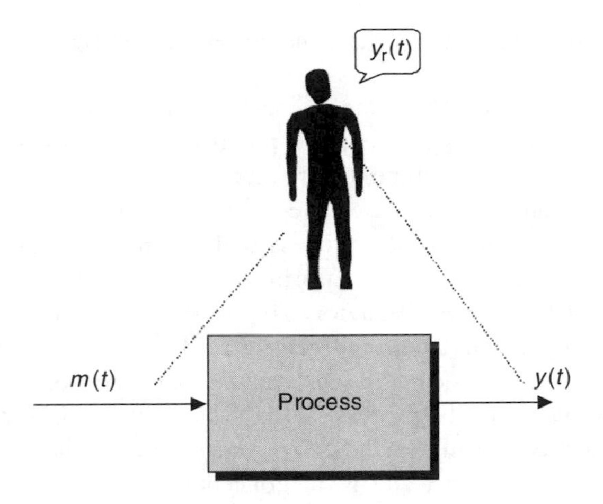


**FIGURE 2.6** Manual control.

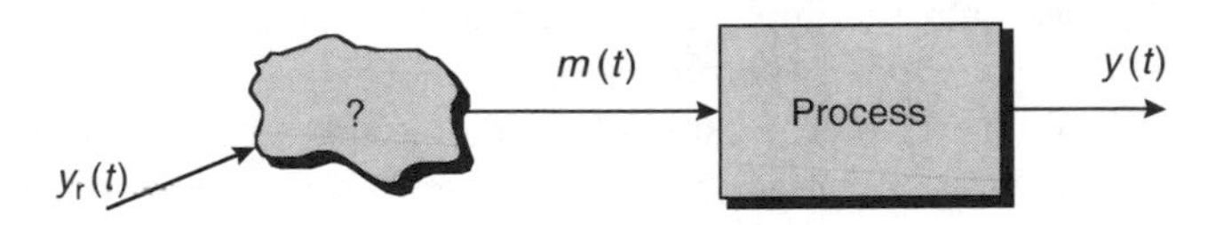


**FIGURE 2.7** Open-loop control.

this model be used to determine $m(t)$, if the desired output, $y_r(t)$, is specified? The control mechanism, then, can simply become an *inverse* model. Indeed, this leads to a perfect control design as we would compute precisely what value of $m(t)$ to send to the process so that $y(t)$ follows $y_r(t)$ exactly. While this appears to be an intuitive solution to our control design problem, there may be practical difficulties in implementing it:

1. In the first place, the assumption of the availability of a perfect process model is never true.
2. It may be computationally difficult or impossible to actually *invert* the model.
3. This approach, which we refer to as open-loop, does not allow the controller to "find out" the actual state of the process, if and when it differs from the model prediction.

The third point becomes crucial when the control mechanism does not have perfect knowledge of the disturbances affecting the process. In reality, the process is subject to disturbances, and the control mechanism needs to be supplied with information regarding the current value of the disturbance. This leads to an alternative control structure, known as the *feedforward control*, if we assume that this disturbance is measured. In this case, the corresponding scheme is as shown in Figure 2.8.

As one can observe, the control mechanism obtains information about changes in the disturbances and helps to produce an anticipative control action $m(t)$ to counteract the effect of the disturbance on the process.

However, in many practical cases, the disturbances will (or can) not be measured; hence our knowledge of the process will be imperfect. Then, it becomes imperative to supply current process information to the control mechanism so that it keeps track of the process behavior. This can be accomplished through the measurement of the output (control) variables as shown in Figure 2.9.

We can see that, in this case, the information obtained from the direct measurement of the output of the process is fed back to the control mechanism. In other words, the mechanism uses the knowledge of the current status of the plant to generate the control signal. This is the concept behind *feedback control.*

Based on the previous discussion, we can summarize four possible configurations:

*Case 0*: The control is implemented through a human operator. This is called a *manual control scheme.*

*Case 1*: The control mechanism acts without current information about the status of the process. This is called an *open-loop control scheme.*

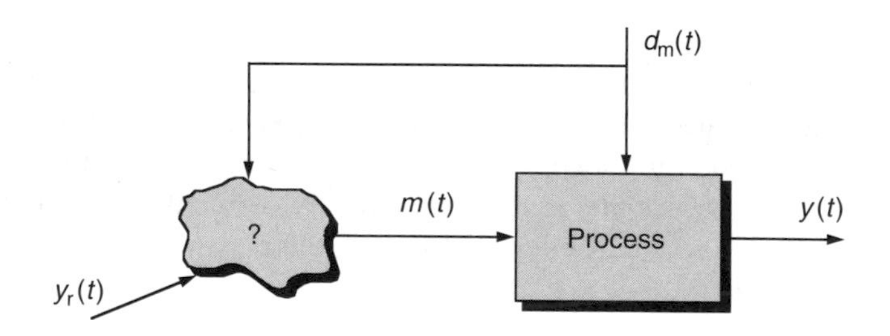


**FIGURE 2.8** The feedforward control scheme.

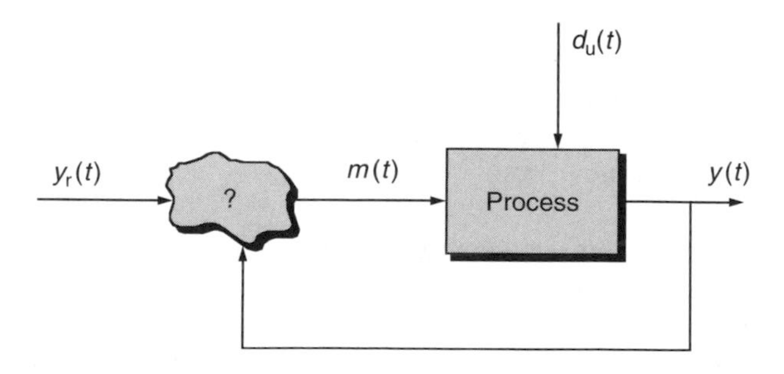


**FIGURE 2.9** The feedback control scheme.

*Case 2*: The control mechanism anticipates the effect of the disturbances, producing a corrective action. This is called anticipative control or a *feedforward control scheme.*

*Case 3*: The control mechanism acts using the information fed back from the measurements. This is called a *feedback control scheme*.

Naturally, there are many more possible alternatives that can be used, as we shall see later in this book and the cases, discussed above, are the most typical ones encountered in practice.

In summary, we can see that control design consists of a series of decision-making steps leading to a comprehensive design strategy.

## 2.3 CONTROL SYSTEM DESIGN

An engineer is expected to approach any design problem in a systematic manner. Accordingly, there is a series of steps to follow during the design of a control system to facilitate the decision-making process of the control engineer. They are outlined below:

1. *Define control objectives* This concerns with defining the operational objectives that a control system is called upon to achieve. The key objective is to maintain the process output variables as close as possible to their targets. By ensuring stability, eliminating disturbances, and optimizing the economic performance of the process, the control system strives to achieve this objective. Naturally, this generic objective statement needs to be translated into specific control objectives for the process in question. This is the task of the control engineer.
2. *Select measured variables* Secondly, we need some means to monitor the performance of the process and observe how it may respond to actions by the control system. This is accomplished by measuring the values of certain process variables (e.g., temperature, pressure, compositions, etc.). It is evident that we directly monitor the variables that represent the control objectives, and this is done whenever possible (we shall discuss cases later showing how a control system is designed when direct measurements are not possible).
3. *Select manipulated variables* Once the control objectives are specified and the measurements are identified, the next question is how we can cause a change in the process. In other words, what are the variables to be manipulated to achieve the control objectives? Usually, we have many options in this selection, and our decision may affect the quality of the process response, i.e., the control performance.
4. *Select control configuration* The control configuration is the information structure used to match the available measurements with the available manipulated variables. Normally, we will have many possible configurations (feedback, feedforward, and their combinations, SISO,

MIMO, etc.). What is best for a given problem is a critical question for the quality of the control system, and is problem-specific.

5. *Tune controller* The controller is the active element that receives the information from the measurements and takes appropriate control actions to adjust the values of the manipulated variables. The question to be answered here is how the information taken from the measurements is used to adjust the values of the manipulated variables. The answer to this question constitutes the *control law*, which is implemented automatically by the controller. As we shall see in later chapters, tuning implies determination of the parameters of the controller.

### Example 2.2

For the stirred-tank heater process in Example 2.1 we can follow the steps below:

1. *Control objective* Maintain the volume (level) and the temperature of the liquid in the tank at their desired (target) values in the presence of variations in the upstream conditions (disturbances), namely fluctuations in the inlet stream temperature and flow rate.
2. *Control variables* Our first attempt is to install measuring devices that will monitor the temperature of the liquid in the tank (remember this is the same temperature as $T_{out}$ due to perfect mixing), and the liquid level, $h$. For this process, this is accomplished by using a thermocouple (for $T_{out}$) and a differential pressure cell (for $h$).
3. *Manipulated variables* We can choose to manipulate the outlet flow rate, $F_{out}$, and the steam flow rate, $F_{st}$. Based on our engineering intuition, we can argue that the steam flow rate will only affect the tank temperature and by varying the outlet flow rate, we can influence the liquid level in the tank.
4. *Control configuration* We could choose either a *feedback control scheme* to deal with the disturbances (inlet stream flow rate and temperature) or even a *feedforward scheme*, since these disturbances can be easily measured. With two control variables and two manipulated variables, this is basically a MIMO control system.
5. *Controller tuning* Depending on the type of function that we suggest relating the control variables to the manipulated variables, we have to determine the parameters of this function. We have to propose relevant criteria for judging the acceptability of a set of parameters. Such criteria typically involve measures of performance for the controlled dynamic behavior of the process.

This systematic approach to control system design needs to be placed in perspective with respect to the inception, execution, and implementation of a plantwide design project.

## 2.4 CONTROL DESIGN PROJECT

As discussed in Chapter 1, the computer control systems became an integral part of enterprise management as they complement other plantwide activities such as information processing, data gathering, on-line optimization, as well as production planning and scheduling. This makes the task of the control engineer much more complex than what it has been in the past, which mostly centered around isolated instrumentation activities. In his or her new role, the control engineer needs to view the needs of the plant from a control standpoint to understand the information management tasks associated with these needs and to make sure that these needs are well balanced with the business goals of the company. Therefore, the control design project is a team effort that moves through a series of stages before reaching its conclusion.

Project teams are brought together to perform either a grassroot design project or a retrofit design project for the plant. While the former aims to build a facility from the ground up, the latter focuses on minor or major improvements in an existing facility. Such a capital project is managed through an integrated approach that takes into account the plant personnel, process technology, as well as the control and information technology.[1] Being part of the latter area, the control engineer needs to maintain a broad perspective in achieving the goals of the control design project.

The plantwide control system project consists of the following activities as discussed in detail by Erickson and Hedrick:[1]

- Preliminary engineering
- Detailed engineering
- Implementation
- Installation
- Commissioning
- Final production start-up and turnover
- Training

### 2.4.1 Preliminary Engineering

The control design project should be considered as part of the overall plant design project from the beginning. Ideally, operability of a plant should be one of the objectives during the conceptual design of a process along with profitability, feasibility, safety, and others. Such an integrated approach would reduce or eliminate costly overdesign factors and operational bottlenecks, and result in a more efficient and flexible plant operation. While the control project traditionally has been started after the plant design is finalized, this trend appears to be changing.

The control engineer typically works with a process flow diagram (PFD) generated during the conceptual plant design project. The PFD associated with a process may show less or greater detail depending on how far the design project has progressed. Generally, the PFD will include major process equipment; their connections and, depending on the design stage, may show material and energy balances and preliminary sizing information. For the control engineer, such a PFD can be the basis for the following activities:

- Develop the process operation description
- Develop the control concept

- Define the preliminary automation and control strategy
- Define the preliminary system architecture
- Prepare the preliminary control technology budget

These activities help to establish the boundaries of the process area, identify control objectives associated with units (as described in Section 2.3), define the extent of automation and the level and type of control system required, decide on the hardware and software platforms to be used for implementation, and a preliminary budget. In most cases, the control engineer works with control system vendors to evaluate the capabilities of available technologies and their relevance to the process control problems articulated at this stage.

### 2.4.2 Detailed Engineering

Once the preliminary control design project is approved, more detailed design activities are conducted. These activities typically result in the following products (deliverables):

- Specification of measurement and final control functions
- Specification of safety instrumented system functions
- Specification of discrete and regulatory control functions
- Specification of procedural control functions
- Specification on process information data models
- Specification of run management functions
- Specification of user-interface functions

These specifications (associated with performance, size, reliability, etc.) are generated to guide control technology development in all areas of hardware, software, and programming.

### 2.4.3 Implementation

At this stage, the hardware for measurement and final control elements are procured. The software associated with these elements are also obtained and tested to ensure that they meet design specifications. Implementation is not complete until the system is verified by these tests.

### 2.4.4 Installation

After the hardware and associated software are successfully tested, they are ready for installation at the site. This involves wiring of interconnections and their verification. The control system is set up, and communication over the local (and if necessary, global) network is tested.

### 2.4.5 Commissioning

This is the stage where each plant system is brought on-line in a systematic fashion and the associated technology is verified to satisfy the overall operational goals of the process. Initial control parameters can be changed (tuned) and various control settings can be experimented with to evaluate their impact on process performance.

### 2.4.6 First Production Startup and Turnover

Once the commissioning step is completed, the plant is ready for its first production. The control engineer (along with some design engineers) oversees the startup and, if this is satisfactory, turns over the plant to the plant personnel (operators). The control design project is completed at this point.

### 2.4.7 Training

Training is a critical activity that continues throughout the operational lifetime of the plant. It starts after the turnover and may focus on system-specific details as well as emergency handling activities. Training is a valuable activity especially for new personnel and can be customized according to their position and duties.

## 2.5 SUMMARY

In this chapter, we introduced the concepts associated with a process control problem, starting with the classification of variables. The control design problem involves the definition of feedback control and how it is different from open-loop and feedforward control configurations. The control system design starts with the definition of the control objective and the selection of control and manipulated variables that target this objective. It concludes with the selection of the control configuration and the determination of parameters in the control law. The control system design is part of the control design project that is, in turn, part of the capital plant project. With this perspective, this chapter also discusses the elements of the control design project from the preliminary design all the way to the commissioning of the control system and the startup and turnover of the plant.

## CONTINUING PROBLEM

A blending process was introduced in Chapter 1 (Figure 1.6). For this process:

- Classify all process variables in terms of the categories discussed in this chapter
- Define the control objectives
- Propose a set of control and manipulated variables and a control configuration for this process

## SOLUTION

### Classification of Variables

- *Input variables*: $F_1, F_2, x_1, x_2$
- *Possible manipulated variables*: $F_1, F_2$
- *Possible disturbances*: $x_1, x_2$
- *Output variables*: $F, x, V$
- *Possible measured variables*: $F, x, V$

If the cross sectional area of the tank is denoted by $A$, the liquid volume in the tank can be expressed as

$$V = Ah$$

We note that the volume changes as the level (height) of the liquid in the tank changes. Furthermore, the exit flow rate $F$ also varies with the level as the flow depends on the static liquid height in the tank. Thus, the actual measured variables to be considered are $h$ and $x$.

### Control Objectives

The control objectives are to produce a product of certain mass fraction $x$ and maintain it at its desired target value, and also to maintain a constant liquid level $h$ in the tank.

### Control Configuration

The level control can be accomplished by measuring the liquid level and manipulating the exit flow rate. The composition control would be accomplished by measuring the blend composition and manipulating one of the feed flow rates.

## REFERENCES

1. Erickson, K.T. and J.L. Hedrick, *Plantwide Process Control*, Wiley, New York, 1999.

# Section I Additional Reading

In Chapter 1, we pointed out that the origins of control applications can be traced back 2000 years to the Middle East, in particular to Egypt, Mesopotamia, and Anatolia. These regions were highly advanced socially and technologically during these years, leading to many early inventions. From the control and automation perspective, the following book, although it may not be easily accessible, presents an intriguing overview of some of the early mechanisms:

Bir, A., *'Kitab al-Hiyal' of Banu Musa Bin Shakir — Interpreted in Sense of Modern System and Control Engineering*, Studies and sources on the history of science, Series No. 4, Research Center for Islamic History, Art and Culture, Istanbul, Turkey, 1990.

The more recent history of control engineering can be found in several books, primarily by S. Bennett:

Bennett, S., *A History of Control Engineering, 1800–1930*, Peter Peregrinus on behalf of the Institution of Electrical Engineers, Stevenage, 1979.

Bennett, S., *A History of Control Engineering, 1930–1955*, Peter Peregrinus on behalf of the Institution of Electrical Engineers, Stevenage, 1993.

Mayr, O., *The Origins of Feedback Control*, MIT Press, Cambridge, MA, 1970.

The reader can find the achievements of some of the pioneers of the field of computer control in the following book:

ISA Ad Hoc Committee on the "Computer Control Pioneers", *The Computer Control Pioneers: A History of the Innovators and their Work*, Instrument Society of America, Research Triangle Park, NC, 1992.

For a more recent treatise on the control implications of human and machine interfaces from the perspective of cybernetics, the reader is referred to

Mindell, D.A., *Between Human and Machine: Feedback, Control, and Computing before Cybernetics*, Johns Hopkins University Press, Baltimore, MD, 2002.

There are several sources where the authors discuss the implications of process control for specific industrial applications. A more recent representative sample is provided below.

Ansari, R.M. and Tadé, M.O., *Nonlinear Model-Based Process Control: Applications in Petroleum Refining*, Springer, New York, NY, 2000.

Cinar, A., Parulekar, S.J., Undey, C., and Birol, G., *Batch Fermentation: Modeling, Monitoring and Control*, Marcel Dekker, New York, NY, 2003.

Erickson, K.T. and Hedrick, J.L., *Plantwide Process Control*, Wiley, New York, NY, 1999.

Johnson, C.D., *Process Control Instrumentation Technology*, Prentice-Hall, Upper Saddle River, NJ, 2000.

Luyben, W.L., Tyréus, B.D., and Luyben, M.L., *Plantwide Process Control*, McGraw-Hill, New York, NY, 1999.

Shinskey, F.G., *Process Control Systems: Application, Design, and Tuning*, McGraw-Hill, New York, NY, 1996.

Wang, L. and Cluett, W.R., *From Plant Data to Process Control: Ideas for Process Identification and PID Design*, Taylor & Francis, London, 2000.

Industrial practitioners and academic researchers meet every 5 years to discuss the current status of the process control field and assess its future, focusing on new technologies and research directions. The proceedings of these meetings are published and are a good source of information. The most recent meeting was held in Tucson, AZ.

Rawlings, J.B., Ogunnaike, B.A., and Eaton, J.W. (Eds.), *Chemical Process Control-VI: Assessment and New Directions for Research, Proceedings of the 6th International Conference on Chemical Process Control*, Cache: American Institute of Chemical Engineers, New York, NY, 2002.

In the last 10 years, we have seen many textbooks being published on the topic of process control. The reader can consult the following references for further reading and more detailed coverage of certain topics.

Bequette, B.W., *Process Control: Modeling, Design, and Simulation*, Prentice-Hall PTR, Upper Saddle River, NJ, 2003.

Chau, P.C., *Process Control: a First Course with MATLAB*, Cambridge University Press, New York, 2002.

Luyben, M.L. and Luyben, W.L., *Essentials of Process Control*, McGraw-Hill, New York, NY, 1997.

Marlin, T.E., *Process Control: Designing Processes and Control Systems for Dynamic Performance*, 2nd ed., McGraw-Hill, New York, NY, 2000.

Ogunnaike, B.A. and Ray, W.H., *Process Dynamics, Modeling, and Control*, Oxford University Press, Oxford, 1994.

Seborg, D.E., Edgar, T.F., and Mellichamp, D.A., *Process Dynamics and Control*, 2nd ed., Wiley, Hoboken, NJ, 2004.

For a complementary textbook on process control in which the key concepts are reviewed in the context of software experiments, the reader is referred to the textbook,

Doyle III, F.J., with Gatzke, E.P. and Parker, R.S., *Process Control Modules: A Software Laboratory for Control Design*, Prentice-Hall PTR, Upper Saddle River, NJ, 2000.

# Section I Exercises

**I.1.** What is the significance of James Watt's flyball governor from the viewpoint of automatic control?

**I.2.** Two liquids are mixed together in a tank and the product is removed through an exit stream at the bottom of the tank. We know that changes in the feedstream flow rates can cause the tank to overflow or completely drain, and this is undesirable. Is it possible to modify the design of the tank to avoid instability? How?

**I.3.** Neutralization of the pH of acidic effluents is a very important problem in wastewater treatment processes. Figure I.1 depicts such a process.

The incoming stream contains mixed acids with varying buffering characteristics and a standard base stream is available for neutralization. Define all the relevant variables for the process as required.

1. State the control objective(s) for this process.
2. What are the possible controlled variables, manipulated variables, and disturbances?
3. Suggest a feedback and a feedforward control strategy for this process and show them schematically.

**I.4.** Frying is one step of an integrated potato chip manufacture line that includes a sequence of operations: cleaning, peeling, slicing, washing, frying, seasoning, and packaging.[1] During frying, starch is gelatinized, water content of the slice is reduced, and absorbing the cooking oil enhances texture and flavor of the chip. For a continuous fryer shown in Figure I.2, draw an I/O diagram, and state all possible control objectives. Classify the output variables, the manipulated variables and the disturbances for this process. Suggest a possible feedback control loop and justify its role with respect to the control objectives.

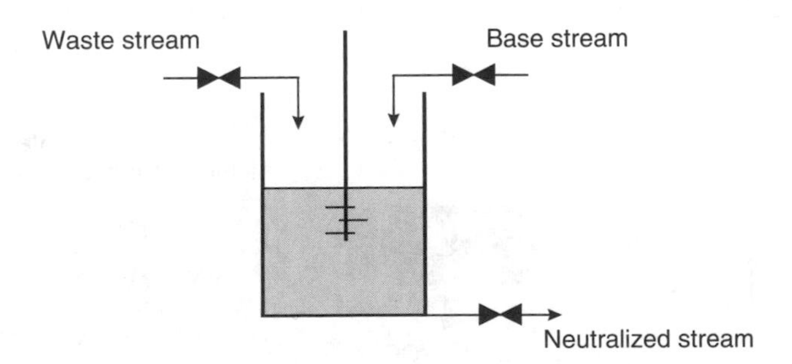


**FIGURE I.1** A pH neutralization process.

**I.5.** To effectively remove volatile organic compounds (VOCs) and hazardous air pollutant emissions, one can use a membrane separation process. This is a viable process when the air stream contains relatively high concentration (10,000 ppm) of vapors. A schematic representation of this process is sketched in Figure I.3.

The VOC-contaminated air stream is first compressed (to about 45–200 psig) and the compressed mixture is sent to the condenser where the organic vapor condenses and is recovered for reuse. The noncondensed air stream (typically 1% organics) proceeds to the membrane unit. To induce selective permeability of the gases, a pressure difference is created across the membrane by a vacuum pump. Organic vapors are enriched on the permeate side and returned upstream. Cleaned gas is vented to atmosphere. For this process,

1. Identify all relevant control objectives and indicate which is the primary one. State all possible disturbances.
2. Identify all controlled variables (outputs) and the available manipulated variables (inputs).
3. Suggest a feedback controller and a feedforward controller to satisfy a control objective identified in (1).

**I.6.** The activated sludge process is the most widely used biological system in wastewater treatment. The process consists of an aeration tank and a settler or clarifier. Wastewater containing some polluting organic substrate (measured by

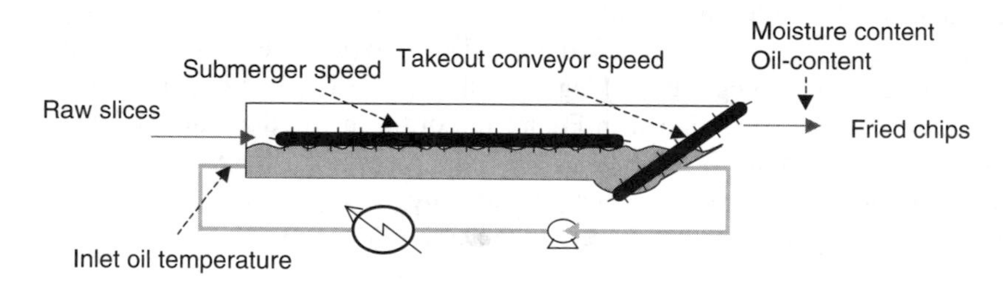


**FIGURE I.2** Schematic representation of frying operation for potato chips.

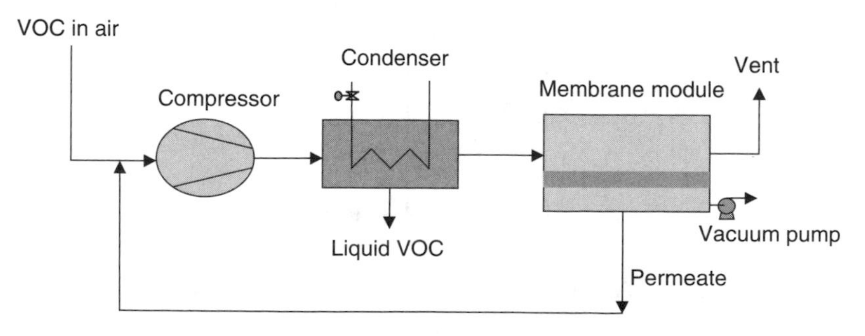


**FIGURE I.3** Schematic representation of a membrane separation process.

its biochemical oxygen demand, BOD) is fed to the aeration basin (operates like a stirred-tank reactor) where the sludge, consisting of various micro-organisms, grows with the consumption of organic substrate (pollutant) and oxygen (supplied as air), thus removing the BOD. The sludge is then passed to the clarifier where it is settled and a part of it is recycled back to the aeration basin. A schematic representation is given in Figure I.4.

For this process,

1. Identify all relevant control objectives and indicate which is the primary one. State possible disturbances.
2. Identify controlled variables (outputs) and the available manipulated variables (inputs).
3. Suggest a feedback controller to satisfy a control objective identified in (1), and draw it schematically.

**I.7.** Solution crystallization is a type of crystallization where there is a solid (solute) separated from a liquid (solvent). A way of achieving this is by cooling the solution to achieve the supersaturation levels that force the solid out of the solution by way of nucleation and growth of crystals. Solution supersaturation is the driving force in any crystallization process and has been recognized as a key variable. Manipulation of the solution supersaturation directly influences nucleation and growth kinetics, which consequently affect the attributes of the crystal product. The size of the crystals formed (actually, one ends up with a large number of different particle sizes) is heavily dependent on the cooling regime. Usually, the downstream processing of the crystal product dictates the objective of the crystallization. For instance, a large crystal size may be demanded where filtration is a post-crystallization operation.

Consider the crystallization of ammonia sulfate from an aqueous solution as shown in Figure I.5. A cooling medium provides the necessary heat transfer through the jacket around the vessel.

For this process,

1. Identify the control objective. Are there any disturbances?

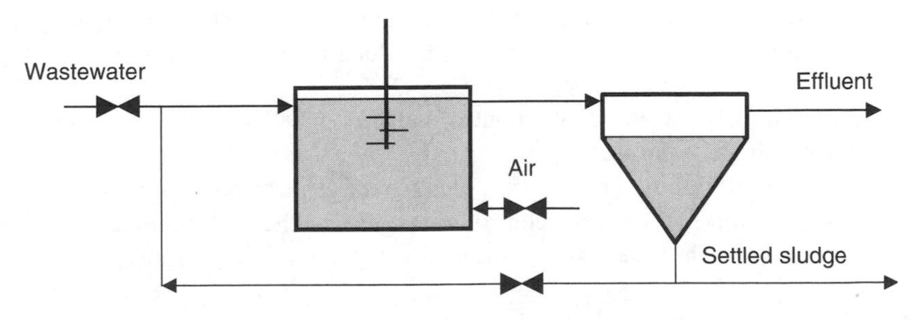


**FIGURE I.4** Schematic representation of an activated sludge process.

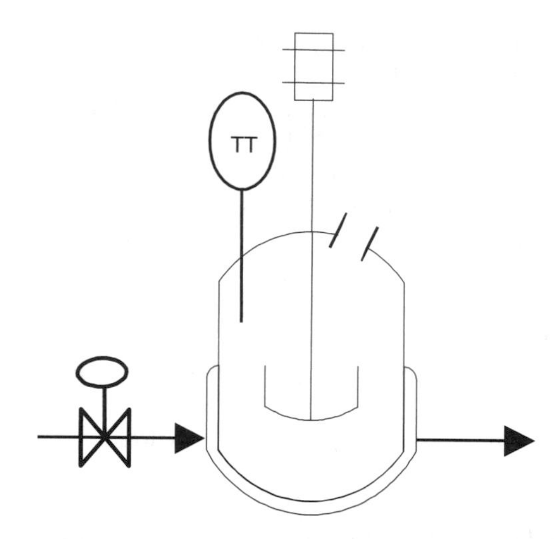


**FIGURE I.5** The sketch of cooling crystallization of ammonia sulfate.

2. Identify the controlled variable (output) and the available manipulated variable (input).
3. Suggest a feedback controller to satisfy the control objective identified in (1), and draw it schematically.
4. Can you suggest a strategy to obtain the optimum crystal properties?

**I.8.** A tank is used to accommodate the difference between the supply (single inlet flow) and demand (single outlet flow) of cooling water as sketched in Figure I.6. It is known that the inlet and outlet flows are subject to magnitude variations according to the power consumption of the pump and the resistance of the valve in the pipeline.

A team of process engineers has been assigned the task to develop a control strategy for the system in Figure I.6.

1. Assume that a control valve is installed in the inlet line, but not in the outlet line:
   - Classify all the process variables from a control point of view.
   - Develop two possible control strategies for the system under study, and sketch these configurations for documentation purposes.

2. Alternatively, assume that a control valve is installed in the outlet line, but not in the inlet line:
   - Classify all the process variables from a control point of view.
   - Develop two possible control strategies for the system under study and sketch these configurations for documentation purposes.

**I.9.** A fired heater (furnace) is used to heat a process stream containing an intermediate product that will be sent to a downstream reactor. It is not only important

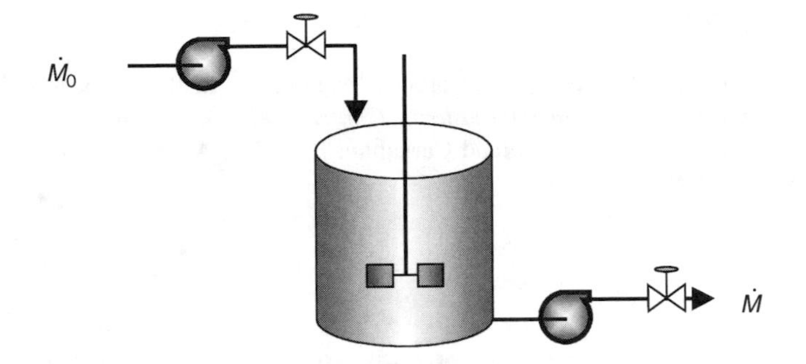


**FIGURE I.6** The surge tank for cooling water.

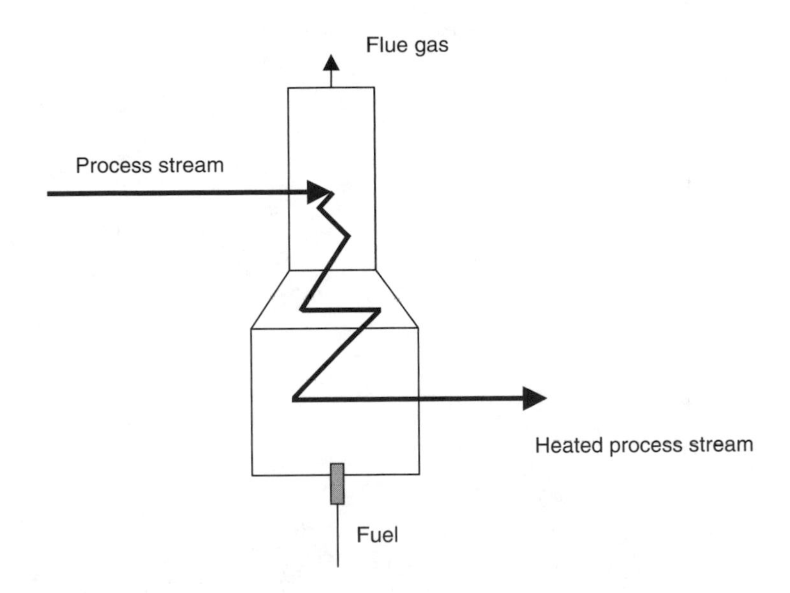


**FIGURE I.7** Schematic representation of a furnace.

to maintain a constant outlet temperature but also keep the furnace temperature below a certain critical value for safety reasons (metallurgical limits of the furnace tubes). A simple schematic representation is shown in Figure I.7.

1. What is the control objective for this system?
2. Identify all external disturbances that will affect the operation of the furnace.
3. Identify all manipulated variables for the control of this unit in the presence of disturbances.
4. Construct a feedback control configuration and a feedforward control configuration that would satisfy the control objectives in the presence of disturbances.

## REFERENCES

1. Nikolaou, M., Computer-aided process engineering in the snack food industry, *Proceedings of the 5th International Conference on Chemical Process Control*, Kantor, J.C., Garcia, C.E., and Carnahan, B. (Eds.), AIChE Symposium Series, Vol. 93, 1997.

# *Section II*

## *Modeling for Control*

# 3 Basic Concepts in Modeling

An engineer must understand the dynamic behavior of a physical system in order to design the equipment, select its operating conditions, and properly implement an automatic control strategy. Such an understanding can be attained by observing the system over time, by making changes in the system, and monitoring the consequences of these changes. This leads to an abstraction of the system's behavior in the engineer's mind, which can be referred to as a *conceptual model* of the system. Indeed, experienced engineers make decisions based on such intuition all the time, demonstrating their knowledge of the system's behavior and their associated expertise. Another expression of the system is the use of mathematical relationships that can explain the physical and chemical phenomena underlying the system's behavior. This gives us *mathematical models* that can be used rigorously to:

1. Improve process understanding
2. Optimize process operation
3. Train personnel
4. Design and evaluate control systems

In more precise terms, Denn[1] provides the following definition:

---

A mathematical model of a process is a system of equations whose solution, given specified input data, is representative of the response of the process to a corresponding set of inputs.

---

We are particularly interested in the evolution of the process over time. Remember that the objective of control is to maintain desired levels of a process variable by manipulating another variable, all in the presence of disturbances. One can observe that all of these variables are time-dependent, hence requiring a *dynamic* model for expressing the behavior of the process.

We should also note that mathematical modeling is still considered an art form, as it relies on past experience and engineering intuition as well as fundamental knowledge.

## 3.1 TYPES OF MODELS

Why model? First and foremost, the objective of modeling is to provide a computational setting within which the behavior of a process can be examined in response to changes in the inputs. For each specific application, the model will be developed to analyze a particular behavior. For example, a model of a distillation column can be used to study flooding phenomena in response to variations in the feed conditions, and their impact on the product quality.

The development of a model and its structure are intimately related to the goals of modeling. In other words, the sophistication of a model should be commensurate with the ultimate application in which it will be used. If one needs to understand precisely the temperature profiles in a gas-phase catalytic reactor to avoid catalyst sintering, the model should try to reflect accurately all physical and chemical phenomena taking place in the gas phase as well as on the catalyst pellets. On the other hand, if one is only interested in finding out the general trending in a blending process to understand the impact of a blend component, a simple relational model may be sufficient.

In this section, we introduce different types of models that represent a variety of process behaviors and are derived from various information sources.

- *Fundamental model* This type of model utilizes the basic laws of physics, chemistry, biology and thermodynamics to arrive at a set of equations that describes the process behavior. A good understanding of the physical and chemical phenomena that underlie the process is required in order to develop the model accurately. For chemical processes, the development of *fundamental models* starts from the use of material, energy and momentum balances.
- *Empirical model I (black-box)* In the absence of a clear understanding of the governing phenomena in a process, a model can still be developed by collecting data during operation (or by planned experiments that reflect the region of operation) and constructing a mathematical relationship among the variables that can explain the observed data. Such models that rely solely on empirical information are referred to as *black-box* models.
- *Empirical model II (gray-box)* Most models will be developed by incorporating empirical knowledge into the fundamental understanding of the process. An example can be the use of mass and energy balances to develop a reactor model in which the rate of the reaction will be based on expressions obtained from laboratory experiments. Such models blending fundamental and empirical knowledge are referred to as *gray-box* models.

## 3.2 CLASSIFICATION OF MODELS

Regardless of whether a process is modeled using empirical or fundamental knowledge, the models can still be classified as a function of certain distinguishing characteristics of their mathematical expressions. These classifications, in a certain way, define *a priori* the degree of difficulty not only in the development and solution of the mathematical model but also the design of the associated control system.

We must keep in mind that the classification of the models is a difficult task and sometimes arbitrary, however, there are general characteristics that can be used to group some of the most commonly used models. The following classifications are offered here:

- Linear vs. nonlinear
- Lumped parameter vs. distributed parameter
- Deterministic vs. stochastic

*Linear vs. nonlinear* Consider a water tank equipped with an immersed heating coil shown in Figure 3.1. One can observe that the water temperature increases as in Figure 3.2a in response to a 10% increase in the heating power (Figure 3.2b). One can also observe that the water temperature decreases as the heating power is turned down by 10%. If the process were linear, one would have observed a mirror image of the variation in the tank temperature. The fact that the system behaves differently when perturbed identically in different directions is one sign that the system indeed has nonlinear characteristics. The most important property of linear systems is that they satisfy the principle of superposition. The principle of superposition establishes that the response produced by the simultaneous application of two different excitatory functions is the sum of the individual responses. A differential (or an algebraic) equation is called nonlinear if it does not satisfy the superposition principle.

The following equations that describe the temporal evolution of an arbitrary variable $x(t)$ are classified as nonlinear:

$$\frac{d^2x}{dt^2} + \left(\frac{dx}{dt}\right)^2 + x = \sin t \tag{3.1}$$

$$\frac{d^2x}{dt^2} + (x^2-1)\frac{dx}{dt} + x = 0 \tag{3.2}$$

A clear sign of nonlinearity in these equations is the presence of multiplication terms involving dependent variables and their derivatives that prevents these equations from satisfying the superposition principle.

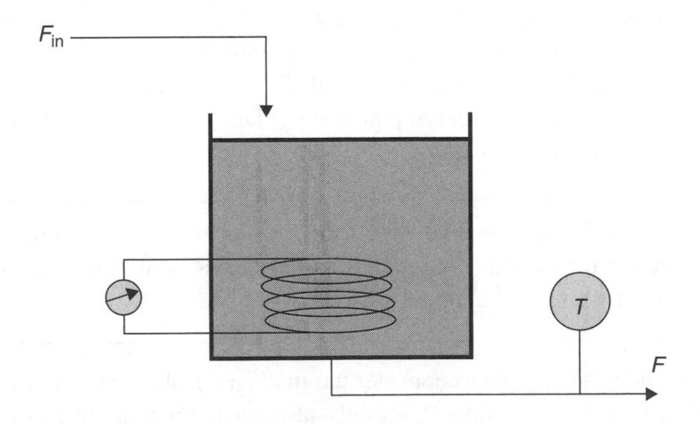


**FIGURE 3.1** A tank with a heating coil.

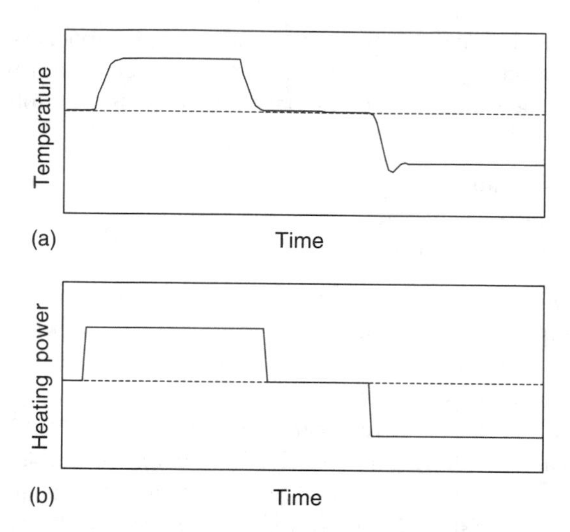


**FIGURE 3.2** Temperature of exit stream (a) in response to changes in heating power (b).

Although most physical systems behave nonlinearly, they are frequently modeled through linear equations. A careful study indicates that most of the so-called linear systems in fact behave linearly only in a limited range of operation. Due to the mathematical difficulty of dealing with nonlinear systems, it is necessary to introduce the idea of *equivalent* linear systems using linearization techniques, as we shall see later.

*Lumped parameter vs. distributed parameter* In a room heated by a space heater (or a fireplace), one notices that the area closer to the heater gets warmer first and the temperature falls as we go away from the heater. This is an indication that the room, as a system, can be classified as a distributed parameter system (DPS) because its temperature not only varies with time but also with spatial position. Lumped parameter systems (LPS) are those in which the spatial variations are ignored (or do not exist) and the dependent variables (concentrations, pressures, temperatures, etc.) are considered uniform within the chosen control volume. The only variations are those with respect to time; thus, they lead to models represented by ordinary differential equations (ODEs) as opposed to DPS, which need to be modeled by partial differential equations (PDEs) due to the presence of both temporal and spatial derivatives.

---

The type of mathematical model will depend on the assumptions initially made to define the system. In general, the more assumptions we make, the simpler the structure of the model will be.

---

To illustrate this point, let us consider the mathematical description of the cooling of a sphere of steel of radius $R$, initially at a given temperature $T_0$ in a current of air at temperature $T_1$. The transient one-dimensional conduction equation governing the temperature of the sphere $T(r, t)$ is

$$\frac{\partial T}{\partial t} = \alpha\left(\frac{\partial^2 T}{\partial r^2} + \frac{2}{r}\frac{\partial T}{\partial r}\right) \tag{3.3}$$

The initial and boundary conditions are given as follows:

$$T(r, 0) = T_0 \quad \text{at } t = 0 \tag{3.4}$$

$$\frac{\partial T}{\partial r} = 0 \quad \text{at } r = 0 \tag{3.5}$$

$$-k\frac{\partial T}{\partial t} = h(T(R, t) - T_1) \quad \text{at } r = R \tag{3.6}$$

where $\alpha = k/\rho c_p$ is the thermal diffusivity of the sphere and $h$ the film heat transfer coefficient. Now, if $\alpha$ is large, then, from a practical point of view, spatial variations of temperature within the sphere would be negligible. Hence, the problem of describing the dynamic behavior of the cooling sphere can be reduced to the solution of the following energy balance equation:

$$\begin{gathered} \rho c_p\left(\frac{4}{3}\pi R^2\right)\frac{\mathrm{d}T}{\mathrm{d}t} = h(4\pi R^2)(T(t) - T_1) \\ T(0) = T_0 \end{gathered} \tag{3.7}$$

which is now a *lumped* parameter model, as opposed to the previous model (Eq. [3.3]), which was a *distributed* parameter model.

*Deterministic vs. stochastic* Deterministic systems are those in which there are no random elements in their mathematical description. In other words, the values of the variables and parameters are fixed numbers and the solution of the mathematical model leads to an exact value of the response. For example, consider again the sphere model (Eq. [3.7])

$$\rho c_p\left(\frac{4}{3}\pi R^2\right)\frac{\mathrm{d}T}{\mathrm{d}t} = h(4\pi R^2)(T(t) - T_1)$$

When $\rho$, $c_p$, $h$, $T_1$, and $R$ are completely specified, the sphere model is called deterministic, as it yields a unique solution for $T(t)$.

Stochastic systems admit random (probabilistic) elements in their mathematical description. In a stochastic model, we may never know the exact value of a quantity but rather the probability associated with the observation of a certain value of the variable. Consequently, the output of a stochastic model is the associated probability rather than a fixed defined number.

For example, in the sphere model, if the air temperature $T_1$ changes in a random fashion, then the model is considered to be stochastic. To study this system, we would need to introduce additional elements for the dependent variables such as their mean values, covariances, and standard deviations, i.e., we would need to know the parameters that can quantify the probability distribution of each variable (see Appendix D).

The theory of control is well developed for linear, deterministic, lumped models. These simple models are clearly not the "best" but will be acceptable for the intended purpose. From the control perspective, we focus on state–space and input–output models.

## 3.3 STATE–SPACE MODELS

---

The state of a dynamic system is the smallest set of variables (called the *state variables*) such that the knowledge of these variables at $t = t_0$, together with the knowledge of the input for $t \geq t_0$, completely determines the behavior of the system for any time $t \geq t_0$.[2]

---

A state–space model explicitly shows the dynamic evolution of state variables associated with the process. In chemical processes, the dynamic models that are developed from material and energy balances are typically first-order differential equations. Hence, a generic state–space model can be expressed as

$$\begin{aligned}
\frac{dx_1}{dt} &= f_1\,(x_1, x_2, \ldots, x_N;\, m_1, m_2, \ldots, m_K;\, d_1, d_2, \ldots, d_L;\, a_1, a_2, \ldots, a_R) \\
\frac{dx_2}{dt} &= f_2\,(x_1, x_2, \ldots, x_N;\, m_1, m_2, \ldots, m_K;\, d_1, d_2, \ldots, d_L;\, a_1, a_2, \ldots, a_R) \quad (3.8) \\
&\vdots \\
\frac{dx_N}{dt} &= f_N\,(x_1, x_2, \ldots, x_N;\, m_1, m_2, \ldots, m_K;\, d_1, d_2, \ldots, d_L;\, a_1, a_2, \ldots, a_R)
\end{aligned}$$

Here, the state variables are $x_i$, the input variables are $m_i$, the disturbance variables are $d_i$, and the model parameters are $a_i$. The model is an $N$-dimensional state–space model with $K$ manipulated inputs, $L$ disturbances, and $R$ parameters.

We can define $J$ output variables, as dictated by $J$ control objectives, which are normally functions of the state as well as the input variables:

$$\begin{aligned}
y_1 &= h_1\,(x_1, x_2, \ldots, x_N;\, m_1, m_2, \ldots, m_K;\, d_1, d_2, \ldots, d_L;\, a_1, a_2, \ldots, a_R) \\
y_2 &= h_1\,(x_1, x_2, \ldots, x_N;\, m_1, m_2, \ldots, m_K;\, d_1, d_2, \ldots, d_L;\, a_1, a_2, \ldots, a_R) \quad (3.9) \\
&\vdots \\
y_J &= h_J\,(x_1, x_2, \ldots, x_N;\, m_1, m_2, \ldots, m_K;\, d_1, d_2, \ldots, d_L;\, a_1, a_2, \ldots, a_R)
\end{aligned}$$

---

State–space models are typically derived from fundamental principles but they can also be obtained from empirical data.

---

## 3.4 INPUT–OUTPUT MODELS

It was discussed before that, in general, a process could be viewed as in Figure 3.3, via its input–output description. This is a more appealing description from the process control viewpoint, as it identifies the most important classification of variables that will be exploited when designing and analyzing control systems. In this case, a convenient conceptual model for control purposes would be expressed as

$$\text{output variables} = f\,(\text{input variables; parameters}) \tag{3.10}$$

or the $j$th output variable can be expressed as a function of all input variables and model parameters.

$$y_j = f_j\,(m_1, m_2, \ldots, m_K;\ d_1, d_2, \ldots, d_L;\ a_1, a_2, \ldots, a_R),\ j = 1, \ldots, J \tag{3.11}$$

---

The input–output model directly represents the cause and effect relationships in a processing system.

---

### Example 3.1

We shall consider the two tanks given in Figure 3.4. With a constant cross-sectional area $A$, the volume of a tank can be expressed as

$$V = Ah$$

where $h$ represents the liquid level. Let us suppose that the control objective is to maintain a constant level in the second tank, while the inlet flow rate to the first tank

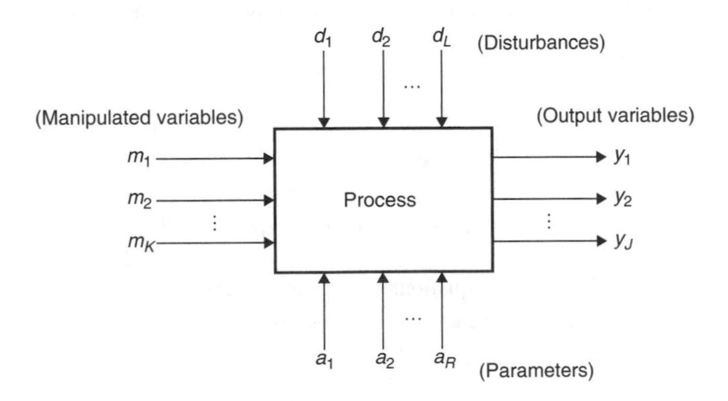


**FIGURE 3.3** Input–output representation of a process.

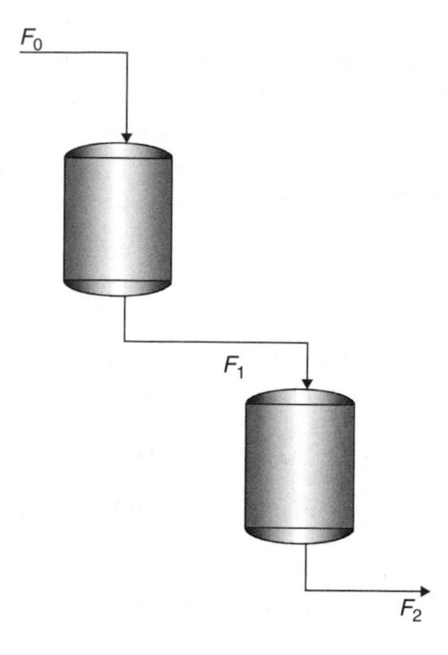


**FIGURE 3.4** Schematic representation of two tanks in series.

is varied. The liquid level is the output (control) variable and the inlet flow rate is the input (manipulated) variable. In Chapter 4, we shall see how a mathematical model for such a process is constructed, but it should suffice to point out here that the dynamic behavior of the process is governed by the transient mass-balance equation for each tank:

$$A_1 \frac{dh_1}{dt} = F_0 - F_1$$
$$A_2 \frac{dh_2}{dt} = F_1 - F_2 \tag{3.12}$$

We shall assume that $F_1 = \alpha_1 h_1$, $F_2 = \alpha_2 h_2$ (for simplicity), capturing the (linear) resistance for the exit piping with the constant $\alpha$, which accounts for the friction effects. Therefore,

$$A_1 \frac{dh_1}{dt} = F_0 - \alpha_1 h_1$$
$$A_2 \frac{dh_2}{dt} = \alpha_1 h_1 - \alpha_2 h_2 \tag{3.13}$$

We shall first place the above equations in the state–space model form. This is done by defining the states, the input and the parameters as follows:

$$x_1 = h_1$$
$$x_2 = h_2$$
$$m = F_0$$

$$a_1 = \alpha_1/A_1$$
$$a_2 = 1/A_1$$
$$a_3 = \alpha_1/A_2$$
$$a_4 = \alpha_2/A_2$$

This leads to the state–space model

$$\frac{dx_1}{dt} = -a_1 x_1 + a_2 m$$
$$\frac{dx_2}{dt} = a_3 x_1 - a_4 x_2 \tag{3.14}$$

To complete the state–space model, the output is expressed as

$$y = x_2 \tag{3.15}$$

Equations (3.14) and (3.15) constitute the state–space model for this process. Next, one can attempt to cast this model in terms of an input–output model. To accomplish this task, we first need to differentiate the second state equation

$$\frac{d^2x_2}{dt^2} = a_3\frac{dx_1}{dt} - a_4\frac{dx_2}{dt} = a_3(-a_1x_1 + a_2m) - a_4\frac{dx_2}{dt} \tag{3.16}$$

and replace the first state equation in Eq. (3.16), and rearrange:

$$\frac{d^2x_2}{dt^2} = a_3\left(-a_1\left(\frac{1}{a_3}\frac{dx_2}{dt} + \frac{a_4}{a_3}x_2\right) + a_2m\right) - a_4\frac{dx_2}{dt} \tag{3.17}$$

By also recognizing Eq. (3.15), this finally results in a single equation describing the effect of the input on the output:

$$\frac{d^2y}{dt^2} + (a_1 + a_4)\frac{dy}{dt} + a_1a_4y = a_2a_3m \tag{3.18}$$

While input–output models, such as Eq. (3.18), are conceptually useful, their form needs to be defined carefully to facilitate their subsequent use in control design and analysis studies. In Chapter 5, we will discuss the Laplace transform, which leads to a very simple and elegant method of solving linear or linearized differential equations resulting from the mathematical modeling of processes, and, thus, will allow us to obtain an algebraic input–output model.

The development of models needs to be placed within the context of the requirements of the control problem. To define the control problem fully, we need to have a sufficient number of relationships. The existence of such relationships (or their lack thereof) determines if the problem is solvable.

## 3.5 DEGREES OF FREEDOM

An important aspect to consider in studying a given process, whose behavior is represented by a series of variables and fundamental relationships that govern them, is the concept of *degrees of freedom.*

---

Degrees of freedom, F are the number of variables that we must specify to completely define a given process.

---

In modeling for control, available degrees of freedom need to be well understood as the control engineer is ultimately responsible for recognizing all variables and the relationships between them so that an effective control strategy can be implemented with desired results.

---

Control of a given process will be feasible when all its degrees of freedom are exhausted.

---

To introduce the concept of degrees of freedom, let us consider the system given in Example 3.1. In this example, we obtained the following expression that relates the input and output variables:

$$\frac{d^2y}{dt^2} + (a_1 + a_4)\frac{dy}{dt} + a_1a_4y = a_2a_3m$$

Can this equation be solved? If it can be solved, how many solutions would it have? To answer these questions, we must consider the number of equations and the number of variables. We know that $a_i$ are known constants; hence, we can observe that

No. of equations: 1
No. of variables: 2 $(y, m)$

In principle, the number of variables that can be arbitrarily specified is given by the difference between the number of variables, $N_v$ and the number of equations, $N_e$. In other words,

$$\text{F (degrees of freedom)} = N_v - N_e \qquad (3.19)$$

---

To specify the problem completely and to obtain a unique solution, F must be equal to zero.

---

We have three different cases according to the value of F, i.e.,

$$\begin{aligned} &F < 0 \\ &F = 0 \\ &F > 0 \end{aligned} \tag{3.20}$$

The first case (F < 0) seldom arises in practice as the number of variables is rarely less than the number of equations. Yet, it may be possible to formulate a problem by introducing extra (redundant) relationships that can lead to an *overdetermined* system. This system will not have a solution until redundant relations are recognized and removed.

Typically, we will have more variables than equations (F > 0) (*underdetermined*), leading to multiple (possibly infinite) solutions of the system of equations.

In the light of this discussion, the model in Eq. (3.18) can be analyzed as follows:

- Yes, there is at least one solution.
- Indeed, there are an infinite number of solutions for each value of $m$.

In other words, if the input $m(t)$ is known (or specified), then there is a unique solution to the output $y(t)$ as a solution of this second-order ODE.

The problem of control is intimately tied to the concept of degrees of freedom, since the presence of a control relationship removes a degree of freedom. This will be discussed later in the book.

## 3.6 MODELS AND CONTROL

While tackling a control design problem, the engineer is basically faced with a given process and must design the remaining part (i.e., the control system) in such a way that the system as a whole satisfies a series of previously determined specifications. Through modeling, we translate a real-world problem into an equivalent mathematical problem. The engineer faces, on the one side, the real physical process, and on the other, its abstract mathematical representation. These worlds are connected by the activities of modeling and implementation (Figure 3.5).

*Modeling* allows us to translate the process and its objectives into the abstract mathematical formulation.

*Implementation* involves all aspects necessary to translate a mathematical model or its results into a physical system.

---

It is important to note that modeling and implementation tasks are always approximations.

---

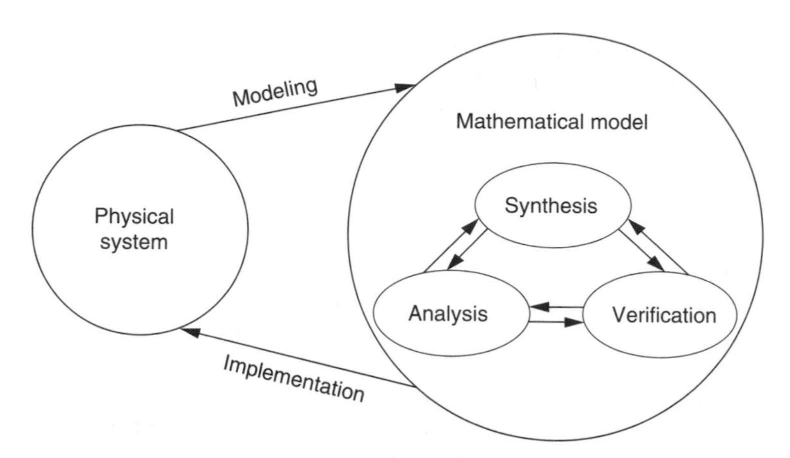


**FIGURE 3.5** Conceptual illustration of the role of models in control development.

Once the design problem is formulated in terms of a model, a mathematical project is carried out that produces the solution of the mathematical version of the design problem. Within the mathematical formulation, we can distinguish three important activities:

1. *Analysis* allows us to infer properties of the physical process through the study of the model.
2. *Synthesis* involves the construction of a function that satisfies the proposed objectives (i.e., the control function).
3. *Verification* is the first test (within the mathematical formulation) of the satisfaction of the objectives. At this stage, we take advantage of computer simulations based on the mathematical model.

It may happen that the proposed solution does not satisfy the proposed objectives. In this case, we have to redesign the control system and complete the corresponding analysis. This process is repeated until a satisfactory design is obtained. The final stage is the implementation on the physical unit or plant where the engineer performs the final test for satisfactory performance.

---

In model-based control strategies, the model itself is part of the solution, i.e., part of the control formulation, as will be seen later in Chapter 16.

---

## 3.7 SUMMARY

This chapter introduced the idea of mathematical modeling as one of the first steps in solving a process control design problem. It is important to understand the types of models one can develop, especially keeping in perspective their final application. As we shall see in subsequent chapters, state–space and input–output models constitute the majority of modeling approaches. Furthermore, we have seen that a process model has a central role in all levels of the development of control systems.

## REFERENCES

1. Denn, M.M., *Process Modelling*, Longman, New York, 1986.
2. Ogata, K., *Modern Control Engineering*, 2nd Ed., Prentice-Hall, Englewood Cliffs, NJ, 1990.

# 4 Development of Models from Fundamental Laws

The first step in any process control study is forming a sufficiently comprehensive understanding of the process to be controlled and its dynamic behavior. Modeling is certainly one of the elements in this activity and perhaps the most important one. A model will help us to create an abstraction of the process and, in turn, will be the source of additional insight into its behavior. In this chapter, we will introduce some of the key aspects of mathematical modeling using the fundamental laws and set the stage for ultimately developing input–output models for control design and analysis purposes.

## 4.1 PRINCIPLES OF MODELING

1. *Basis* In this chapter, mathematical models are based on the fundamental laws, such as the laws of conservation of mass, energy, and momentum. To study process dynamics, we will express them in their unsteady-state (transient) form.
2. *Assumptions* In creating an abstraction of the process, an engineer has to exercise sound judgment regarding what simplifying assumptions can be made and validated. Obviously, a rigorous model that encompasses every phenomenon down to the microscopic detail would be so complex that it would require extensive time and effort to develop and might be impractical to solve. An engineering compromise between a rigorous process description and an answer that is good enough for the intended purpose (in this case control) is always preferred.
3. *Consistency* Once all the model equations have been determined, it is considered prudent, particularly with large, complex systems of equations, to make sure that the number of unknown or unspecified variables equals the number of equations. The so-called *degrees of freedom* of the system must be zero in order to obtain a unique solution. If this is not the case, the system may be underdetermined or overdetermined and the formulation of the problem would be inconsistent.
4. *Solution* The available solution techniques and tools must be kept in mind as the mathematical model is developed. A model that is unsolvable analytically or numerically is of very little value.
5. *Verification* An important but often neglected part of developing a mathematical model is proving that the model describes the real situation accurately. For this purpose, one has to verify that the model can satisfactorily

predict the process behavior by comparing the predictions with data collected from the actual process.

## 4.2 MODELS BASED ON FUNDAMENTAL LAWS

In order to characterize a processing system (tank heater, batch reactor, distillation column, heat exchanger, etc.) and its behavior, we need:

1. A set of fundamental dependent quantities whose values describe the natural state of the system | State Variables
2. A set of equations that describe how the natural state of the given system changes with time | State Equations

The state equations are derived from application of the conservation laws:

1. Total mass balance
2. Component mass balance
3. Energy balance
4. Momentum balance

The conservation law for a quantity ***S***, within a defined control volume, establishes that:

Accumulation of ***S*** per unit of time = Amount of ***S*** in per unit of time
− Amount of ***S*** out per unit of time
+ Amount of ***S*** generated per unit of time
− Amount of ***S*** consumed per unit of time

where ***S*** could be total mass, mass of individual components, total energy, or momentum.

The application of conservation laws will yield a set of *differential equations* with the state variables as dependent variables and time as the independent variable. The solution of these differential equations will determine how the state variables evolve with time; i.e., it will characterize the *dynamic behavior* of the process.

### Example 4.1

Consider a liquid surge tank with constant cross-sectional area depicted in Figure 4.1 where the control objective is to maintain the liquid level at a desired reference value.

The fundamental quantity that provides the information about the state of the tank is the total mass of the liquid in the tank.

Let us start with the law of conservation of mass, assuming that the liquid properties are uniform within the control volume of the tank. The total mass in the tank is expressed as

$$\text{Total mass} = \rho V = \rho A h \tag{4.1}$$

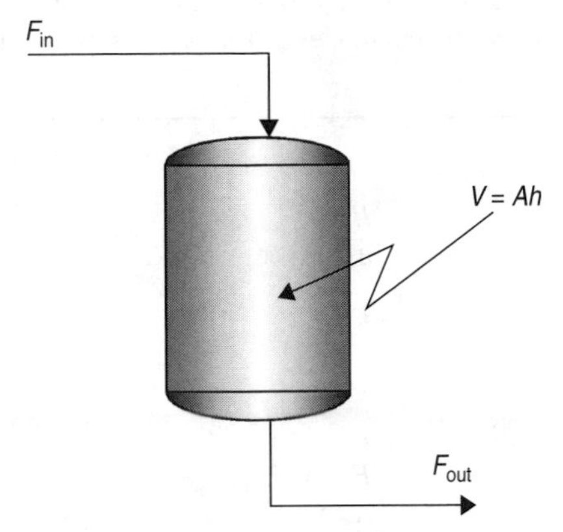


**FIGURE 4.1** A liquid surge tank.

where $\rho$ is the density of the liquid, $V$ the volume of the liquid in the tank, $A$ the cross sectional area of the tank, and $h$ the level of the liquid in the tank.

We can observe that the state variable and the parameters for the system are:

- State variable: $h$
- Constant parameters: $\rho$ and $A$

---

We assume that the density $\rho$ is constant.

---

Applying the conservation law to the fundamental quantity, the total mass becomes

$$\frac{[\text{Accumulation of total mass}]}{\text{time}} = \frac{[\text{Input of total mass}]}{\text{time}} - \frac{[\text{Output of total mass}]}{\text{time}} \tag{4.2}$$

or

$$\frac{d(\rho A h)}{dt} = \rho F_{in} - \rho F_{out} \tag{4.3}$$

In this equation, $F_{in}$ and $F_{out}$ denote the volumetric flow rates of the inlet stream and the outlet stream, respectively.

Eliminating density from Eq. (4.3) and recognizing that the cross-sectional area is constant, we have the model equation that describes the change of level in the tank as a function of the inlet and outlet flow rates.

$$A\frac{dh}{dt} = F_{in} - F_{out} \tag{4.4}$$

This state equation along with all the variables involved (state, input, and parameters) constitutes the mathematical model of the process as summarized below:

---

*State Equation*:

$$A\frac{dh}{dt} = F_{in} - F_{out}$$

| | |
|---|---|
| *State Variable*: | $h$ |
| *Output Variable*: | $h$ |
| *Input Variables*: | |
| *Disturbance variable*: | $F_{in}$ |
| *Manipulated variable*: | $F_{in}$ (or $F_{out}$) |
| *Parameters*: | $A, \rho$ |

---

### *Steady-State Condition*

A very important concept associated with the process model is the steady-state condition of the system.

---

*Steady-state* is the state at which the state variables do not change with time. This is reflected in the mathematical model by having the accumulation term vanish.

---

The application of the steady-state condition leads to a set of algebraic equations, the solution of which yields the value of the state variables at steady-state, which is also referred to as the *equilibrium* state. It may happen that a system has more than one steady-state solution. In this case, we say that the system has multiplicity or multiple steady-states.

## Example 4.2

Consider the model of the tank level obtained in Example 4.1:

$$A\frac{dh}{dt} = F_{in} - F_{out}$$

At steady-state, we expect that the level $h$ no longer changes with time. Hence, its derivative with respect to time vanishes, leading to

$$A\frac{dh_s}{dt} = 0 = F_{in,s} - F_{out,s} \qquad (4.5)$$

The subscript s denotes the steady-state condition. Equation (4.5) simply indicates that at steady-state, the inflow to the tank is equal to the outflow from the tank.

As the phenomenon that governs the system behavior becomes more complex, the model equations also become more complex. To model a chemical reaction occurring in a reactor, one requires information on the kinetic rates, explaining the generation and consumption of various chemical species involved in the reaction. If the ambient temperature is different from the temperature of a liquid flowing in a pipe, one has to express the heat flux across the tube surface through the appropriate resistance equations. Modeling a two-phase mixture of multiple chemical components demands the availability of gas–liquid and possibly liquid–liquid equilibria information, such as equations of state and nonideal equilibrium relations.

---

Process modeling effort brings together many facets of chemical engineering knowledge to arrive at a model that can represent the actual process behavior to the extent commensurate with the ultimate use of the model.

---

**Example 4.3**

For the stirred-tank heater in Figure 2.5, the fundamental quantities that provide the information about the dynamic behavior of the heater are:

- The total mass of the liquid in the tank
- The total energy of the liquid in the tank

Of course, it is assumed that the momentum of the heater is constant and- thus no momentum balance is required. Let us now identify the state variables of the process, starting with the definition of total mass from Example 4.1.

*Total mass in the tank*

$$\text{Total mass} = \rho V = \rho A h \tag{4.6}$$

*Total energy of the liquid in the tank*

$$\text{Total energy} = E = U + K + P \tag{4.7}$$

Here, $U$, $K$, and $P$ represent the internal, kinetic, and potential energies of the system, respectively. Since the liquid in the tank can be considered stationary, we have

$$\frac{\mathrm{d}K}{\mathrm{d}t} = \frac{\mathrm{d}P}{\mathrm{d}t} = 0 \quad \text{and} \quad \frac{\mathrm{d}E}{\mathrm{d}t} = \frac{\mathrm{d}U}{\mathrm{d}t} \tag{4.8}$$

For liquid systems, one can assume that

$$\frac{\mathrm{d}U}{\mathrm{d}t} \cong \frac{\mathrm{d}H}{\mathrm{d}t} \tag{4.9}$$

where $H$ denotes the total enthalpy of the liquid in the tank. Furthermore,

$$H = \rho V c_p (T - T_{ref}) = \rho A h c_p (T - T_{ref}) \tag{4.10}$$

where $c_p$ is the heat capacity of the liquid in the tank and $T_{ref}$ is a reference temperature where the specific enthalpy of the liquid is assumed to be zero. We can conclude that the state variables and the parameters for the system are:

- State variables: $h$ and $T$
- Constant parameters: $\rho$, $A$, $c_p$, and $T_{ref}$

---

It is assumed that the density $\rho$ and the heat capacity $c_p$ are constant and independent of temperature.

---

We will now apply the conservation principle on the two fundamental quantities: the total mass and the total energy

*Total mass balance*

As in Example 4.1, we obtain the variation in the height of the level in the tank with time as

$$A \frac{dh}{dt} = F_{in} - F_{out} \tag{4.11}$$

This equation describes the dynamic behavior of the level in the tank for changes in the inlet and outlet flow rates.

*Total energy balance*

$$\frac{[\text{Accumulation of total energy}]}{\text{time}} = \frac{[\text{Input of total energy}]}{\text{time}} - \frac{[\text{Output of total energy}]}{\text{time}} + \frac{[\text{Energy supplied by steam}]}{\text{time}}$$

or

$$\frac{d(\rho A h c_p (T - T_{ref}))}{dt} = \rho F_{in} c_p (T_{in} - T_{ref}) - \rho F_{out} c_p (T - T_{ref}) + Q \tag{4.12}$$

---

Given that the tank is well stirred, the temperature of the contents of the tank is equal to the temperature of the outlet stream.

---

$Q$ represents the amount of heat supplied by steam per unit of time. Assuming $T_{ref} = 0$ and since $A$, $\rho$, and $c_p$ are constants, we have

$$\frac{d(hT)}{dt} = h \frac{dT}{dt} + T \frac{dh}{dt} \tag{4.13}$$

After some algebra, we can rewrite the energy balance equation (Eq. [4.12]) as

$$Ah\frac{\mathrm{d}T}{\mathrm{d}t} = F_{\text{in}}(T_{\text{in}} - T) + \frac{Q}{\rho c_p} \tag{4.14}$$

Furthermore, the amount of heat $Q$ supplied by the steam to the liquid in the tank heater is given by the following heat transfer rate equation:

$$Q = U_{\text{t}}A_{\text{t}}(T_{\text{st}} - T) \tag{4.15}$$

where, $U_{\text{t}}$ is the overall heat transfer coefficient, $A_{\text{t}}$ the total area of heat transfer, and $T_{\text{st}}$ the temperature of the steam in the coil.

As a result, we have the following *state equations*:

$$A\frac{\mathrm{d}h}{\mathrm{d}t} = F_{\text{in}} - F_{\text{out}} \tag{4.16}$$

$$Ah\frac{\mathrm{d}T}{\mathrm{d}t} = F_{\text{in}}(T_{\text{in}} - T) + \frac{U_{\text{t}}A_{\text{t}}(T_{\text{st}} - T)}{\rho c_p} \tag{4.17}$$

These constitute a set of differential equations and fully characterize the dynamic behavior of the process. The *steady-state* behavior of the process can be expressed by setting the accumulation terms equal to zero in the state equations as follows:

$$0 = F_{\text{in, s}} - F_{\text{out, s}} \tag{4.19}$$

$$0 = F_{\text{in,s}}(T_{\text{in}} - T_{\text{s}}) + \frac{U_{\text{t}}A_{\text{t}}(T_{\text{st}} - T_{\text{s}})}{\rho c_p} \tag{4.20}$$

The subscript s again denotes the steady-state value of the corresponding variable. The solution of this set of algebraic equations, for given values of the parameters, will provide the steady-state operating conditions for the tank level and the temperature.

---

*State equations*:

$$A\frac{\mathrm{d}h}{\mathrm{d}t} = F_{\text{in}} - F_{\text{out}}$$

$$Ah\frac{\mathrm{d}T}{\mathrm{d}t} = F_{\text{in}}(T_{\text{in}} - T) + \frac{U_{\text{t}}A_{\text{t}}(T_{\text{st}} - T)}{\rho c_p}$$

| | |
|---|---|
| *State variables*: | $h$ and $T$ |
| *Output variables*: | $h$ and $T$ |
| *Input variables*: | |
| *Disturbances*: | $T_{\text{in}}$ and $F_{\text{in}}$ |
| *Manipulated variables*: | $T_{\text{st}}$ and $F_{\text{out}}$ |
| *Parameters*: | $A$, $\rho$, $c_p$, $U_{\text{t}}$, and $A_{\text{t}}$ |

---

## 4.3 ROLE OF PROCESS SIMULATORS IN MODELING

In the last decade, interest in integrating computer modeling and process simulation in the design of chemical processes has grown exponentially due to the apparent economical, environmental, and operational benefits. Process simulation packages like HYSYS™, ASPEN PLUS™, gProms™, and PRO/II™ also provide a valuable basis for the design and overall evaluation of advanced process control applications.

In general, the simulators are used to illustrate the flow of information between the units of a designed process flowsheet in a transparent way, and result in a significant reduction of the required time for process development. Moreover, they provide a consistent basis for comparing process alternatives where a number of different ideas can be generated and evaluated very quickly.

In terms of plantwide control considerations, simulation of highly integrated processes points process design and control engineers toward the salient interactions among the process units. Practically, the design and optimization of chemical processes involve the study of both steady-state and dynamic behaviors. Thus, steady-state and dynamic simulation models help in the process development by analyzing and validating the design and ideas before their implementation to avoid costly modifications and to ensure safe operation.

Seider et al.[1] and Luyben[2] illustrate the use of simulation tools for both steady-state and dynamic models in process creation, flowsheeting, optimization and evaluation of the controllability, and resilience of processes. They also underscore the importance of the process simulations in educating and training future engineers.

Steady-state simulation of chemical processes has been a standard process engineering activity for many years. Steady-state models can perform energy and material balances and evaluate different plant scenarios. Therefore, the simulation serves to optimize the designed process in terms of various objectives such as economical, environmental, operational, social, etc. However, chemical plants are never truly at steady-state, and a dynamic simulation of the designed processes is required to help in understanding the overall plant performance through its complex dynamic behavior.

Process and control engineers are quite familiar with dynamic evaluations to assess the process control concepts, investigate variable interactions, and design control strategies. The concept of dynamic modeling for typical chemical processes is introduced in control textbooks, such as Seborg et al.,[3] Luyben,[4] and Bequette.[5] Dynamic simulation offers many benefits over steady-state simulation where it allows the design and process engineers to investigate the time dependent behavior of a process. All variables can be observed regardless of instrumentation and the process model can be taken well beyond the safe limits of operation. Using a dynamic model, individual and plantwide control strategies can be designed and tested and even the control loops can be tuned before choosing one that may be suitable for implementation.

Therefore, the dynamic analysis is an essential step, as it provides valuable insight and complements the steady-state analysis by identifying specific areas in a plant that may have difficulty achieving the steady-state objective. Having said that, dynamic simulation models still have somewhat limited applications due to their high computational overhead and often substantial development times.

### 4.3.1 New Trends in Process Simulations

Today, with advances in computational speed and memory allocation, more and more engineering problems are being solved with the help of modeling or simulation software packages that are used to facilitate the design process and enable the engineer to focus on problem definition and analysis. Examples of such software include spreadsheet-based packages like MS Excel, analysis and programming languages or environments such as MATLAB, Speedup, or gPROMS, equation solvers such as Polymath and fluids packages, and process modeling packages, such as HYSYS. It is possible to use one or more of these tools to solve a given problem or to aid in the modeling of a given process.

However, most design problems often need to be broken into smaller (sub)problems, and these problems are then solved sequentially or sometimes in parallel. In some cases, the entire process model can be written in a single software package, such as gPROMS. There are many modeling problems, however, which may need the concurrent use of two or more software packages to arrive at a meaningful solution. This is especially the case where a specific problem requires the use of, for instance, both a physical properties package, such as SuperPro and a programing language, such as gPROMS. Each program offers unique advantages and specialties, but there is to date no single program that incorporates all of the data and methods, which may possibly be used to solve every problem. Nor will there ever be an all-encompassing process solver that would be a solution to all chemical engineering modeling situations. As new hypotheses and theories are developed, process modeling tools need to be restructured and improved.

A possible approach to overcome this shortcoming is to develop techniques to interconnect different software packages, enabling data communication across established links.[6] This leads to increased accuracy and speed of calculation while increasing process-modeling and problem-solving capabilities. The vision is to create an open system wherein the different programs can import and export data and communicate with each other freely. This interconnectivity will lead to the ability to create and maintain flexible and powerful hybrid models of processes and systems, essentially harnessing the powers of multiple tools. The hybrid architecture, illustrated in Figure 4.2, facilitates combining the strengths of each particular program allowing powerful simulations to be developed.

The current standard for open architecture is the Global CAPE-Open organization that comprises software developers, users, and researchers whose goal is to develop standard interfaces that can connect different software tools in chemical engineering.[7]

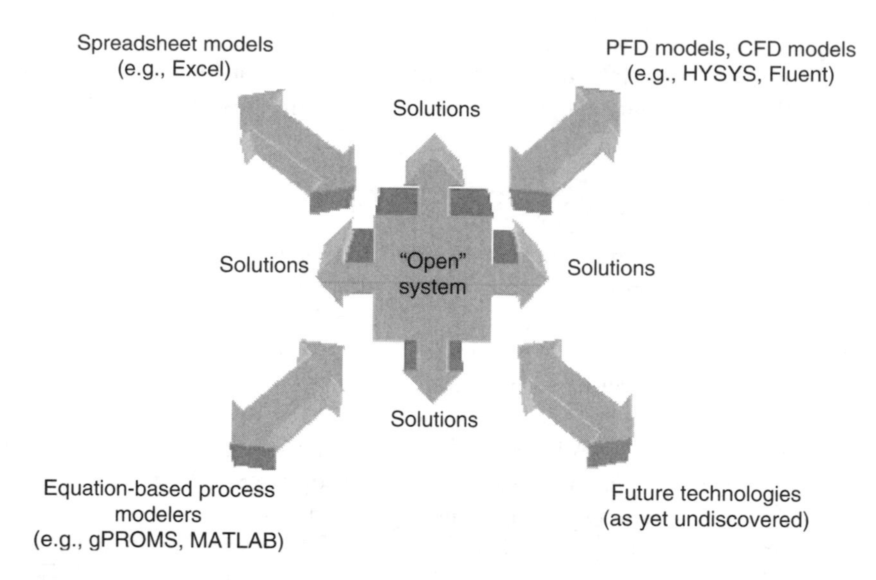


**FIGURE 4.2** The open software vision.

## Example 4.4

A pilot-scale crystallization facility, which resides in the Department of Chemical Engineering at the University of Sydney, is described briefly.[6] The central unit operation of the pilot facility is the crystallizer, which is heated or cooled by an oil stream flowing through its jacket. The heat exchange oil is circulated through hot and cold heat exchangers by a pump. A boiler and a cooling tower are used to heat and cool the oil at the heat exchangers, respectively. Cooling is used to generate the required supersaturation. As the solution is cooled, the solute will crystallize in accordance with the saturation concentration at that temperature.

Detailed modeling of this process is a complex undertaking not only because there are numerous unit operations present but also because of the complexity in their dynamics. Particularly, the crystallization phenomenon represents a difficult modeling task due to its highly nonlinear nature. The importance of having a model of this plant will become evident later in this book (Chapter 21) when the model develo-ped will be used within an on-line environment for optimal operation of the plant. Figure 4.3 shows the breakdown of responsibilities among different software.

HYSYS is responsible for the modeling of the utilities section, including all heating and cooling flows, temperatures, and pressures. The HYSYS flowsheet representing the utilities unit of the pilot facility is shown in Figure 4.4.

gPROMS provides solutions for the crystallizer temperature and more significantly it provides the crystal size distribution (CSD) output from the distributed population balance equations. MS Excel, while acting as a data link and used for data management and visualization, is also utilized for the calculation of the crystallizer jacket heat transfer model.

A graphical user interface (GUI) is developed to facilitate the execution of the simulation, utilizing the MS Excel Visual Basic (VBA) environment. The main window is shown in Figure 4.5. Using this interface, gPROMS and HYSYS are

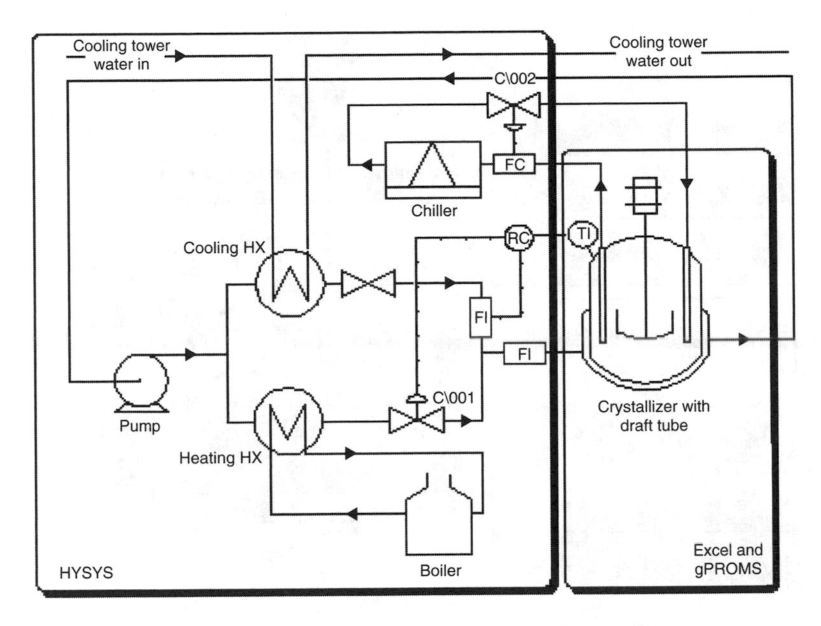


**FIGURE 4.3** Breakdown of responsibilities among the different software.

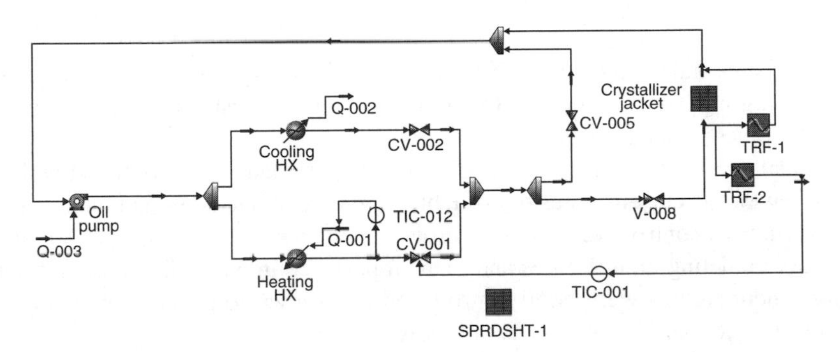


**FIGURE 4.4** HYSYS flowsheet for the utilities unit of the crystallization pilot plant facility.

executed in the background and all necessary visualization is performed in the MS Excel environment.

## 4.4 SUMMARY

In this Chapter, we introduced some of the key aspects of mathematical modeling using fundamental laws. Specifically, we discussed the principles of modeling, i.e., basis, assumptions, consistency, solution, and verification. Also, we introduced the fundamental conservation laws as tools for constructing dynamic models and finally, we have shown some applications to typical chemical processes.

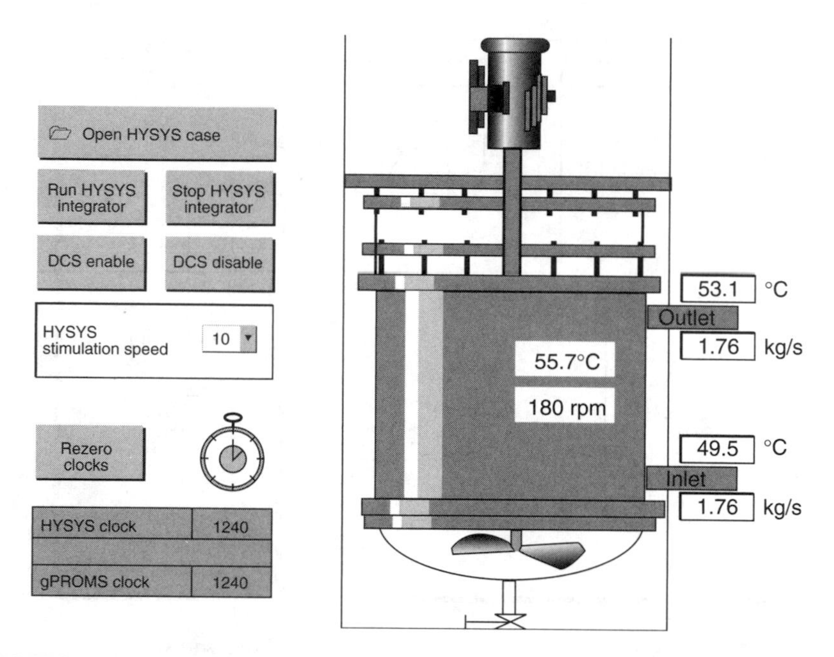


**FIGURE 4.5** The GUI where connection and simulation execution can be performed.

The concepts and the models considered in this chapter set the stage for ultimately developing input–output models for control design and analysis purposes, which are the topics of subsequent chapters.

Finally, in this chapter, the role of a process simulator was discussed with an overview of the current status of available commercial packages and their importance within a control system design project. Furthermore, some new trends such as open modeling architectures and their importance in expanding current simulator functionalities were briefly introduced and shown to provide endless communication among different software tools.

## CONTINUING PROBLEM

For the blending process

- Define the state variables
- Develop the state-space model
- State clearly the key assumptions during the modeling process

### SOLUTION

As pointed out in Section 4.2, the state variables are a set of fundamental dependent quantities, which describe the natural state of the system. Therefore, the state variables in this problem are $x$ (mass fraction of product in the exit stream) and $h$ (height of liquid in the tank). Furthermore, we can state the following assumptions:

1. Density, $\rho$, is constant throughout and independent of temperature (if we are dealing with dilute aqueous solutions, this is a good assumption)
2. The tank is well mixed, thus, the concentration in the tank is the same as the outlet stream concentration
3. The concentrations of feedstreams are constant
4. The heat of mixing is negligible

First, the total mass balance of the system is written as

$$\frac{\text{accumulation}}{\text{time}} = \frac{\text{input}}{\text{time}} - \frac{\text{output}}{\text{time}}$$

$$\begin{aligned}\frac{\mathrm{d}m}{\mathrm{d}t} &= \rho A\frac{\mathrm{d}h}{\mathrm{d}t} = IN - OUT = (\rho F_1 - \rho F_2) - \rho F \\ &= \rho(F_1 + F_2 - F)\end{aligned} \tag{4.21}$$

where $m$ denotes the mass of liquid in the tank and $V = Ah$. This equation simplifies to

$$A\frac{\mathrm{d}h}{\mathrm{d}t} = F_1 + F_2 - F \tag{4.22}$$

Next, a component balance is performed. The mass of the product in the tank is given as $mx = \rho Vx$, thus,

$$\rho\frac{\mathrm{d}(Vx)}{\mathrm{d}t} = \rho(F_1x_1 + F_2x_2 - Fx) \tag{4.23}$$

and as we have $\mathrm{d}(Vx)/\mathrm{d}t = V(\mathrm{d}x/\mathrm{d}t) + x(\mathrm{d}V/\mathrm{d}t)$, and after canceling the density term, we get

$$V\frac{\mathrm{d}x}{\mathrm{d}t} + x\frac{\mathrm{d}V}{\mathrm{d}t} = F_1x_1 + F_2x_2 - Fx \tag{4.24}$$

We also know that $V = Ah$, thus, $\mathrm{d}V/\mathrm{d}t = A(\mathrm{d}h/\mathrm{d}t)$ that results in

$$V\frac{\mathrm{d}x}{\mathrm{d}t} = F_1x_1 + F_2x_2 - Fx - xA\frac{\mathrm{d}h}{\mathrm{d}t} \tag{4.25}$$

Substituting Eq. (4.21) into the above equation, we get

$$Ah\frac{\mathrm{d}x}{\mathrm{d}t} = F_1(x_1 - x) + F_2(x_2 - x) \tag{4.26}$$

The state-space model for this system is, thus, defined by the equations

$$A\frac{dh}{dt} = F_1 + F_2 - F \tag{4.27}$$

$$Ah\frac{dh}{dt} = F_1(x_1 - x) + F_2(x_2 - x) \tag{4.28}$$

Furthermore, the outlet flow rate is typically a function of the static height in the tank, and this dependence is often represented through the relationship

$$F = \beta\sqrt{h} \tag{4.29}$$

where $\beta$ is a constant that represents the inverse of the resistance to the flow. Finally, the model equations are given as

$$A\frac{dh}{dt} = F_1 + F_2 - \beta\sqrt{h} \tag{4.30}$$

$$Ah\frac{dx}{dt} = F_1(x_1 - x) + F_2(x_2 - x) \tag{4.31}$$

## REFERENCES

1. Seider, W.D., Seader, J.D., and Lewin, D.R., *Product and Process Design Principles*, 2nd Ed., McGraw-Hill, New York, 2004.
2. Luyben, W.L., *Plantwide Dynamic Simulators in Chemical Processing and Control*, Marcel Dekker, New York, 2002.
3. Seborg, D.E., Edgar, T.F., and Mellichamp, D.A., *Process Dynamics and Control*, 2nd Ed., Wiley, Hoboken, NJ, 2004.
4. Luyben, W.L., *Process Modeling, Simulation and Control for Chemical Engineers*, 2nd Ed., McGraw-Hill, New York, 1990.
5. Bequette, B.W., *Process Control – Modeling, Design and Simulation*, Prentice-Hall, NJ, 2003.
6. Abbas, A., Guevara, V., and Romagnoli, J.A., Hybrid modelling: Architecture for the solution of complex process systems, Proceedings of ESCAPE-12, The Hague, Netherlands, 2002.
7. The CAPE-OPEN Laboratories Network, http://www.co-lan.org, 2004.

# 5 Input–Output Models: The Transfer Function

Most chemical processes are known to exhibit nonlinear behavior; thus, the modeling techniques discussed in the previous chapter will lead to a set of nonlinear ordinary differential equations (ODEs). Yet, such models are often intractable and pose fundamental challenges in obtaining closed-form (analytical) solutions. Moreover, we have already seen in Chapter 3 that *simple* input–output representations help the control engineer visualize the control problem and devise solutions very effectively.

Historically, control theory has been well established for *linear* systems, and numerous analytical and synthetic tools are available for control system design and evaluation. Can the control engineer afford a compromise here? There are many instances where nonlinear processes remain in the vicinity of a particular operating point, such that a linear approximation of the process model in this region may be sufficiently accurate. Such local models can provide significant intuition and insight into the problem and lead to very effective control strategies. In this chapter, we will start with a discussion of the linear version for process models described by nonlinear differential equations.

To facilitate the construction of simple input–output (I/O) relationships from linear models, we will introduce the Laplace transform. A major benefit is that this transformation converts the linear differential equations into algebraic equations, thus simplifying the mathematical manipulations required to develop an I/O model. Such an algebraic model, called the *transfer function*, is used exclusively in control system design and analysis.

## 5.1 LINEAR (LINEARIZED) MODEL

In exploring the dynamic behavior of a chemical process, its fundamental model plays a key role, as we have seen in Chapter 4. Yet, most process models are often classified as *nonlinear*, and there is no general mathematical theory for the analytical solution of nonlinear ODEs. Therefore, a *global* study of the process behavior, explaining the process characteristics throughout the whole operating envelope is infeasible. A compromise can be reached, however, by focusing on the *local* behavior of a process, described by a linear model, only valid in the vicinity of a specific (and desired) operating point. One method of obtaining such approximate models is to use a series expansion. Consider a process described by the following ODE:

$$\frac{dx}{dt} = f(x) \tag{5.1}$$

The nonlinear function $f(x)$ can be expanded around a reference point $(x_0)$ using an infinite Taylor series expansion (Appendix A). If the expansion terms are truncated after the first-order term, we obtain a *linear* equation

$$\frac{dx}{dt} = f(x_0) + K(x - x_0) \tag{5.2}$$

where $K$ is a constant that corresponds to the partial derivative of the function with respect to the variable of interest (see Appendix A). Equation (5.2) is a linear *approximation* of Eq. (5.1) and can be used confidently as long as the domain of $x$ is restricted sufficiently close to the reference point $x_0$.

### 5.1.1 Deviation Variables

Let us assume that $x_s$ represents the steady-state value of the dependent variable. At steady-state, we expect the following condition to hold,

$$\frac{dx_s}{dt} = 0 = f(x_s) \tag{5.3}$$

If we consider $x_s$ as the reference point around which Eq. (5.1) is linearized (i.e., $x_0 = x_s$), we have

$$\frac{dx}{dt} \approx f(x_s) + \left.\frac{df}{dx}\right|_{x_s} (x - x_s) = f(x_s) + K(x - x_s) \tag{5.4}$$

Now, we can subtract the steady-state equation (5.3) from the dynamic equation (5.4), and obtain

$$\frac{d(x - x_s)}{dt} = K(x - x_s) \tag{5.5}$$

The *deviation variable* is defined as

$$\bar{x} = x - x_s$$

The final expression is

$$\frac{d\bar{x}}{dt} = K\bar{x} \tag{5.6}$$

which is a linear equation in terms of the *deviation variable* $\bar{x}$.

The notion of a deviation variable is very useful in process control where we are always interested in tracking the variables of interest as they move away from or come close to an equilibrium (steady-state) point. A set of deviation variables provides an intuitive basis for explaining this dynamic behavior with the appropriate reference to the desired operating point.

**Example 5.1**

We shall consider the tank system given in Figure 4.1. The state equation found from the total mass balance (Example 4.1) is

$$A\frac{\mathrm{d}h}{\mathrm{d}t} = F_{\mathrm{in}} - F_{\mathrm{out}}$$

If the outlet flow rate, $F_{\mathrm{out}}$, is a function of the static pressure head in the tank (liquid level), with the typical square-root dependence, we have

$$F_{\mathrm{out}} = \beta\sqrt{h}, \quad \beta = \text{constant}$$

Then, the resulting total mass balance yields a nonlinear dynamic model as

$$A\frac{\mathrm{d}h}{\mathrm{d}t} + \beta\sqrt{h} = F_{\mathrm{in}} \tag{5.7}$$

Rearranging and expanding the equation using Taylor expansion (Appendix A),

$$\begin{aligned}\frac{\mathrm{d}h}{\mathrm{d}t} &= -\frac{\beta\sqrt{h}}{A} + \frac{F_{\mathrm{in}}}{A} \\ &= \left[-\frac{\beta\sqrt{h_{\mathrm{s}}}}{A} + \frac{F_{\mathrm{in,s}}}{A}\right] + \frac{\partial}{\partial h}\left[-\frac{\beta\sqrt{h}}{A} + \frac{F_{\mathrm{in}}}{A}\right]_{h_{\mathrm{s}}, F_{\mathrm{in,s}}}(h - h_{\mathrm{s}}) \\ &\quad + \frac{\partial}{\partial F_{\mathrm{in}}}\left[-\frac{\beta\sqrt{h}}{A} + \frac{F_{\mathrm{in}}}{A}\right]_{h_{\mathrm{s}}, F_{\mathrm{in,s}}}(F_{\mathrm{in}} - F_{\mathrm{in,s}})\end{aligned}$$

We can define the following *deviation variables*:

$$\bar{h} = h - h_{\mathrm{s}}, \quad \bar{F}_{\mathrm{in}} = F_{\mathrm{in}} - F_{\mathrm{in,s}}$$

Finally, we have a linear model in terms of the *deviation variables* given by

$$\frac{\mathrm{d}\bar{h}}{\mathrm{d}t} = a\bar{F}_{\mathrm{in}} - b\bar{h} \tag{5.8}$$

The constants are defined as

$$a = \frac{1}{A}, \quad b = \frac{1}{A}\left[\frac{\mathrm{d}(\beta\sqrt{h})}{\mathrm{d}h}\right]_{h=h_{\mathrm{s}}} = \frac{\beta}{2A\sqrt{h_{\mathrm{s}}}}.$$

Usually, we will be concerned with maintaining the value of a process variable (temperature, concentration, etc.) at some desired steady-state. Consequently, the steady-state becomes a natural candidate point around which the approximate linearized model is developed. In such cases, the deviation variable describes directly the magnitude of the displacement of a system from the desired level of operation. Furthermore, if the controller of the given process has been designed well, it will not allow the process variable to move far from the desired steady-state value. So, the approximate linearized model will be satisfactory to describe the dynamic behavior of the process near the steady-state.

### 5.1.2 Higher Dimensional Equations

The linearization technique can be easily generalized (as shown in Appendix A) for the case of higher dimensional state-space models. Let us demonstrate this with the following two nonlinear ODEs that consist of two state and two input variables:

$$\frac{dx_1}{dt} = f_1(x_1, x_2, u_1)$$
$$\frac{dx_2}{dt} = f_2(x_1, x_2, u_2) \tag{5.9}$$

The Taylor expansion is carried out around the steady-state operating point captured by the equations

$$0 = f_1(x_{1,s}, x_{2,s}, u_{1,s})$$
$$0 = f_2(x_{1,s}, x_{2,s}, u_{2,s})$$

We can define the deviation variables,

$$\bar{x}_1 = x_1 - x_{1,s}; \quad \bar{x}_2 = x_2 - x_{2,s}; \quad \bar{u}_1 = u_1 - u_{1,s}; \quad \bar{u}_2 = u_2 - u_{2,s}$$

and arrive at the linear set of differential equations:

$$\frac{d\bar{x}_1}{dt} = K_{11}\bar{x}_1 + K_{12}\bar{x}_2 + K_{13}\bar{u}_1$$
$$\frac{d\bar{x}_2}{dt} = K_{21}\bar{x}_1 + K_{22}\bar{x}_2 + K_{23}\bar{u}_2 \tag{5.10}$$

As before, the constants correspond to the partial derivative of the function with respect to the variable of interest, e.g.,

$$K_{12} = \left.\frac{\partial f_1}{\partial x_2}\right|_{x_{1,s}, x_{2,s}, u_{1,s}, u_{2,s}}$$

**Example 5.2**

Consider the stirred-tank heater discussed in Example 4.3. The state equations developed from the total mass balance and the energy balance are

$$A\frac{dh}{dt} = F_{in} - F_{out} \tag{5.11}$$

$$Ah\frac{dT}{dt} = F_{in}(T_{in} - T) + \frac{UA_t(T_{st} - T)}{\rho c_p} \tag{5.12}$$

The set of linear ODEs representing the dynamic model of this process is given by (see Appendix A)

$$\frac{d\bar{h}}{dt} = a\bar{F}_{in} - b\bar{h} \tag{5.13}$$

$$\frac{d\bar{T}}{dt} = c\bar{T} + d\bar{h} + e\bar{F}_{in} + f\bar{T}_{st} + g\bar{T}_{in} \tag{5.14}$$

We note that Eq. (5.13) is essentially the same as in Example 5.1. The calculation of the constants in Eq. (5.14) is left as an exercise for the reader.

### 5.1.3 Linear State-Space Model

Recalling the definition of nonlinear state-space models (Eq. [3.8]), the linearized state-space equations can be rewritten in a compact notation as follows:

$$\frac{d\boldsymbol{x}}{dt} = A\boldsymbol{x} + B\boldsymbol{u} \tag{5.15}$$

Here, the boldface letters indicate the vector quantities,

$$\boldsymbol{x} = [x_1, \quad x_2 \quad \cdots \quad x_N]^T, \quad \boldsymbol{u} = [u_1, \quad u_2 \quad \cdots \quad u_K]^T$$

The matrices $A$, often referred to as the *Jacobian*, and $B$ are defined as

$$\begin{aligned} A &= \begin{bmatrix} a_{11} \ldots a_{1N} \\ a_{N1} \ldots a_{NN} \end{bmatrix}, \quad a_{ij} = \left.\frac{\partial f_l}{\partial x_j}\right|_{\substack{x=x_s \\ u=u_s}}, \quad j = l, \ldots, N, l = 1, \ldots, N \\ B &= \begin{bmatrix} b_{11} \ldots b_{1K} \\ b_{N1} \ldots b_{NK} \end{bmatrix}, \quad b_{ij} = \left.\frac{\partial f_l}{\partial x_j}\right|_{\substack{x=x_s \\ u=u_s}}, \quad j = l, \ldots, K \end{aligned} \tag{5.16}$$

This is the so-called linear *state-space* representation of the system. In general, the vector of output variables can be expressed in terms of the vector of state variables as

$$\boldsymbol{y} = C\boldsymbol{x} \tag{5.17}$$

We must note that there may be cases where the outputs also depend on the process inputs, i.e.,

$$\boldsymbol{y} = C\boldsymbol{x} + D\boldsymbol{u}$$

In Eq. (5.17), we observe that the outputs are linear combinations of the state variables. The linear state-space representation of the system in terms of input, state, and output variables is summarized below (considering $D = 0$).

$$\begin{aligned} \dot{\boldsymbol{x}} &= \frac{dx}{dt} = A\boldsymbol{x} + B\boldsymbol{u} \quad \text{(state equation)} \\ \boldsymbol{y} &= C\boldsymbol{x} \qquad\qquad\qquad\quad \text{(output equation)} \end{aligned} \tag{5.18}$$

In the particular case where the state variables are also the outputs of the system, i.e., $C = I$ (the identity matrix), the state-space equations simplify to

$$\frac{\mathrm{d}\boldsymbol{y}}{\mathrm{d}t} = A\boldsymbol{y} + B\boldsymbol{u} \tag{5.19}$$

This is an *input–output* representation of the system when there are multiple outputs and multiple inputs.

### Example 5.3

Let us consider the stirred-tank heater studied in Example 4.3. The modeling equations are also used in Example 5.2:

$$\frac{\mathrm{d}\bar{h}}{\mathrm{d}t} = a\bar{F}_{\text{in}} - b\bar{h} \tag{5.13}$$

$$\frac{\mathrm{d}\bar{T}}{\mathrm{d}t} = c\bar{T} + d\bar{h} + e\bar{F}_{\text{in}} + f\bar{T}_{\text{st}} + g\bar{T}_{\text{in}} \tag{5.14}$$

We can define the state-space model for this process as

$$\frac{\mathrm{d}\boldsymbol{x}}{\mathrm{d}t} = A\boldsymbol{x} + B\boldsymbol{u}$$

where we have $\boldsymbol{x} = [\bar{h} \quad \bar{T}]$ and $\boldsymbol{u} = [\bar{F}_{\text{in}} \quad \bar{T}_{\text{st}} \quad \bar{T}_{\text{in}}]$. The state matrix and the input matrix become

$$A = \begin{bmatrix} -b & 0 \\ d & c \end{bmatrix}; \quad B = \begin{bmatrix} a & 0 & 0 \\ e & f & g \end{bmatrix}$$

In this case, we can assume that both the level and the temperature are measured, thus, the output equation becomes

$$\boldsymbol{y} = \begin{bmatrix} y_1 \\ y_2 \end{bmatrix} = I\boldsymbol{x} = \begin{bmatrix} 1 & 0 \\ 0 & 1 \end{bmatrix} \begin{bmatrix} \bar{h} \\ \bar{T} \end{bmatrix}$$

## 5.2 CONCEPT OF TRANSFER FUNCTION

Previously, we have described techniques to develop mathematical models that explain the dynamic operation of a process. We noted that solving such models (i.e., obtaining input–output relationships) requires either analytical or numerical integration of the corresponding differential equations. Complicating this effort is the nonlinear nature of these models. To overcome this problem, we introduced a technique that allows the representation of a nonlinear model as a linear approximation around a certain operating point. Such a linear representation is the starting point of many analytical techniques in process control.

In this section, we will introduce the Laplace transform technique (see Appendix B for a formal definition and details) that will help us develop linear I/O process models in a straightforward manner from linear differential equations.

A major benefit is that this transformation converts the differential equations into algebraic equations, thus simplifying the mathematical manipulations required to construct an I/O model. Such an algebraic model, called a *transfer function*, can be used very effectively in control system design and analysis.

**Example 5.4**

We recall that a model expressing the relationship between an input $m(t)$ and an output $y(t)$ for a two-tank system was developed in Example 3.1

$$\frac{d^2y}{dt^2} + (a_1 + a_4)\frac{dy}{dt} + a_1a_4y = a_2a_3m \tag{5.20}$$

We observe that this is a linear equation and the steady-state equation dictates the following relationship:

$$a_1a_4y_s = a_2a_3m_s \tag{5.21}$$

The I/O model can be put into the usual deviation variable form by performing the algebraic manipulations as discussed before and recognizing $\bar{y} = y - y_s$, $\bar{m} = m - m_s$,

$$\frac{d^2\bar{y}}{dt^2} + (a_1 + a_4)\frac{d\bar{y}}{dt} + a_1a_4\bar{y} = a_2a_3\bar{m} \tag{5.22}$$

Let us now take Laplace transform of both sides of Eq. (5.22),

$$(s^2\bar{y}(s) - s\bar{y}(0) - \bar{y}^1(0)) + (a_1 + a_4)(s\bar{y}(s) - \bar{y}(0)) + a_1a_4\bar{y}(s) = a_2a_3\bar{m}(s) \tag{5.23}$$

By definition, we have $\bar{y}(0) = \bar{y}^1(0) = 0$. Collecting terms, and solving for $\bar{y}(s)$, we get

$$(s^2 + (a_1 + a_4)s + a_1a_4)\bar{y}(s) = a_2a_3\bar{m}(s) \tag{5.24}$$

This is the representation of the I/O model in Eq. (3.18) in the $s$-domain. We note that this relationship is now algebraic in nature. If the input function $\bar{m}(s)$ is known, one can even go further and compute the output $\bar{y}(s)$ in the Laplace domain using only algebraic manipulations and use the inverse Laplace transformation to finally obtain the time-domain solution of Eq. (5.20) for this specific input.

## 5.3 TRANSFER FUNCTIONS OF SISO PROCESSES

---

The transfer function of a linear dynamic process is defined as the ratio of the output variable to the input variable in the Laplace domain.

---

Consider the process represented in Figure 5.1, where we consider the input signal $u(t)$ and the output signal, $y(t)$.

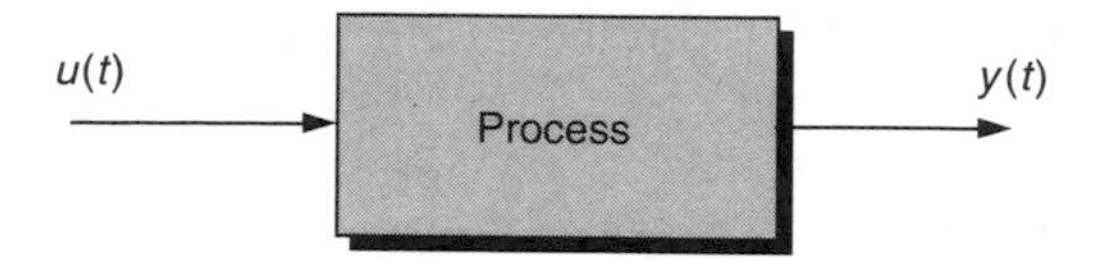


**FIGURE 5.1** I/O representation of a process in time domain.

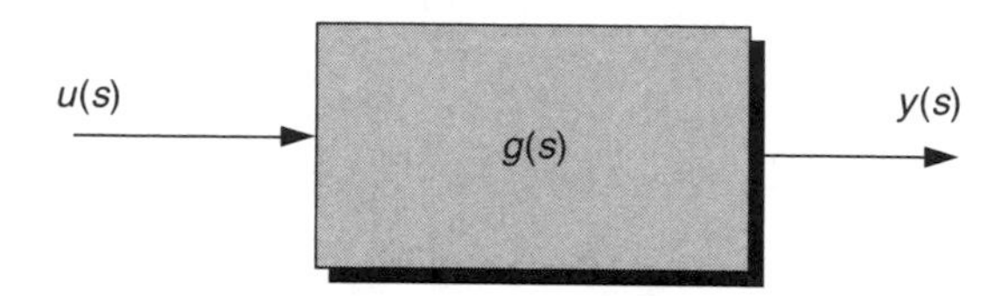


**FIGURE 5.2** I/O representation of a process in Laplace domain.

The transfer function $g(s)$ is defined as:

$$g(s) = \frac{\text{Laplace transform of the output}}{\text{Laplace transform of the input}} \tag{5.25}$$

Schematically, this can be represented as in Figure 5.2.

The following general differential equation describes a linear time-invariant process,

$$a_0 y^{(n)} + a_1 y^{(n-1)} + \cdots + a_n y = b_0 u^{(m)} + b_1 u^{(m-1)} + \cdots + b_m u \tag{5.26}$$

where $n \geqslant m$, and $y^{(n)}$ represents the $n$th-order derivative of $y(t)$. For convenience, all initial conditions are assumed to be equal to zero (which arises naturally when deviation variables are used). The model is called *time-invariant* because the coefficients ($a_i$, $b_j$) are assumed not to vary with time. Also, we should note that while in chemical processes, $n$ never exceeds 2, there are practical benefits in considering this general form, as will be obvious later.

We obtain the transfer function representation of Eq. (5.26) by applying the Laplace transformation to both sides of the equation.

$$a_0 y(s)s^n + a_1 y(s)s^{n-1} + \cdots + a_n y(s) = b_0 u(s)s^m + b_1 u(s)s^{m-1} + \cdots + b_m u(s) \tag{5.27}$$

Next, we collect terms by factoring out $y(s)$ and $u(s)$,

$$(a_0 s^n + a_1 s^{n-1} + \cdots + a_n)y(s) = (b_0 s^m + b_1 s^{m-1} + \cdots + b_m)u(s) \tag{5.28}$$

Finally, by taking the ratio, we have

$$g(s) = \frac{y(s)}{u(s)} = \frac{b_0 s^m + \cdots + b_m}{a_0 s^n + \cdots + a_n} \tag{5.29}$$

This is the transfer function of the process represented by Eq. (5.26). The highest power of $s$ in the denominator is called the order of the process.

---

- The transfer function model provides a simpler I/O model than in the time domain.
- It completely describes the dynamic behavior of the output variable when the corresponding input variable is specified.
- The transfer function is a system property and thus independent of the form of the input.
- We cannot form the transfer function with nonzero initial conditions. The transfer function is only defined for deviation variables.

---

### 5.3.1 Asymptotic Theorems

When the process model is expressed in the Laplace domain, the dynamic behavior of the output variable in response to a change in the input variable can be analyzed by two asymptotic theorems. These theorems exploit the algebraic nature of the transfer function to predict initial and final conditions of the process output response.

#### *Initial Value Theorem*

The initial value of a function $f(t)$ in the time domain can be calculated from the following expression:

$$\lim_{t\to 0} f(t) = \lim_{s\to\infty}[sF(s)] \tag{5.30}$$

#### *Final Value Theorem*

A very useful asymptotic property of a function is its value as time grows and a steady-state (if it exists) is approached.

$$\lim_{t\to\infty} f(t) = \lim_{s\to 0}[sF(s)] \tag{5.31}$$

The latter is a very useful theorem as it indicates the location of the new process steady-state in response to a change in the input variable. For control purposes, we would be frequently interested in determining this steady-state location as a key process performance measure.

**Example 5.5**

The level dynamics in a tank was studied in Example 5.1 and the model, in deviation variable form, was expressed as follows:

$$\frac{\mathrm{d}\bar{h}}{\mathrm{d}t} + b\bar{h} = a\bar{F}_{\text{in}} \tag{5.32}$$

We can take the Laplace transform of Eq. (5.32) and rearrange the terms, which results in

$$\frac{\bar{h}(s)}{\bar{F}_{in}(s)} = \frac{a}{s+b} \tag{5.33}$$

or

$$\frac{\bar{h}(s)}{\bar{F}_{in}(s)} = \frac{k}{\tau s+1} \tag{5.34}$$

where the constants are $k = a/b$, $\tau = 1/b$. The transfer function for this process is given as

$$g(s) = \frac{k}{\tau s+1} \tag{5.35}$$

We note that the denominator polynomial is first-order and this process is said to be of *first-order dynamics*. We shall see later in Chapter 8 the precise meaning.

Let us now use the asymptotic theorems to see what we can learn from their implementation. First, let us use the Initial Value Theorem, considering the output variable as our function.

$$\lim_{s\to\infty}[s\bar{h}(s)] = \lim_{s\to\infty}\left[s\frac{k}{\tau s+1}\,F_{in}(s)\right] = 0 \tag{5.36}$$

This confirms that the initial condition for the level was indeed zero. Now, consider the input variable to be a unit step function whose Laplace transform is expressed as

$$F_{in}(s) = \frac{1}{s} \tag{5.37}$$

Let us explore the new steady-state level in the tank when the input flow rate is changed in this manner.

$$\lim_{s\to 0}[s\bar{h}(s)] = \lim_{s\to 0}\left[sg(s)\frac{1}{s}\right] = \lim_{s\to 0}\left[\frac{k}{\tau s+1}\right] = k \tag{5.38}$$

This indicates that the level will increase to a new steady-state value given by $k$.

## 5.4 PROPERTIES OF TRANSFER FUNCTIONS

Recalling the SISO case, we have shown that, in general, a *rational* transfer function can be represented as a ratio of two polynomials, i.e.,

$$g(s) = \frac{y(s)}{u(s)} = \frac{b_0 s^m + \cdots + b_m}{a_0 s^n + \cdots + a_n} = \frac{z(s)}{p(s)} \tag{5.39}$$

Here, $z(s)$ and $p(s)$ denote the numerator and denominator polynomials, respectively.

All physical systems need to be *proper* and *causal*. Simply, if the order of the numerator polynomial is less than or equal to the denominator polynomial ($n \geqslant m$), the system is referred to as proper, and improper, if $m > n$ (i.e., containing differentiators). A process is called causal if its output depends only on past inputs, and noncausal, if its output depends also on the future inputs (i.e., requiring prediction).

*Characteristic Equation* The denominator polynomial $p(s)$ when equated to zero is called the characteristic equation, i.e.,

$$p(s) = a_0 s^n + a_1 s^{n-1} + \cdots + a_n = 0 \tag{5.40}$$

The roots of the polynomials $p(s)$ and $z(s)$ play an important role in shaping and explaining the dynamic behavior of the process. We introduce two fundamental concepts:

---

The roots of the polynomial $z(s)$ are called the *zeros of the transfer function* or the *zeros of the process.*

---

When evaluated at the zeros, the transfer function becomes zero.

---

The roots of the polynomial $p(s)$ are called the poles of the transfer function or the poles of the process.

---

When evaluated at the poles, the transfer function becomes unbounded (indeterminate).

Both the poles and the zeros of the process play a fundamental role in the analysis of the dynamic behavior of the process and in the design of its control system as will be shown later.

### Example 5.6

Consider the two-tank process discussed in Example 5.3. The I/O relationship was found to be

$$(s^2 + (a_1 + a_4)s + a_1 a_4)\overline{y}(s) = a_2 a_3 \overline{m}(s) \tag{5.41}$$

We can express the transfer function for this process as follows:

$$\frac{\overline{y}(s)}{\overline{m}(s)} = g(s) = \frac{a_2 a_3}{s^2 + (a_1 + a_4)s + a_1 a_4} \tag{5.42}$$

It can be observed that this process does not have any zeros,

$$z(s) = a_2 a_3 = \text{constant}$$

and it has two poles located at the roots of the characteristic equation,

$$p(s) = s^2 + (a_1 + a_4)s + a_1 a_4 = 0$$

The poles are given by

$$p_1 = -a_1$$
$$p_2 = -a_4$$

By referring to Example 3.1, we can observe that the pole locations depend on the process parameters, such as the tank cross-sectional area and the outlet valve constant. Thus, by changing the design conditions, we can cause this process to exhibit a different dynamic behavior. We will analyze such systems in detail in Chapter 8.

## 5.5 NONRATIONAL TRANSFER FUNCTIONS

There are many instances where the measurement of the output variable may be delayed, hence, the information used to assess process status may be old. These instances may include the following:

- *Sensor response* Some sensors, such as those based on chromatographic separation, may have significant delays in producing the measurement. Therefore, product composition measurements may be available only after a certain period of time has elapsed.
- *Physical proximity of sensor* If a sensor is placed at a location farther downstream than the actual location of interest, this measurement will only reflect the delayed condition of the process due to transport delays in the flow stream. An example is a pH sensor placed 5 m downstream of the actual point of mixing due to physical constraints of the plant layout.

Similarly, the inputs influencing the behavior of a process may also be delayed, depending on the actuator type and location.

- *Actuator response* If one is using an electric heater as an actuator to affect the temperature in a bath, there may be a time delay between the time a command is sent to the heater and the actual time that the energy is made available to the bath.
- *Physical proximity of actuator* If a valve used to adjust the flow rate of an input stream is placed farther upstream than the location at which the change is demanded, the transport time will cause a delay in the implementation of this change on the process.

In modeling processes where delays are present, the delayed variables are represented by explicitly incorporating the delay information. A linear model in deviation variables given by,

$$\frac{\mathrm{d}y(t)}{\mathrm{d}t} + ay(t) = bm(t - t_{\mathrm{D}}) \tag{5.43}$$

indicates that the input variable *m(t)* is delayed by a period of $t_{\mathrm{D}}$ time units. In constructing the I/O model in the Laplace domain, we exploit the *translation property* of the Laplace transformation, and obtain

$$sy(s) + by(s) = bm(s)\mathrm{e}^{-t_{\mathrm{D}}s} \tag{5.44}$$

This leads to the transfer function

$$\frac{y(s)}{u(s)} = g(s) = \frac{b}{s + b}\mathrm{e}^{-t_{\mathrm{D}}s} \tag{5.45}$$

This is an *nonrational* transfer function due to the presence of the transcendental exponential term.

---

The time delays have a strong influence on the control system and may severely degrade its performance as will be discussed later.

---

### Example 5.7

Consider the stirred-tank heater process redrawn in Figure 5.3, where the thermocouple used to measure the temperature of the outlet stream is placed at some distance

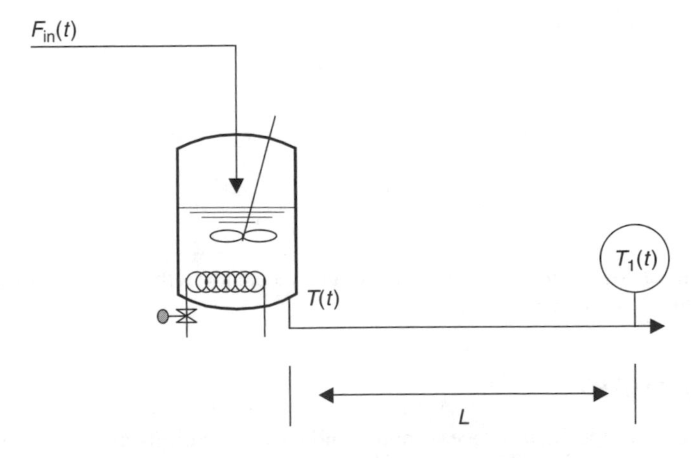


**FIGURE 5.3** Measuring the temperature of a fluid leaving a tank.

from the tank. We are interested in constructing the transfer function relating the measured temperature, $T_1$, to the inlet flow rate, $F_{in}$.

The following assumptions are made:

1. The tank level and the inlet flow rate are constant
2. Inlet temperature and the steam temperature are kept constant
3. The exit piping between the tank outlet and the measurement point is well insulated
4. The flow of the liquid through the exit piping is ideal plug flow

Under these assumptions, the model developed in Example 5.2 reduces to the following simple energy balance:

$$\frac{\mathrm{d}\overline{T}}{\mathrm{d}t} = c\overline{T} + \mathrm{e}\overline{F}_{in} \tag{5.46}$$

Applying Laplace transformation, we obtain the following transfer function model:

$$\frac{\overline{T}(s)}{\overline{F}_{in}(s)} = \frac{\mathrm{e}}{s + c} \tag{5.47}$$

We note that this model does not account for the fact that the temperature is actually measured farther downstream. The transport delay can be expressed by

$$t_{\mathrm{D}} = \frac{A_c L}{F_{out}} \tag{5.48}$$

where $A_c$ is the cross-sectional area of the pipe and $L$ the length of pipe. The measured temperature is now related to the exit temperature in the following manner:

$$T_1(t) = T(t - t_{\mathrm{D}}) \tag{5.49}$$

In the Laplace domain, we have

$$\frac{T_1(s)}{T(s)} = \mathrm{e}^{-t_{\mathrm{D}}s} \tag{5.50}$$

Finally, we can obtain the transfer function between $T_1(s)$ and $F_{in}(s)$:

$$\frac{T_1(s)}{F_{in}(s)} = \frac{T_1(s)}{T(s)} \frac{T(s)}{F_{in}(s)} = \frac{\mathrm{e}}{s + c}\mathrm{e}^{-t_{\mathrm{D}}s} \tag{5.51}$$

This model now correctly captures the dynamic behavior of the process as observed by the plant personnel.

## 5.6 SUMMARY

How to construct a linear process model and its implications are discussed in this chapter. The definition of the transfer function results from our desire to express the process dynamics in terms of an input–output model in the Laplace domain. The properties of this representation are also discussed, specifically the poles and zeros of the transfer function. The transfer function is also extended to include an nonrational element, such as a time delay.

## CONTINUING PROBLEM

For the blending process:

- Linearize the nonlinear model equations
- Develop the transfer functions between the level and the feed flow rates
- Develop the transfer functions between the outlet composition and the feed flow rates

### SOLUTION

To linearize the nonlinear differential equations, they will be expanded in a Taylor series around the steady-state operating point, and all terms after the first-order derivatives will be neglected.

First, we linearize the total mass balance equation (4.30):

$$A\frac{\mathrm{d}h}{\mathrm{d}t} = F_1 + F_2 - \beta\sqrt{h}$$

Here, the term $\beta\sqrt{h}$ is nonlinear and $\beta$ is a constant. We expand this term using Taylor series around the point $h_s$,

$$\beta\sqrt{h} = \beta\sqrt{h_s} + \left[\frac{\mathrm{d}(\beta\sqrt{h})}{\mathrm{d}h}\right]_{h=h_s}(h - h_s) + \cdots \tag{5.52}$$

Neglecting second- and higher-order terms in the series, we have the state equation

$$A\frac{\mathrm{d}h}{\mathrm{d}t} + \beta\sqrt{h_s} + \left[\frac{\mathrm{d}(\beta\sqrt{h})}{\mathrm{d}h}\right]_{h=h_s}(h - h_s) = F_1 + F_2 \tag{5.53}$$

The steady-state equation is given by

$$A\frac{\mathrm{d}h_s}{\mathrm{d}t} = 0 = F_{1,s} + F_{2,s} - \beta\sqrt{h_s} \tag{5.54}$$

Subtracting (5.54) from (5.53) we obtain

$$A\frac{\mathrm{d}(h - h_s)}{\mathrm{d}t} + \left[\frac{\mathrm{d}(\beta\sqrt{h})}{\mathrm{d}h}\right]_{h=h_s}(h - h_s) = (F_1 + F_{1,s}) + (F_2 - F_{2,s}) \tag{5.55}$$

We define the *deviation variables* as $\bar{h} = h - h_s$, $\bar{F}_1 = F_1 - F_{1,s}$, $\bar{F}_2 = F_2 - F_{2,s}$. Then, we obtain the linear model in terms of derivation variables:

$$\frac{d\bar{h}}{dt} = \frac{\bar{F}_1}{A} + \frac{\bar{F}_2}{A} + a_1\bar{h} \tag{5.56}$$

where

$$a_1 = \frac{1}{A}\left[\frac{d(\beta\sqrt{h})}{dh}\right]_{h=h_s} = -\frac{\beta}{2A\sqrt{h_s}}$$

Next, we study the component balance equation:

$$Ah\frac{dx}{dt} = F_1(x_1 - x) + F_2(x_2 - x) \tag{5.57}$$

$$\begin{aligned}\frac{dx}{dt} &= \frac{F_1x_1}{Ah} - \frac{F_1x}{Ah} + \frac{F_2x_2}{Ah} - \frac{F_2x}{Ah} \\ &= f(F_1, F_2, h, x_1, x_2, x)\end{aligned} \tag{5.58}$$

After linearization and some algebra, we arrive at,

$$\frac{dx}{dt} = f(F_1, F_2, h, x_1, x_2, x)|_s$$

$$+a_2(F_1-F_{1,s})+a_3(F_2-F_{2,s})+a_4(h-h_s)+a_5(x-x_s)+a_6(x_1-x_{1,s})+a_7(x_2-x_{2,s}) \tag{5.59}$$

where

$$a_2 = \left.\frac{df}{dF_1}\right|_s = \left[\frac{x_1}{Ah} - \frac{x}{Ah}\right]_s = \left[\frac{1}{Ah}(x_1 - x)\right]_s$$

$$a_3 = \left.\frac{df}{dF_2}\right|_s = \left[\frac{1}{Ah}(x_2 - x)\right]_s$$

$$a_4 = \left.\frac{df}{dh}\right|_s = \left[-\frac{1}{Ah^2}(F_1x_1 - F_1x + F_2x_2 - F_2x)\right]_s = 0$$

$$a_5 = \left.\frac{df}{dx}\right|_s = \left[-\frac{F_1}{Ah} - \frac{F_2}{Ah}\right]_s = -\frac{F_s}{Ah}$$

$$a_6 = \left.\frac{df}{dx_1}\right|_s = \left[-\frac{F_1}{Ah}\right]_s = -\frac{F_{1,s}}{Ah}$$

$$a_7 = \left.\frac{df}{dx_2}\right|_s = \left[\frac{F_2}{Ah}\right]_s = \frac{F_{2,s}}{Ah}$$

The steady-state component balance equation is given by

$$\frac{dx_s}{dt} = 0 = \left[ \frac{F_1 x_1}{Ah} - \frac{F_1 x}{Ah} + \frac{F_2 x_2}{Ah} - \frac{F_2 x}{Ah} \right]_s \quad (5.60)$$

Subtracting Eq. (5.60) from Eq. (5.59) and defining the usual deviation variables, we have

$$\frac{d\bar{x}}{dt} = a_2 \bar{F}_1 + a_3 \bar{F}_2 + a_5 \bar{x} + a_6 \bar{x}_1 + a_7 \bar{x}_2 + \cdots \quad (5.61)$$

Hence, the set of linear equations that represent the dynamic model of the blending process can be summarized as

$$\begin{aligned} \frac{d\bar{h}}{dt} &= \frac{\bar{F}_1}{A} + \frac{\bar{F}_2}{A} + a_1 \bar{h} \\ \frac{d\bar{x}}{dt} &= a_2 \bar{F}_1 + a_3 \bar{F}_2 + a_5 \bar{x} + a_6 \bar{x}_1 + a_7 \bar{x}_2 \end{aligned} \quad (5.62)$$

Applying the Laplace transform to both equations, we obtain

$$\begin{aligned} s\bar{h} &= \frac{\bar{F}_1}{A} + \frac{\bar{F}_2}{A} + a_1 \bar{h} \\ s\bar{x} &= a_2 \bar{F}_1 + a_3 \bar{F}_2 + a_5 \bar{x} + a_6 \bar{x}_1 + a_7 \bar{x}_2 \end{aligned} \quad (5.63)$$

Grouping the terms, we have

$$\begin{aligned} s\bar{h} - a_1 \bar{h} &= \bar{h}(s - a_1) = \frac{\bar{F}_1}{A} + \frac{\bar{F}_2}{A} \\ s\bar{x} - a_5 \bar{x} &= \bar{x}(s - a_5) = a_2 \bar{F}_1 + a_3 \bar{F}_2 + a_6 \bar{x}_1 + a_7 \bar{x}_2 \end{aligned} \quad (5.64)$$

Solving for $\bar{h}$ and $\bar{x}$ and after algebraic manipulations, we obtain

$$\begin{aligned} \bar{h} &= \frac{1/A}{(s - a_1)} \bar{F}_1 + \frac{1/A}{(s - a_1)} \bar{F}_2 \\ \bar{x} &= \frac{a_2}{(s - a_5)} \bar{F}_1 + \frac{a_3}{(s - a_5)} \bar{F}_2 + \frac{a_6}{(s - a_5)} \bar{x}_1 + \frac{a_7}{(s - a_5)} \bar{x}_2 \end{aligned} \quad (5.65)$$

or

$$\begin{aligned} \bar{h} &= \frac{k_1}{(\tau_1 s + 1)} \bar{F}_1 + \frac{k_1}{(\tau_1 s + 1)} \bar{F}_2 \\ \bar{x} &= \frac{k_2}{(\tau_2 s + 1)} \bar{F}_1 + \frac{k_3}{(\tau_2 s + 1)} \bar{F}_2 + \frac{k_4}{(\tau_2 s + 1)} \bar{x}_1 + \frac{k_5}{(\tau_2 s + 1)} \bar{x}_2 \end{aligned} \quad (5.66)$$

where we have, $k_1 = -1/a_1A$, $k_2 = -a_2/a_5$, $k_3 = -a_3/a_5$, $k_4 = -a_6/a_5$, $k_5 = -a_7/a_5$, and $\tau_1 = -1/a_1$, $\tau_2 = -1/a_5$. Hence, Eq. (5.66) can be summarized using the transfer functions:

$$\begin{aligned}\bar{h}(s) &= g_{11}(s)\bar{F}_1(s) + g_{12}(s)\bar{F}_2(s)\\ \bar{x}(s) &= g_{21}\bar{F}_1(s) + g_{22}\bar{F}_2(s) + g_{23}x_1(s) + g_{24}(s)x_2(s)\end{aligned} \tag{5.67}$$

where the transfer functions are defined as

$$\begin{aligned} g_{11}(s) &= \frac{k_1}{\tau_1 s+1}, \quad g_{12}(s) = \frac{k_1}{\tau_1 s+1},\\ g_{21}(s) &= \frac{k_2}{\tau_2 s+1}, \quad g_{22}(s) = \frac{k_3}{\tau_2 s+1},\\ g_{23}(s) &= \frac{k_4}{\tau_2 s+1}, \quad g_{24}(s) = \frac{k_5}{\tau_2 s+1}.\end{aligned} \tag{5.68}$$

The following model parameters are defined: $A = 5\ \mathrm{m}^2$, $\beta = 2\ \mathrm{m}^{5/2}/\mathrm{min}$. We also define the steady-state operating point as $F_{1,s} = 1\ \mathrm{m}^3/\mathrm{min}$, $F_{2,s} = 1\ \mathrm{m}^3/\mathrm{min}$, $x_{1,s} = 0.4$, $x_{2,s} = 0.8$, $h_s = 1$ m, $F_s = 2\ \mathrm{m}^3/\mathrm{min}$, $x_s = 0.6$. This leads to the following transfer functions for the blending process:

$$\begin{aligned} g_{11}(s) &= \frac{1}{5s+1}, \quad g_{12}(s) = \frac{1}{5s+1},\\ g_{21}(s) &= \frac{-0.1}{2.5s+1}, \quad g_{22}(s) = \frac{0.1}{2.5s+1},\\ g_{23}(s) &= \frac{0.5}{2.5s+1}, \quad g_{24}(s) = \frac{0.5}{2.5s+1}.\end{aligned} \tag{5.69}$$

# 6 Models from Process Data

In Chapter 4, we introduced the general principles of developing dynamic models to represent mathematically the unsteady (transient) behavior of a process. An alternative approach to developing models based on fundamental laws is to design and conduct plant experiments and use the data obtained from these experiments to construct dynamic relationships between the process inputs and the process outputs (Figure 6.1). Such methods are typically referred to as *black box* models since the input–output behavior is learned but the inner workings of the process cannot be uncovered. These empirical models are also tailor-made for process control purposes as they relate the control objectives (outputs) directly to the available manipulated variables (inputs). This implies that a model developed as such explains the behavior of all equipment physically present between the input and the output, such as valves and sensors, and not just the process of interest. This is not necessarily a drawback and perhaps even an advantage, as the model would now contain all the relevant information for control design and analysis.

## 6.1 DEVELOPMENT OF EMPIRICAL MODELS

The development of models from plant data requires careful planning and execution, as it may involve a number of plant personnel and interrupt normal operation. The engineer needs to minimize the impact of modeling experiments on the

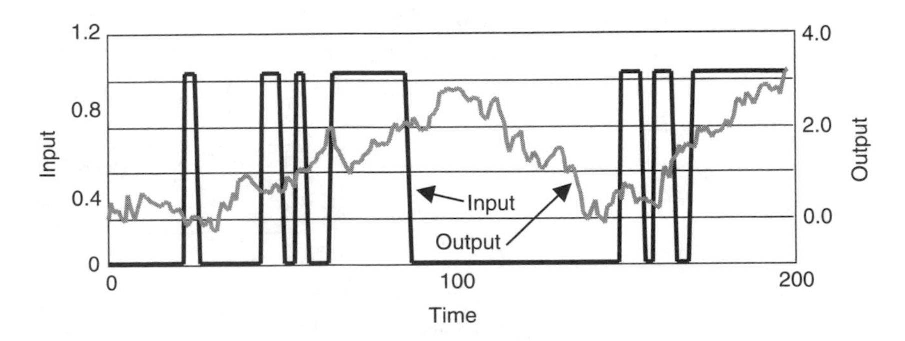


**FIGURE 6.1** Output transient data collected from a process subject to a well-defined input sequence.

plant operation, yet has to make sure that the data collected are sufficiently *rich* in information. This is a key trade-off and can be confronted very effectively with a well-reasoned plan that relies on existing process knowledge and modeling fundamentals. Developing a model using data collected from the process (Figure 6.2) requires the following key steps.

### 6.1.1 Performing the Experiments

As a first step in the design of experiments, one has to identify the variables that are essential for the modeling study. The engineer needs to assure the presence of reliable sensors for all such variables and determine how frequently they are measured. Next, the base operating conditions need to be determined. As we have seen in the previous chapter, linear models explain process behavior around local operating conditions and hence, it is imperative that such conditions are well documented to indicate the region of validity for the model to be developed.

To collect information on how the process responds to changes in the designated inputs, a series of perturbations needs to be introduced to those input variables. The experimental design should clearly indicate the magnitude of these perturbations (the process needs to stay within the linear regime) as well as their duration (the process needs to reach its natural steady-state). The magnitude and duration of these perturbations should also be kept to a minimum to accommodate quality and safety concerns in the plant.

The key to the success of plant experiments is in keeping all other variables constant while observing the effect of a designated input on the output(s). This may be rather challenging due to the presence of unidentified disturbances in the plant, but, nevertheless, the model should only reflect the dynamics associated with the input and output variables of interest.

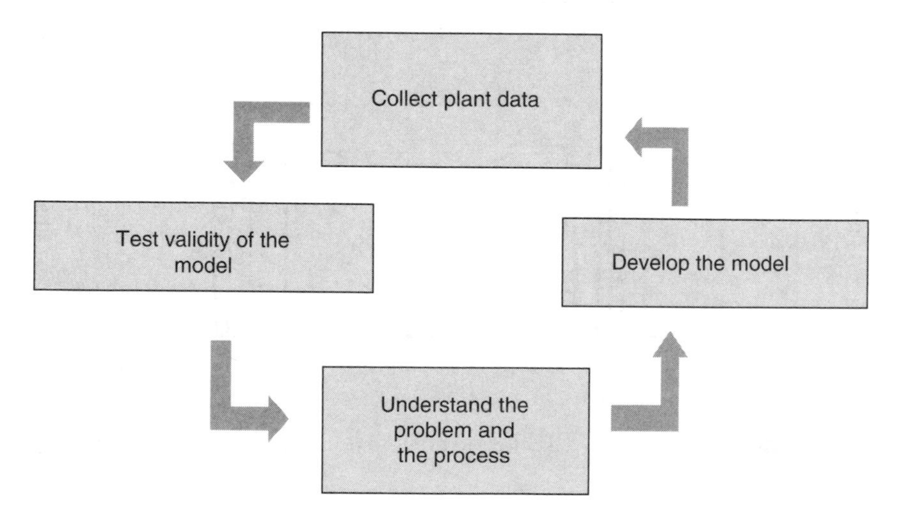


**FIGURE 6.2** Key steps in conceptual model development.

### 6.1.2 Developing the Model

The engineer is responsible for suggesting the *form* of the model and determining its *parameters* so that the model explains the observed data well. While the model form clearly depends on the intended use, there has to be an acceptable trade-off between complexity and accuracy.

---

It is worth pointing out that most chemical processes can be reasonably described by first-order models with time delay.

---

Typically, a model form is proposed and its success in explaining the data is evaluated using a performance criterion. Other forms may be tried if the performance is not found satisfactory. This underscores the *iterative* nature of empirical model development. Iterations may be conducted in a more intelligent manner if the engineer is familiar with the process under study and also by observing key features of the data that can point toward specific model forms and structure.

Depending on the form of the model, there may be a number of parameters that needs to be computed so that the model provides a reasonable fit to the data. For example, a first-order model with time delay would have three parameters to be determined (see Example 5.6). Such a determination can be made using either graphical or algorithmic methods, and we will illustrate both in this chapter.

### 6.1.3 Evaluating the Model

The most important element in model evaluation is the choice of the performance criterion. A simple visual inspection can be made by plotting the output variable data collected when the input perturbation is introduced against the output variable predicted by the model when it is subjected to the same input perturbation. Then, the engineer can decide if the *predicted* output sufficiently captures the key features observed in the *actual* output data.

---

If this evaluation is made with data not used in model development, it is referred to as *cross-validation*.

---

The cross-validation step is crucial because we would like to assess the predictive capabilities of the model beyond the experimental conditions used for model development. Yet, it should be clear that the model can only represent what it has seen already. Hence, the engineer should recognize the fact that the extrapolation capabilities of linear models are limited.

A common performance measure used in model evaluation is the integral square error (ISE) criterion. It is based on the error between the observed data and the model prediction

$$\text{ISE} = \sum_{i=1}^{N}(y_{i,\text{ observed}} - y_{i,\text{ predicted}})^2 \tag{6.1}$$

Here, $N$ refers to the number of observations. This is especially effective in comparing the performance of various choices for the model form and can help the engineer narrow down the alternative models. Smaller the ISE, closer is the predicted output response to the actual output response.

## 6.2 PROCESS REACTION CURVE

Owing to its simplicity and intuitive nature, this is the most common method in practice. Its drawbacks are often counterbalanced by the amount of conceptual information that it can provide with minimal effort and process interruption.

The method takes its name from the fact that the output response of a process to a step change in the input is referred to as the *process reaction curve* (PRC). We must note that including the effects of the measurement noise as well as unaccounted disturbances, the overall shape of the output response can be quite complicated. Yet, in many practical instances, this shape can be well approximated using a first-order-plus-time-delay (FOPTD) model, expressed in the *standard* form as

$$\frac{y(s)}{u(s)} = g(s) = \frac{ke^{-t_D s}}{\tau s + 1} \tag{6.2}$$

In this form, $k$ is the process gain, $\tau$ is the process time constant and $t_D$ the time delay (or dead time). We defer a full discussion of these terms to Chapter 8. The task here is to determine three model parameters, $k$, $\tau$, and $t_D$.

The process is subjected to a step change of magnitude $\hat{u}$ in the input, and the output variable response is recorded over time. A typical process reaction curve is displayed in Figure 6.3b. Implicit in this method is the expectation that the process output will generally conform to the shape in Figure 6.3b. To determine the model parameters, the following three steps are followed

1. The process gain $k$ is found by calculating the ratio of the change in the steady-state value of $y(t)$ ($\Delta y$) to the change in the steady-state value of $u(t)$ ($\Delta u$). If the input and the output are expressed in *deviation variables*, then $\Delta y$ and $\Delta u$ correspond to the final steady-state value of the output and the final steady-state value of the input, respectively:

$$k = \frac{\Delta y}{\Delta u} = \frac{y(t\rightarrow\infty)}{u(t\rightarrow\infty)} \tag{6.3}$$

In Figure 6.3, these points can be identified as follows:

$$k = \frac{\hat{y}}{\hat{u}} \tag{6.4}$$

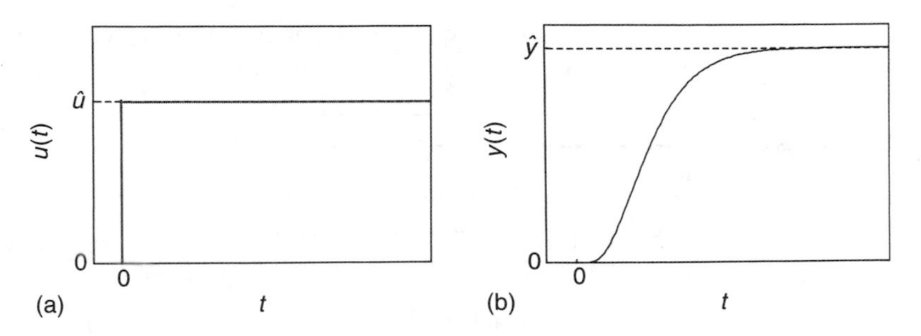


**FIGURE 6.3** Process reaction curve (b) in response to a step change in the input variable (a).

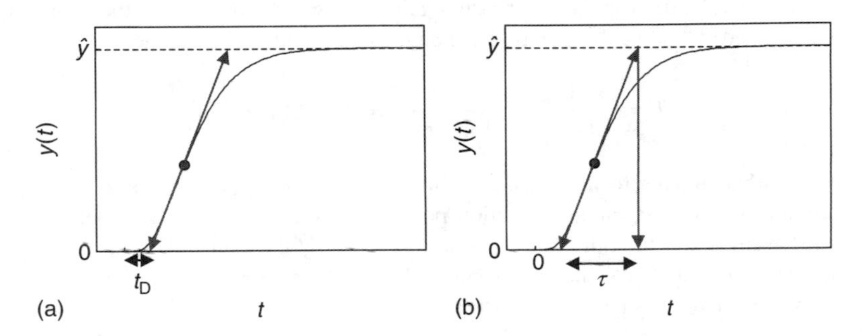


**FIGURE 6.4** Calculation of the time delay (a) and the time constant (b).

2. Next, we need to locate the inflection point of the process reaction curve, and draw a tangent along it (Figure 6.4a). The intersection of the tangent line and the time axis ($y = 0$) corresponds to the estimate of the time delay, $t_D$.
3. If the tangent line is extended to intersect the steady-state response line ($y = \hat{y}$) (Figure 6.4b), the point of intersection corresponds to $t = t_D + \tau$, thus $\tau$ can be calculated by inspection.

## Example 6.1

A linear model will be determined using the process reaction curve method for a heat exchanger shown in Figure 6.5. The shell-and-tube heat exchanger reduces the temperature of a process stream flowing in the tubes by using cooling water in the shell side. The model will represent the input–output relationship between the exit temperature of the process stream $T_2(t)$ and the flow rate of the cooling water $F_3(t)$, where both are expressed in deviation variables.

The process reaction curve (Figure 6.6a) is obtained in response to a unit step change (1 Liter/min) in the cooling water flow rate. The first observation is that the output response is not smooth due to the presence of noise in the thermocouple recordings. This makes the use of the process reaction curve method rather difficult as it may become harder to pin down the key graphical features like the inflection point.

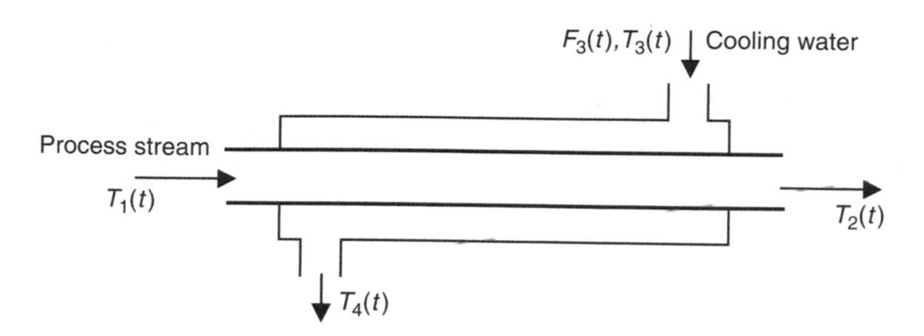


**FIGURE 6.5** Schematic representation of the shell-and-tube heat exchanger.

Let us start by determining the process gain. As discussed previously, the gain can be computed by finding the ratio of the change in the output to the change in the input.

$$k = \frac{\Delta T_2(t)}{\Delta F_3(t)} = \frac{2.5°\text{C}}{1\,\text{Liter/min}} = 2.5°\text{C/Liter/min} \tag{6.5}$$

We illustrate another method of determining the remaining parameters. Since it is difficult to draw an accurate inflection point through the noisy measurements, we propose to exploit the time domain solution of the FOPDT model for a unit step change in the input. We shall leave this solution as an exercise to the reader (see Chapter 8), and just supply the result here

$$y(t) = k(1 - \mathrm{e}^{-(t-t_\mathrm{D})/\tau}) \tag{6.6}$$

If we rearrange Eq. (6.6), we obtain

$$\ln\left(\frac{k-y}{k}\right) = -\frac{t-t_\mathrm{D}}{\tau} = -\frac{1}{\tau}t + \frac{t_\mathrm{D}}{\tau} \tag{6.7}$$

This equation should be valid for all points along the response curve, hence by suitably choosing two points (Figure 6.6b), we can have two equations from which the two unknowns $\tau$ and $t_\mathrm{D}$ can be solved. For the choices depicted in Figure 6.6b, the parameters are computed as

$$\tau = 19 \text{ min}$$

$$t_\mathrm{D} = 13 \text{ min}$$

resulting in the final model:

$$g(s) = \frac{2.5\mathrm{e}^{-13s}}{19s + 1} \tag{6.8}$$

We must note that this method is also to a certain extent subjective as by choosing different pairs, one may get different values for the unknown parameters. While it is important to recognize such modeling *uncertainties*, incorporation of this information into the final design of the control system is nontrivial.

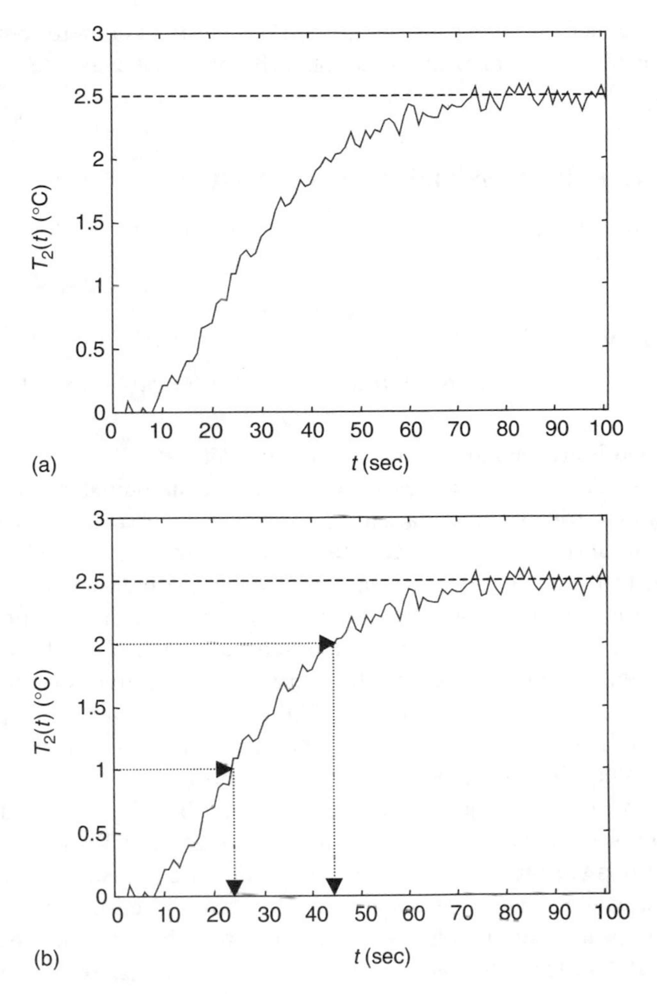


**FIGURE 6.6** Process reaction curve for the heat-exchanger (a) and application of the method (b).

If the process exhibits nonlinear characteristics, the dynamic response will be a function of the input magnitude and direction. Care should be exercised to perturb the process with sufficiently small inputs to avoid amplifying the nonlinear effects. It may also be prudent to change the direction of the input and evaluate the symmetry of the response. If different (qualitatively and quantitatively) responses are observed, a practical solution may be to average the parameters of the models thus obtained to arrive at an *average* model that can provide a reasonable accuracy within the region of operation. Again, the control engineer needs to acknowledge the uncertainties (or systematic errors) resulting from this experiment.

If a FOPDT model does not represent the data sufficiently well, one can use a second-order model with two poles. Depending on the type of dynamics involved, the methods of Harriot or Smith may be used.[1] For more complex models, one can resort to numerical estimation techniques that exploit the principle of regression.

## 6.3 LINEAR REGRESSION IN MODELING

Least-squares estimation is a method used for constructing algebraic input–output models of the form:[2]

$$y(i+1) = -a_1 y(i) - a_2 y(i-1) - \cdots - a_n y(i-n+1) + b_1 u(i) + b_2 u(i-1) + \cdots + b_m u(i-m+1) \tag{6.9}$$

where, $a_j$ and $b_j$ are the model parameters to be calculated.

This is a *discrete-time* representation of the input–output model, and such representations arise from the fact that the variables are sampled (measured) at discrete time instants and, in reality, are not available on a continuous basis. Supposing that the current time instant is $i$, Eq. (6.9) indicates that the output variable at the next time instant $i + 1$ depends on the measurement of the output variable at the current and previous time instants as well as the value of the input variable at the current and previous time instants. The complexity of the model (captured by the choice of $n$ and $m$) depends on the *memory* of the process, in other words, how far back in time, one must go to find input and output measurements influencing the output at the next time instant.

Typically, an input sequence $\{u(1), u(2), ..., u(N)\}$ is applied to the process and the corresponding sequence of outputs $\{y(1), y(2), ..., y(N)\}$ is recorded. This data set is then used to find the parameters in an optimal fashion. Let us denote by $\hat{y}$ the estimated value of the output variable from Eq. (6.9). The idea is to formulate the optimization problem in such a way that the difference between the recorded value of the output and its model prediction by Eq. (6.9) is minimized. This leads to the objective function

$$J = \frac{1}{2}\sum_{i=1}^{N} e^2(i) \tag{6.10}$$

where $e(i) = y(i) - \hat{y}(i)$. With the model defined as in Eq. (6.9), the error at the time instant $i$ can be expressed as

$$e(i) = y(i) - \hat{y}(i) = y(i) - \phi\theta \tag{6.11}$$

The vectors in Eq. (6.11) are given as

$$\begin{aligned} \phi(i+1) &= [-y(i) \cdots -y(i-n+1)\ u(i) \cdots u(i-m+1)] \\ \theta &= [a_1 \cdots a_n \quad b_1 \cdots b_m]^T \end{aligned} \tag{6.12}$$

Given the data set with $N$ input–output pairs, we can also construct the following vectors:

$$Y = \begin{bmatrix} y(2) \\ \vdots \\ y(N) \end{bmatrix}, \quad \Phi = \begin{bmatrix} \phi(i+1) \\ \vdots \\ \phi(N) \end{bmatrix}, \quad E = \begin{bmatrix} e(2) \\ \vdots \\ e(N) \end{bmatrix} \tag{6.13}$$

The least-squares problem can then be formulated as,

$$\text{minimize } J(\theta) = \frac{1}{2} E^{\mathrm{T}} E \tag{6.14}$$

It is shown[2] that the minimum of Eq. (6.14) is reached for the *optimal* parameter values $\theta^*$:

$$\theta^* = (\Phi^{\mathrm{T}}\Phi)^{-1}\Phi^{\mathrm{T}} Y \tag{6.15}$$

Note that the right-hand side of Eq. (6.15) consists of known vectors, hence the parameters of the model, optimal in the sense of least squares, can be computed in a straightforward manner. The following example will demonstrate the use of this method.

### Example 6.2

We collected 100 pairs of input–output data from a process with a sampling time of 1 minute. Figure 6.7 displays the input and output sequences. We point out here that this input sequence is referred to as a pseudo-random binary signal (PRBS), which is a common input sequence for developing linear models. The most notable characteristics of a PRBS are its two-valued (binary) amplitude and the switching frequency that follows a normal probability distribution (Appendix D).

We will determine if a first-order model would represent the dynamics of this process sufficiently well. Recall that a first-order process can be represented by the differential equation

$$\tau \frac{\mathrm{d}y(t)}{\mathrm{d}t} = ku(t) - y(t) \tag{6.16}$$

In terms of the model form in Eq. (6.9), a first-order process would have $n = 1$ and $m = 1$, hence,

$$y(i) = -a_1 y(i-1) + b_1 u(i-1) \tag{6.17}$$

To help build some insight into this representation, let us consider developing this model by starting from Eq. (6.16). We will approximate the derivative term in Eq. (6.16) by the finite-difference formula

$$\frac{\mathrm{d}y}{\mathrm{d}t} \cong \frac{y(i+1) - y(i)}{\Delta t} \tag{6.18}$$

This indicates that the derivative is calculated by knowing the values of two consecutive data points and dividing the difference by the time taken between the

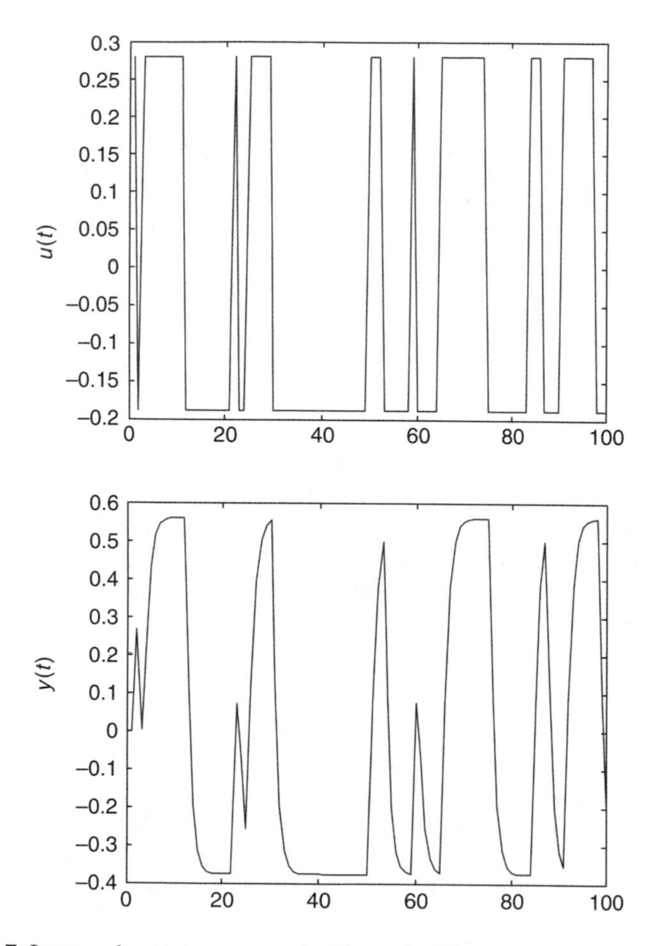


**FIGURE 6.7** Input and output sequences for Example 6.2.

points. In practice, data are collected at fixed sampling points and $\Delta t$ corresponds to the sampling time. The smaller the sampling time, the more accurate is the approximation in Eq. (6.18). We can rewrite Eq. (6.16) for the time instant $i$ as

$$y(i) = k\left(\frac{\Delta t}{\tau}\right)u(i-1) - \left(\frac{\Delta t}{\tau} - 1\right)y(i-1) \tag{6.19}$$

This is exactly the same equation as in Eq. (6.17) with $a_1 = [(\Delta t/\tau) - 1]$ and $b_1 = k(\Delta t/\tau)$.

This model can be used to determine if the plant behavior can be explained by first-order dynamics by fitting the coefficients $a_1$ and $b_1$. The objective is to minimize the error between the observed data points and the model prediction.

$$\text{minimize } J(a_1, b_1) = \frac{1}{2}\sum_{i=1}^{100} e^2(i) \tag{6.20}$$

The error is given as

$$e(i) = y(i) - \hat{y}(i) = y(i) + a_1 y(i-1) - b_1 u(i-1) \tag{6.21}$$

Using the notation in Eq. (6.12), we can define the following vectors:

$$\begin{aligned} \phi(i+1) &= [-\,y(i) \quad u(i)] \\ \theta &= [a_1 \quad b_1]^{\mathrm{T}} \end{aligned} \tag{6.22}$$

Using the data set associated with Figure 6.7, Eq. (6.15) yields the following result

$$\begin{bmatrix} a_1 \\ b_1 \end{bmatrix} = (\Phi^{\mathrm{T}}\Phi)^{-1}\Phi^{\mathrm{T}} Y = \begin{bmatrix} -0.5162 \\ 1.0135 \end{bmatrix} \tag{6.23}$$

The collected data are then plotted in Figure 6.8 against the model prediction and one can observe the goodness of fit visually. For this experiment, ISE can be calculated as 0.084. The conclusion is that the model may be considered adequate for explaining the process dynamics. However, further tests for cross-validation would be required to evaluate the prediction capabilities of the model better.

---

Note that the first-order model parameters $k$ and $\tau$ can be determined through the definitions in Eq. (6.19) as $k = 2.1$ and $\tau = 2.06$ min.

---

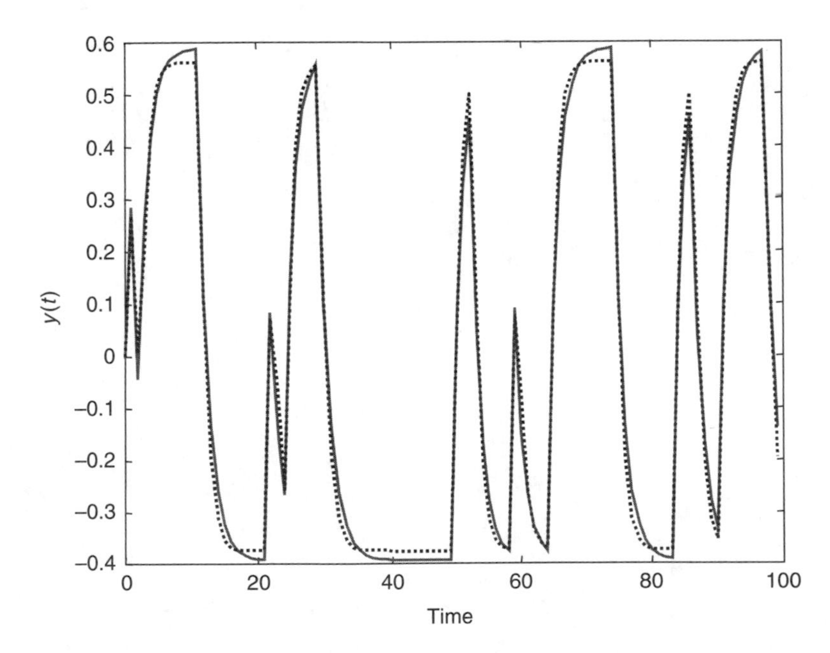


**FIGURE 6.8** Comparison of process data and model prediction. Dotted line represents the actual data points, and the solid line is the model prediction.

## 6.4 SUMMARY

An empirical model of a process can be built from transient data collected by varying process inputs and observing the process output response. We discussed two approaches in determining the model. In the PRC approach, data are fit into a prespecified model, namely the FOPDT structure. In the second approach, using the concept of linear regression, one can build models with more general structures that can explain a variety of dynamic behaviors.

## REFERENCES

1. Seborg, D.E., T.F. Edgar, and D.A. Mellichamp, *Process Dynamics and Control*, 2nd Edition, Wiley, NY, 2004.
2. Astrom, K.J. and B. Wittenmark, *Computer-Controlled Systems – Theory and Design*, 2nd Ed., Prentice Hall, Englewood Cliffs, NJ, 1990.

# Section II Additional Reading

An excellent treatise on modeling that discusses its etymology, as well as its implications for physicochemical systems is

Aris, R., *Mathematical Modeling Techniques*, Dover, New York, NY, 1994.

A more recent book by Aris focuses on the specific modeling strategies for chemical processes and includes a number of his historical papers on mathematical modeling. The book by Hangos and Cameron is an excellent source for not only the philosophy and principles of fundamental model development, but also for empirical model building and diagnosis.

Aris, R., *Mathematical Modeling: A Chemical Engineer's Perspective*, Academic Press, San Diego, CA, 1999.

Hangos, K. and Cameron, I., *Process Modeling and Model Analysis*, Academic Press, New York, 2001.

The text by Franks introduces modeling strategies from the viewpoint of computer simulations and is considered to be the first one of its kind:

Franks, R.G.E., *Mathematical Modeling in Chemical Engineering*, Wiley, New York, NY, 1967.

Fundamental model building (specifically, dynamic models) for chemical process systems is often covered in dynamics and control textbooks. These books also have chapters on empirical modeling strategies:

Bequette, B.W., *Process Dynamics: Modeling, Analysis, and Simulation*, Prentice-Hall, Upper Saddle River, NJ, 1998.

Luyben, W.L., *Process Modeling, Simulation and Control for Chemical Engineers*, 2nd ed., McGraw-Hill, New York, NY, 1990.

Ogunnaike, B.A. and Ray, W.H., *Process Dynamics, Modeling, and Control*, Oxford University Press, Oxford, 1994.

In building fundamental models of processes, sometimes, the observed results contradict our intuition. Often, the reason for this is the poor understanding of assumptions that are made to derive the model equations. The following article explores the source of incorrect interpretations of the dynamic behavior of a stirred-tank heater, a common process used in many textbooks, including ours.

Romagnoli, J.A., Palazoglu, A., and Whitaker, S., Dynamics of a stirred-tank heater: Intuition and analysis, *Chem. Eng. Educ.*, 35(1), 46–49, 2001.

Some insight into the modeling of biological systems can be found in the following texts:

Bailey, J.E. and Ollis, D.F., *Biochemical Engineering Fundamentals*, 2nd ed., McGraw-Hill, New York, NY, 1986.

Shuler, M.L. and Kargi, F., *Bioprocess Engineering: Basic Concepts*, 2nd ed., Prentice-Hall, Upper Saddle River, NJ, 2001.

Additional information on empirical modeling of dynamic systems can be found in a number of books on identification and control.

Eykhoff, P., *System Identification: Parameter and State Estimation*, Wiley-Interscience, London, 1974.

Ramirez, W.F., *Process Control and Identification*, Academic Press, Boston, MA, 1994.

Seinfeld, J.H. and Lapidus, L., *Process Modeling, Estimation, and Identification*, Prentice-Hall, Englewood Cliffs, NJ, 1974.

Soderstrom, T. and Stoica, P., *System Identification*, Prentice-Hall, New York, NY, 1989.

Zhu, Y., *Multivariable System Identification for Process Control*, Pergamon Press, Amsterdam, 2001.

# Section II Exercises

**II.1.** Consider a continuous blending process where the water is mixed with slurry to give slurry of the desired consistency (Figure II.1). They are mixed in a constant volume ($V$) blending tank, and the mass fraction of the solids in the inlet slurry stream is given as $x_s$, with a volumetric flow rate of $q_s$. Since $x_s$ and $q_s$ vary, the water make-up mass flow rate $w$ is changed to compensate for these variations. Write an unsteady-state model for this blender that can be used to predict the dynamic behavior of the mass fraction of solids in the exit stream $x_e$ for changes in $x_s$, $q_s$, or $w$. What is the number of degrees of freedom for this process?

**II.2.** A binary mixture at the saturation point is fed to a single-stage flash unit (Figure II.2), where the mixture is heated at an unknown rate ($Q$). The feed flow rate and feed mole fractions are known and may vary with time. Assume that $x$ represents the mole fraction of the more volatile component (e.g., $x_f$ is the mole fraction of the more volatile component in the feedstream) and the molar heat of vaporization is the same for both components. Flow rate is given in moles per unit time. $H$ represents the molar liquid holdup.

1. Derive the modeling equations for this system. State your assumptions clearly and explicitly.
2. Derive the transfer function between the overhead mole fraction of the more volatile component and its feed mole fraction. (**Hint:** Assume constant molar holdup.)

**II.3.** Most separation processes in the chemical industry consist of a sequence of stages. For example, sulfur dioxide present in combustion gas may be removed by the use of a liquid absorbent (such as dimethylalanine) in a multistage absorber. Consider the three-stage absorber displayed in Figure II.3.

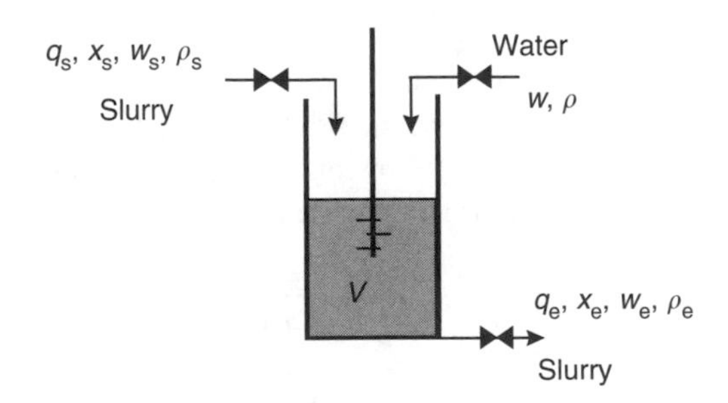


**FIGURE II.1** Schematic of slurry blending process.

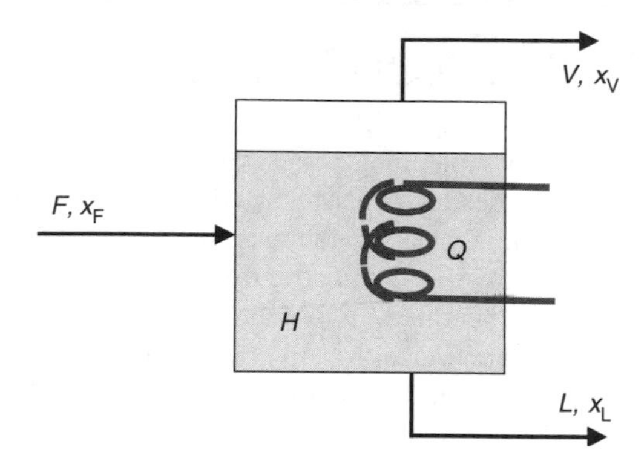


**FIGURE II.2** Schematic of a flash unit.

This process is modeled through the following equations:[1]

$$\tau \frac{dx_1}{dt} = K(y_f - b) - (1 + S)x_1 + x_2$$

$$\tau \frac{dx_2}{dt} = Sx_1 - (1 + S)x_2 + x_3$$

$$\tau \frac{dx_3}{dt} = Sx_2 - (1 + S)x_3 + x_f$$

where $H$ is the liquid holdup in each stage and assumed to be constant, and $x$ and $y$ represent liquid and vapor compositions, respectively. Also, $\tau = H/L$ is the liquid residence time, $S = aG/L$ is the stripping factor, $K = G/L$ is the gas-to-liquid ratio, and $A$ and $b$ are constants.

1. How many variables are there? How many equations (relationships)? What are the degrees of freedom?
2. Is this system underdetermined or overdetermined? Why?
3. What additional relationships, if necessary, can you suggest to reduce the degrees of freedom to zero?

**II.4.** Consider a liquid chromatography unit for the separation of a mixture containing $N$ components. Assuming that the process is isothermal, and there are no radial concentration gradients, the following governing equations for solute $j$ in the mobile phase and on the adsorbent can be obtained:

$$u_0 \frac{\partial c_j}{\partial z} + \varepsilon_t \frac{\partial c_j}{\partial t} + (1 - \varepsilon)\frac{\partial q_j}{\partial t} = \varepsilon D_L \frac{\partial^2 c_j}{\partial z^2}$$

$$\frac{\partial q_j}{\partial t} = k_{a,j} c_j q_{m,j}\left(1 - \sum_{i=1}^{N} \frac{q_i}{q_{m,i}}\right) - k_{d,j} q_j$$

In this model, $c$ is the concentration of solute in the mobile phase, $q$ the adsorbate concentration, $u_0$ the superficial velocity, $\varepsilon$ and $\varepsilon_t$ are column void fraction and

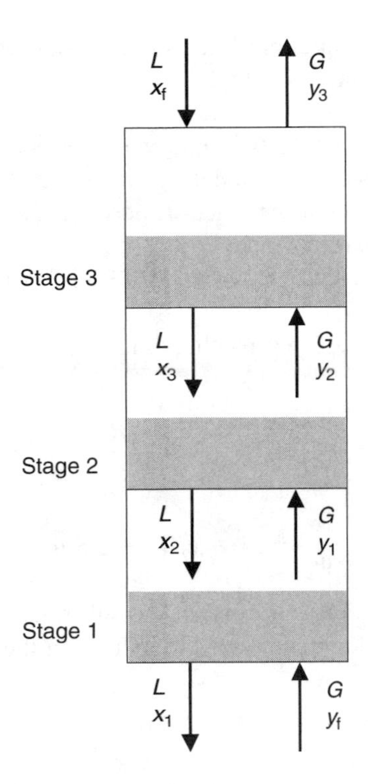


**FIGURE II.3** Schematic of a three-stage absorber.

total void fraction, respectively, $D_L$ is the axial dispersion coefficient, $q_m$ is the maximum adsorbate concentration, and $k_{a,j}$ and $k_{d,j}$ are the adsorption and desorption rate constants for solute $j$, respectively:

1. How would you classify this system of equations? Why?
2. How many variables are there? How many equations (relationships)? What is the number of degrees of freedom?
3. Is this system underdetermined or overdetermined? Why?
4. What additional relationships, if necessary, can you suggest to reduce the degrees of freedom to zero?

**II.5.** A liquid-phase isothermal reaction takes place in a continuously stirred-tank reactor. The reaction is second-order and the outlet flow rate depends linearly on the liquid volume in the tank. The model equations are given as

$$\frac{dC}{dt} = \frac{F_{in}}{V}(C_{in} - C) - kC^2$$

$$\frac{dV}{dt} = F_{in} - F = F_{in} - \beta V$$

where $C$ is the tank concentration, $V$ the tank volume, $F$ the flow rate, the subscript in refers to inlet conditions, and $k$ and $\beta$ are constants:

1. What are the state variable(s), input variable(s), and output variable(s)? Obtain a linear state-space model for this system.
2. Derive the process transfer function between the outlet (tank) concentration and the inlet flow rate.
3. What are the poles and zeros of this transfer function?

**II.6.** A bioreactor is represented by the following model that uses the Monod kinetics:

$$\frac{dx_1}{dt} = (\mu - D)x_1$$

$$\frac{dx_2}{dt} = (4 - x_2)D - 2.5\,\mu x_1$$

where $x_1$ is the biomass concentration, $x_2$ the substrate concentration, and $D$ the dilution rate. The specific growth rate $\mu$ depends on the substrate concentration as follows:

$$\mu = \frac{0.53x_2}{0.12 + x_2}$$

In this system, we are interested in controlling the biomass concentration using the dilution rate, at the steady-state defined by $x_{1s} = 1.4523$, $x_{2s} = 0.3692$, and $D_s = 0.4$.

1. Obtain a linear state-space model for this system.
2. Derive the process transfer function for this system.
3. What is the order of this process?

**II.7.** Consider the following model:

$$\frac{d^2x}{dt^2} + a\frac{dx}{dt} + x = 1$$

with the initial conditions, $x(0) = x'(0) = 0$.

1. Using Laplace transformation, find the solutions of this model when $a = 1$ and $a = 3$.
2. Plot the solutions on one graph and discuss the effect of the parameter $a$ on the solutions.

**II.8.** An oil stream is heated as it passes through two well-mixed tanks in series. Heat is supplied in the first tank through a heating coil, and the volumes of both tanks are constant. Also assuming constant physical properties, the following model equations are developed through energy balances around each tank:

$$\rho c_p V_1 \frac{dT_1}{dt} = \rho c_p F(T_{in} - T_1) + Q$$

$$\rho c_p V_2 \frac{dT_2}{dt} = \rho c_p F(T_1 - T_2)$$

Develop the transfer functions among the second tank temperature (output) $T_2$, the heat input (manipulated variable) $Q$, and the flow rate (disturbance) $F$. $T_{in}$ can be assumed as constant (what if it is not?).

**II.9.** Consider the equation:

$$\frac{d^3y(t)}{dt^3} + a\frac{d^2y(t)}{dt^2} + 8\frac{dy(t)}{dt} + 4y(t) = 3m(t)$$

We can have either $a = 5$ or $a = 1$. The output variable $y(t)$ and the input variable $m(t)$ are both in deviation variable form.

1. Express this equation in the form of a state-space model, i.e.,

$$\dot{x} = Ax + Bm$$

$$y = Cx + Dm$$

2. Plot the response of the output $y(t)$ variable to a unit-step change in the input variable $m(t)$ for both cases of the parameter $a$.

**II.10.** A process is modeled by the following equations:

$$2\frac{dx_1}{dt} = -2x_1 + \exp(-x_1) - 3u_1x_2$$

$$\frac{dx_2}{dt} = -x_2 + \frac{2x_1}{1 + x_2} + 4u_2$$

The control objectives dictate the following output equations:

$$y_1 = x_1$$

$$y_2 = x_2$$

1. Find the four transfer functions relating the outputs $(y_1, y_2)$ to the inputs $(u_1, u_2)$.

2. Solve the equations with the conditions, $u_1(t) = 1$, $u_2(t) = 1$, $y_1(0) = 0$, and $y_2(0) = 0$.
3. Plot the output responses. What is the steady-state reached by the outputs?

**II.11.** Consider the state-space model

$$\begin{aligned} x &= Ax + Bu \\ y &= Cx \end{aligned}, A = \begin{bmatrix} -2 & 1 & 0 \\ -3 & 0 & 3 \\ -1 & 0 & -3 \end{bmatrix}, B = \begin{bmatrix} 0 \\ 3 \\ 1 \end{bmatrix}, C = [1 \quad 0 \quad 0]$$

Determine the eigenvalues of the state matrix. Also, find the transfer function model for this system. Report the poles and zeros of this transfer function. Obtain the response of this model to a step change in the input.

**II.12.** Consider the following state-space model:

$$\begin{bmatrix} \dot{x}_1 \\ \dot{x}_2 \end{bmatrix} = \begin{bmatrix} -2.405 & 0 \\ 0.833 & -2.238 \end{bmatrix} \begin{bmatrix} x_1 \\ x_2 \end{bmatrix} + \begin{bmatrix} 7 \\ -1.117 \end{bmatrix} u$$

$$y = [0 \quad 1] \begin{bmatrix} x_1 \\ x_2 \end{bmatrix}$$

Find the transfer function $g(s)$ where $y(s) = g(s)u(s)$. Determine the poles and zeros.

**II.13.** A chemical reactor has been operating at steady-state for a long time with the feed flow rate $F_{in}$ kept constant at 3.5 m$^3$/min. To handle a projected increase in upstream capacity, the operator decides to increase the feedflow rate suddenly by 10%, resulting in a change in the outlet stream composition recorded in Table II.1. Using the process reaction curve method, obtain an empirical transfer function model for this process.

**TABLE II.1**
**Composition Data in Response to a Change in Flow Rate**

| Time (min) | Change (gmol/m$^3$) | Time (min) | Change (gmol/m$^3$) |
|---|---|---|---|
| 0 | 0 | 1.6 | 0.35 |
| 0.2 | 0 | 1.8 | 0.5 |
| 0.4 | 0 | 2 | 0.55 |
| 0.6 | 0.02 | 3 | 0.7 |
| 0.8 | 0.1 | 4 | 0.9 |
| 1 | 0.15 | 5 | 0.95 |
| 1.2 | 0.2 | 6 | 1 |
| 1.4 | 0.3 | 8 | 1 |

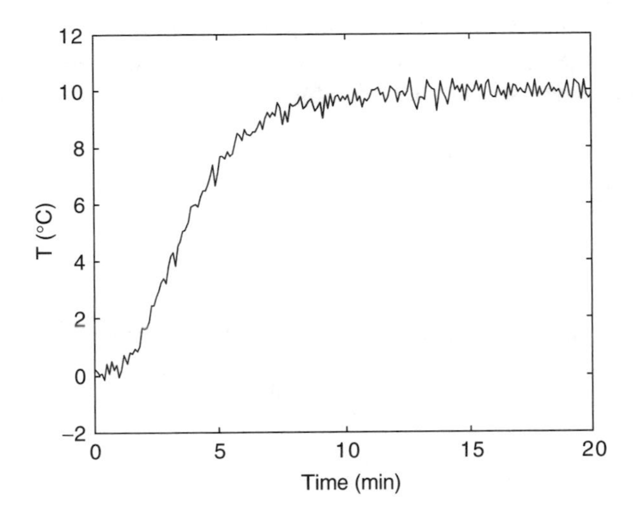


**FIGURE II.4** Temperature response of the outlet process stream.

**II.14.** An experiment was performed on a shell-and-tube heat exchanger that heats a process stream with medium pressure steam. In the experiment, the steam valve was opened an additional 5% in a stepwise manner. The resulting temperature response of the process outlet stream is given in Figure II.4. Determine the process model parameters using the reaction curve method, and estimate the inaccuracies due to the data and calculation methods.

**II.15.** Bioreactors are used to produce a variety of pharmaceuticals and food products. A simple bioreactor model involves biomass and substrate. The biomass consists of cells that consume the substrate. The following material balance equations are derived for a bioreactor,[2]

$$\frac{dx_1}{dt} = (\mu - D)x_1$$

$$\frac{dx_2}{dt} = D(x_{2f} - x_2) - \frac{\mu x_1}{Y}$$

where $D$ is the dilution rate, $x_1$ the biomass concentration, $x_2$ the substrate concentration, $x_{2f}$ the feed concentration of the substrate, and $Y$ the yield with $Y = 0.4$. The specific growth rate $\mu$ for a system with substrate inhibition is given as

$$\mu = \frac{\mu_{max} x_2}{k_m + x_2 + k_1 x_2^2}$$

The constants are given as $k_m = 0.12$ g/L, $k_1 = 0.4545$ L/g, and $\mu_{max} = 0.53\ h^{-1}$. The steady-state values of the input and the state variables are $D_s = 0.3\ h^{-1}$, $x_{1s} = 1.5302$, and $x_{2s} = 0.1745$.

Construct a simulation of this process in MATLAB/Simulink. By introducing small step changes in the dilution rate, observe the biomass response. Develop a discrete-time model (see Eq. [6.9]) for this system using the step-response data.

## REFERENCES


1. Seborg, D.E., Edgar, T.F., and Mellichamp, D.A., *Process Dynamics and Control*, 2nd ed., Wiley, New York, 2004.
2. Bequette, B.W., *Process Dynamics: Modeling, Analysis and Simulation*, Prentice-Hall, Englewood Cliffs, NJ, 1998.

# Section III

## Process Analysis

# 7 Stability

As pointed out before, the first step in system analysis is to develop a mathematical model for the process of interest. Once a mathematical model is available, there are various analytical techniques that can be used to study the dynamic behavior of the process.

The objective of this analysis step is the determination of process characteristics, such as stability and transient response. However, knowing whether a process is absolutely stable or not is insufficient information for most practical purposes. For instance, if a process is known to be stable, we would also like to know how close it might be to becoming unstable. Furthermore, for a stable process, we may still want to know the shape of its dynamic response, such as the presence of oscillations, settling time, etc., so that the dynamic performance can be quantified.

Certainly, the direct (often numerical) solution of the model equations may be employed to analyze the dynamics of the process in response to various input perturbations. Yet, such simulations only provide information for specific operating conditions and are difficult to generalize. We would like to focus on analytical techniques that do not require the solution of a model and hence offer the possibility of a more generic account of process behavior.

## 7.1 STABILITY OF LINEAR SYSTEMS

An intuitive definition of stability means that the process states must not grow without bound in response to a bounded input or an initial condition. To explore what this means in the context of a linear dynamic system, let us start with the statespace representation (5.18), and focus on the state equation

$$\dot{\boldsymbol{x}}(t) = \frac{\mathrm{d}\boldsymbol{x}(t)}{\mathrm{d}t} = A\boldsymbol{x}(t) + B\boldsymbol{u}(t) \tag{7.1}$$

Note that the boldfaced lowercase letters indicate vectors, and the uppercase letters denote matrices (see Appendix C). The integration of this linear differential equation, with the vector of initial conditions $\boldsymbol{x}(t = 0) = \boldsymbol{x}_0$, results in

$$\boldsymbol{x}(t) = H(t)\boldsymbol{x}_0 + \int_0^t H(\tau)B\boldsymbol{u}(\tau)\mathrm{d}\tau \tag{7.2}$$

Here, $H(t)$ is called the *state transition matrix* and is given by

$$H(t) = \mathrm{e}^{At} \tag{7.3}$$

Since we are only interested in bounded inputs (Figure 7.1) and bounded, non-zero initial conditions for most practical problems, the stability of the states can be inferred from the characteristics of the state transition matrix only.

We observe that the dynamic evolution of the states depends directly on the *eigenvalues* of the matrix $H(t)$, and it can be shown[1] that the eigenvalues $\alpha_i$ of $H(t)$ are related to the eigenvalues $\lambda_i$ of the state matrix $A$ by

$$\alpha_i = e^{\lambda_i t} \tag{7.4}$$

If the eigenvalue of the state matrix $A$ is expressed as a complex number (for generality), $\lambda_i = \beta_i \pm j\omega_i$, then we arrive at the simple conditions depicted in Table 7.1 for the stability of linear processes expressed by Eq. (7.1).

---

A linear dynamic system with constant coefficients is *unstable* if its state matrix has eigenvalues with positive real parts.

---

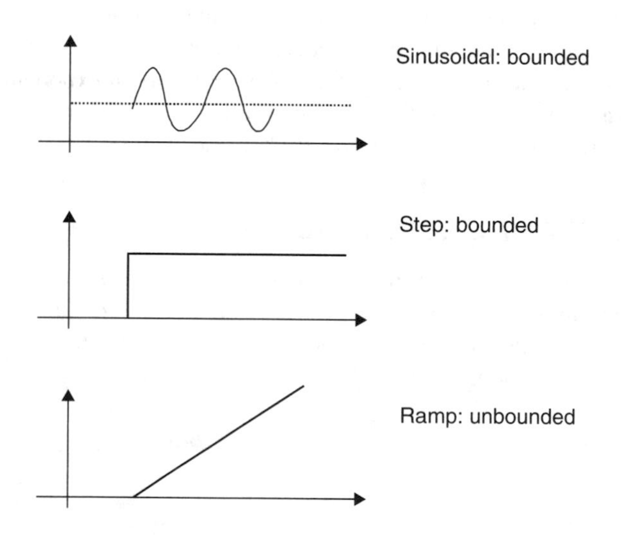


**FIGURE 7.1** Bounded and unbounded inputs for a process.

**TABLE 7.1**
**Stability Conditions for Systems Described by Eq. (7.1)**

| Stability Condition | Eigenvalue Condition |
|---|---|
| Unstable | $\beta_i > 0$ |
| Critically stable | $\beta_i = 0$ |
| Asymptotically stable | $\beta_i < 0$ |

### Example 7.1

Let us consider the state equations for a process as a function of a constant design parameter $a$:

$$\dot{\boldsymbol{x}}(t) = \begin{bmatrix} \dot{x}_1(t) \\ \dot{x}_2(t) \end{bmatrix} = \begin{bmatrix} -2.5-a & -1.3 \\ 0.8 & -0.5-a \end{bmatrix} \begin{bmatrix} x_1(t) \\ x_2(t) \end{bmatrix} + \begin{bmatrix} 1 \\ 0 \end{bmatrix} u(t) \tag{7.5}$$

The initial condition vector is given as

$$\begin{bmatrix} x_1(0) \\ x_2(0) \end{bmatrix} = \begin{bmatrix} 0 \\ 0 \end{bmatrix}$$

The input is considered to be a unit step (see Figure 7.1), hence it is bounded. Table 7.2 summarizes the process stability characteristics for three cases, and Figure 7.2 demonstrates the salient response characteristics of both states.

As pointed out in Table 7.1, the type of stability is determined solely by the real part of the eigenvalues of the state matrix, $\beta_i$:

1. When $\beta_i$ is negative, the states reach the new condition in an exponential manner (Figure 7.2a).
2. When $\beta_i$ is positive, the states grow unbounded in an exponential manner (Figure 7.2b).
3. When $\beta_i$ is zero, the states oscillate continuously, neither decaying nor growing, thus creating a critically (marginally) stable condition (Figure 7.2c).

While the system stability is strictly determined by the *real* part of the eigenvalues of the state matrix, the *imaginary* part $\omega_i$ also plays a role in shaping the state evolution over time. If $\omega_i = 0$, then the state response solely consists of exponential terms, but if $\omega_i \neq 0$, then the process response will contain *sinusoidal* terms with *frequencies* $\omega_i$. The magnitude of $\omega_i$ determines the degree of the oscillatory behavior of the states. Figure 7.3 displays the response of the states for the previous example for three different sets of eigenvalues. Note that the responses are stable but the frequency of oscillations grows with the magnitude of the imaginary part of the eigenvalues.

**TABLE 7.2**
**Stability Conditions for System Described by Eq. (7.5)**

| Parameter $a$ | Stability Condition | Eigenvalues $\lambda_{1,2}$ |
|---|---|---|
| 2.5 | Asymptotically stable | $-4 \pm 0.2j$ |
| $-2.5$ | Unstable | $1 \pm 0.2j$ |
| $-1.5$ | Critically stable | $\pm 0.2j$ |

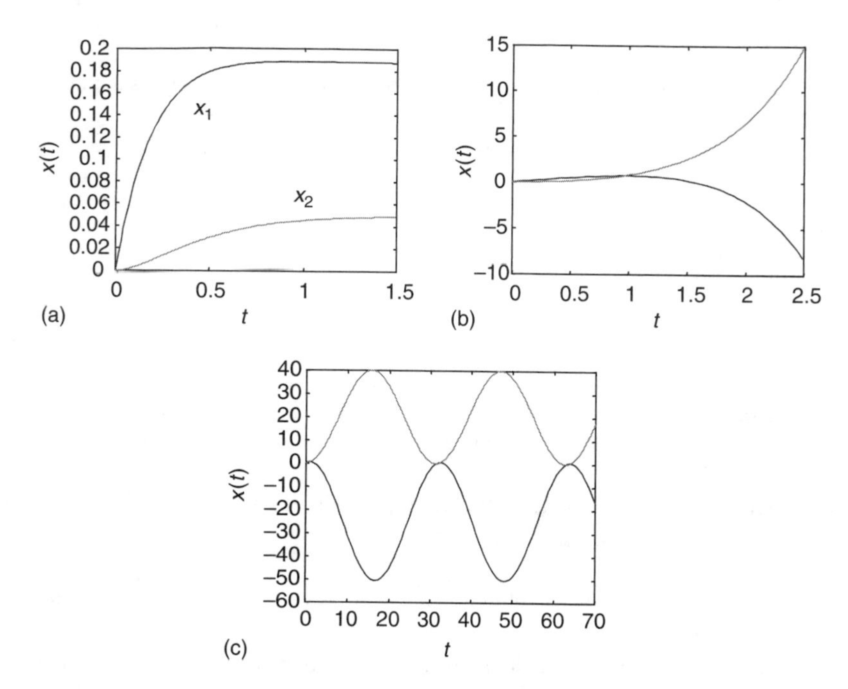


**FIGURE 7.2** Evolution of states for a step input: (a) $a = 2.5$, (b) $a = -2.5$, (c) $a = -1.5$.

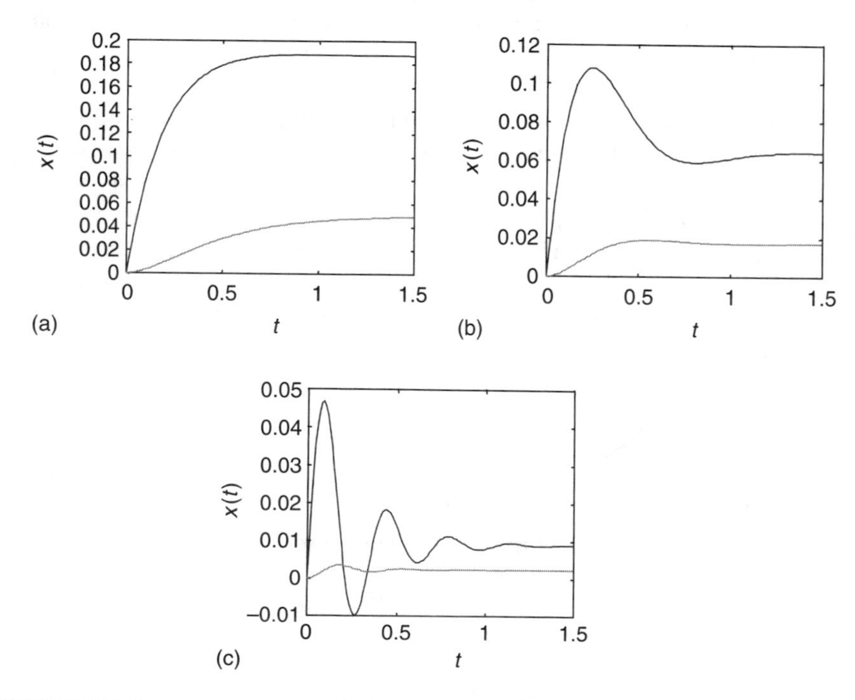


**FIGURE 7.3** Process response with eigenvalues (a) $\lambda_{1,2} = -4 \pm 0.2j$, (b) $\lambda_{1,2} = -4 \pm 5.59j$, and (c) $\lambda_{1,2} = -4 \pm 17.86j$.

## 7.2 INPUT–OUTPUT STABILITY

The state-space equations (5.18) can be used to construct an input–output model in the Laplace domain. We start with the equations

$$\dot{x} = \frac{dx}{dt} = Ax + Bu$$
$$y = Cx \tag{7.6}$$

and proceed to take their Laplace transform, keeping in mind that the variables are expressed in *deviation variable* form. This results in

$$sx(s) = Ax(s) + Bu(s)$$
$$y(s) = Cx(s) \tag{7.7}$$

As we will be focusing largely on SISO processes, for this particular case, the input–output model, or the process transfer function can be expressed as

$$\frac{y(s)}{u(s)} = g(s) = C[sI - A]^{-1}B \tag{7.8}$$

Here, all matrices should have consistent dimensions and $I$ denotes the identity matrix. Equation (7.8) illustrates how one would go from the state-space representation to the transfer function representation of a process.

It can be shown easily that the characteristic equation of the transfer function in Eq. (7.8) is given by the following relationship:

$$p(s) = \text{determinant}(sI - A) = 0 \tag{7.9}$$

Thus, one can conclude that the poles of a transfer function are equivalent to the eigenvalues of the corresponding state matrix. This observation facilitates the stability analysis significantly for processes expressed by transfer functions. The location of the poles of a transfer function determines the *bounded input–bounded output* (BIBO) stability of a process.

---

If the transfer function of a dynamic process has a pole with a positive real part, the process is unstable. If the real part is zero, then the process is critically stable.

---

We can analyze the stability of a dynamic process graphically through the complex plane (or $s$-plane). Since $s$ is a complex variable, it has a real and an imaginary part.

$$s = \beta \pm j\omega$$

We can represent the complex variable $s$ by a point in the $s$-plane using the real and imaginary coordinates as shown in Figure 7.4.

We assert that the dynamic process is stable if and only if all its poles are located on the left-hand side of the complex $s$-plane (Figure 7.5). If the poles are located *on* the imaginary axis, then we have a critically stable condition.

### Example 7.2

Consider the same set of state equations as in Example 7.1. For the case of $a = 2.5$, we have

$$\dot{\mathbf{x}}(t) = \begin{bmatrix} \dot{x}_1(t) \\ \dot{x}_2(t) \end{bmatrix} = \begin{bmatrix} -5 & -1.3 \\ 0.8 & -3 \end{bmatrix} \begin{bmatrix} x_1(t) \\ x_2(t) \end{bmatrix} + \begin{bmatrix} 1 \\ 0 \end{bmatrix} u(t) \quad (7.10)$$

The output equation is given as

$$y(t) = [1 \quad 0] \begin{bmatrix} x_1(t) \\ x_2(t) \end{bmatrix} \quad (7.11)$$

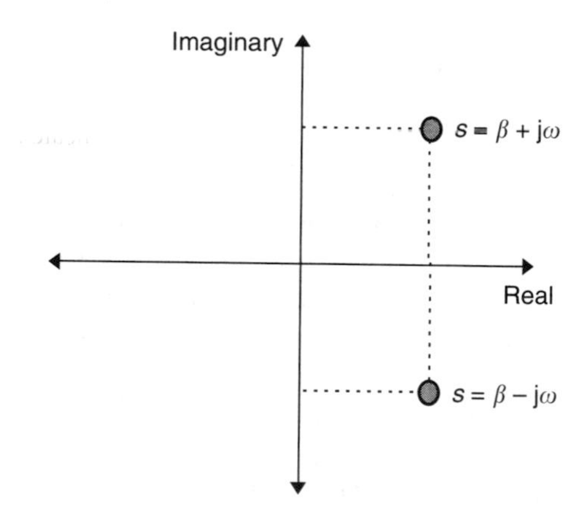


**FIGURE 7.4** The $s$-plane and the point $s = \beta \pm j\omega$.

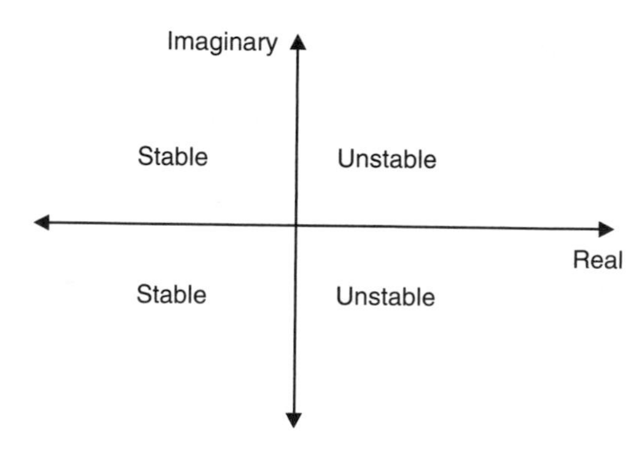


**FIGURE 7.5** Regions of stability and instability in the $s$-plane.

This results in the transfer function, as specified in Eq. (7.8),

$$g(s) = C[sI - A]^{-1}B = [1 \quad 0]\begin{bmatrix} s+5 & 1.3 \\ -0.8 & s+3 \end{bmatrix}^{-1}\begin{bmatrix} 1 \\ 0 \end{bmatrix} \tag{7.12}$$

Noting that

$$\begin{bmatrix} s+5 & 1.3 \\ -0.8 & s+3 \end{bmatrix}^{-1} = \frac{\text{adjoint}\begin{bmatrix} s+5 & 1.3 \\ -0.8 & s+3 \end{bmatrix}}{\text{determinant}\begin{bmatrix} s+5 & 1.3 \\ -0.8 & s+3 \end{bmatrix}} = \frac{\begin{bmatrix} s+3 & 1.3 \\ 0.8 & s+5 \end{bmatrix}}{(s+3)(s+5) - (1.3)(-0.8)}$$

we have the transfer function expressed as

$$g(s) = \frac{s+3}{s^2+8s+16.04} \tag{7.13}$$

The characteristic equation and the poles are given as

$$p(s) = s^2+8s+16.04 \Rightarrow p_{1,2} = -4\pm0.2j$$

We note that the poles are the same as the eigenvalues and indicate a stable process.

---

We need to be cautious when we derive a transfer function model from a state-space model, because a zero (or zeros) may cancel a pole (or poles). This becomes especially important if the cancelled pole is unstable, which means that that mode of the process would be hidden from us.

---

Several methods are available for determining the response characteristics of linear time-invariant systems. We will focus on two methods that are simpler and more direct than the time-domain methods for practical linear stability analyses. These methods also allow for a nice visual interpretation of process stability and its dependence on process parameters. They are: Routh's Criterion, and Root-Locus method.

## 7.3 ROUTH'S CRITERION

Routh's Criterion helps in determining the location of the poles, without the need to calculate them explicitly by studying the coefficients of the pole polynomial $p(s)$.

The procedure is as follows:

***Step 1*** Express the polynomial $p(s)$ in the form

$$a_0s^n + a_1s^{n-1} + \cdots + a_n = 0 \tag{7.14}$$

where $a_n \neq 0$ (i.e., we eliminate the possibility of a root at zero).

***Step 2*** There are two possible cases:

1. *If a coefficient is zero or negative* in the presence of at least one positive coefficient, then there is at least one root with a positive real part. In this case, the system is unstable. If we are only interested in absolute stability, there is no need to carry the procedure any further and we stop.
2. *If all coefficients are positive*, then we cannot conclude anything about the stability of the process since the positivity condition is *necessary* for stability but *not sufficient*. In this case, we need to proceed with Step 3.

***Step 3*** If all coefficients are positive, then we group the coefficients in the *Routh Array* as shown in Table 7.3. The array has as many rows as the degree of the polynomial $p(s)$. The unknown coefficients are calculated in the following manner:

$$b_1 = \frac{a_1 a_2 - a_0 a_3}{a_1} \qquad c_1 = \frac{b_1 a_3 - a_1 b_2}{b_1} \qquad d_1 = \frac{c_1 b_2 - b_1 c_2}{c_1}$$

$$b_2 = \frac{a_1 a_4 - a_0 a_5}{a_1} \qquad c_2 = \frac{b_1 a_5 - a_1 b_3}{b_1} \qquad d_2 = \frac{c_1 b_3 - b_1 c_3}{c_1}$$

$$b_3 = \frac{a_1 a_6 - a_0 a_7}{a_1} \qquad c_3 = \frac{b_1 a_7 - a_1 b_4}{b_1} \qquad \vdots$$

$$\vdots \qquad\qquad \vdots$$

This procedure is continued until the $n$th row is completed. The final Routh Array will have an upper triangular structure.

The Routh Criterion establishes that the number of roots with positive real parts is equal to the number of sign changes in the coefficients of the second column of the array. Thus, *necessary* and *sufficient* conditions for stability are: (1) all original coefficients are positive, and (2) all terms in the second column of the Routh Array have a positive sign.

**TABLE 7.3**
**The Routh Array**

| | | | | | |
|---|---|---|---|---|---|
| $s^n$ | $a_0$ | $a_2$ | $a_4$ | $a_6$ | . |
| $s^{n-1}$ | $a_1$ | $a_3$ | $a_5$ | $a_7$ | . |
| $s^{n-2}$ | $b_1$ | $b_2$ | $b_3$ | $b_4$ | . |
| $s^{n-3}$ | $c_1$ | $c_2$ | $c_3$ | | |
| $s^{n-4}$ | $d_1$ | $d_2$ | | | |
| $\vdots$ | $\vdots$ | $\vdots$ | | | |
| $s^1$ | $e_1$ | | | | |
| $s^0$ | $f_1$ | | | | |

### Example 7.3

Let the characteristic equation be given by

$$s^4 + 2s^3 + 3s^2 + 4s + 5 = 0 \tag{7.15}$$

Following the procedure indicated above, the Routh Array in Table 7.4 is generated. In this example, the number of sign changes in the first column is equal to two. This means that there are two roots of the polynomial with positive real parts. Hence, the process is unstable.

### Example 7.4

Consider the following characteristic equation, which is a function of a constant parameter $k$:

$$s^4 + 3s^3 + 3s^2 + 2s + k = 0 \tag{7.16}$$

From the Routh Array given in Table 7.5, the following observations can be made:

1. For stability, $k$ must be positive. This is the first requirement in Routh's Criterion.
2. To maintain stability, $k$ should always be less than 14/9.
3. When $k = 14/9$, the process exhibits oscillations with constant amplitude, as we will have a pair of complex poles on the imaginary axis.

**TABLE 7.4**
**Routh Array for Example 7.3**

| | | | |
|---|---|---|---|
| $s^4$ | 1 | 3 | 5 |
| $s^3$ | 2 | 4 | 0 |
| $s^2$ | 1 | 5 | |
| $s^1$ | −6 | | |
| $s^0$ | 5 | | |

**TABLE 7.5**
**The Routh Array for Example 7.5**

| | | | |
|---|---|---|---|
| $s^4$ | 1 | 3 | $k$ |
| $s^3$ | 3 | 2 | 0 |
| $s^2$ | 7/3 | $k$ | |
| $s^1$ | 2–9/7 $k$ | | |
| $s^0$ | $k$ | | |

## 7.4 ROOT-LOCUS METHOD

Evans[2] developed a simple method to find the roots of the characteristic polynomial. This method is called the Root-Locus method and consists of traversing, in the complex plane, the roots of the characteristic equation for all values of a system parameter as it varies from zero to infinity.

Let the characteristic equation be represented by

$$p(s, k) = r(s) + kl(s) = 0 \tag{7.17}$$

where $k$ is a constant system parameter. The Root-Locus concept is founded on tracing the locus of the roots of this equation as $k$ is varied from zero to infinity. One can then observe the evolution of the system stability as the roots move around the $s$-plane.

### Example 7.5

We shall study the characteristic equation given by

$$p(s, k) = s^2 + s + k = 0 \tag{7.18}$$

Note that $r(s) = s^2 + s$, $l(s) = 1$. To demonstrate the concept behind the Root-Locus diagram, let us find the roots analytically in terms of $k$ and then vary $k$ from zero to infinity. In this case, $p(s, k)$ is a quadratic equation, hence the roots are

$$p_1 = -\frac{1}{2} + \frac{1}{2}\sqrt{1 - 4k}, \quad p_2 = -\frac{1}{2} - \frac{1}{2}\sqrt{1 - 4k} \tag{7.19}$$

We observe that the roots are real for $k \leq 1/4$ and complex for $k > 1/4$.

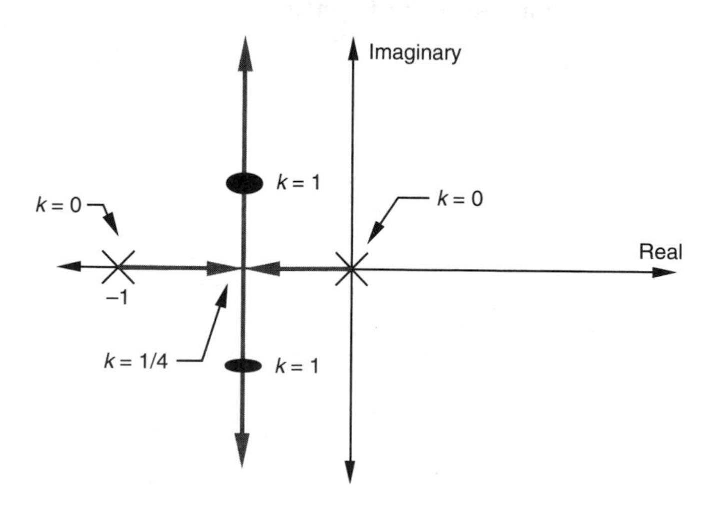


**FIGURE 7.6** The Root-Locus diagram for Eq. (7.18).

The Root-Locus diagram for this process is displayed in Figure 7.6 for all values of $k$ (the arrows indicate the direction of movement of the roots as $k$ increases). Once the diagram is prepared, one can immediately observe the value of $k$ that produces a root (or a pole) at a desired location.

From Figure 7.6, we can observe the behavior of the system for different values of $k$:

1. When $k = 0$, then $s_1 = 0$ and $s_2 = -1/2$.
2. Increasing $k$ from zero to 1/4, the poles move toward the point $(-1/2, 0)$. They move along the real axis, hence they are all real.
3. At $k = 1/4$, the two real poles are equal.
4. For $k > 1/4$, the poles become complex, and since the real part becomes independent of $k$, they move along the vertical line $s = -1/2$.

It is interesting to note that this process never becomes unstable for any value of $k$.

From the Root-Locus diagram one can observe the effect of varying $k$ on the dynamic response of the system. Since the method is a graphical procedure to trace the roots of the characteristic equation, it provides a visual medium to analyze the behavior of the roots of any polynomial equation arising in the study of a physical system.

The Root-Locus method is especially important in the analysis and design of control systems where the parameter $k$ is associated with the controller, as we shall see later.

## 7.5 SUMMARY

We have seen in this chapter that the stability of a dynamic process is captured by the eigenvalues of its state matrix. We then pointed out that the eigenvalues of the state matrix are equivalent to the poles of the corresponding transfer function model. We stated that if the real parts of all the eigenvalues of the state matrix lie in the left half of the complex plane, then the system was stable. We introduced two methods for stability analysis. In Routh's Criterion, the goal was to determine the stability condition (stable or unstable) of a system without actually computing the poles of the transfer function model. In Root-Locus analysis, we traced the roots of the characteristic function in the complex plane as a function of a system parameter, thus providing insight toward the extent of stability.

## REFERENCES


1. Brogan, W.L., *Modern Control Theory*, 3rd Ed., Prentice-Hall, Englewood Cliffs, NJ, 1990.
2. Evans, W.R., Control system synthesis by Root-Locus method, *AIEE Trans. II*, 69, 66–69, 1950.

# 8 Dynamic Performance

In this chapter, we investigate in detail the dynamics of processes with specific structures in an attempt to identify key performance characteristics. We start with the simplest case of a first-order process and continue with the dynamic behavior of second-order processes. Finally, we discuss the dynamic response of higher order systems as well as the impact of time delays and zeros on process performance.

We have seen in the previous chapter that stability is a prerequisite for acceptable dynamic performance. Now, we will focus on the qualitative response characteristics and generalize specific process features. We need to articulate what constitutes satisfactory dynamic performance and understand how this performance can be influenced by certain elements of the process model.

## 8.1 FIRST-ORDER PROCESSES

In Example 5.1, we have shown that the model for a process tank,

$$\frac{d\bar{h}(t)}{dt} = a\bar{F}_{in}(t) - b\bar{h}(t) \tag{8.1}$$

resulted in the transfer function

$$\frac{\bar{h}(s)}{\bar{F}_{in}(s)} = g(s) = \frac{k}{\tau s + 1} \tag{8.2}$$

with the parameters $k = a/b$ and $\tau = 1/b$. Two related observations indicate that this is a *first-order* process:

1. Equation (8.1) is a *first-order* differential equation in $\bar{h}(t)$.
2. The transfer function in Eq. (8.2) has *one pole* located at $s = -1/\tau$.

Indeed, Eq. (8.2) is in the standard form with two characteristic parameters $k$ and $\tau$ shown explicitly. These parameters have important interpretations for explaining process performance.

---

*Process gain k*: This parameter can be interpreted as the ultimate value of the response (new steady-state) for a unit step change in the input. In other words, $k$ indicates how much the process amplifies or attenuates the input signal after reaching steady-state.

*Time constant* $\tau$: The time constant of a process is a measure of time needed for the process to adjust to a change in the input. The magnitude of $\tau$ indicates whether the process reacts in a rapid or a slow manner when the input is changed.

---

Consider how a first-order process responds to a step change of magnitude $\hat{u}$ ($\overline{F}_{in}(s) = \hat{u}/s$).

$$\overline{h}(s) = \frac{k}{\tau s + 1}\frac{\hat{u}}{s} \tag{8.3}$$

Using the partial fraction expansion method (see Appendix B), we can calculate the time-domain solution that describes the dynamic performance of the tank level in response to a step change in the inlet flow rate.

$$\overline{h}(t) = \hat{u}k(1 - e^{-t/\tau}) \tag{8.4}$$

Equation (8.4) reveals many features of a first-order process:

1. The ultimate (steady-state) value of the response is equal to $\hat{u} \cdot k$ for a step change of magnitude $\hat{u}$; or $k$, when $\hat{u} = 1$ (unit step change).
2. We observe that when the time is equal to the process time constant ($t = \tau$), the system reaches 63.2% of its final response, $\overline{h}(t = \tau) = \hat{u} \cdot k(0.632)$. After approximately $3\tau$, the transient response is essentially at steady-state.

---

When the inlet flow rate increases, the liquid level goes up. However, as the level increases the hydrostatic pressure also increases, in turn, the outlet flow rate, and the system eventually reaches a new steady-state. This feature of stable first-order processes is referred to as *self-regulation*.

---

### Example 8.1

We recall that the process tank model in Eq. (8.1) has the following parameters derived in Example 5.1:

$$a = \frac{1}{A}, \quad b = \frac{1}{A}\left[\frac{d(\beta\sqrt{h})}{dh}\right]_{h=h_s} = \frac{\beta}{2A\sqrt{h_s}}$$

We can easily show that the first-order process parameters are

$$k = \frac{a}{b} = \frac{2\sqrt{h_s}}{\beta} = \frac{2h_s}{\beta\sqrt{h_s}} \tag{8.5}$$

$$\tau = \frac{1}{b} = \frac{2A\sqrt{h_s}}{\beta} = \frac{2Ah_s}{\beta\sqrt{h_s}} = \frac{2V_s}{\beta\sqrt{h_s}} \tag{8.6}$$

The following general observations can be made regarding the impact of process characteristics on dynamic performance:

1. Equation (8.5) indicates that the process gain is proportional to the resistance to flow through the outlet valve ($1/\beta\sqrt{h_s}$). The higher the resistance, the more sensitive the tank level to changes in the inlet flow rate.
2. Equation (8.6) shows that the time constant of the process is directly proportional to the steady-state storage capacity of the tank. The larger the capacity, the slower the change in the level when the inlet flow rate changes.
3. The time constant is also proportional to the resistance to flow through the outlet valve. The tank level responds faster if the resistance is smaller; in other words, the new steady-state is reached more quickly if the outlet valve resistance is diminished.

These characteristics are further explored through the simulations depicted in Figure 8.1. The resistance is changed by varying the value of $\beta$ and the storage capacity is altered by changing the cross-sectional area $A$ of the tank. Simulations clearly support the observations made above. When the resistance is higher (value of $\beta$ lower) (Figure 8.1a), the input variable is more amplified due to the higher gain, and the response is slower as the time constant also increases. On the other hand, when we have a larger tank capacity (higher cross-sectional area) (Figure 8.1b), the gain is unchanged, yet the response becomes more sluggish. The inset in Figure 8.1b reveals the characteristic exponential rise of a first-order process during its initial transient response.

---

Since most linear models arise from the approximation of nonlinear processes, the model parameters in most cases depend on the operating conditions of the process. This will produce variations in the transfer function parameters as the process changes operating conditions; hence they need to be adapted every time.

---

### *Special Case: Integrating Processes*

What if the liquid from the tank is not drawn through a valve but maintained constant by a pump (Figure 8.2)? The model based on the overall mass balance, again, is given by

$$A\frac{dh}{dt} = F_{in} - F_{out} \tag{8.7}$$

At steady-state, the inlet flow rate has to be equal to the outlet flow rate,

$$0 = F_{in,s} - F_{out,s} \tag{8.8}$$

The transfer function between the inlet flow rate and the tank level can be easily derived, assuming deviation variables $\bar{h} = h - h_s$ and $\bar{F}_{in} = F_{in} - F_{in,s}$ as

$$\frac{\bar{h}(s)}{\bar{F}_{in}(s)} = g(s) = \frac{k}{s} \tag{8.9}$$

with $k = 1/A$. Equation (8.9) depicts a *pure integrating* (or a pure capacitative) process.

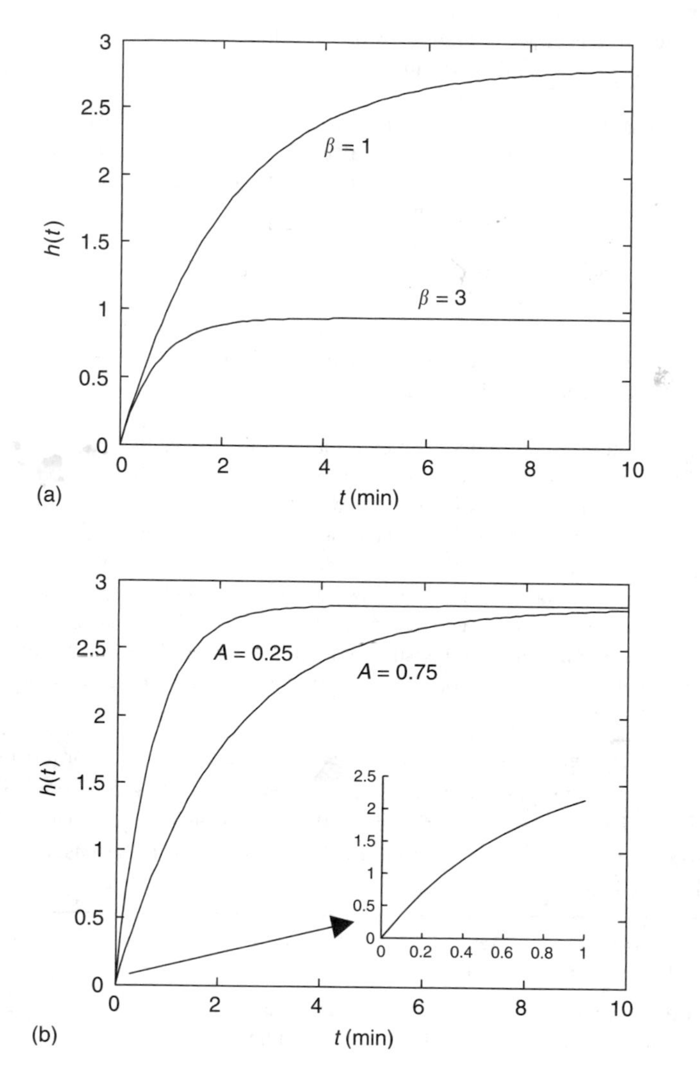


**FIGURE 8.1** Response of the tank level ($h_s = 2$ m) with changes in (a) flow resistance, with $A = 0.75$ m$^2$ and (b) cross sectional area, with $\beta = 1$ m$^{5/2}$/min.

An integrating process is a special case of first-order processes where the pole is located at the origin. Owing to this characteristic, the output grows unbounded when the input is changed as a step function. For the tank in Figure 8.2, one can observe that as soon as the inlet flow rate is changed from its steady-state value where it is equal to the outlet flow rate dictated by the pump, the tank will either drain or overflow. It can be concluded that such processes are *not* self-regulating.

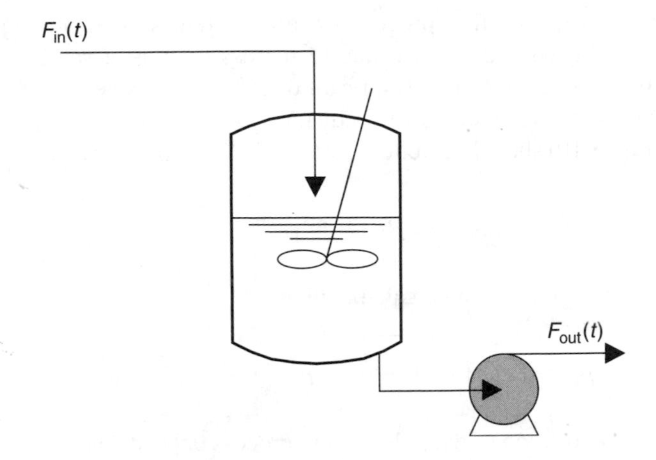


**FIGURE 8.2** A process tank with a pump.

## 8.2 SECOND-ORDER PROCESSES

Similar to Eq. (8.2) representing the standard form of a first-order process, the following transfer function representation is the standard form for a second-order process:

$$g(s) = \frac{k}{\tau^2 s^2 + 2\xi\tau s + 1} \tag{8.10}$$

Analogously, such transfer functions arise from input–output models expressed through second-order differential equations. Second-order processes have three characteristic parameters, $k$, $\tau$, and $\xi$.

---

*Process gain k*: This is the ultimate value of the response (new steady-state) for a unit step change in the input. The gain has the same physical interpretation as in first-order processes.

*Time constant or natural period* $\tau$: The time constant of a second-order process is the measure of time necessary for it to adjust to a change in the input. Again, this parameter has the same physical interpretation as in first-order processes. However, this parameter is also referred to as the natural period of oscillation for second-order processes that exhibit oscillatory behavior.

*Damping factor* $\xi$: The value of this parameter, as will be shown later, will determine the presence and the level of oscillatory behavior in the system. This parameter is unique to second-order processes.

---

Very rarely will one find processes with inherent second- or higher order dynamics. The higher orders arise due to the association of several first-order capacities (see Section 8.2.1) or as a result of the control system.

An analysis of the characteristic equation and the associated poles for the process in Eq. (8.10) should be quite informative. With the characteristic equation expressed as

$$p(s) = \tau^2 s^2 + 2\xi\tau s + 1 = 0 \tag{8.11}$$

The poles of this process can be calculated as

$$p_1 = -\frac{\xi}{\tau} + \frac{\sqrt{\xi^2 - 1}}{\tau}, \quad p_2 = -\frac{\xi}{\tau} - \frac{\sqrt{\xi^2 - 1}}{\tau} \tag{8.12}$$

We note that the location of the poles of the process depends on the parameters $\xi$ and $\tau$. For actual processes that are inherently second-order, the time constant and the damping factor are typically nonzero, thus indicating that such processes are always stable.

The dependency on $\xi$ is important because according to the value of this parameter, the characteristic equation will have either *real* or *complex* roots. We observe three possible cases:

1. When $\xi > 1$; two distinct and real poles (*overdamped* process)
2. When $\xi = 1$; two equal poles (*critically damped* process)
3. When $\xi < 1$; two complex conjugate poles (*underdamped* process)

Let us examine how a second-order process responds to a unit step change in the input, i.e.,

$$y(s) = \frac{k}{\tau^2 s^2 + 2\xi\tau s + 1}\frac{1}{s} \tag{8.13}$$

---

*Overshoot* represents the degree with which the dynamic response exceeds its ultimate (steady-state) value.

---

Figure 8.3 depicts the response of a second-order process for a variety of model parameters. The following observations are made:

1. Responses exhibiting oscillation and overshoot are obtained only for $\xi < 1$, as expected from the analysis of the location of the poles of the process as a function of the damping factor.
2. For values of $\xi > 1$, we obtain a sluggish (slow) response, but without oscillations and overshoot. This again can be predicted from the previous analysis of the pole locations.

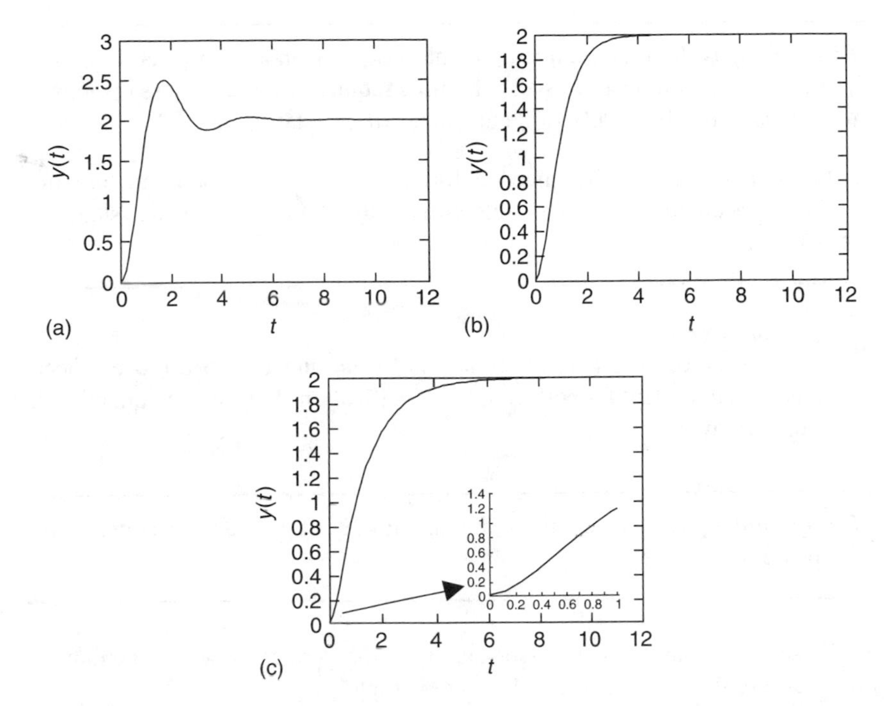


**FIGURE 8.3** Step response of a second-order process ($k = 2$, $\tau = 0.5$): (a) $\xi = 0.4$, (b) $\xi = 1.0$, (c) $\xi = 1.4$.

3. The fastest response without exceeding the ultimate value of the response (no overshoot) is obtained for the critically damped case, i.e., $\xi = 1$.
4. The inset in Figure 8.3c contrasts the initial transient response of a second-order process with that of a first-order process (Figure 8.1b). We note the initial slow response followed by an exponential rise, common to all second-order processes.

## Special Case: Step Response of Underdamped Processes

An underdamped process frequently arises in the design of control systems where the performance specifications are often phrased through its characteristic features. This will be discussed further in Chapter 12. Here, we will introduce these characteristic features to establish an intuitive understanding of the performance of such processes.

The underdamped processes are recognized by the oscillatory behavior they exhibit, and it is noted that these oscillations decay in time. *Overshoot* already quantifies the magnitude of the initial peak of this response. Yet, it is important for the engineer to also capture the initial speed of this response as well as to determine a time limit when the process can be viewed as to have *practically* reached its steady-state value.

*Rise time* $t_R$ is the time required for the process response to reach its final value for the first time, or it is also the time required for the process response to increase from 10 to 90% of its ultimate (steady-state) value.

*Settling time* $t_S$ is the time needed for the response to reach and remain within a specified percentage (frequently 2 to 5%) of its ultimate (steady-state) value.

As we observed before, the oscillations for an underdamped process decay over time. It is important for performance specifications to be able to quantify the rate of this decay.

*Decay ratio* represents the ratio of the magnitudes of two successive peaks in the process response.

These definitions, typically associated with the step response of underdamped processes, are illustrated in the following example.

### Example 8.2

Consider a second-order process with the transfer function

$$\frac{y(s)}{u(s)} = g(s) = \frac{2}{s^2 + s + 1} \tag{8.14}$$

We can extract the following key parameters from this transfer function:

$$k = 2$$

$$\tau = 1 \text{ min}$$

$$\xi = 0.5$$

The unit step response of this underdamped process is displayed in Figure 8.4. The following characteristics of this response are identified:

1. The ultimate (steady-state) value of the output is 2.0, which is equal to the process gain.
2. The percent overshoot can be calculated as follows:

$$\%\text{overshoot} = \frac{3.05 - 2.0}{2.0} 100 = 52.5\%$$

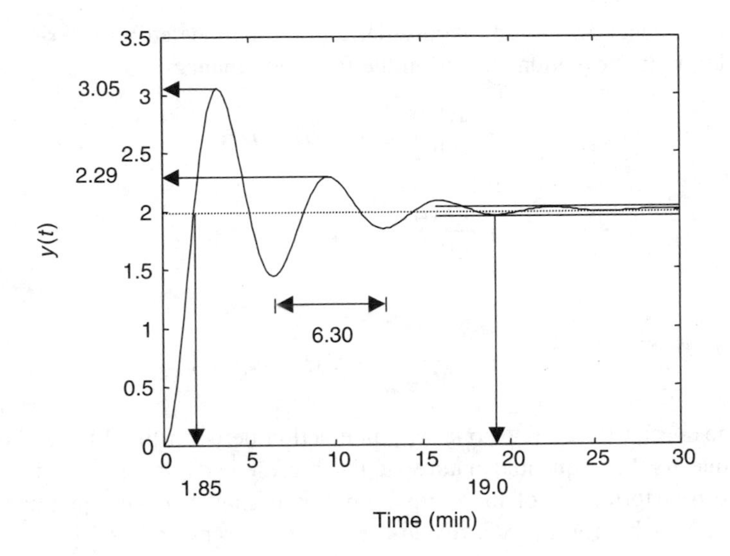


**FIGURE 8.4** The unit-step response of the process in Eq. (8.14).

- The decay ratio is given by

$$\text{decay ratio} = \frac{2.29 - 2.0}{3.05 - 2.0} = 0.28$$

- The rise time is observed to be 1.85 min, and the settling time (based on 2% of the ultimate value) appears to be 19 min.

---

*Period of oscillation* is the time span between two successive peaks (or two successive valleys) of the process response.

---

- From Figure 8.4, we can also determine the period of oscillation as 6.30 min.

These observations fully quantify the step response of this underdamped process.

## 8.3 MULTI-CAPACITY PROCESSES

Higher order transfer functions arise owing to the presence of multiple capacity processes that affect the dynamic behavior of an output variable. A good example of this is the process of two tanks in series discussed in Examples 3.1 and 5.3, where it was shown that the resulting process is second-order due to the fact that we have two first-order processes in series. In that example, the mass storage capacity of each tank gave rise to first-order process characteristics.

For the general case, let us consider a process modeled by a series of first-order differential equations linked in the following manner:

$$\tau_1 \frac{dy_1(t)}{dt} = -y_1(t) + u(t)$$
$$\tau_2 \frac{dy_2(t)}{dt} = -y_2(t) + y_1(t) \qquad (8.15)$$
$$\vdots$$
$$\tau_n \frac{dy_n(t)}{dt} = -y_n(t) + y_{n-1}(t)$$

Note the special structure of one-way interaction between the differential equations, due to the sequential nature of the represented process. By taking the Laplace transformation of these equations (assuming that the equations are in deviation variable form), we arrive at the following transfer functions:

$$\frac{y_1(s)}{u(s)} = \frac{1}{\tau_1 s + 1}$$
$$\frac{y_2(s)}{y_1(s)} = \frac{1}{\tau_2 s + 1} \qquad (8.16)$$
$$\vdots$$
$$\frac{y_n(s)}{y_{n-1}(s)} = \frac{1}{\tau_n s + 1}$$

The transfer function between the input $u(s)$ and the output $y_n(s)$ is expressed as

$$\frac{y_n(s)}{u(s)} = g(s) = \frac{1}{\tau_1 s + 1} \frac{1}{\tau_2 s + 1} \cdots \frac{1}{\tau_n s + 1}$$
$$= \frac{1}{(\tau_1 s + 1)(\tau_2 s + 1) \cdots (\tau_n s + 1)} \qquad (8.17)$$

This is an $n$th-order process, characterized by $n$ poles. One can observe the following:

- Multi-capacity processes resulting from capacities in series yield transfer functions that are concatenations of the individual capacities.
- The poles of the process are real numbers and equal to the individual poles and also equal to the inverse of the individual time constants ($p_i = -1/\tau_i$).
- If the individual time constants are equal, then the poles are equal.
- If $n = 2$, then we can conclude that this will always result in an overdamped or critically damped response.

**Example 8.3**

Consider a process stream being heated as it passes through three tanks in series (Figure 8.5). We are interested in determining the relationship between the temperature of this stream as it leaves the third tank, $T_3(t)$, and its temperature as it enters the first tank, $T_{in}(t)$. This is a case where the models represent the capacity of each tank for energy storage. We make the following assumptions:

- The tank levels and the inlet flow rate, $F_{in}(t)$, are constant
- The steam temperature is maintained as constant

The model for this process can be developed as follows (see Example 5.2):

$$\frac{d\bar{T}_1(t)}{dt} = -c_1\bar{T}_1(t) + g_1\bar{T}_{in}(t)$$

$$\frac{d\bar{T}_2(t)}{dt} = -c_2\bar{T}_2(t) + g_2\bar{T}_1(t) \tag{8.18}$$

$$\frac{d\bar{T}_3(t)}{dt} = -c_3\bar{T}_3(t) + g_3\bar{T}_2(t)$$

The transfer function model can be constructed in a straightforward manner:

$$\frac{\bar{T}_3(s)}{\bar{T}_{in}(s)} = \frac{g_1}{s+c_1}\,\frac{g_2}{s+c_2}\,\frac{g_3}{s+c_3} = \frac{k}{(\tau_1 s+1)(\tau_2 s+1)\cdots(\tau_3 s+1)} \tag{8.19}$$

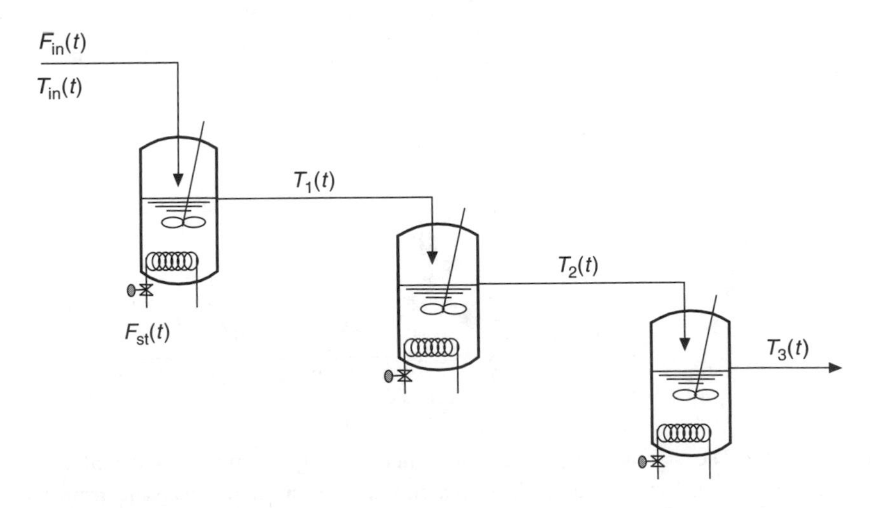


**FIGURE 8.5** Three heating tanks in series.

Here, the parameters are defined as follows:

$$k = \frac{g_1 g_2 g_3}{c_1 c_2 c_3}$$

$$\tau_1 = 1/c_1$$

$$\tau_2 = 1/c_2$$

$$\tau_3 = 1/c_3$$

Figure 8.6 shows the general overdamped characteristics of the unit-step response of this process and the initial transient behavior.

The previous example illustrates the fact that increasing the number of lags (capacities) increases the sluggishness of the response and the flatness of the initial transient (Figure 8.6). This is the reason why FOPDT models (see Chapter 6) usually provide a satisfactory representation of complex, multicapacity processes as encountered in typical industrial applications.

### *Special Case: Interacting Processes*

Although the capacities may be physically placed in series, there may be material or energy recycles between them to make the modeling more complex. Potentially, the

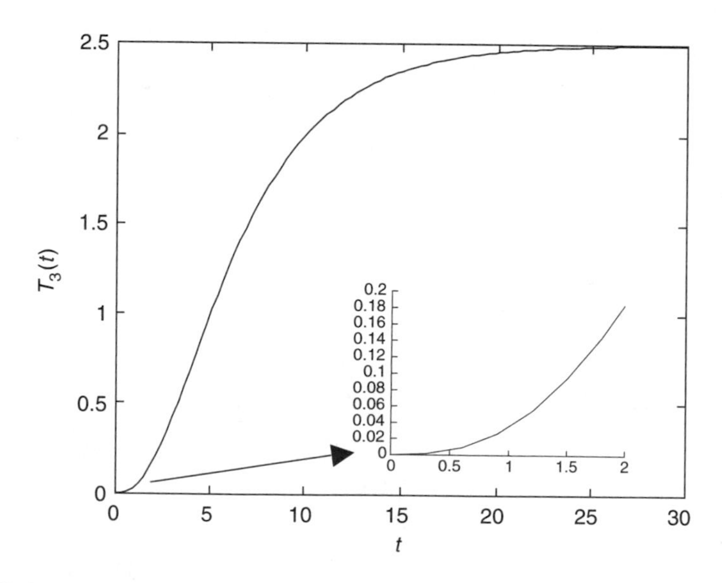


**FIGURE 8.6** The unit-step response of the process in Eq. (8.19) with $k = 2.5$, $\tau_1 = 1$, $\tau_2 = 2$, and $\tau_3 = 4$. The inset shows the typical initial transient for multicapacity processes.

equations may appear as

$$
\begin{aligned}
\tau_1 \frac{dy_1(t)}{dt} &= -y_1(t) + y_2(t) + \cdots + y_n(t) + u(t) \\
\tau_2 \frac{dy_2(t)}{dt} &= y_1(t) - y_2(t) + \cdots + y_n(t) \\
&\vdots \\
\tau_n \frac{dy_n(t)}{dt} &= y_1(t) + y_2(t) + \cdots - y_n(t)
\end{aligned}
\tag{8.20}
$$

An immediate observation is that while Eq. (8.15) can be solved sequentially for a given input, Eq. (8.20) represents a *coupled* set of differential equations that can only be solved simultaneously. This naturally has implications for numerical simulations and analysis of such systems. One can also appreciate the algebraic difficulties in obtaining transfer function models for such systems starting from differential equations.

More importantly, processes modeled by equations, such as in Eq. (8.20), have certain dynamic performance features that cannot be generalized from simple multicapacity processes. One such feature is that the time constants of interacting processes may no longer be directly associated with the time constants of individual capacities. The following example studies a typical interacting process.

### Example 8.4

Consider a process consisting of two tanks as shown in Figure 8.7. The feed to the second tank is introduced at the bottom, thus its flow rate is influenced by the difference in the static heights in both tanks. Subscripts 1 and 2 will refer to the first and second tanks, respectively.

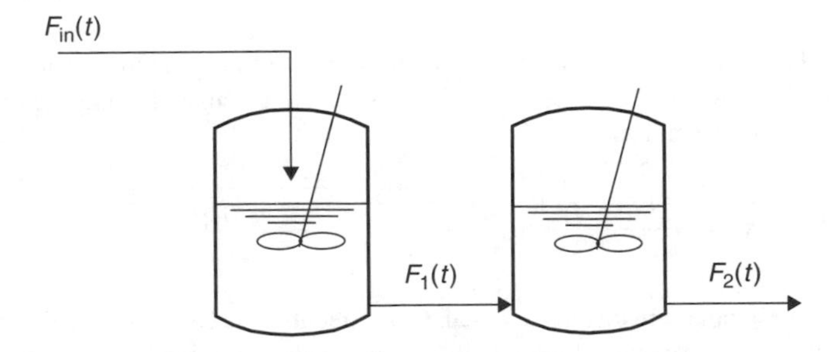


**FIGURE 8.7** The process of two interacting tanks.

For simplicity, let us assume that the resistance of outlet piping is linearly dependent on the static height (e.g., $F_2 = \beta_2 h_2$). A total mass balance in each tank leads to

$$A_1 \frac{dh_1(t)}{dt} = F_{in}(t) - \beta_1(h_1 - h_2) \tag{8.21}$$

$$A_2 \frac{dh_2(t)}{dt} = \beta_1(h_1 - h_2) - \beta_2 h_2 \tag{8.22}$$

Equations (8.21) and (8.22) constitute a set of *coupled differential equations* and they need to be solved simultaneously. This indicates the mutual effect of the two tanks.

Defining deviation variables and using the Laplace transformation, we obtain a set of coupled algebraic equations. Solving these equations for $\bar{h}_1$ and $\bar{h}_2$, one has the following transfer functions:

$$\begin{aligned} \bar{h}_1(s) &= \frac{(\tau_2/\beta_1)s + (1/\beta_1 + 1/\beta_2)}{\tau_1\tau_2 s^2 + (\tau_1 + \tau_2 + A_1/\beta_2)s + 1}\bar{F}_{in}(s) \\ \bar{h}_2(s) &= \frac{1}{\tau_1\tau_2 s^2 + (\tau_1 + \tau_2 + A_1/\beta_2)s + 1}\bar{F}_{in}(s) \end{aligned} \tag{8.23}$$

where, we have $\tau_1 = A_1/\beta_1$ and $\tau_2 = A_2/\beta_2$ as the time constants of the individual tanks. We note the following:

- Equations (8.23) show that the dynamic response of the two interacting tanks follows a second-order behavior.
- We can compare Eq. (8.23) with the transfer function model (left as an exercise to the reader) one would obtain for two noninteracting tanks (see Example 3.1).

$$\bar{h}_2(s) = \frac{1/\beta_2}{\tau_1\tau_2 s^2 + (\tau_1 + \tau_2)s + 1}\bar{F}_{in}(s) \tag{8.24}$$

These transfer functions differ primarily in the coefficient of $s$ in the denominator by the factor $A_1/\beta_2$. This term indicates the degree of interaction between the tanks.

- The larger the value of $A_1/\beta_2$, the more influential is the interaction.
- The poles of the process (Eq. [8.24]) can be calculated using the quadratic formula as:

$$p_{1,2} = \frac{-(\tau_1 + \tau_2 + A_1/\beta_1) \pm \sqrt{(\tau_1 + \tau_2 + A_1/\beta_2)^2 - 4\tau_1\tau_2}}{2\tau_1\tau_2} \tag{8.25}$$

- These poles are distinct and real. Consequently, we can conclude that the dynamic response of the two interacting tanks is always overdamped.
- It is also noted that the time constants of the overall system (inverse of the poles) are not equal to the individual time constants.

## 8.4 EFFECT OF ZEROS

We have discussed in Chapter 7 that the poles unequivocally determine the stability characteristics of the process response. Here, we will explore the effects the zeros may have in shaping this response. To illustrate our point, let us consider the following transfer function:

$$\frac{y(s)}{u(s)} = g(s) = \frac{s + \gamma}{(s + \sigma_1)(s + \sigma_2)} \tag{8.26}$$

with $\gamma \neq \sigma_1 \neq \sigma_2 \neq 0$. We already know that the locations of the poles, $p_1 = -\sigma_1$ and $p_2 = -\sigma_2$, determine the stability features of the output response. A simple way to uncover the effect of the zero, $z = -\gamma$, would be to develop the partial fraction expansion of this system for an arbitrary input. Indeed, the partial fraction expansion (see Appendix B) reveals the simplest dynamic phenomena that, when joined together, identically describe the overall process performance.

$$\begin{aligned} y(s) &= g(s)u(s) \\ &= \frac{s + \gamma}{(s + \sigma_1)(s + \sigma_2)} u(s) = \frac{A}{s + \sigma_1} u(s) + \frac{B}{s + \sigma_2} u(s) \\ &= A g_1(s) u(s) + B g_2(s) u(s) \\ &= y_1(s) + y_2(s) \end{aligned} \tag{8.27}$$

Equation (8.27) states that this second-order process can be expressed by a combination of two first-order processes ($g_1$, $g_2$) whose individual contributions to the overall process performance are dictated by the constant coefficients $A$ and $B$. Using partial fraction expansion rules, we can calculate the following coefficients

$$A = \frac{-\sigma_1 + \gamma}{-\sigma_1 + \sigma_2}$$
$$B = -\frac{-\sigma_2 + \gamma}{-\sigma_1 + \sigma_2} \tag{8.28}$$

We observe that the location of the zero (along with the pole locations) influences the relative contribution of each term in Eq. (8.27).

**Example 8.5**

To illustrate the effect of a zero, let us consider the following transfer function:

$$\frac{y(s)}{u(s)} = g(s) = \frac{s + \gamma}{(s + 1.0)(s + 1.5)} \tag{8.29}$$

We will study two zero locations: (1) $z = -\gamma = -1.25$; (2) $z = -\gamma = -0.1$.

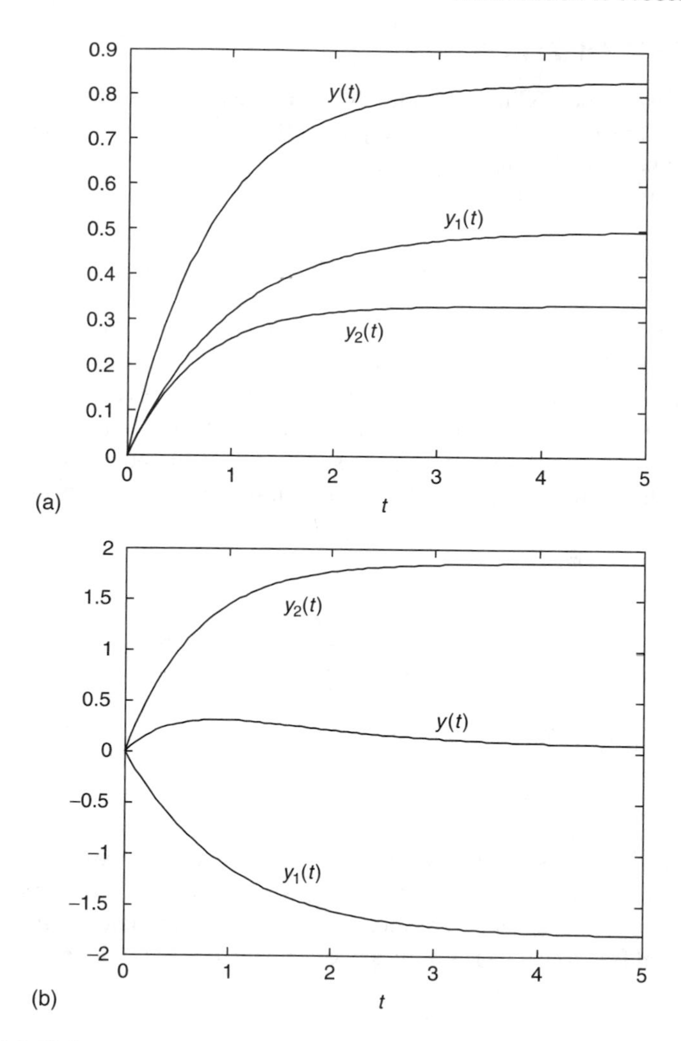


**FIGURE 8.8** Unit-step response of a process in Eq. (8.29), (a) $\gamma = 1.25$, (b) $\gamma = 0.1$.

The constants $A$ and $B$ can be calculated (Eq. [8.28]) and the unit-step response of the process (Eq. [8.29]) is analyzed as follows:

- $A = 0.5$ and $B = 0.5$: In this case, the individual first-order transfer functions have an equal dynamic contribution to the overall response (Figure 8.8a).
- $A = -1.8$ and $B = 2.8$: Here, two first-order processes have *opposing* contributions as delineated by the difference in sign. The overall response given in Figure 8.8b shows that the initial transient is dominated by the faster process ($g_2$), but the other one ($g_1$) dominates later and brings down the overall process response.

These zeros are located on the left half-plane (LHP). One can generalize that when processes have zeros on the LHP, the initial transient response will be in the direction of the ultimate (steady-state) value of the process.

*Special Case: Right Half-Plane Zeros*

When the zeros are located on the right half-plane (RHP), the process response exhibits a special dynamic behavior. Let us modify Eq. (8.29) as follows:

$$\frac{y(s)}{u(s)} = g(s) = \frac{s - 1.25}{(s + 1.0)(s + 3.0)} \tag{8.30}$$

with the zero $z = 1.25$. This yields the output response below, after partial fraction expansion as

$$y(s) = \left[-\frac{1.125}{s + 1} + \frac{2.125}{s + 3}\right]u(s) = [-1.125g_1(s) + 2.125g_2(s)]u(s) \tag{8.31}$$

The unit-step response of this process is depicted in Figure 8.9. We can arrive at the following generalizations:

1. The process with a RHP zero always consists of individual dynamic phenomena that oppose each other.
2. The initial transient response always moves in the opposite direction of the ultimate (steady-state) value of the process response.

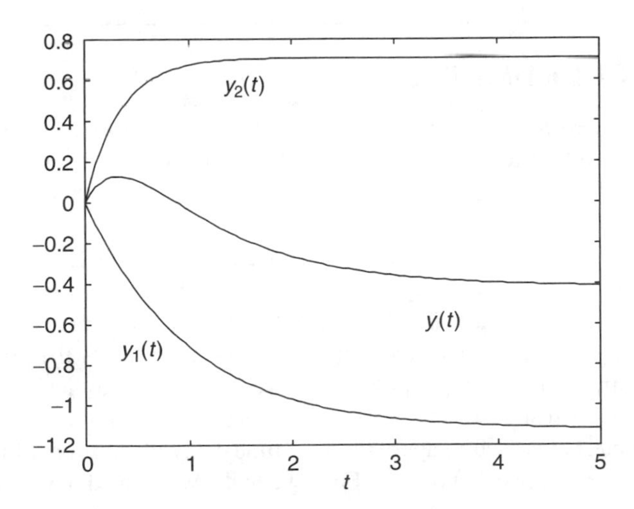


**FIGURE 8.9** Unit-step response of a process with inverse response behavior.

This special dynamic behavior is referred to as *inverse response* (or wrong-way behavior), and can be exhibited by various physical processes:

- *Distillation columns when the steam pressure to the reboiler is suddenly changed.* An increase in the steam pressure will eventually decrease the reboiler level by boiling more liquid. However, initially, the amount of frothing on the trays above the reboiler will increase, thus causing an initial increase in the reboiler liquid level, which will be later eliminated by the increase in vapor boil-up.
- *Drum boilers when the flow rate of the cold feed water is increased.* We would expect that the liquid level in the drum will increase. However, initially the level decreases for a short period, since the cold water causes the temperature to drop, which eventually decreases the volume of the entrained vapor bubbles, leading to an initial decrease in the liquid level of the boiling water. With a constant heat supply, the steam production remains constant and the liquid level, after the initial drop, will start to increase.
- *Tubular catalytic reactors with exothermic chemical reaction when feed temperature is increased.* Initially, the exit temperature decreases due to an increase in conversion at the entrance of the reactor, which depletes reactants toward the end of the reactor, allowing products to cool. After the accompanying decrease in temperature toward the exit, higher reaction rates along the reactor eventually lead to higher exit temperatures.

---

Inverse response is the result of two physical effects that act on the process output variable in opposite ways and with different time scales.

---

## 8.5 EFFECT OF TIME DELAYS

In Chapter 5, we introduced the concept of time delay and its associated transfer function. The only effect of time delay on the process response was to translate (shift) the response along the time axis for a period of time that corresponds to the delay.

When time delays are part of the model, we can no longer represent the transfer function as a ratio of two polynomials in $s$, since the exponential term is a nonrational element. Thus, there is an incentive to find rational approximations to the exponential term in the delay, since this will allow us to factor the process transfer function in terms of simple poles and zeros and use analytical techniques to study the process response.

The simplest approach to approximate a time delay by a rational function is to use a Taylor series expansion of $e^{-t_D s}$. First, we will rewrite the delay term as a ratio,

$$e^{-t_D s} = \frac{e^{-(t_D/2)s}}{e^{(t_D/2)s}} \tag{8.32}$$

The Taylor expansion of an exponential term around $s = 0$ is given as

$$e^{-as} \cong 1 - as + \frac{a^2s^2}{2!} - \frac{a^3s^3}{3!} + \cdots \tag{8.33}$$

Truncating this series after the first-order term, Eq. (8.32) becomes

$$e^{-t_D s} \approx \frac{1 - \frac{t_D}{2}s}{1 + \frac{t_D}{2}s} \tag{8.34}$$

This expression should serve as a reasonable approximation to the delay term, especially for small delays.

The approximation given in Eq. (8.34) is called the *first-order Padé approximation*, and it consists of first-order polynomials in the numerator and the denominator. Note that we have a stable pole at $p = -1/(t_D/2)$ and an RHP zero at $z = 1/(t_D/2)$. Basically, the wrong-way behavior precipitated by the RHP zero helps to create the appearance of a time delay by delaying the response in the "right" direction by a certain period of time.

## 8.6 SUMMARY

In this chapter, we focused on analyzing the qualitative response characteristics and generalized the specific features of typical processes. Starting from the simpler representation of first-order models and continuing with the dynamic response of higher order models, the key parameters that explain the process behavior were identified. The effect of interaction among capacities was also investigated and its consequence on the overall system performance was quantified. Finally, the impact of the time delays and system zeros on the overall process system behavior was discussed. It was demonstrated that the presence of RHP zeros generates a special dynamic behavior referred to as inverse response (or wrong-way) behavior.

# 9 Frequency Response

The aim of process analysis is to uncover dynamic features that would help an engineer to build a sound understanding of the process operation. It is clear that computer simulations accomplish this task by creating a visual image of the process output in response to various stimuli. Yet, we cannot expect to generate simulations for all process behaviors, and simulations do not lend themselves to rigorous generalizations. In the previous chapters, we have studied transfer functions by focusing our attention on their poles and zeros and the implications of the pole-zero locations for dynamic performance. Naturally, we were able to develop generalizable observations for first- and second-order processes, as well as processes that consist of multiple capacities.

This chapter introduces an alternative way to analyze process dynamics through *frequency response*. As it will be shown later, this procedure yields a powerful tool not just for analyzing dynamic processes but also for designing and tuning controllers.

## 9.1 WHAT IS FREQUENCY RESPONSE?

The control engineer needs all the tools available for probing a process to uncover its key response characteristics. A special approach to extracting this key information from the process is to perturb the process with a well-characterized input signal and look for *patterns* in the output signal to match known features. The input selected for this analysis has a sinusoidal shape with a known frequency and amplitude.

When a stable, linear process is subjected to a sinusoidal input with constant frequency and amplitude, the *ultimate* output response (i.e., after a long period of time) is also sinusoidal (Figure 9.1). Furthermore, one can observe that the output signal has the same frequency as the input signal. There are two key features in the output response that reveals the underlying process characteristics:

1. The amplitude of the output signal may be different from the amplitude of the input signal and
2. The output signal appears to lag behind the input signal, as the input and output sinusoids are out of phase.

These differences capture unique process features. The *amplification* or the *attenuation* of the input signal, as it passes through the process, reveals information regarding the process gain. Moreover, the *shift* in the phase should be related to the time constants (capacities) associated with the process.

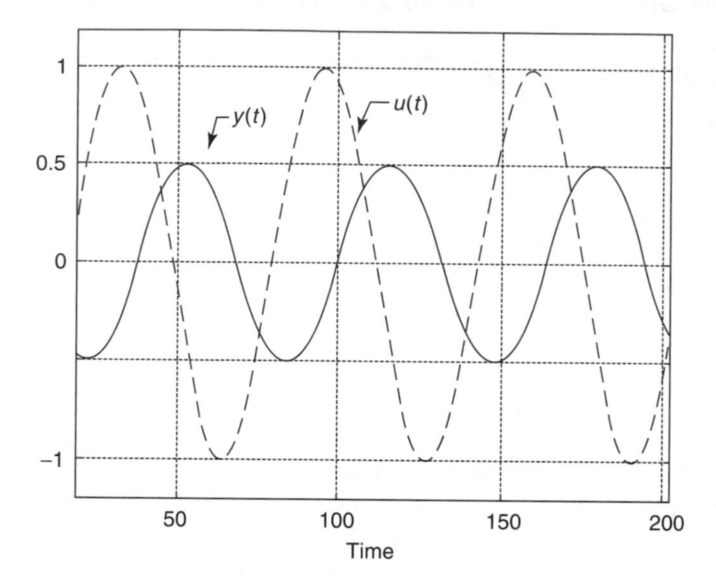


**FIGURE 9.1** A typical long-term sinusoidal response to a sinusoidal input.

If this analysis is performed for a large set of frequencies, we have the *process frequency response* captured by two key factors:

$$\text{amplitude ratio} = AR(\omega) = \frac{\text{output signal magnitude}}{\text{input signal magnitude}} \tag{9.1}$$

$$\text{phase angle} = \phi(\omega) = \text{phase difference between the input and the output} \tag{9.2}$$

In this chapter, we will introduce a general procedure for *stable*, linear processes with constant parameters, which will allow characterization of the frequency response from the knowledge of the transfer function.

The algebra involves the use of complex numbers, as the transfer functions are expressed in the complex (Laplace) domain. The polar coordinates facilitate the visualization of complex numbers and provide an intuitive setting for their operations. A brief discussion is provided in the next section.

## 9.2 COMPLEX NUMBERS IN POLAR COORDINATES

Consider a complex number given by

$$s = \alpha + \mathrm{j}\beta \tag{9.3}$$

This number can be represented in the complex plane with the coordinates representing the real and imaginary axes (Figure 9.2a). The representation of this point in polar coordinates takes the form

$$s = \delta \mathrm{e}^{-\mathrm{j}\phi} \tag{9.4}$$

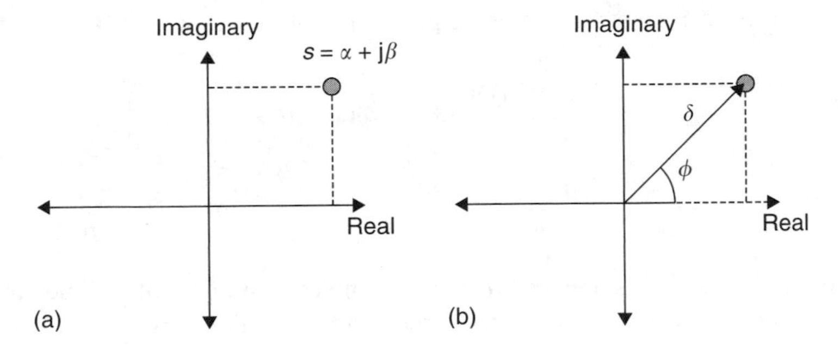


**FIGURE 9.2** Visualization of a complex number as a vector.

The two key quantities are defined as:

$$\delta = \text{magnitude of } s = |s| = \sqrt{\alpha^2 + \beta^2} \tag{9.5}$$

$$\phi = \text{argument of } s = \arg(s) = \tan^{-1}\left[\frac{\beta}{\alpha}\right] \tag{9.6}$$

These quantities are depicted in Figure 9.2b. The magnitude represents the *length* of the vector, and the argument is the *angle* that it makes with the real axis. Such a representation greatly facilities the algebraic operations between complex numbers such as multiplication,

$$\begin{aligned} s = s_1 \cdot s_2 &= \left(\delta_1 e^{-j\phi_1}\right)\left(\delta_2 e^{-j\phi_2}\right) \\ &= \left(\delta_1 \delta_2\right) e^{-j(\phi_1+\phi_2)} \\ &= \delta e^{-j\phi} \end{aligned} \tag{9.7}$$

Hence, the new complex number has a magnitude that corresponds to the *multiplication* of the individual magnitudes, and an angle that is equal to the *addition* of the individual angles.

## 9.3 CONSTRUCTION OF THE FREQUENCY RESPONSE

Consider a stable, linear process with the transfer function

$$\frac{\bar{y}(s)}{\bar{u}(s)} = g(s) \tag{9.8}$$

For a sinusoidal input $\bar{u}(t) = A\sin\omega t$, the output will be given by

$$\bar{y}(s) = g(s)\frac{A\omega}{s^2 + \omega^2} \tag{9.9}$$

Equation (9.9) can be expanded using partial fractions (Appendix B)

$$\bar{y}(s) = g(s)\left[\frac{A\omega}{(s+\mathrm{j}\omega)(s-\mathrm{j}\omega)}\right]$$
$$= \frac{a}{s+\mathrm{j}\omega} + \frac{a'}{s-\mathrm{j}\omega} + \frac{b_1}{s-p_1} + \frac{b_2}{s-p_2} + \cdots + \frac{b_n}{s-p_n} \tag{9.10}$$

where $a$ and $b_i$ are constants and $a'$ is the complex conjugate of $a$. Poles of $g(s)$ are given by $p_i$. The time-domain response can be calculated as

$$\bar{y}(t) = a\mathrm{e}^{-\mathrm{j}\omega t} + a'\mathrm{e}^{\mathrm{j}\omega t} + b_1\mathrm{e}^{-p_1 t} + b_2\mathrm{e}^{-p_2 t} + \cdots + b_n\mathrm{e}^{-p_n t} \tag{9.11}$$

Since the process is assumed to be stable, the exponential terms containing the poles will vanish. This is true even if the poles are not distinct. Therefore, the ultimate (steady-state) response will be given by

$$\bar{y}_{ss}(t) = a\mathrm{e}^{-\mathrm{j}\omega t} + a'\mathrm{e}^{\mathrm{j}\omega t} \tag{9.12}$$

From Eq. (9.10), we can compute the constants $a$ and $a'$:

$$a = g(s)\frac{A\omega}{s^2+\omega^2}(s+\mathrm{j}\omega)\bigg|_{s=-\mathrm{j}\omega} = -\frac{Ag(-\mathrm{j}\omega)}{2\mathrm{j}} \tag{9.13}$$

$$a' = g(s)\frac{A\omega}{s^2+\omega^2}(s-\mathrm{j}\omega)\bigg|_{s=\mathrm{j}\omega} = \frac{Ag(\mathrm{j}\omega)}{2\mathrm{j}} \tag{9.14}$$

Therefore, Eq. (9.12) becomes

$$\bar{y}_{ss}(t) = -\frac{Ag(-\mathrm{j}\omega)}{2\mathrm{j}}\mathrm{e}^{-\mathrm{j}\omega t} + \frac{Ag(\mathrm{j}\omega)}{2\mathrm{j}}\mathrm{e}^{\mathrm{j}\omega t} \tag{9.15}$$

---

The long-term response is purely sinusoidal with frequency $\omega$; therefore we are concerned only with the case where we have pure imaginary poles (or eigenvalues) of the response. This allows us to evaluate the resulting complex numbers only at $s = \mathrm{j}\omega$, which also corresponds to the stability boundary.

---

Since $g(\mathrm{j}\omega)$ is a complex number, we can now express it in polar coordinates as follows:

$$g(\mathrm{j}\omega) = |g(\mathrm{j}\omega)|\mathrm{e}^{\mathrm{j}\phi}, \quad \phi = \arg(g(\mathrm{j}\omega)) \tag{9.16}$$

One can also show that

$$g(-\mathrm{j}\omega) = |g(-\mathrm{j}\omega)|\mathrm{e}^{-\mathrm{j}\phi} = |g(\mathrm{j}\omega)|\mathrm{e}^{-\mathrm{j}\phi} \tag{9.17}$$

We can now replace $g(-j\omega)$ and $g(j\omega)$ in Eq. (9.15)

$$\begin{aligned}\bar{y}_{ss}(t) &= -\frac{A|g(j\omega)|}{2j}e^{-j(\omega t+\phi)} + \frac{A|g(j\omega)|}{2j}e^{j(\omega t+\phi)} \\ &= A|g(j\omega)|\frac{e^{-j(\omega t+\phi)} - e^{j(\omega t+\phi)}}{2j}\end{aligned} \tag{9.18}$$

By recognizing the trigonometric identity, we finally have

$$\bar{y}_{ss}(t) = A|g(j\omega)|\sin(\omega t + \phi) \tag{9.19}$$

The key result is that the process frequency response can be constructed directly from the knowledge of the process transfer function, $g(s)$, evaluated at $s = j\omega$:

---

1. The amplitude ratio

$$AR = \frac{A|g(j\omega)|}{A} = |g(j\omega)|$$

2. The phase shift is the angle

$$\phi = \text{argument of } g(j\omega)$$

---

## 9.4 EVALUATION OF THE FREQUENCY RESPONSE

Given the transfer function of a process, the frequency response can be computed via the following steps:

1. Start by evaluating $g(s)$ at $s = j\omega$.
2. Express $g(j\omega)$ as a complex number.
3. Calculate the magnitude (modulus) and the argument of the complex number $g(j\omega)$.
4. Calculate $AR(\omega)$ and $\phi(\omega)$.

We can demonstrate the use of this method by the following examples.

### Example 9.1

Consider the first-order transfer function,

$$g(s) = \frac{k}{\tau s + 1}$$

First, we need to evaluate the transfer function at $s = j\omega$.

$$g(j\omega) = \frac{k}{\tau j\omega + 1} \tag{9.20}$$

To express the transfer function in the standard form of a complex number, we multiply Eq. (9.20) by the complex conjugate of the denominator as well as divide by it:

$$g(\mathrm{j}\omega) = \frac{k}{\mathrm{j}\tau\omega + 1}\frac{(-\mathrm{j}\tau\omega + 1)}{(-\mathrm{j}\tau\omega + 1)} = \frac{k(-\mathrm{j}\tau\omega + 1)}{\omega^2\tau^2 + 1}$$
$$= \frac{k}{\omega^2\tau^2 + 1} + \mathrm{j}\frac{(-\tau\omega)k}{\omega^2\tau^2 + 1} = \alpha + \mathrm{j}\beta \tag{9.21}$$

where

$$\alpha = \frac{k}{\omega^2\tau^2 + 1}, \quad \beta = \frac{(-\tau\omega)k}{\omega^2\tau^2 + 1} \tag{9.22}$$

The modulus can be calculated based on Eq. (9.5):

$$|g(\mathrm{j}\omega)| = \sqrt{\alpha^2 + \beta^2} = \sqrt{\left(\frac{k}{\tau^2\omega^2 + 1}\right)^2 + \left(\frac{(-\tau\omega)k}{\tau^2\omega^2 + 1}\right)^2}$$
$$= \sqrt{\frac{k^2(\tau^2\omega^2 + 1)}{(\tau^2\omega^2 + 1)^2}} = \frac{k}{\sqrt{\tau^2\omega^2 + 1}} \tag{9.23}$$

and the argument is based on Eq. (9.6):

$$\arg(g(\mathrm{j}\omega)) = \tan^{-1}\left[\frac{\beta}{\alpha}\right] = \tan^{-1}\left[\frac{k(-\tau\omega)/(\tau^2\omega^2 + 1)}{k/(\tau^2\omega^2 + 1)}\right]$$
$$= \tan^{-1}(-\tau\omega) \tag{9.24}$$

This yields the amplitude ratio and the phase shift for a first-order process:

$$AR = |g(\mathrm{j}\omega)| = \frac{k}{\sqrt{\omega^2\tau^2 + 1}}$$
$$\phi = \arg(g(\mathrm{j}\omega)) = \tan^{-1}(-\tau\omega) = -\tan^{-1}(\tau\omega) \tag{9.25}$$

## Example 9.2

Let us consider a second-order process transfer function given by

$$g(s) = \frac{2}{s^2 + s + 1} \tag{9.26}$$

Again, following the procedure as we have,

$$g(\mathrm{j}\omega) = \frac{2}{(\mathrm{j}\omega)^2 + \mathrm{j}\omega + 1} = \frac{2}{-\omega^2 + \mathrm{j}\omega + 1} = \frac{2}{(1 - \omega^2) + \mathrm{j}\omega}$$
$$= \frac{2}{(1 - \omega^2) + \mathrm{j}\omega}\frac{(1 - \omega^2) - \mathrm{j}\omega}{(1 - \omega^2) - \mathrm{j}\omega} \tag{9.27}$$

$$g(\mathrm{j}\omega) = \frac{2(1 - \omega^2) - 2\mathrm{j}\omega}{(1 - \omega^2)^2 - \omega^2} = \frac{2(1 - \omega^2)}{(1 - \omega^2)^2 + \omega^2} - \mathrm{j}\frac{2\omega}{(1 - \omega^2)^2 + \omega^2}$$

The modulus and the argument of $g(j\omega)$ yield

$$
\begin{aligned}
|g(j\omega)| &= \sqrt{\frac{4(1-\omega^2)^2}{[(1-\omega^2)^2+\omega^2]^2}+\frac{4\omega^2}{[(1-\omega^2)^2+\omega^2]^2}} \\
&= 2\sqrt{\frac{(1-\omega^2)^2+\omega^2}{[(1-\omega^2)^2+\omega^2]^2}} \\
&= 2\sqrt{\frac{1}{[(1-\omega^2)^2+\omega^2]^2}} = \frac{2}{\sqrt{(1-\omega^2)^2+\omega^2}}
\end{aligned}
\tag{9.28}
$$

$$
\begin{aligned}
\arg(g(j\omega)) &= \tan^{-1}\left[\frac{2\omega}{(1-\omega^2)^2+\omega^2}\,\frac{(1-\omega^2)^2+\omega^2}{2(1-\omega^2)}\right] \\
&= \tan^{-1}\left[\frac{-\omega}{1-\omega^2}\right]
\end{aligned}
\tag{9.29}
$$

Finally, we have

$$
\begin{aligned}
AR &= |g(j\omega)| = \frac{2}{\sqrt{(1-\omega^2)^2+\omega^2}} \\
\phi &= \arg(g(j\omega)) = \tan^{-1}\left[\frac{-\omega}{1-\omega^2}\right]
\end{aligned}
\tag{9.30}
$$

As shown before, the frequency response of a system is characterized by its amplitude ratio and its phase angle with the frequency as the independent variable. A graphical representation of these relationships is very useful for system analysis and design. There are two types of plots commonly used to visualize the frequency response of a process:

- *Bode diagrams* or logarithmic diagrams
- *Nyquist diagrams* or polar diagrams

## 9.5 BODE DIAGRAMS

The Bode diagrams (or Bode plots) consist of two complementary graphs:

- The magnitude of $g(j\omega)$ or the amplitude ratio $AR$ versus $\omega$.
- The argument of $g(j\omega)$ or the phase shift $\phi$ versus $\omega$.

Both plots are expressed as a function of frequency $\omega$. A logarithmic scale is used for the $\omega$-axis (abscissa) because the magnitude and the phase angle can be plotted over a greater range of frequencies than with a linear frequency axis. Furthermore, all frequencies are equally emphasized and such graphs often result in straight lines.

---

The $AR$ may be plotted on a logarithmic scale in decibel (db) units:

$$\text{db} = 20\log_{10}|g(j\omega)|$$

---

The magnitude vs. log $\omega$ plot is called the *Bode magnitude plot*, and the phase angle versus log $\omega$ plot is called the *Bode phase plot*. The Bode magnitude plot is sometimes referred to as the *log-modulus plot* in the control literature.

**Example 9.3**

Consider a first-order process given by the transfer function model

$$g(s) = \frac{2}{3s + 1} \tag{9.31}$$

Using Eq. (9.25), we can express the *AR* and the $\phi$ as follows:

$$\begin{aligned} AR &= |g(j\omega)| = \frac{2}{\sqrt{9\omega^2 + 1}} \\ \phi &= \arg(g(j\omega)) = \tan^{-1}(-3\omega) \end{aligned} \tag{9.32}$$

The corresponding Bode plot is given in Figure 9.3. We note the following features on this plot:

- At low frequencies (approaching steady-state), the *AR* plot *asymptotically* approaches 2, which represents the gain of the process.
- At high frequencies, the *AR* approaches 0 (in the logarithmic scale) with a slope of –1. The slope corresponds to the difference between the numerator and denominator polynomial degrees of the process transfer function.

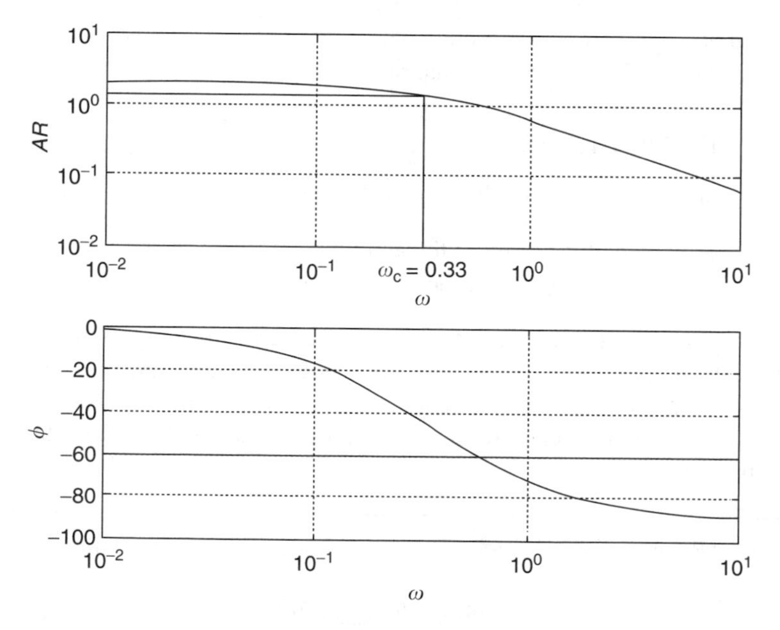


**FIGURE 9.3** Bode plot of a first-order process in Example 9.3.

- The frequency at which the $AR$ reaches 63.2% of the process gain corresponds to the crossover frequency, $\omega_c$. This provides critical information about process dynamics because, by definition, $\omega_c = 1/\tau$. For this example, note that $\omega_c = 0.33 \text{ min}^{-1}$.
- The frequency range $[0, \omega_c]$ is defined as the process *bandwidth*.
- The phase plot indicates that the maximum possible phase shift for a first-order process is 90°, and it is actually a phase *lag* as the output signal trails the input signal. This lag is typical for all causal processes.
- A first-order process is also referred to as a *low-pass filter*, as it allows low frequency signals to pass through relatively unchanged, but significantly attenuates (filters) high-frequency inputs. (Note that measurement noise typically has high frequency components.)
- In a low-pass filter, the output follows the sinusoidal input faithfully at low frequencies. However, as the input frequency is increased, the output can no longer follow the input faithfully because a certain amount of time is necessary for the process to overcome its inertia. Therefore, at high frequencies, the $AR$ approaches zero and the $\phi$ becomes –90°.

What if the first-order process in Eq. (9.31) had a time delay, $t_D = 0.5$ min ? The equations for $AR$ and $\phi$ are modified as follows:

$$AR = |g(j\omega)| = \frac{2}{\sqrt{9\omega^2 + 1}}$$
$$\phi = \arg(g(j\omega)) = -\tan^{-1}(3\omega) - (0.5\omega) \tag{9.33}$$

Note that the difference is only in the $\phi$, and the $AR$ is not affected. The corresponding Bode diagram for this case is given in Figure 9.4. For this case, one observes that the output signal can have a large phase lag depending on the frequency of the input signal.

## 9.6 NYQUIST DIAGRAMS

A Nyquist diagram (or a polar plot) is an alternative medium to visualize the frequency response characteristics of a linear dynamic process. The coordinates for this plot are the real and imaginary axes. As discussed in Section 9.2, given a specific value of the frequency $\omega$, a vector anchored at the origin has a length corresponding to the $AR$ and makes an angle $\phi$ with the Real axis, corresponding to the $\phi$ for the process transfer function. This procedure is repeated for a range of frequencies to obtain the Nyquist plot (Figure 9.5).

On this plot, we can make the following observations:

- The point A has a corresponding frequency $\omega_A$.
- The distance from point A to the origin (0,0) is the $AR$ at $\omega_A$.
- The angle made with the real axis is the $\phi$ (lag) at the frequency $\omega_A$.
- The Nyquist plot displays the real and imaginary parts of the complex variable $g(j\omega)$.

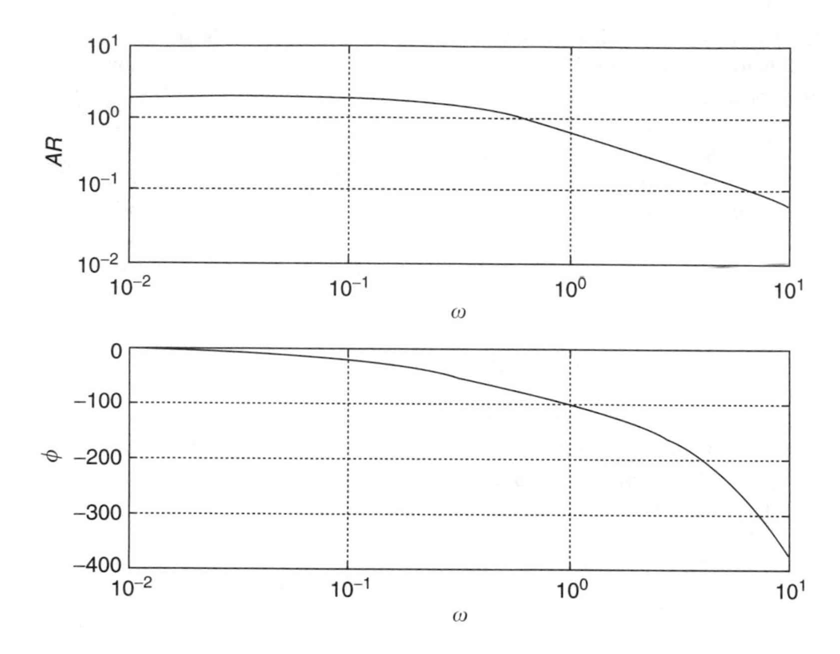


**FIGURE 9.4** Bode plot of a first-order process with time delay in Example 9.3.

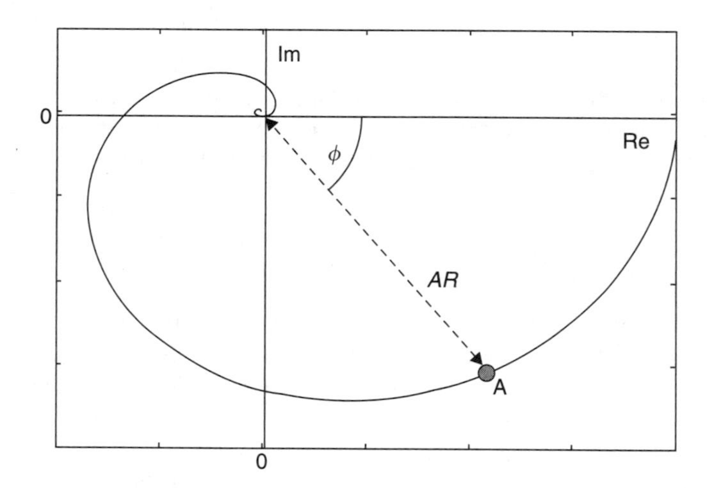


**FIGURE 9.5** A Nyquist plot with point A indicating the value at frequency $\omega_A$.

It should also be noted that the Nyquist plot starts on the positive real axis corresponding to the low-frequency range ($\omega = 0$), and as the frequency increases from 0 to $\infty$, we trace the whole length of the Nyquist plot, converging to the origin for all causal processes.

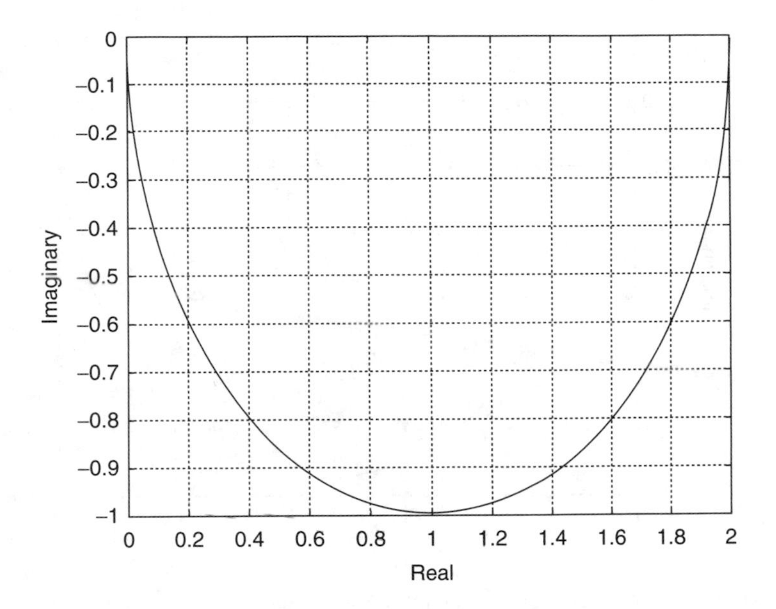


**FIGURE 9.6** The Nyquist plot for Example 9.5.

---

The Nyquist plot contains the same information as the Bode plot for the same system.

---

### Example 9.5

Consider a first-order process given in Example 9.3 and the corresponding frequency response equations for the *AR* and the $\phi$ as given in Eq. (9.32).

The Nyquist diagram for this system is shown in Figure 9.6. Note that the plot originates from point (2,0) and meets the origin at high frequencies. The entire plot is confined to the lower-right quadrant as we recall that the maximum possible $\phi$ is 90° for first-order processes, regardless of the *AR*.

Consider again the first-order system in Eq. (9.31), but with an associated time delay, recalling the *AR* and $\phi$ as in Eq. (9.33). The Nyquist plot for this case is given in Figure 9.7.

---

With the addition of a delay term, the *AR* remains the same; however, the $\phi$ is significantly affected. The net effect of the delay is to alter the phase characteristics of the process, which results in the circling of the origin at high frequencies with a decreasing radius.

---

A number of control system design and analysis methods are based on frequency response characteristics and will be discussed later in Chapter 11.

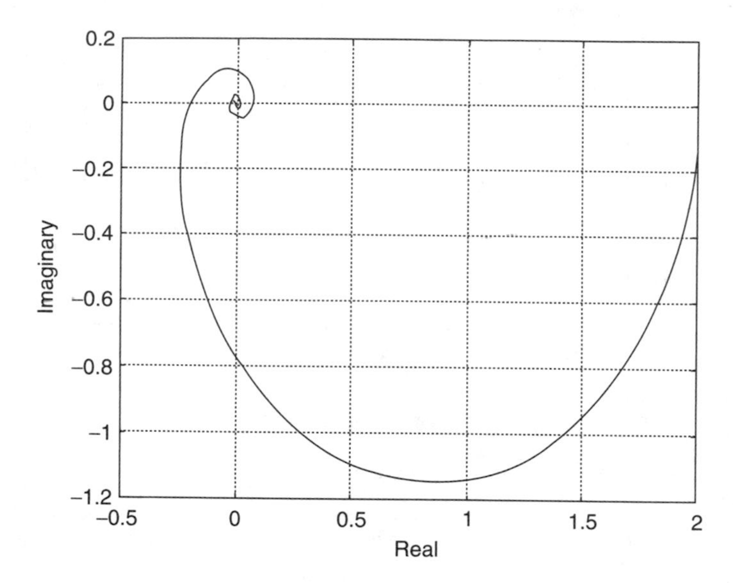


**FIGURE 9.7** The Nyquist plot of Example 9.5 with time delay.

## 9.7 SUMMARY

In this chapter, we introduced the basic elements of frequency response analysis starting with the representation of complex numbers in polar form. This was followed by the development of frequency responses for first- and second-order processes to explain the mechanics of the derivations. Frequency response of processes is best expressed through Bode and Nyquist diagrams that can offer a graphical representation of the process dynamics in the frequency domain.

# Section III Additional Reading

In his comprehensive article on the historical development of stability theory, Michel points to a number of milestones that influence the evolution of the theory of feedback control:

Michel, A.N., Stability: The common thread in the evolution of feedback control, *IEEE Control System Mag.*, 16, 50–60, 1996.

Early studies by mathematicians on the algebraic criteria for stability of dynamic systems enabled the control engineers with the appropriate tools for analyzing and ensuring the stability of closed-loop designs.These required the availability of the equations of motion:

Routh, E.J., *A Treatise on the Stability of a Given State of Motion*, Macmillan, London, 1877.

Furthermore, around 1940s, with the advent of electronic amplifiers in phone communications, Nyquist and Bode developed frequency-domain criteria that studied the stability of negative feedback amplifier systems:

Bode, H.W., *Network Analysis and Feedback Amplifier Design*, Van Nostrand, New York, NY, 1945.

Nyquist, H., Regeneration theory, *Bell Syst. Tech. J.*, 11, 126–147, 1932.

The implications of frequency response methods for controlling process systems were studied in the following book:

Caldwell, W.I., Coon, G.A., and Zoss, L.M., in *Frequency Response for Process Control*, Zoss, L.M. (Ed.), McGraw-Hill, New York, 1959.

The greater challenge lies, even today, in the dynamic analysis of nonlinear systems. Nonlinear systems offer a very rich set of dynamic behaviors, ranging from input and output multiplicities to bifurcations and chaos:

Guckenheimer, J. and Holmes, P.H., *Nonlinear Oscillations, Dynamical Systems, and Bifurcations of Vector Fields*, Springer, New York, 1990.

The stability of nonlinear dynamic systems was first studied by Liapunov, whose methods still remain as the cornerstone of the analysis techniques studied or proposed today.

Hahn, W., *Theory and Application of Liapunov's Direct Method* (English ed., prepared by Siegfried H. Lehnigk), Prentice-Hall, Englewood Cliffs, NJ, 1963.

Freeman, R.A. and Kokotović, P.V., *Robust Nonlinear Control Design: State-Space and Lyapunov Techniques*, Birkhäuser, Boston, 1996.
Sastry, S., *Nonlinear Systems: Analysis, Stability, and Control*, Springer, New York, 1999.

In chemical engineering, early studies on system stability focused on the dynamics of chemical reactors, and largely the CSTRs.

Aris, R. and Amundson, N.R., An analysis of chemical reactor stability and control. Parts I-III, *Chem. Eng. Sci.*, 7, 121–155, 1958.
Perlmutter, D., *Stability of Chemical Reactors*, Prentice-Hall, Englewood Cliffs, NJ, 1972.
Friedly, J.C., *Dynamic Behavior of Processes*, Prentice-Hall, Englewood Cliffs, NJ, 1972.

Nonlinear differential equations that represent the energy and material balances for the CSTR exhibit a wide range of dynamic behaviors and their classification as a function of process parameters have been analyzed in an often-cited article by Ray and his co-workers:

Uppal, A., Ray, W.H., and Poore, A.B., On the dynamic behavior of continuous stirred tank reactors, *Chem. Eng. Sci.*, 29, 967–985, 1974.

A more recent text that offers an in-depth dynamic analysis of reactors is given below:

Elnashaie, S.S.E.H. and Elshishini, S.S., *Dynamic Modeling, Bifurcation and Chaotic Behavior of Gas-Solid Catalytic Reactors*, Gordon and Breach, Amsterdam, 1996.

# Section III Exercises

**III.1.** Investigate each of the transfer functions below for BIBO stability and asymptotic stability:

1. $g(s) = \dfrac{K}{(\tau s + 1)(s^2 + \omega^2)}$

2. $g(s) = \dfrac{K}{(s^2 + 4s + 3)}$

3. $g(s) = \dfrac{K(s + 1)}{s(s^2 + 4s + 3)}$

**III.2.** For the system described by two first-order systems in parallel,

$$g(s) = \frac{-2}{3s + 1} + \frac{1}{s + 1}$$

The input signal is described by the following transfer function:

$$u(s) = \frac{1}{s - 1}$$

Is this a bounded input? Is the system output bounded? Why? Discuss the implications of this behavior.

**III.3.** Find the poles of and classify the following processes as being stable (asymptotically or critically), or unstable:

$$g_1(s) = \frac{1}{(3s + 1)(s + 1)}$$

$$g_2(s) = \frac{1}{(3s + s)(s + 2)}$$

$$g_3(s) = \frac{1}{3s^3 + s^2 + s + 1}$$

$$g_4(s) = \frac{1}{s^4 + 4s^3 + 4s^2 + 2s + 1}$$

Explain your decisions.

**III.4.** Use the Routh's Criterion to evaluate the stability of the process with the following characteristic equation:

$$p(s) = 4s^3 + 2s^2 + s + 1$$

**III.5.** A system is being studied for stability, and its characteristic equation is given as follows:

$$p(s) = 5s + 1 + 2Ke^{-2s}$$

1. Apply the Routh's Criterion to this problem. What can you conclude about the system when $K = 4$.
2. Determine the range of $K$ that would maintain stability. Is this a guaranteed result?

**III.6.** A process has the following transfer function:

$$g(s) = \frac{4(as + 1)}{(5s + 1)(s + 1)}$$

For the following values of the parameter $a$:

- $a = 10$
- $a = 2$
- $a = 0.5$
- $a = -1$

compute the responses for a step-change of magnitude 0.5 and plot them in a single figure. What conclusions can you draw concerning the zero location? Is the location of the pole corresponding to $\tau_2 = 5$ important?

**III.7.** For each of the processes with transfer functions given below, (1) identify the poles and zeros of the transfer function, as well as the process gain and (2) sketch the response to a unit step input change, clearly showing the key response characteristics.

$$g(s) = \frac{10}{0.1s + 1} - \frac{5}{0.04s + 1}$$

$$g(s) = \frac{10}{0.2s + 1} - \frac{5}{0.3s + 1}$$

**III.8.** Consider a second-order process represented by the transfer function,

$$\frac{y(s)}{u(s)} = \frac{1}{4s^2 + s + 2}$$

Introduce a step change of magnitude 5 into the system and find, (1) % overshoot, (2) decay ratio, (3) maximum value of $y(t)$, (4) ultimate value of $y(t)$, (5) rise time, and (6) period of oscillation.

**III.9.** A time-delay element is basically a distributed system. One approximate way to get the dynamics of distributed systems is to lump them into a number of perfectly mixed sections. Prove that a series of $N$ mixed tanks is equivalent to a pure time delay as $N$ tends to infinity. (**Hint:** Keep the volume of the total system constant as more and more sections are used).

**III.10.** A step change of magnitude 4 is introduced into a process having the transfer function:

$$\frac{y(s)}{u(s)} = \frac{10}{s^2 + 1.6s + 4}$$

Obtain the output response by simulation and determine:

- Percent overshoot
- Rise time
- Settling time
- Period of oscillation
- Ultimate value of $y(t)$

**III.11.** A process consists of five perfectly stirred tanks in series. The volume in each tank is 50 L, and the volumetric flow rate through the system is 5 L/min. At some particular time, the inlet concentration of the nonreacting species is changed from 0.7 to 0.83 (mass fraction) and held there. Write an expression for $c_5$ (the concentration leaving the fifth tank) as a function of time, and determine exit concentrations from all tanks at $t = 30$ min.

**III.12.** A two-tank flow-surge system is given in Figure III.1, along with the block diagram representing the approximate dynamics of the system.

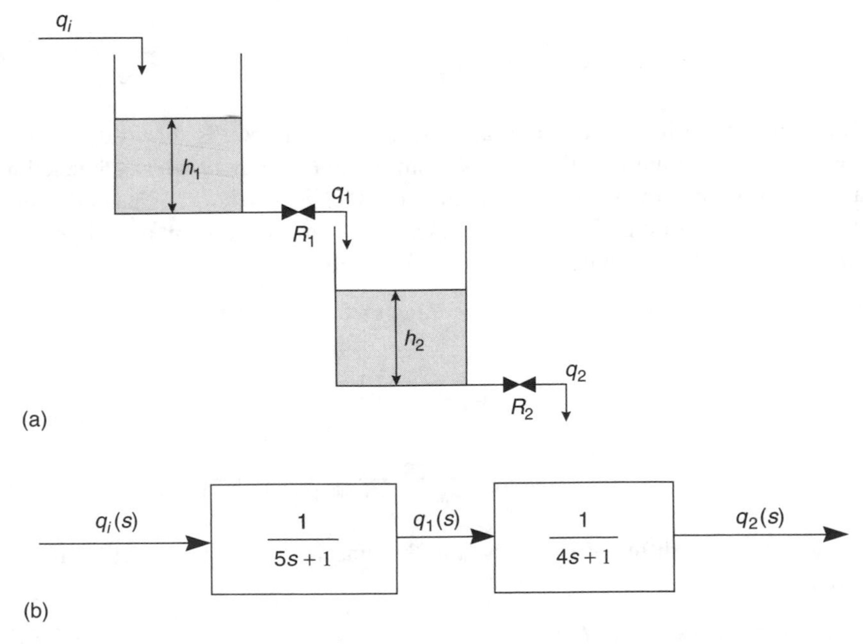


**FIGURE III.1** A two-tank surge process (a) and its transfer function model (b).

1. What is the response $q_2(t)$ to a step input of magnitude 0.5 m$^3$/min in $q_i(t)$ if the system is initially at steady-state corresponding to $\bar{q}_i = \bar{q}_1 = \bar{q}_2 = 1$ m$^3$/min. The head-flow relations for the tanks are

$$q_1 = \frac{5\text{m}^3/\text{min}}{\text{m}} h_1, \quad q_2 = \frac{2\text{m}^3/\text{min}}{\text{m}} h_2$$

2. What are the ultimate values of each tank level after 1 m$^3$ of liquid is suddenly added to the first tank? Why?

**III.13.** The dynamic behavior of a drum boiler can be represented by the following transfer function between the liquid level and the cold water feed:

$$\frac{H'(s)}{Q'_{\text{in}}(s)} = \frac{(\tau - 1)s + 1}{s(\tau s + 1)}$$

Plot the poles and zeros of this transfer function in the complex plane for two cases:

1. $\tau = 0.5$
2. $\tau = 2.0$

For each case *sketch* the level response to a step change in the cold water flow rate.

**III.14.** An exothermic chemical reactor has the following transfer function relationship between the inlet flow rate and the reactor temperature.

$$g(s) = \frac{2(-2.5s + 1)}{9s^2 + 3s + 1}$$

The units of the input are L/min and the output is in °C. Is this system overdamped or underdamped? Does this system exhibit overshoot when subjected to a unit-step change in the input? If yes, what is the % overshoot? What is the ultimate value of the output? Sketch the system response for this input and provide an explanation for the dynamic behavior observed.

**III.15.** Draw the Bode diagrams of the transfer functions below,

$$g_1(s) = \frac{1}{(3s + 1)(s + 1)}$$

$$g_2(s) = \frac{1}{(3s + 1)(s + 1)(0.1s + 1)}$$

Comment on the differences observed in the diagrams and their practical implications.

# Section IV

## Feedback Control

# 10 Basic Elements of Feedback Control

In previous chapters, we acquired many tools for modeling and analyzing process systems. We are now in a position to consider the elements of feedback control, starting with the basic definitions and continuing in the next chapters with analysis and design of feedback controllers.

## 10.1 FEEDBACK CONTROL PROBLEM

In principle, a control problem is formulated to ensure that a process operates at its design specifications, satisfying some predetermined dynamic performance criteria. The most common structure is feedback, as this structure can monitor variations in the process and successfully compensate for unwanted excursions in a manner consistent with the performance objectives. In process industries, one encounters two classes of control problems:

1. *Set-point tracking* The role of the feedback controller, in this case, is to ensure satisfactory tracking of the set-point (reference) value of the output variable. This arises in processes where the customer's demand for a product varies in terms of quality and or the quantity of the product (output variable). For example, the process may have to produce different grades of a product at different times during its continuous operation and the plant operators expect a smooth and rapid transition between these changes. Sometimes, this problem is also referred to as the *servo problem*, dating back to the *servomechanisms*.[1]
2. *Disturbance rejection* The most common problem in process industries is to maintain steady operation around a set-point value in the face of ever-present disturbances. The role of the feedback controller is to react to the variations in the process operation caused by the disturbances and recover the desired plant operation in a timely and smooth manner. Disturbances may prove costly for plant operation if they are not compensated for in an appropriate manner. If product quality deteriorates due to a disturbance, the product may have to be discarded or reprocessed, which is both economically and environmentally unattractive. This problem is sometimes referred to as the *regulator* (or the *load*) *problem*.

How does a feedback control system function to respond to these problems? Let us start by expressing a process through its input–output representation as in

Figure 10.1, where we can identify as the inputs, the manipulated variable $m$ and the disturbance variable $d$, along with the output, the controlled variable $y$. The process transfer function is denoted as $g(s)$, where the disturbance transfer function is given by $g_d(s)$.

The feedback control problem starts with the measurement of the controlled variable. Figure 10.2 depicts the measurement step where a measuring element (sensor) provides the value of the controlled variable for subsequent calculations. The sensor could be a thermocouple (TC) for temperature measurements or differential pressure cell for flow rate measurements, to cite a few examples. The measured variable $y_m$ is the output of the measuring element transfer function $g_m(s)$.

The measured value of the controlled variable is next compared with the desired set-point value $y_{sp}$ to generate the tracking error $e$ (Figure 10.3). The tracking error is fed through the control law, $g_c(s)$, to generate the control action $c$. This part of the loop takes place in what we shall call the *controller*, or the control mechanism. The role of the feedback controller is to produce a control action based on the set-point tracking error that, when implemented through the manipulated variable on the process, helps the controlled variable to approach the desired set-point.

The control action is implemented in the process through a final control element (or an actuator), $g_f(s)$, as shown in Figure 10.4. Often, the final control element is a valve that regulates the flow rate of a process stream.

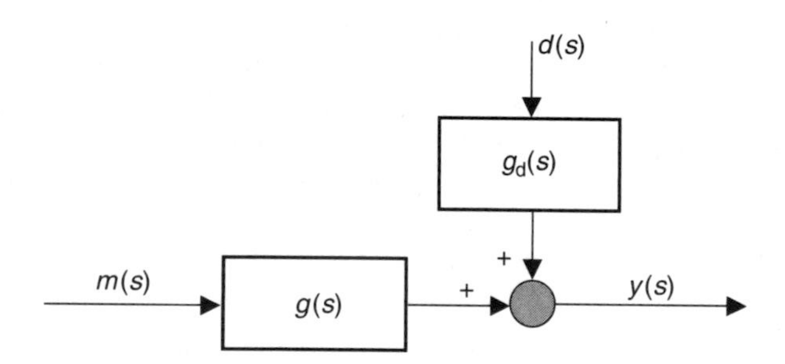


**FIGURE 10.1** Input–output representation of a process.

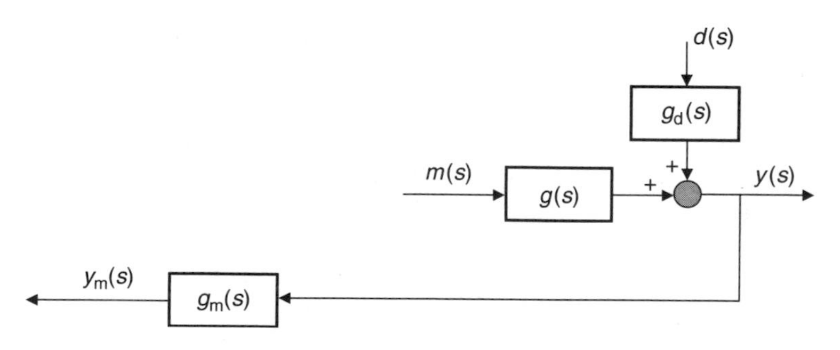


**FIGURE 10.2** Process and the measuring element.

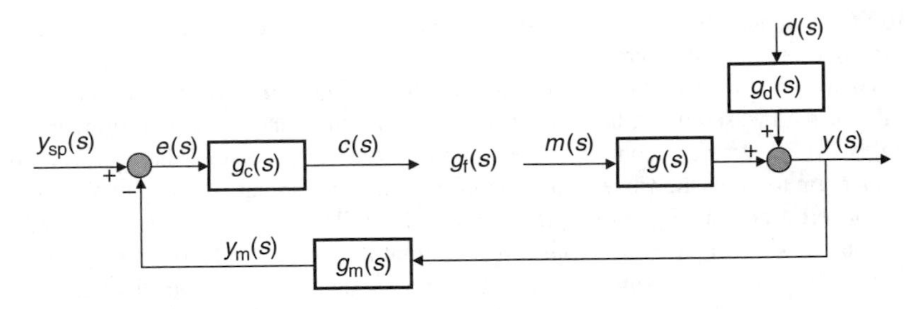


**FIGURE 10.3** Calculation of the tracking error and the control action.

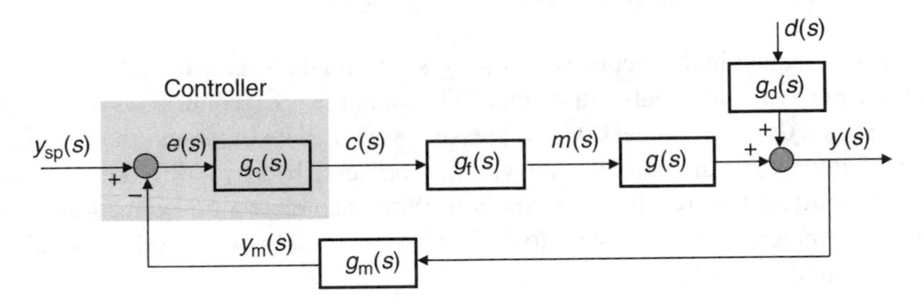


**FIGURE 10.4** Final control element closes the feedback loop.

The system in Figure 10.1 is considered in *open-loop*, whereas the feedback-controlled system of Figure 10.4 is considered to be in *closed-loop*.

The most important feature of a feedback control system is the fact that it *learns* the process behavior through continuous measurements of the output, and feeds this information back to the controller that commands a certain change in the manipulated variable. The effect of this change on the process is *learned* by the subsequent measurement, and this loop is followed indefinitely.

### Example 10.1

Consider the heat exchanger studied in Example 6.1 and Figure 6.5, where a process stream is cooled by circulating cooling water in the shell side. We shall define the following control problem:

Maintain the design value of the exit process stream temperature $T_2$ at its set-point $T_{2,sp}$ in the presence of variations in the inlet process stream temperature $T_1$, by manipulating the cooling water flow rate, $F_3$.

We can now identify all the elements of a feedback control system that could deal with the control problem as stated.

First, we need to measure the output variable $T_2$, and a TC can accomplish this task. Next, the cooling water stream would require a control valve to help change its flow rate, $F_3$. In most cases, sensors are equipped with a transmitter (TT for a temperature transmitter) to deliver a current (or sometimes a voltage) signal that corresponds to the measurement made. In other words, the sensor is calibrated such that a certain measurement value corresponds to a current level. The current range for most electronic sensors is 4–20 mA. If our TC is calibrated in the range 0 to

100°C, then one can immediately conclude that a current signal of 12 mA indicates a temperature measurement of 50°C.

Today, most industrial controllers are electronic, and as they receive current signals for the measurement, they also produce a current signal to be sent to the control valve. Yet, most control valves operate with air pressure that acts on a diaphragm to move the valve stem. Therefore, another element is typically required to convert a current signal to a pressure signal. I/P Transducers convert 4–20 mA signals to 3–15 psi pressure signals, thus facilitating the use of the proper signal.

The temperature set-point $T_{2,sp}$ is specified in degrees but the controller needs to convert it internally to a current signal $T_{2,sp,m}$ using a TT to establish the error signal. Figure 10.5a illustrates these elements on the process, and Figure 10.5b shows the corresponding conceptual feedback control diagram.

Each element in the feedback control system should be considered as a physical system with an input–output pair. For example, the control valve can be viewed as a dynamic process with a pressure signal as its input and the flow rate as its output. Consequently, its behavior can be described by a differential equation or equivalently by a transfer function. With each element described as such, the development of closed-loop block diagrams is greatly facilitated, as we shall see later in this chapter.

## 10.2 CONTROL LAW

The control signal $c(t)$ is calculated, given the value of the error $e(t)$, through a predefined functional relationship.

$$c(t) = C[e(t)] \tag{10.1}$$

The function $C[\cdot]$ constitutes the *control law*. By specifying $C[\cdot]$, we are, in effect, establishing the manner with which the error information is utilized by the controller. The most common functional form is the *three-mode* proportional-integral-derivative (PID) control law.

In this book, with a slight abuse of terminology, the PID control law will also be referred to as the PID *controller* since, in practice, that is its most common use.

Each mode of the PID control law will be analyzed next to provide some insight into their individual roles in shaping the control signal.

### 10.2.1 PROPORTIONAL MODE

This mode produces a control signal that is proportional to the error

$$c(t) = k_c e(t) + c_b \tag{10.2}$$

$k_c$ represents the *proportional gain* of the controller, and defines how sensitive the controller is to errors present in the system. $c_b$ is a bias signal that corresponds to the value of the control signal when the error is zero. The bias signal can also be interpreted as the steady-state value of the control signal. Thus, defining the

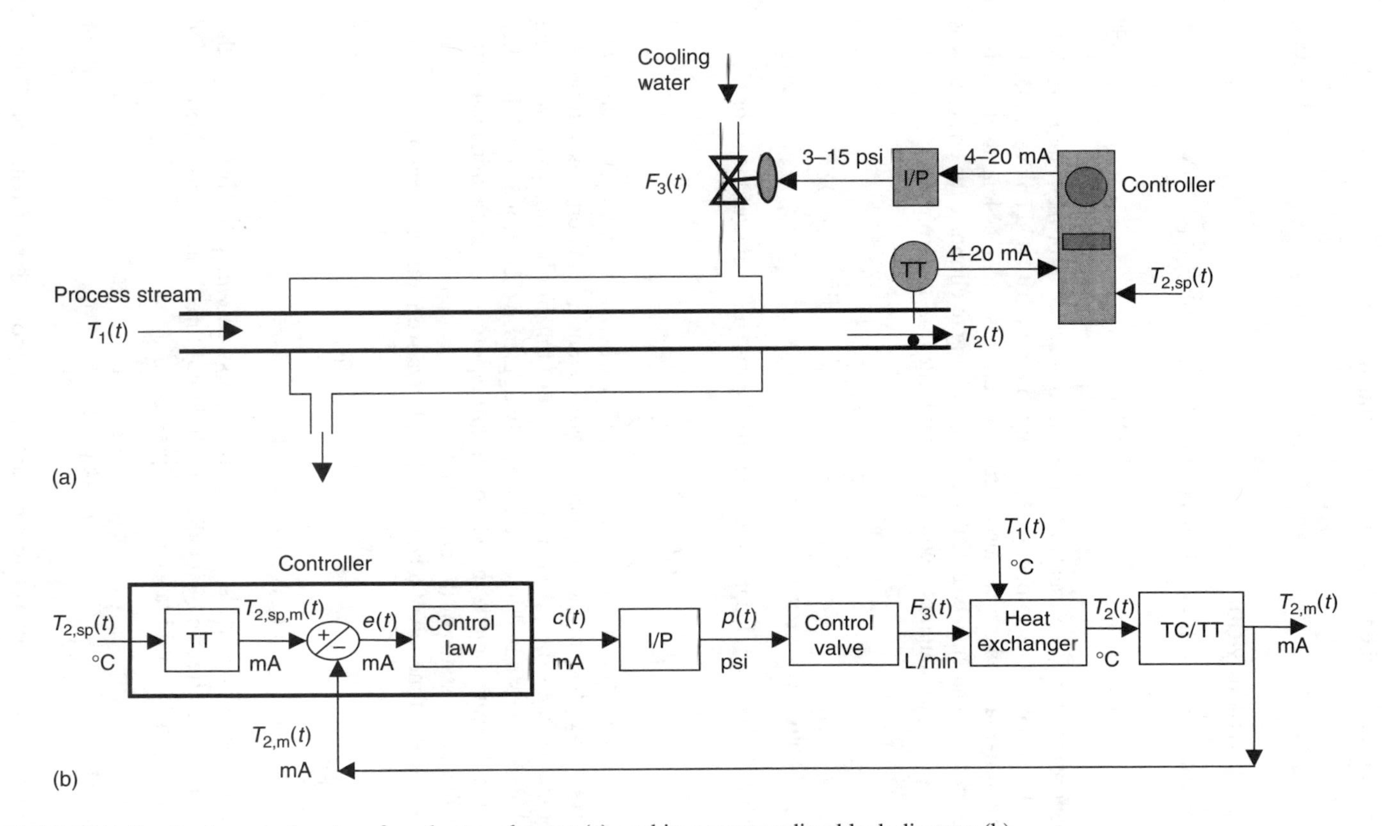


**FIGURE 10.5** Feedback control system for a heat exchanger (a), and its corresponding block diagram (b).

deviation variable $\bar{c}(t) = c(t) - c_b$, and recognizing that by definition $\bar{e}(t) = e(t)$, Eq. (10.2) results in the following transfer function:

$$\frac{\bar{c}(s)}{\bar{e}(s)} = g_c(s) = k_c \tag{10.3}$$

### 10.2.2 Integral Mode

The control signal for this mode is produced by the integral equation,

$$c(t) = \frac{k_c}{\tau_I} \int e(t)\, dt + c_b \tag{10.4}$$

The new parameter $\tau_I$ represents the *integral time constant* or the *reset time*. With this mode, the controller responds effectively to errors that build up over time. This is a very important feature because even if the error is small, as long as it persists, a large control signal may be calculated, thus helping to eliminate the error quickly.

The transfer function of a controller with integral mode only is,

$$\frac{\bar{c}(s)}{\bar{e}(s)} = g_c(s) = k_c\left(\frac{1}{\tau_I s}\right) \tag{10.5}$$

### 10.2.3 Derivative Mode

In this mode, the control signal responds to the rate of change of the error signal,

$$c(t) = k_c \tau_D \frac{de(t)}{dt} + c_b \tag{10.6}$$

A new parameter $\tau_D$ is introduced as the *derivative time constant*. The role of this mode is to judge the change in the error. For instance, if the error is still present but not increasing as fast, the controller may use this information to decrease the control signal, thus possibly avoiding overly aggressive control actions. In other words, the derivative mode introduces an *anticipatory* control action as it extrapolates the future status of the error.

The transfer function of a controller in derivative mode is given by

$$\frac{\bar{c}(s)}{\bar{e}(s)} = g_c(s) = k_c(\tau_D s) \tag{10.7}$$

### 10.2.4 Three-Mode Controller (PID)

The *PID control law* yields a three-term expression where the behavior of the controller can be affected by a judicious choice of three parameters. The transfer function of a PID control law can be expressed as

$$\frac{\bar{c}(s)}{\bar{e}(s)} = g_c(s) = k_c\left(1 + \frac{1}{\tau_I s} + \tau_D s\right) \tag{10.8}$$

Common forms of the PID controller such as the Proportional Controller (P-controller) and the proportional-integral controller (PI controller) can be easily obtained by setting $\tau_I = \infty$, $\tau_D = 0$, and $\tau_D = 0$, respectively. The PI controller is sometimes referred to as the proportional-plus-reset-controller.

In Chapter 12, we shall study methods of *controller tuning* where the best combination of these parameters is calculated satisfying specific performance objectives.

### Special Case: Discrete PID

As noted in Chapter 6, in industrial applications, the information flow between the process and the process control computers is not in analog (continuous) but in digital (discrete) form. Therefore, continuous control laws such as PID need to be discretized for practical applications. Consider the integrodifferential form of the PID control law,

$$c(t) = k_c\left[1 + \frac{1}{\tau_I}\int e(t)\,dt + \tau_D\frac{de(t)}{dt}\right] + c_b \tag{10.9}$$

Using a first-order finite difference formula for the derivative and a rectangular integration to replace the integral, Eq. (10.9) can be expressed as follows:

$$c_k = k_c\left[e_k + \frac{\Delta t}{\tau_I}\sum_{i=1}^{k} e_i + \frac{\tau_D}{\Delta t}(e_k - e_{k-1})\right] + c_b \tag{10.10}$$

Equation (10.10), referred to as the *position form* of PID, calculates the control signal at the sampling instant $k$ by using information on the error from the previous sampling instances, where $\Delta t$ is the sampling time. An alternative expression is the *velocity form* of PID,

$$c_k - c_{k-1} = \Delta c = k_c\left[(e_k - e_{k-1}) + \frac{\Delta t}{\tau_I}e_k + \frac{\tau_D}{\Delta t}(e_k - 2e_{k-1} + e_{k-2})\right] + c_b \tag{10.11}$$

This form calculates the *change* in the control signal from the previous value and is easier to implement algorithmically.

## 10.3 CLOSED-LOOP TRANSFER FUNCTIONS

For analysis and design of feedback systems, we need to establish the algebra for the closed-loop structure. The input–output blocks and the transfer functions for each feedback element constitute the key pieces for this algebra. First, we start with the closed-loop *block diagram* that will help us visualize the feedback system and its elements (Figure 10.4).

The closed-loop block diagram depicted in Figure 10.4 is to some extent a simplified representation, but it captures the most important feedback principles and will be used as such in the sequel. For each element shown, we can write the corresponding transfer function relating its output to its input, but we neglect any dynamics that the transmission lines may have.

From Figure 10.4, we observe that the process output can be expressed as a sum of two transfer functions,

$$y(s) = g_p(s)m(s) + g_d(s)d(s) \tag{10.12}$$

Our goal is to derive an input–output relationship for the closed-loop process. We note that the inputs to the closed-loop process are the set-point $y_{sp}(s)$ and the disturbance $d(s)$ variables. The output of interest is $y(s)$. We recognize that the manipulated variable is produced by the actuator (control element),

$$m(s) = g_f(s)c(s) \tag{10.13}$$

However, the control signal is expressed as

$$c(s) = g_c(s)e(s) \tag{10.14}$$

and the error signal can be calculated as follows:

$$e(s) = y_{sp}(s) - y_m(s) = y_{sp}(s) - g_m(s)y(s) \tag{10.15}$$

We can replace Eq. (10.15) in Eq. (10.14) and the resulting expression can be then substituted in Eq. (10.13) to obtain

$$m(s) = g_f(s)g_c(s)[y_{sp}(s) - g_m(s)y(s)] \tag{10.16}$$

If we replace the manipulated variable in Eq. (10.12) by Eq. (10.16), and collect the terms, we obtain the following expression:

$$y(s) = \frac{g_p(s)g_f(s)g_c(s)}{1 + g_p(s)g_f(s)g_c(s)g_m(s)}y_{sp}(s) + \frac{g_d(s)}{1 + g_p(s)g_f(s)g_c(s)g_m(s)}d(s) \tag{10.17}$$

We can now define the *closed-loop transfer functions* from Eq. (10.17) as follows:

$$y(s) = G_{sp}(s)y_{sp}(s) + G_d(s)d(s) \tag{10.18}$$

It is easy to see that the transfer function $G_{sp}(s)$ embodies the dynamics of the set-point response, i.e., it explains how the process output will respond when the set-point is changed. On the other hand, $G_d(s)$ indicates how the process output will respond when a disturbance enters the process. These responses can be shaped by the choice of the controller, and they also critically depend on the dynamics of the actuator as well as the sensor.

---

Denominators of the closed-loop transfer functions are the same, indicating that they share the same stability characteristics.

---

In the next section, we focus on the dynamic performance of the closed-loop process in the presence of a PID controller, demonstrating the effect of each PID mode on the closed-loop response in more detail.

## 10.4 CLOSED-LOOP PERFORMANCE

The influence of a PID controller on the closed-loop process behavior can be analyzed in terms of three key response characteristics:

- *Speed of response* We expect that the closed-loop process would respond quicker than its open-loop counterpart. In effect, we often introduce feedback to speed up an otherwise sluggish process.
- *Asymptotic behavior* The process output should reach the desired set-point in an asymptotic manner. In other words, at steady-state, there shall be no difference between the process output and the set-point.
- *Damping behavior* For processes that exhibit an underdamped response, the degree of damping can be adjusted by the choice of the controller parameters. For instance, we should be able to shape the output response to attain an acceptable level of overshoot.

### 10.4.1 PROPORTIONAL MODE

For the case of a P-Controller, the transfer function is

$$g_c(s) = k_c \tag{10.19}$$

To demonstrate the closed-loop performance, let us consider a first-order process,

$$y(s) = \frac{k_p}{\tau_p s + 1} m(s) + \frac{k_d}{\tau_p s + 1} d(s) \tag{10.20}$$

To simplify the analysis, in the remainder of this chapter, let us assume that $g_f = 1$ and $g_m = 1$. For the disturbance rejection problem with $y_{sp} = 0$, the closed-loop transfer function can be determined as

$$y(s) = G_d(s)d(s) = \frac{k_d}{\tau_p s + 1 + k_p k_c} d(s) \tag{10.21}$$

After rearranging, we obtain

$$y(s) = \frac{[k_d/(1 + k_p k_c)]}{[\tau_p/(1 + k_p k_c)]s + 1} d(s) \tag{10.22}$$

We can make the following observations:

1. The closed-loop process is still first-order (i.e., the controller does not affect the order of the original process).
2. The time constant of the closed-loop process is smaller than that of the process without control.
3. The closed-loop gain is smaller than the disturbance (open-loop) gain.

### Example 10.2

Consider the following process transfer function,

$$y(s) = \frac{2}{5s + 1} m(s) + \frac{1}{5s + 1} d(s) \tag{10.23}$$

The closed-loop transfer function for the disturbance rejection problem, with a P controller, becomes

$$y(s) = G_d(s)d(s) = \frac{[1/(1 + 2k_c)]}{[5/(1 + 2k_c)]s + 1} d(s) \tag{10.24}$$

Figure 10.6 displays the output response for a unit-step change in the disturbance for various values of the proportional gain. We can make the following observations:

1. The closed-loop process response is indeed faster in comparison to the open-loop response, and the speed of response increases with the value of the proportional gain.
2. The response shows the characteristic features of a first-order process.
3. For perfect disturbance rejection, we would like the output response to return asymptotically to its set-point value ($y_{sp} = 0$). However, as one can see there is a steady-state error regardless of the value of the proportional gain, although the error appears to be decreasing with increasing $k_c$.

---

*Offset* is defined as the difference between the ultimate (steady-state) value of the process response and its set-point.

---

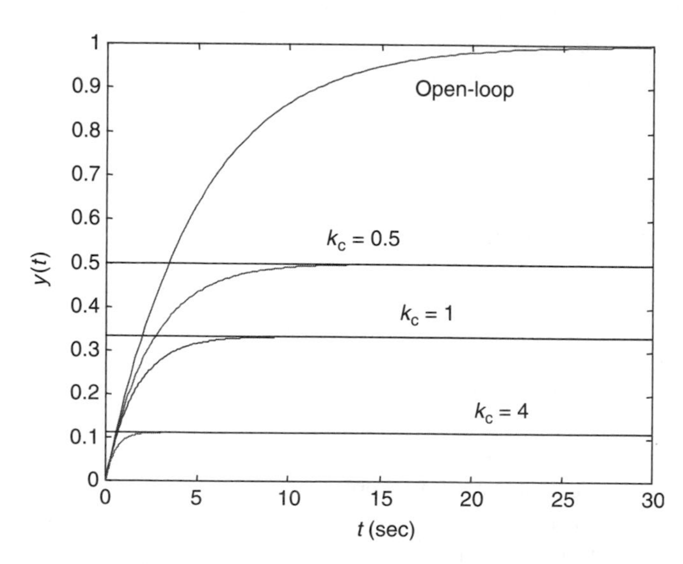


**FIGURE 10.6** Closed-loop response for process in Eq. (10.23) for various $k_c$ values.

The offset for this problem can be calculated analytically, using the Final Value Theorem discussed in Chapter 5:

$$\text{Offset} = \frac{1}{1 + 2k_c} - 0 = \frac{1}{1 + 2k_c} \qquad (10.25)$$

It is clear that there will always be an offset for this closed-loop process, and it will decrease as the proportional gain is increased.

---

A feedback control system under P control exhibits a steady-state error (offset). Is this true for all processes under P control?

---

We note that the offset would be equal to zero only for an *infinite* controller gain, which is clearly infeasible. Furthermore, using a high proportional gain to diminish the offset may not be advisable for high order processes where stability may be adversely affected.

### 10.4.2 Integral Mode

Focusing only on the integral mode of the PID controller, the controller transfer function becomes

$$g_c(s) = \frac{k_c}{\tau_I s} \qquad (10.26)$$

We shall consider the same process model in Eq. (10.20), but, this time, study the set-point tracking problem, assuming that there are no disturbances ($d = 0$). The closed-loop transfer function can be determined as

$$y(s) = G_{sp}(s)y_{sp}(s) = \frac{1}{[(\tau_I \tau_p)/(k_c k_p)]s^2 + [\tau_I/(k_c k_p)]s + 1} y_{sp}(s) \qquad (10.27)$$

For a first-order process under PID control using integral mode only, we can make the following immediate observations:

- The order of the closed-loop process is increased by one, compared to the open-loop process.
- The integral time affects both the closed-loop time constant (speed of response) and the damping behavior.

**Example 10.3**

We will again study the process given by Eq. (10.23) but this time, assume no disturbances, $d = 0$:

$$y(s) = \frac{2}{5s + 1} m(s) \qquad (10.28)$$

The closed-loop transfer function for the set-point tracking problem is expressed as

$$y(s) = \frac{1}{(5\tau_I/2k_c)s^2 + (\tau_I/2k_c)s + 1} y_{sp}(s) \tag{10.29}$$

One can immediately notice that the effect of the proportional gain is simply to increase the speed of response, but it would also affect the damping. For this example, let us assume that $k_c = 2.5$ and focus on the effect of the integral time only. The time constant and the damping factor can now be calculated from the standard form (Eq. [8.10]) as,

$$\begin{aligned} \tau &= \sqrt{\tau_I} \\ \xi &= \frac{1}{2}\sqrt{\frac{\tau_I}{25}} = \frac{1}{10}\sqrt{\tau_I} \end{aligned} \tag{10.30}$$

Figure 10.7 illustrates the output response to a unit-step change in the set-point for various values of $\tau_I$. We can make the following observations:

- The closed-loop process responds slower as the value of the integral time increases.
- The response shows the characteristic features of an underdamped second-order process.
- As the integral time is decreased, the rise time decreases and the overshoot increases.
- For all values of the integral time, the output response asymptotically reaches the new set-point, without any offset.

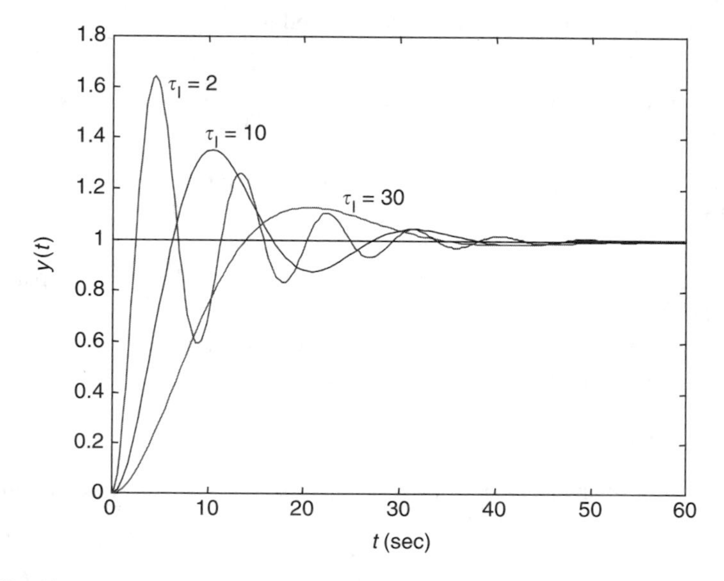


**FIGURE 10.7** Closed-loop response for process in Eq. (10.25) for various $\tau_I$ values.

The offset for this problem can be calculated analytically also, using the Final Value Theorem discussed in Chapter 5.

$$\text{Offset} = 1 - 1 = 0 \tag{10.31}$$

We note that the offset is zero regardless of the values of the controller parameters. The mere presence of the integral mode eliminates the offset for this process.

With the integral mode, the speed of response is increased at the expense of a more underdamped (larger overshoot) output response. This is a key *trade-off* in determining closed-loop performance specifications.

### 10.4.3 Derivative Mode

With only the derivative mode of the PID controller, the transfer function becomes

$$g_c(s) = k_c \tau_D s \tag{10.32}$$

We shall consider the same process model in Eq. (10.20), and study the disturbance rejection problem with $y_{sp} = 0$. The closed-loop transfer function becomes

$$y(s) = G_d(s)d(s) = \frac{k_d}{(\tau_p + k_p k_c \tau_D)s + 1} d(s) \tag{10.33}$$

The derivative mode is never used by itself, thus not many meaningful conclusions can be drawn from the study of Eq. (10.33). However, the following observations are generalizable and will be illustrated in the next example.

- The order of the closed-loop process remains the same as the open-loop process.
- The derivative time effectively slows down the closed-loop process compared to the open-loop process.

#### Example 10.4

A combination of all three modes of the PID controller is used in most applications. In this example, we will use the PID controller with the process transfer function given in Eq. (10.23), and study the disturbance rejection problem. The PID controller is given as

$$g_c(s) = k_c \left( 1 + \frac{1}{\tau_I s} + \tau_D s \right) \tag{10.34}$$

We shall assume that $k_c = 2$ and $\tau_I = 5$. The closed-loop transfer function becomes,

$$y(s) = \frac{1.25s}{[(25 + 20\tau_D)/4]s^2 + 6.25s + 1} d(s) \tag{10.35}$$

One can also calculate

$$\tau = \sqrt{\frac{25 + 20\tau_D}{4}}$$
$$\xi = \frac{6.25}{\sqrt{25 + 20\tau_D}} \qquad (10.36)$$

Figure 10.8 displays the output response for a unit-step change in the disturbance for various values of the derivative time.

We can clearly observe that the derivative time increases the settling time of the closed-loop process and it slows down its speed of response. Another observation for this problem is that the damping is also reduced, creating more oscillatory responses as the derivative time is increased. The closed-loop responses exhibit no offset due to the presence of integral action.

### 10.4.4 Integral Windup

When a controller has integral action, a persistent error will cause the integral term to increase (or decrease) to a value of large magnitude, thereby saturating the control valve action either at fully open or fully closed. In essence, the control is lost and cannot be recovered unless special provisions are built into the control system. This situation is referred to as *integral windup*, and the integral term does not diminish until the sign of the error changes. Many commercial controllers limit the value of the integral term and modify the integral action when this situation is encountered. This can be accomplished by temporarily decreasing the integral time constant as the magnitude of the error decreases but before

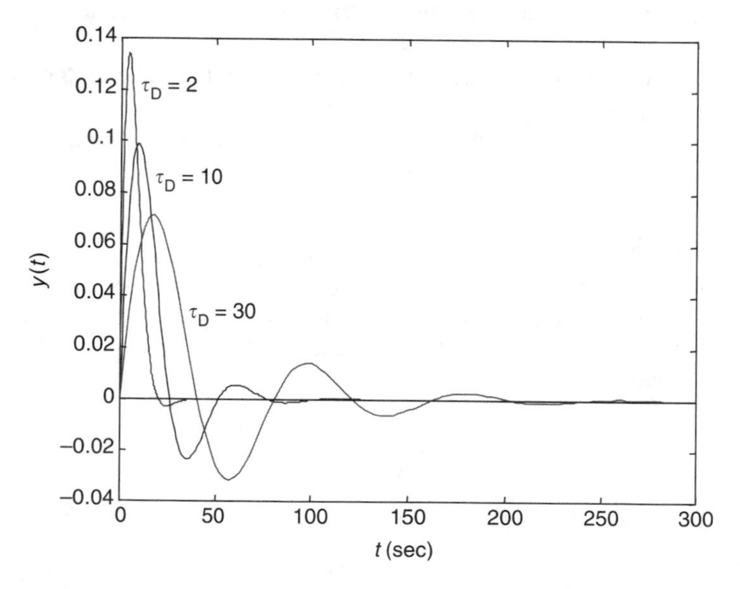


**FIGURE 10.8** Closed-loop response for process in Eq. (10.35) for various $\tau_D$ values.

the error changes sign. These provisions are called antiwindup schemes and are generally available on most industrial controllers.

We note that the velocity form of the PID controller (Eq. [10.11]) is inherently protected against integral windup.

## 10.5 SUMMARY

In this chapter, the concept of feedback control is introduced. With the aid of practical examples, such as temperature control of a heat exchanger, we illustrated the mechanism of feedback control. This was followed by a discussion on the types of conventional feedback controllers. These included the P, PI and PID controllers. The dynamics of these controllers in the closed-loop system were also studied, demonstrating the effect of each mode on the performance.

The closed-loop equations, along with the definitions of the control mechanisms, have now established the basic framework for the analysis of closed-loop systems that will be presented in the next chapter.

## CONTINUING PROBLEM

For the blending process, we shall determine the closed-loop transfer functions between the output and the set-point as well as the output and the disturbance. The following specifications are given:

- The manipulated variable is the flow rate of stream 2, $F_2$, to control outlet composition, $x$.
- The disturbance is the flow rate of stream 1, $F_1$.
- The feed compositions are assumed to be constant.
- There is a proportional controller, $g_c(s) = k_c$.
- The dynamics of the actuators and the sensors will be accounted for by pure dead-time elements, modifying the model transfer functions (Eq. [5.69]) as follows:

$$\bar{h}(s) = \frac{e^{-0.1s}}{5s+1}\bar{F}_1(s) + \frac{e^{-0.1s}}{5s+1}\bar{F}_2(s)$$

$$\bar{x}(s) = \frac{-0.1e^{-s}}{2.5s+1}\bar{F}_1(s) + \frac{0.1e^{-s}}{2.5s+1}\bar{F}_2(s) + \frac{0.5e^{-s}}{2.5s+1}\bar{x}_1(s) + \frac{0.5e^{-s}}{2.5s+1}\bar{x}_2(s)$$

Perform closed-loop simulations for unit-step changes in the disturbance and the set-point, with different values of the proportional gain.

### SOLUTION

The general closed-loop equation is given by Eq. (10.17). For the blending process, using the transfer functions as defined above, we have the set-point

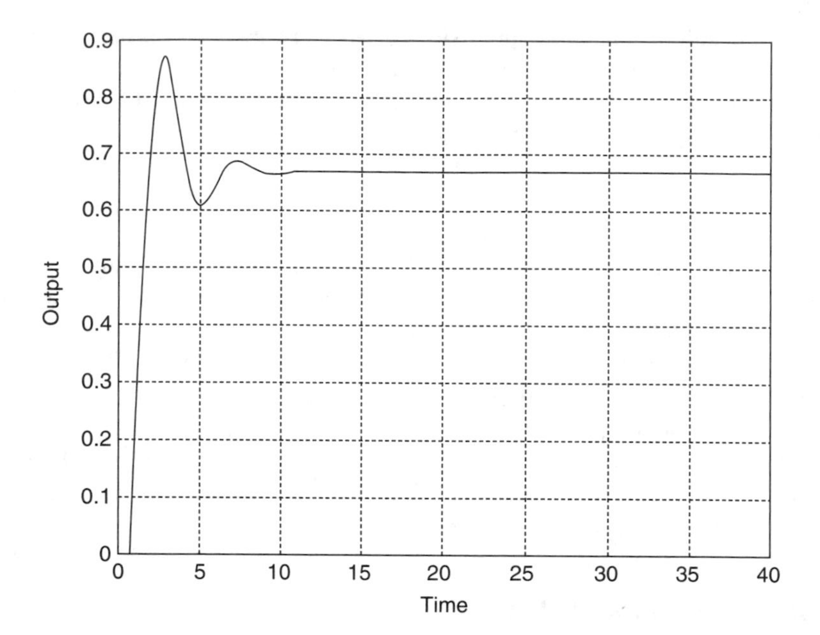


**FIGURE 10.9** Closed-loop response to a change in the set-point for the blending process using a P controller, $k_c = 20$.

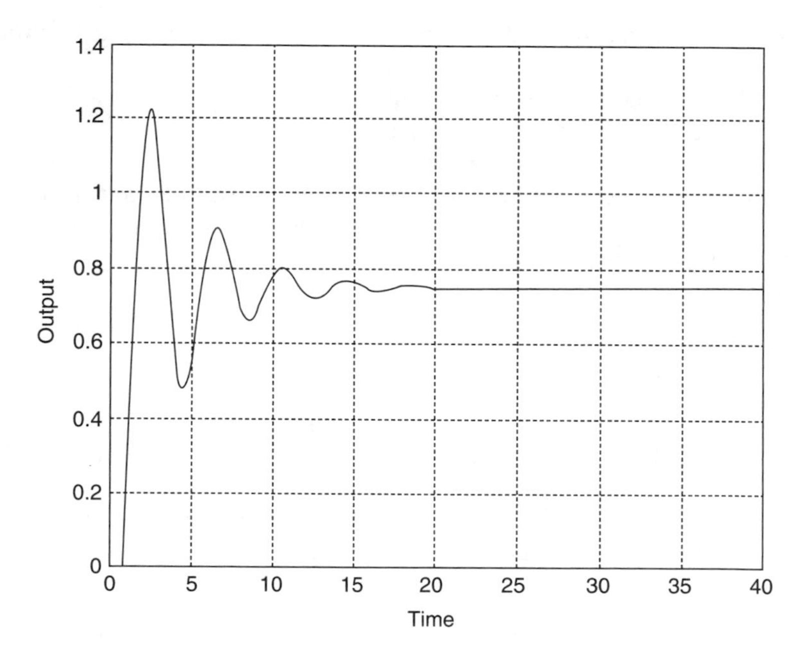


**FIGURE 10.10** Closed-loop response to a change in the set-point for the blending process using a P controller, $k_c = 30$.

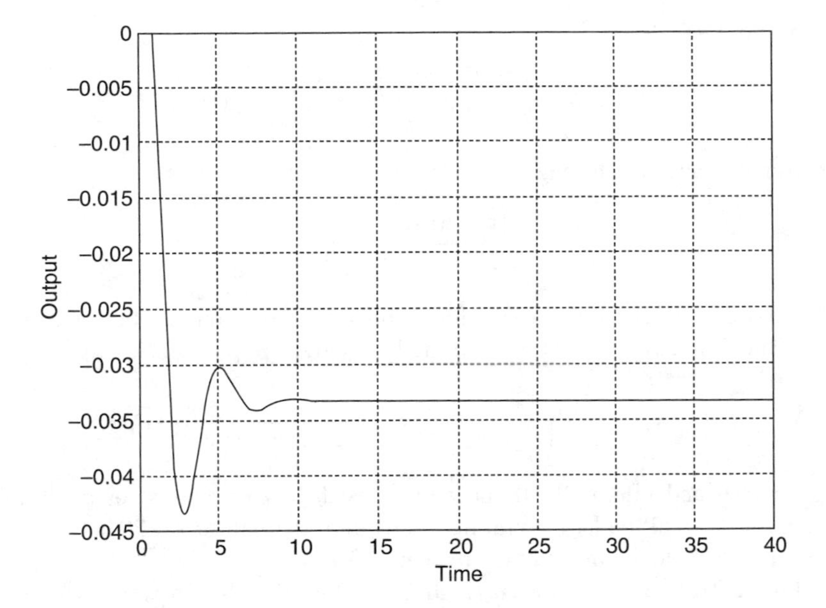


**FIGURE 10.11** Closed-loop response to a change in the disturbance for the blending process using a P controller, $k_c = 20$.

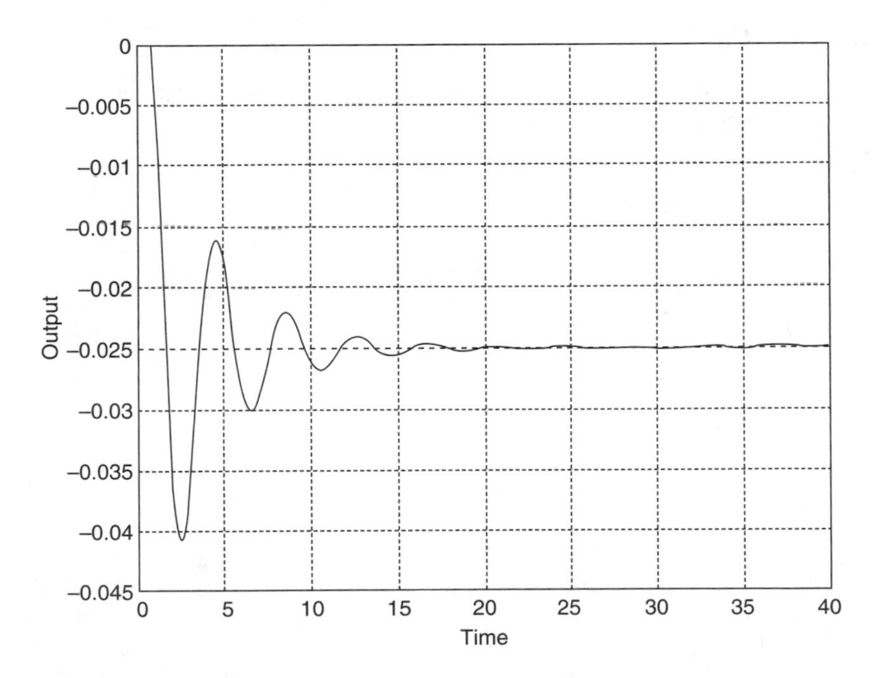


**FIGURE 10.12** Closed-loop response to a change in the disturbance for the blending process using a P controller, $k_c = 30$.

transfer function as

$$\bar{x}(s) = \frac{[0.1e^{-s}/(2.5s + 1)]k_c}{1 + [0.1e^{-s}/(2.5s + 1)]k_c}\bar{x}_{sp}(s)$$

Rearranging and simplifying,

$$\bar{x}(s) = \frac{(0.1e^{-s})k_c}{2.5s + (0.1e^{-s}k_c + 1)}\bar{x}_{sp}(s)$$

The closed-loop transfer function for disturbance rejection can be found similarly. The final closed-loop transfer function is given by

$$\bar{x}(s) = \frac{0.1e^{-s}k_c}{2.5s + 1 + e^{-s}k_c}\bar{x}_{sp}(s) + \frac{-0.1e^{-s}k_c}{2.5s + 1 + e^{-s}k_c}\bar{F}_1(s)$$

Figure 10.9 and Figure 10.10 show the closed-loop responses for a set-point change ($\bar{F}_1(s) = 0$) under P controller with controller gains $k_c = 20$ and $k_c = 30$, respectively. Note the offset in the responses.

Figure 10.11 and Figure 10.12 illustrate the closed-loop responses of the process to a unit-step change in the disturbance ($\bar{x}_{sp}(s) = 0$) with the same settings as given above. Again, the offset is noticeable in the responses.

## REFERENCES

1. Thaler, G.J. and Brown, R.G., *Servomechanism Analysis*, McGraw-Hill, New York, 1953.

# 11 Stability Analysis of Closed-Loop Processes

The basic elements and properties of feedback control systems were introduced in Chapter 10. This chapter focuses on the stability aspects of the closed-loop process and describes the methodologies for stability analysis.

Knowing whether a process is absolutely stable or not is insufficient for most practical purposes. If a process is shown to be stable, we still would like to know how close it is to being unstable. We need to determine the *relative* stability of a system and this is the topic of this chapter. As we learn more about the stability of the closed-loop process, we will also gain information about its anticipated dynamic performance.

## 11.1 CLOSED-LOOP STABILITY

We established in Chapter 7 that the BIBO stability of a process expressed by a transfer function is inferred from the location of its poles. In a similar manner, the stability characteristics of the closed-loop process can also be determined through the poles of its transfer functions $G_{sp}(s)$ and $G_d(s)$. As noted before, these poles are common to both transfer functions which are based on the closed-loop block diagram of Figure 10.4 as given by the solution of the equation,

$$1 + g_p(s)g_f(s)g_c(s)g_m(s) = 0 \tag{11.1}$$

This is the *closed-loop characteristic equation* and includes the controller transfer function. This will allow us to influence the stability and shape the response of the closed-loop process by the choice and tuning of the controller.

---

A closed-loop process is stable if all the roots of its characteristic equation have negative real parts.

---

### Example 11.1

Consider a blending process depicted in Figure 11.1. The process consists of two tanks where a product blending operation takes place. We measure the composition at the exit stream of Tank 2 and manipulate the addition of pure component A to meet the target product composition. The closed-loop block diagram for this process is given in Figure 11.2.

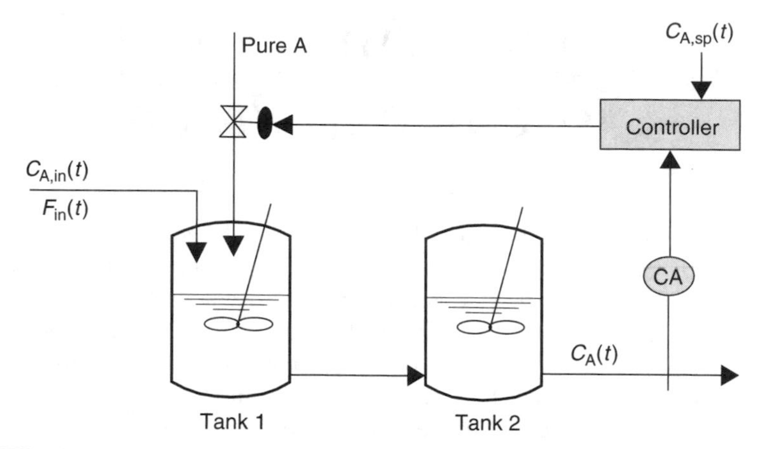


**FIGURE 11.1** Schematic representation of the blending process in Example 11.1.

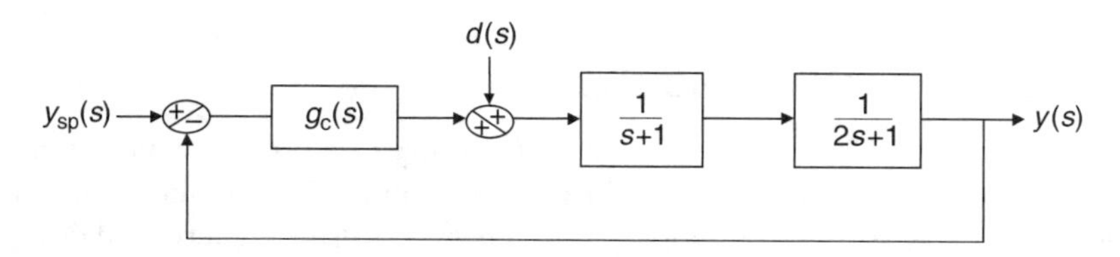


**FIGURE 11.2** Closed-loop block diagram of Example 11.1.

Let us attempt to control the process using a *PI controller*, and as usual for simplicity, we assume that $g_m(s) = g_f(s) = 1$. The characteristic equation of the closed-loop process is

$$p(s) = 1 + k_c\left(1 + \frac{1}{\tau_I s}\right)\left(\frac{1}{s+1}\right)\left(\frac{1}{2s+1}\right) = 0 \tag{11.2}$$

On rearranging we have,

$$\begin{aligned} p(s) &= \tau_I s(s+1)(2s+1) + k_c(\tau_I s + 1) = 0 \\ &= 2\tau_I s^3 + 3\tau_I s^2 + (k_c\tau_I + \tau_I)s + k_c = 0 \end{aligned} \tag{11.3}$$

It can be observed that the location of the poles will critically depend on the values of the controller parameters chosen. Table 11.1 displays three such selections and the resulting poles.

The selection of the controller parameters can result in an unstable closed-loop process that is naturally unacceptable. Other choices may result in asymptotically stable or critically stable closed-loop behaviors.

This example shows the significance of the location of the closed-loop process poles in determining the performance and stability of the controlled system as a function of the controller parameters.

**TABLE 11.1**
**Stability of the Closed-Loop Process of Figure 11.2**

| Parameters | Stability Condition | Poles |
|---|---|---|
| $k_c = 0.1$, $\tau_I = 1.0$ | Asymptotically stable | $p_1 = -1, p_2 = -0.36, p_3 = -0.14$ |
| $k_c = 3$, $\tau_I = 0.5$ | Critically stable | $p_1 = -1.5, p_{2,3} = \pm 1.414j$ |
| $k_c = 2$, $\tau_I = 0.1$ | Unstable | $p_1 = -2.5, p_{2,3} = 0.5 \pm 1.937j$ |

There are several methods for the analysis of the closed-loop poles and the effect of controller parameters on the stability of the closed-loop process. Two often-used methods that were introduced in Chapter 7 are

1. *Routh's Criterion* This method tests if any of the closed-loop poles are located to the right of the imaginary axis, without the calculation of the actual values of the poles.
2. *Root-Locus method* This method plots the roots of the characteristic equation in the complex plane as the gain of the controller ($k_c$) tends from zero to infinity.

Another set of methods originates from the frequency response analysis discussed in Chapter 9:

1. *Bode stability criterion* This criterion establishes *phase* and *gain* margins that quantify the distance of the process to instability.
2. *Nyquist stability criterion* A very versatile tool, Nyquist stability criterion checks closed-loop stability by finding if a critical point is encircled by the Nyquist contour.

## 11.2 ROUTH'S CRITERION

For stability, we need to check if any root of the characteristic equation of the closed-loop process is located in the complex RHP, or, in other words, if any of the closed-loop poles have positive real parts.

Expanding the closed-loop characteristic equation as a polynomial, we obtain

$$1 + g_p g_f g_c g_m = a_0 s^n + a_1 s^{n-1} + \cdots + a_n = 0 \tag{11.4}$$

Once the characteristic equation is expressed as in Eq. (11.4), we can apply the Routh's Criterion as described in Chapter 7 to check the stability of the closed-loop process.

**Example 11.2**

Consider the closed-loop characteristic equation from Example 11.1

$$p(s) = 2\tau_I s^3 + 3\tau_I s^2 + (k_c \tau_I + \tau_I)s + k_c = 0 \tag{11.5}$$

**TABLE 11.2**
**Routh Array for Example 11.2**

| | | |
|---|---|---|
| $s^3$ | $1$ | $0.5(k_c + 1)$ |
| $s^2$ | $1.5$ | $k_c$ |
| $s^1$ | $\dfrac{0.75(k_c + 1) - k_c}{1.5}$ | |
| $s^0$ | $k_c$ | $0$ |

First, we note that all coefficients are positive, so no immediate conclusions can be drawn. We shall analyze the case where the integral time is $\tau_I = 0.5$. This results in the characteristic equation

$$p(s) = s^3 + 1.5s^2 + 0.5(k_c + 1)s + k_c = 0 \tag{11.6}$$

The Routh Array can now be formed as in Table 11.2. All elements in the *first* column are guaranteed to be greater than zero except for the third row entry ($b_1$), which is not obvious, and depends on the value of the controller gain. By rearranging, that entry becomes

$$b_1 = 0.5 - 0.167k_c \tag{11.7}$$

This term can be either positive or negative depending on the value of $k_c$. For example, consider the following cases:

- If $k_c = 5$, then the first column entry becomes
  $b_1 = -0.333 < 0 \Rightarrow$ ***Unstable***
- If $k_c = 1$, then the first column entry becomes
  $b_1 = 0.333 > 0 \Rightarrow$ ***Stable***
- In general, the system will be stable if

$$k_c < 3 \tag{11.8}$$

We note $k_c = 3$ as the critical stability condition.

## 11.3 ROOT-LOCUS METHOD

We have seen that the stability characteristics of a closed-loop system depend on the values of the gain. In the complex plane, this dependency can be represented as a Root-Locus plot.

*Root-Locus plot* traces, in the complex plane, all the roots of the characteristic polynomial (poles of the closed-loop process) as the controller gain varies from zero to infinity.

The construction of the Root-Locus plot and the basis of the method have been discussed in Chapter 7 for a given polynomial. Here, we will apply the method to the closed-loop characteristic polynomial.

In general, the so-called open-loop transfer function is given as $g_{OL} = g_m g_p g_f g_c$, thus the characteristic equation can be expressed as

$$1 + g_{OL} = 0 \tag{11.9}$$

If we also express the open-loop transfer function as a ratio of two polynomials, $g_{OL} = k[N(s)/D(s)]$ with a gain $k$, then the characteristic equation can be expressed as an equation, linear in the parameter $k$,

$$D(s) + kN(s) = 0 \tag{11.10}$$

We recall the following rules that are derived from an asymptotic analysis:

- $k = 0 \quad \Rightarrow$ Root-Locus starts at the open-loop system poles
- $k \to \infty \quad \Rightarrow$ Root-Locus ends at the zeros of the open-loop system

This allows an approximate representation of the Root-Locus for some simple systems. For more complex systems, there are various control softwares (such as MATLAB) that can construct the Root-Locus diagram using the information about the individual transfer functions in the closed-loop as data.

**Example 11.3**

We again consider the control problem introduced in Example 11.1. In this example, three different controllers will be used to construct Root-Locus diagrams.

*P Control*

The open-loop transfer function is given by

$$g_{OL}(s) = \frac{k_c}{(2s + 1)(s + 1)} = \frac{k_c/2}{(s + 0.5)(s + 1)} = \frac{k}{(s + 0.5)(s + 1)} \tag{11.11}$$

The characteristic equation becomes

$$D(s) + kN(s) = (s + 1)(s + 0.5) + k = 0 \tag{11.12}$$

The Root-Locus plot for this system is shown in Figure 11.3.

The roots start from the poles at –1 and –0.5, travel across the real axis until they meet, and then they become a complex conjugate pair. Thus, from the Root-Locus plot, we conclude that the system, under proportional control, is always stable for any value of the controller gain.

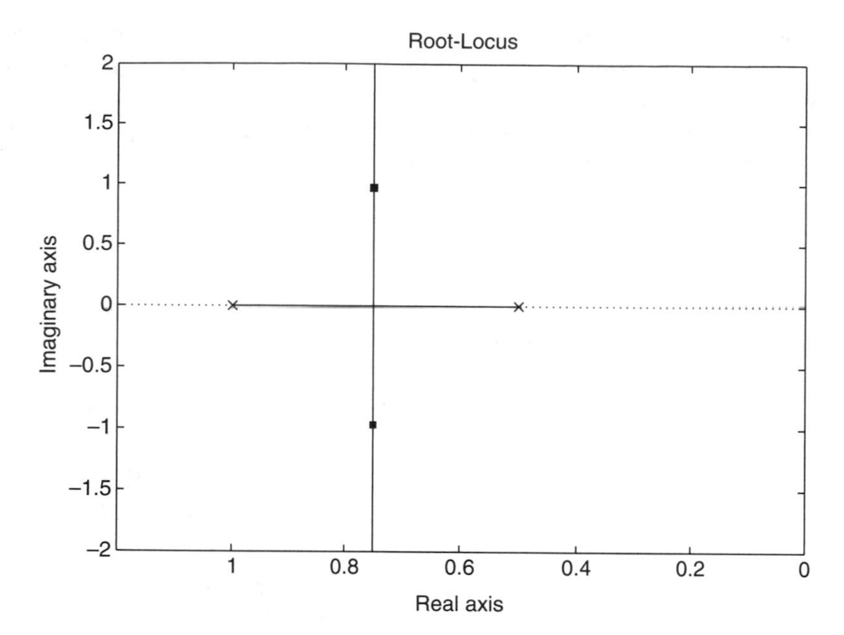


**FIGURE 11.3** The Root-Locus plot of Example 11.3 with P controller.

*PI Control*

In this case, $g_{\mathrm{OL}}(s)$ is given by

$$g_{\mathrm{OL}} = \frac{k_c}{(2s+1)(s+1)}\left(1+\frac{1}{\tau_{\mathrm{I}} s}\right) = \frac{(k_c/2)\left(s+\dfrac{1}{\tau_{\mathrm{I}}}\right)}{s(s+0.5)(s+1)} = \frac{k\left(s+\dfrac{1}{\tau_{\mathrm{I}}}\right)}{s(s+0.5)(s+1)} \qquad (11.13)$$

We can see that the controller now introduces a new pole at the origin, and a zero at $s = -1/\tau_{\mathrm{I}}$. The characteristic equation becomes

$$s(s+0.5)(s+1) + k\left(s+\frac{1}{\tau_{\mathrm{I}}}\right) = 0 \qquad (11.14)$$

The Root-Locus plot for this system is shown in Figure 11.4 for $\tau_{\mathrm{I}} = 0.5$.

We observe that the Root-Locus now moves to the complex RHP for certain values of the proportional gain. We can see that with the PI controller, the system may become unstable if $\tau_{\mathrm{I}} = 0.5$ and $k > 1.5$.

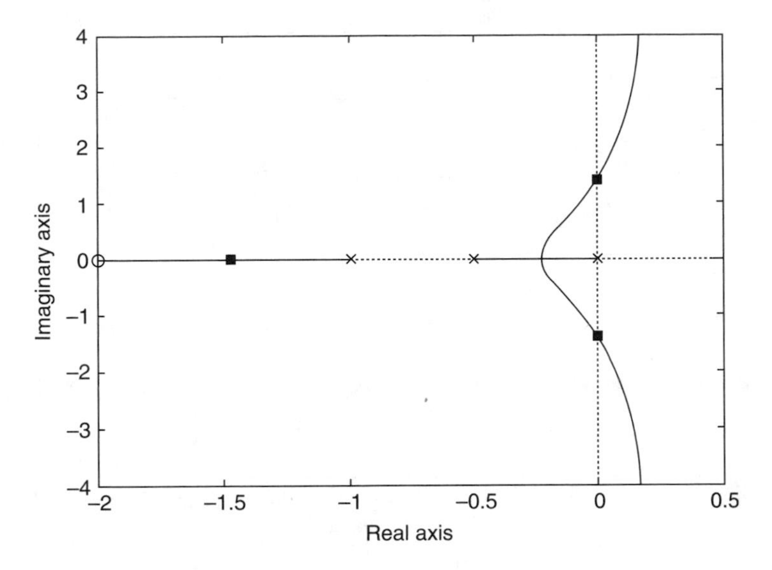


**FIGURE 11.4** The Root-Locus plot with PI controller.

*PID Control*

The transfer function $g_{OL}(s)$ is given by

$$g_{OL} = \frac{k_c}{(2s+1)(s+1)}\left(1+\tau_D s+\frac{1}{\tau_I s}\right)$$
$$= \frac{(k_c/2)(\tau_I\tau_D s^2+\tau_I s+1)/(\tau_I s)}{(s+0.5)(s+1)} = \frac{k(\tau_D s^2+s+1/\tau_I)}{s(s+0.5)(s+1)} \tag{11.15}$$

Again, we have the same definition of the gain as before. We can see that the controller introduces a new pole at the origin and two zeros at $s=-1\pm j$ for $\tau_I = 1.0$ and $\tau_D = 0.5$. The characteristic equation becomes,

$$s(s+0.5)(s+1)+k(\tau_D s^2+s+1/\tau_I)=0 \tag{11.16}$$

The Root-Locus plot for this system is shown in Figure 11.5.

We can see that the system, for these values of the integral and derivative time constants, is always stable for any value of the proportional gain.

*Zeros at Infinity*

On some Root-Locus plots (e.g., Figure 11.4), one can observe that while part of the locus terminates at the open-loop zeros of the system, there may be other parts of the locus that extend to infinity. Indeed, we again assert in these situations that the

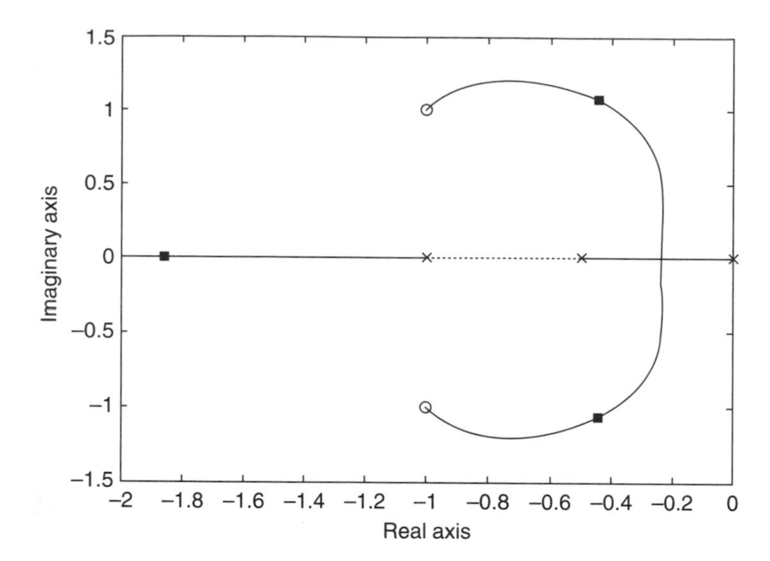


**FIGURE 11.5** Root-Locus plot for PID controller in Example 11.3.

Root-Locus terminates at the open-loop zeros, but qualify that statement by indicating that these are *zeros at infinity*. By definition, a system has as many zeros at infinity as the difference between the number of open-loop poles of $g_{OL}(s)$ and the number of open-loop zeros of $g_{OL}(s)$. Thus, it is easy to see, after inspecting Eq. (11.13) that while one of the poles would be traced to an open-loop zero, the other two poles will approach to zeros at infinity as shown in Figure 11.4.

## 11.4 FREQUENCY RESPONSE METHODS

The Bode plots introduced in Chapter 9 can be used to provide stability information for *minimum phase* processes (processes with no open-loop poles or zeros in the complex RHP). Recall that to construct Bode plots, the transfer functions are evaluated at the stability boundary, $s = j\omega$. Before we proceed, we also define the *crossover frequency* as the frequency at which the phase shift becomes 180°.

To visualize the Bode stability criterion, imagine a sinusoidal wave injected at some point along the feedback loop and traveling along it. The frequency of the wave is so chosen that the wave is shifted by a full period (180°) as it passes through all the transfer functions within the loop. Consider that all other inputs to the closed-loop process are constant and the "open-loop" transfer function is given by

$$g_{OL} = g_m g_p g_f g_c$$

If the gain of $g_{OL}$ is 1, the injected wave will be completely annihilated by the wave coming through the loop. However, if the gain of $g_{OL}$ is greater than 1, the wave will be enhanced by each pass through the loop, becoming unstable, otherwise, the wave will be attenuated, thus stable. The Bode stability criterion can now be stated:

---

The closed-loop process is stable if the *AR* of the corresponding open-loop transfer function is smaller than 1 at the crossover frequency.

---

### Example 11.4

We shall study the open-loop transfer function obtained in Example 11.3 for a PI controller

$$g_{\mathrm{OL}} = \frac{k_c}{(2s+1)(s+1)}\left(1 + \frac{1}{\tau_{\mathrm{I}} s}\right) = \frac{k\left(s + \dfrac{1}{\tau_{\mathrm{I}}}\right)}{s(s+0.5)(s+1)} \tag{11.17}$$

Figure 11.6 displays the Bode plot with the controller parameters, $k = 2$ and $\tau_{\mathrm{I}} = 0.5$. We can observe that at the crossover frequency ($\omega = 0.75$), the amplitude ratio becomes larger than 1, thus indicating instability.

If the Bode plot is redrawn with a new set of controller parameters, $k = 0.5$, $\tau_{\mathrm{I}} = 0.5$, then we can conclude that the system is stable (Figure 11.7), because at the crossover frequency the *AR* is 0.32.

One of the advantages of Bode plots is that they provide *relative* stability information through the definition of a gain margin and a phase margin.

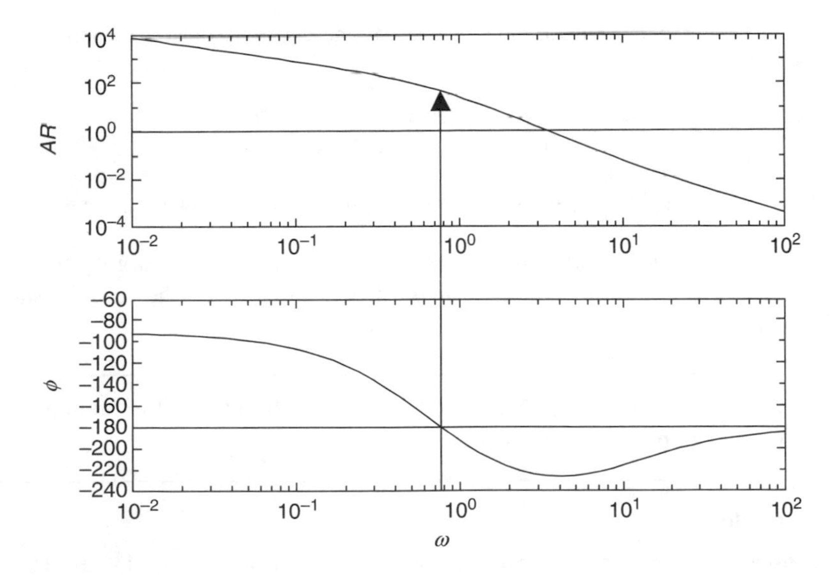


**FIGURE 11.6** The Bode plot for Example 11.4 with the unstable closed-loop process.

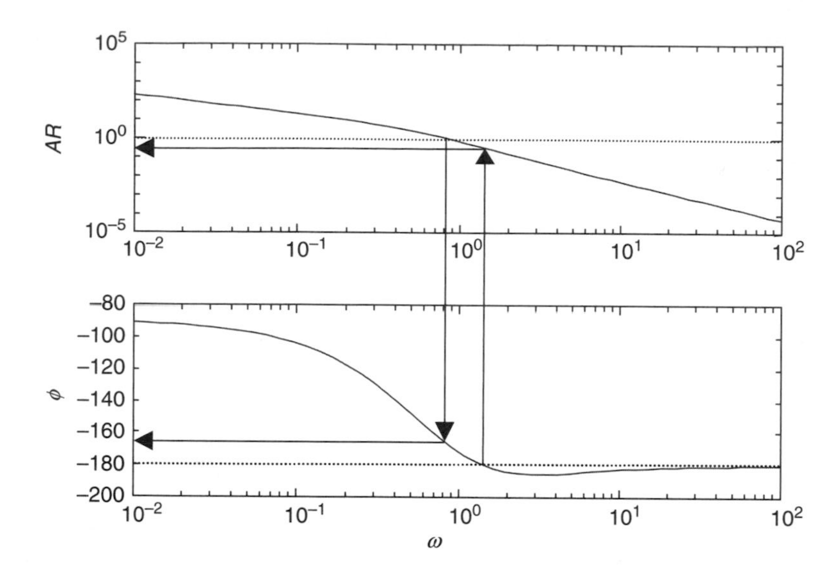


**FIGURE 11.7** The Bode plot for Example 11.4 with the stable closed-loop process.

---

Gain margin is defined as $GM = 1/AR_c$, where $AR_c$ is the $AR$ at the crossover frequency.

Phase margin is defined as $PM = 180° - \phi_c$, where $\phi_c$ is the phase shift corresponding to an $AR$ of 1.

---

Given these definitions, Figure 11.7 also reveals these margins that result in the following:

$$\text{GM} = 1/0.32 = 3.125$$
$$\text{PM} = 180° - 167° = 13°$$

The larger the margins, the farther the closed-loop process would be from instability.

The Nyquist plots defined in Chapter 9 offer a more general stability criterion as it can be applied to *nonminimum phase* processes as well. The Nyquist stability criterion is stated as

---

A closed-loop process is stable if the Nyquist contour of $g_{OL}$ does not encircle the critical point $(-1,0)$. It is unstable, otherwise.

---

**Example 11.5**

Consider the open-loop transfer function studied in Example 11.4 (Eq. [11.13]). The Nyquist plots for the same set of controller parameters are depicted in Figure 11.8.

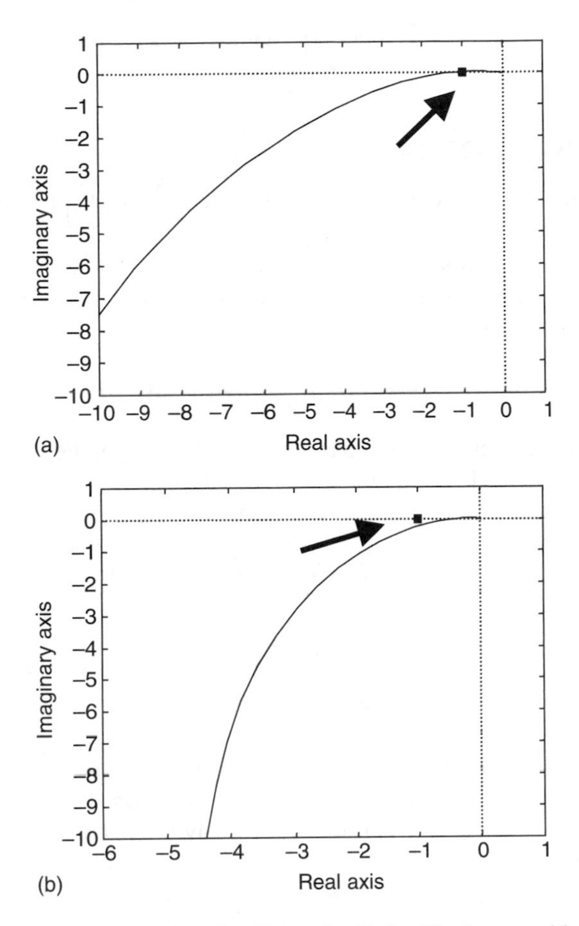


**FIGURE 11.8** The Nyquist plots for Example 11.5 with the unstable (a) and stable (b) closed-loop processes.

It can be seen that Figure 11.8a depicts an encirclement of the critical point, while Figure 11.8b does not. Hence, the stability results of the previous example are confirmed. Figure 11.8b can also reveal the phase and gain margins as before. This is left as an exercise to the reader.

## 11.5 SUMMARY

In this chapter, we have discussed the stability of the closed-loop system. The main governing equation of the dynamics of the closed-loop system was introduced and the closed-loop characteristic equation was defined to explore closed-loop stability through its roots. Two analysis methods, Routh's Criterion and Root-Locus, were revisited to determine the stability characteristics of the closed-loop system. The impact of different control actions on closed-loop stability was investigated. We also introduced two frequency response criteria for stability. Bode and Nyquist plots can be used to evaluate closed-loop stability and also determine how close a system is to instability (or far from instability).

## CONTINUING PROBLEM

For the blending process, with the conditions defined in Chapter 10,

- Write the characteristic equation,
- Analyze the range of controller gains that makes the closed-loop system stable using Root-Locus.

### SOLUTION

The closed-loop transfer function, considering a P Controller, between the set-point and the output variable is given by

$$\bar{x}(s) = \frac{(0.1e^{-s})k_c}{2.5s + (0.1e^{-s}k_c + 1)}\bar{x}_{sp}(s) \tag{11.18}$$

The characteristic equation is expressed as

$$2.5s + 1 + 0.1e^{-s}k_c = 0 \tag{11.19}$$

This is not a rational transfer function because of the exponential term. To use the Root-Locus method, we need to transform this equation to a polynomial form. Using first-order Padé approximation for the delay (Chapter 8), we obtain

$$(2.5s + 1) + 0.1\frac{(1 - 0.5s)}{(1 + 0.5s)}k_c = 0 \tag{11.20}$$

or

$$(2.5s + 1)(1 + 0.5s) + 0.1(1 - 0.5s)k_c = 0$$

This is now a polynomial equation. Expressing this equation in the standard form yields,

$$1.25s^2 + (3 - 0.05k_c)s + (1 + 0.1k_c) = 0 \tag{11.21}$$

For the Routh's Criterion, all coefficients in the characteristic polynomial need to be positive. Since $k_c$ is positive, the condition for stability, then, becomes

$$(3 - 0.05k_c) > 0 \tag{11.22}$$

Or, we can show that

$$k_c < 3/0.05 = 60$$

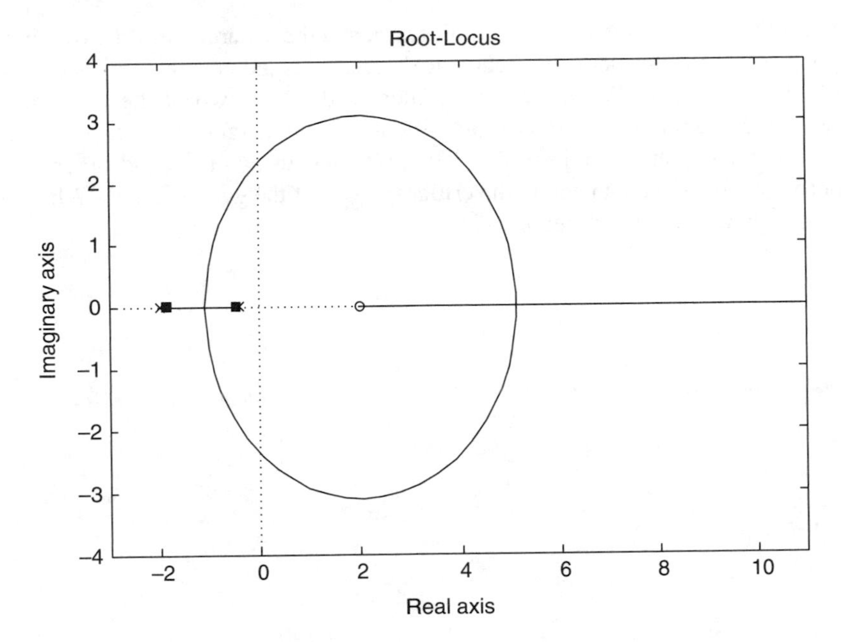


**FIGURE 11.9** Root-Locus for the blending process, squares show the root location when $k_c = 1$.

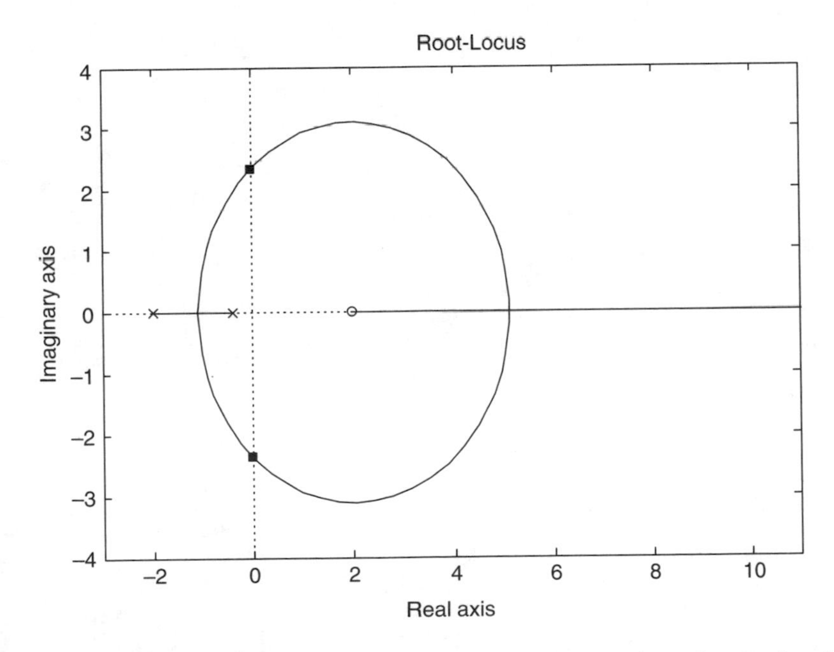


**FIGURE 11.10** Root-Locus for the blending process, squares show the root location when $k_c = 59$.

Figure 11.9 displays the Root-Locus diagram, and the squares show the root locations when the controller gain is equal to 1. The roots indicate that this is a stable closed-loop process. When the gain is increased to 60, two of the roots move toward the imaginary axis, indicating critical stability (Figure 11.10).

The reader should repeat these calculations using a second-order Padé approximation and comment on the critical values of the gain obtained. What can one conclude about the differences?

# 12 Feedback Control Design

This chapter is devoted to stating explicitly the design objectives in terms of the performance criteria, hence developing quantitative measures of closed-loop performance, which allow us to compare alternative controller designs. The design techniques that are based on open- and closed-loop responses are discussed. We elaborate on the practical aspects of feedback controller selection.

## 12.1 DESIGN OBJECTIVES

At some stage during the design of a control system, we face the question of which type of controllers to use (i.e., P, PI, PID) and what the controller parameters should be, i.e., the issue of controller tuning. To answer these questions, we need to establish quantitative measures of closed-loop system behavior (*performance criteria*) that, in turn, will allow us to evaluate alternative designs.

The basic goal of the control system design is to meet desired *performance specifications*. These specifications are essentially the constraints placed on the characteristics of the system response. Generally, they take the form of:

1. *Time-domain specifications* They are expressed in terms of the time response of the closed-loop system.
2. *Frequency-response specifications* They are described in terms of the frequency response of the closed-loop system.

In this chapter, we focus on the time-domain specifications. In general, these specifications deal with important properties of the system dynamics, such as the *speed of response* and the *response accuracy* or allowable tracking error.

---

Stability (or relative stability) is the most important measure of system performance. Since stability, in turn, defines a range of admissible controller parameters, we first concentrate on achieving desired performance within the allowable stability range.

---

In any process, the dynamic response first goes through a transient period before it reaches its steady-state (equilibrium) point (for stable processes, naturally). Consequently, we need to consider response characteristics related to the steady-state behavior as well as characteristics exhibited during the transient.

*Steady-state performance* is expressed by the ability of the process to reach its equilibrium point and is usually specified as zero-tracking error at steady-state. This is naturally related to the presence or lack of *offset* in the closed-loop response as discussed in Chapter 10 when we analyzed the effect of different controllers on closed-loop characteristics.

*Transient performance* is based on important features of the closed-loop transient response to an input disturbance, usually in the form of a unit-step change. The most typical characteristics are overshoot, decay ratio, rise time, settling time, and the period of oscillation (Chapter 8) for typical second-order processes.

The designer of the control system has the freedom to choose from any of the above performance characteristics as a criterion for selecting the controller type and its tuning parameters.

1. Different performance criteria lead to different control designs.
2. It should be emphasized that a single response characteristic does not describe the dynamic response completely and probably multiple criteria have to be adopted.
3. The performance criteria are conflictive in nature, for example, increasing the speed of response will also increase the overshoot.

---

The control design problem is a *trade-off* among conflicting objectives and the designer's task is to balance the conflicting characteristics to ensure a satisfactory outcome.

---

### 12.1.1 Error-Based Criteria

The criteria described so far are based on a few points of the complete dynamic response of the closed-loop system and thus are simple to formulate and observe. There are other criteria that are based on the shape of the complete closed-loop response. They are the so-called *time-integral performance criteria* or *error-based criteria.*

For a closed-loop process we define the error as $e(t) = y_{sp}(t) - y(t)$. The time-integral criteria focus on the quantification of this error in time. The simplest and the most common is the ISE.

$$\text{ISE} = \int_0^\infty e^2(t)\,\mathrm{d}t \tag{12.1}$$

This criterion can be minimized as a function of the controller parameters, i.e., an optimization problem can be formulated to yield the optimal controller parameters. The error can be formulated for either a change in the set-point or a change in the disturbance, thus, possibly resulting in two different sets of optimal controller parameters for set-point tracking and disturbance rejection problems.

Another criterion can be formulated by focusing only on the absolute value of the error, integral of the absolute error (IAE)

$$\text{IAE} = \int_0^\infty |e(t)|\,\mathrm{d}t \tag{12.2}$$

Hence, a more practical criterion is the integral of the time-weighted absolute error (ITAE):

$$\text{ITAE} = \int_0^\infty t\,|e(t)|\,\mathrm{d}t \tag{12.3}$$

This is useful because it gives more importance to errors that persist in time.

### Example 12.1

Consider a first-order system under PI control as shown in Figure 12.1. As before, it is assumed that $g_f = g_m = 1$, and we have $k_p = 0.1$ and $\tau_p = 10$.

A popular performance specification is the quarter-decay criterion that basically sets the decay ratio (Chapter 8) to be 1/4. In this example, let us compare two sets of controller parameters found to satisfy the quarter-decay criterion:

1. $k_c = 10$ and $\tau_I = 0.464$
2. $k_c = 50$ and $\tau_I = 0.258$

Figure 12.2 shows the closed-loop responses. We can clearly see that, although they are based on the same criterion, the overall response is quite different.

Table 12.1 displays the integral-time criteria for these two sets. One can observe that the second set appears to have slightly better performance based on ISE and IAE criteria. However, ITAE criterion indicates better integral performance for the first set.

In the previous example, we considered controller tuning based on set-point tracking. It should be clear that if we design the controller to provide a rapid and smooth response to a set-point change, it might result in a sluggish control for disturbances rejection and *vice versa*. Thus, a *trade-off* is required again in selecting the controller settings that are satisfactory for both problems.

Control tuning is the adjustment procedure of the feedback controller parameters to obtain a desired or specified closed-loop response. As in everything else, the difficulty increases with the number of parameters that must be adjusted. A P controller, for example, has just one parameter to be adjusted, thus simple tuning procedures can be implemented. On the other hand, a PID controller (three-mode

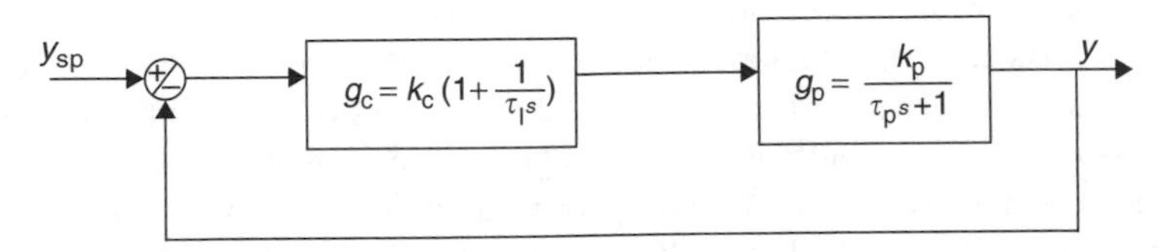


**FIGURE 12.1** Closed-loop process with a PI controller.

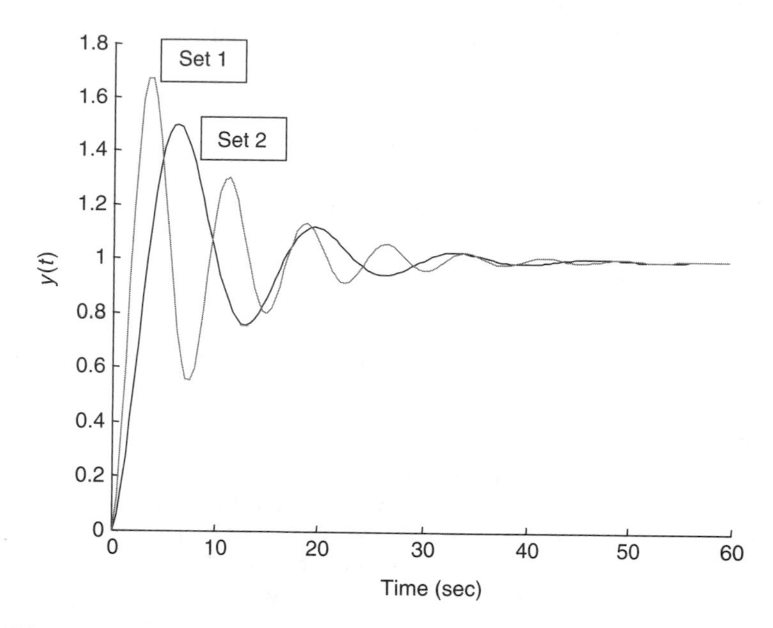


**FIGURE 12.2** Closed-loop response for two sets of quarter-decay designs.

**TABLE 12.1**
**Integral-Time Criteria for Two Sets of PI Controller Tunings**

| Criterion | ISE | IAE | ITAE |
|---|---|---|---|
| Set 1 | 6.355 | 14.752 | 128.83 |
| Set 2 | 6.090 | 14.697 | 130.22 |

controller) requires the tuning of three parameters, thus creating a more complex combinatorial problem. Nevertheless, one can still use some of the ideas discussed in the previous section to pursue the objective.

In Sections 12.2 and 12.3, we will study two tuning methods, the step-testing (or the open-loop tuning) method and the on-line (or the closed-loop) tuning method, respectively.

## 12.2 OPEN-LOOP TUNING (COHEN–COON) METHOD

This is a popular tuning technique, also known as the PRC method.[1]

The first step is to develop a process model that expresses the relationship between the manipulated variable (input) and the process variable (output) response. In open-loop, a step change of known magnitude in manipulated variable $c(t)$ is introduced, and the values of the process variable, $y_m(t)$, are recorded over time.

The curve $y_m(t)$ is called the PRC of the process and depends not only on the process dynamics but also on the dynamics of the final control element as well as the dynamics of the measurement device. For most processing units, the PRC has a *sigmoid shape* as shown in Figure 6.3 and can be fit to a FOPTD model.

$$g_{\mathrm{PRC}}(s) = \frac{y_{\mathrm{m}}(s)}{c(s)} \cong \frac{ke^{-t_{\mathrm{D}}s}}{\tau s + 1} \tag{12.6}$$

The model parameters can be calculated following the procedures outlined in Chapter 6. Cohen and Coon reported design relations, using the approximate model in Eq. (12.6), that are aimed at providing closed-loop responses with a decay ratio of 1/4. These relations are shown in Table 12.2.

The Cohen–Coon method has several advantages and disadvantages.

| **Advantages** | **Disadvantages** |
|---|---|
| Only a single experiment is necessary | The experimental test must be performed under open-loop conditions |
| It does not require a trial-and-error procedure | It may be difficult to determine the slope at the inflection point accurately, especially for noisy processes |
| The controller settings are calculated easily | The method is based on the assumption that the FOPTD model is a good approximation of the real process |

## 12.3 CLOSED-LOOP TUNING (ZIEGLER–NICHOLS) METHOD

Ziegler and Nichols[2] proposed the continuous cycling method many years ago, but it is still among the popular methods for tuning PID controllers. This is a closed-loop tuning methodology also known as the ultimate-gain method since

**TABLE 12.2**
**Cohen–Coon Controller Settings**

| Controller | Settings | Cohen–Coon |
|---|---|---|
| P | $k_{\mathrm{c}}$ | $\frac{1}{k}\frac{\tau}{t_{\mathrm{D}}}\left(1 + \frac{t_{\mathrm{D}}}{3\tau}\right)$ |
| PI | $k_{\mathrm{c}}$ | $\frac{1}{k}\frac{\tau}{t_{\mathrm{D}}}\left(0.9 + \frac{t_{\mathrm{D}}}{12\tau}\right)$ |
| | $\tau_{\mathrm{I}}$ | $t_{\mathrm{D}}\frac{\left(30 + 3\frac{t_{\mathrm{D}}}{\tau}\right)}{9 + 20\frac{t_{\mathrm{D}}}{\tau}}$ |
| PID | $k_{\mathrm{c}}$ | $\frac{1}{k}\frac{\tau}{t_{\mathrm{D}}}\left(\frac{3t_{\mathrm{D}} + 16\tau}{12\tau}\right)$ |
| | $\tau_{\mathrm{I}}$ | $t_{\mathrm{D}}\frac{\left(32 + 6\frac{t_{\mathrm{D}}}{\tau}\right)}{13 + 8\frac{t_{\mathrm{D}}}{\tau}}$ |
| | $\tau_{\mathrm{D}}$ | $\frac{4t_{\mathrm{D}}}{11 + 2\frac{t_{\mathrm{D}}}{\tau}}$ |

the main goal is to obtain the closed-loop system ultimate gain ($k_{cu}$). The concept of ultimate gain is very important in control design and is introduced below.

---

*Ultimate gain* ($k_{cu}$) is the largest value of the controller gain $k_c$ that maintains closed-loop stability when a P controller is used.

---

The first step in the method is to obtain $k_{cu}$ experimentally, which can be done through the following procedure:

***Step 1*:** Eliminate integral and derivative actions on the PID, i.e., keep a P controller.

***Step 2*:** Set $k_c$ at a low value and place the controller in automatic mode.

***Step 3*:** Increase the controller gain $k_c$ by small increments until the closed-loop system response exhibits continuous cycling or sustained oscillations with constant amplitude. The period of the sustained oscillations is referred to as the ultimate period, $p_u$ .

***Step 4*:** The PID controller settings are then calculated from the knowledge of $k_{cu}$ and $p_u$, using Z–N tuning relations given in Table 12.3.

The continuous cycling method offers several significant advantages and disadvantages as explained below.

| Advantages | Disadvantages |
|---|---|
| Closed-loop experiments are used, compared to the open-loop approach of the PRC method | Continuous cycling may be objectionable, since the process is pushed to the stability limit |
| The controller settings are easily calculated | Caution should be exercised for open-loop unstable processes, since they tend to be unstable for small and large gains and only stable for intermediate gains |
| Does not assume a previous knowledge of the model of the process | Some simple processes that can be represented exactly as a first- or second-order without delay do not possess an ultimate gain |

---

It should be emphasized that the settings provided by the techniques described in this chapter have to be regarded as *good* first estimates and subsequent fine-tuning should be considered.

---

**TABLE 12.3**
**Z–N Settings**

| Controller | $k_c$ | $\tau_I$ | $\tau_D$ |
|---|---|---|---|
| P | $0.5\ k_{cu}$ | — | — |
| PI | $0.45\ k_{cu}$ | $p_u/1.2$ | — |
| PID | $0.6\ k_{cu}$ | $p_u/2$ | $p_u/8$ |

### Example 12.2

Consider a process with the transfer functions

$$g_p(s) = \frac{2e^{-s}}{(10s+1)(5s+1)}, \quad g_f(s) = 1, \quad g_m(s) = 1 \tag{12.7}$$

We will use the techniques described above to design PI and PID controllers.

*Cohen–Coon Method*

An experiment is carried out to obtain the open-loop reaction curve given in Figure 12.3. Based on this curve, a FOPTD model is obtained,

$$g_{PRC}(s) \cong \frac{2e^{-4s}}{(13s+1)} \tag{12.8}$$

The comparison of the step response of the approximate model and the actual response of the second-order system is also given in Figure 12.3. Based on the FOPDT approximation, the controller parameters in Table 12.4 are obtained using the Cohen–Coon settings as in Table 12.2.

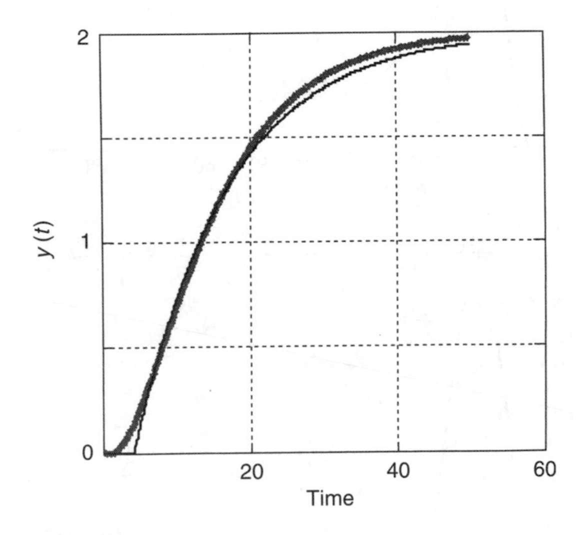


**FIGURE 12.3** PRC for Example 12.2. Thin line represents the model.

**TABLE 12.4**
**Cohen–Coon Settings for Example 12.2**

| Controller | $k_c$ | $\tau_I$ | $\tau_D$ |
|---|---|---|---|
| PI | 1.50 | 8.16 | — |
| PID | 2.29 | 8.75 | 1.37 |

The closed-loop responses for two types of controllers and for both set-point changes and disturbances are shown in Figure 12.4 and Figure 12.5.

*Z–N Method*

In the Z–N method, the first step is to obtain the ultimate gain from experiments on the second-order model. Sustained oscillations were obtained for a value of the controller gain, $k_c = 8$ , using P control only (see Figure 12.6).

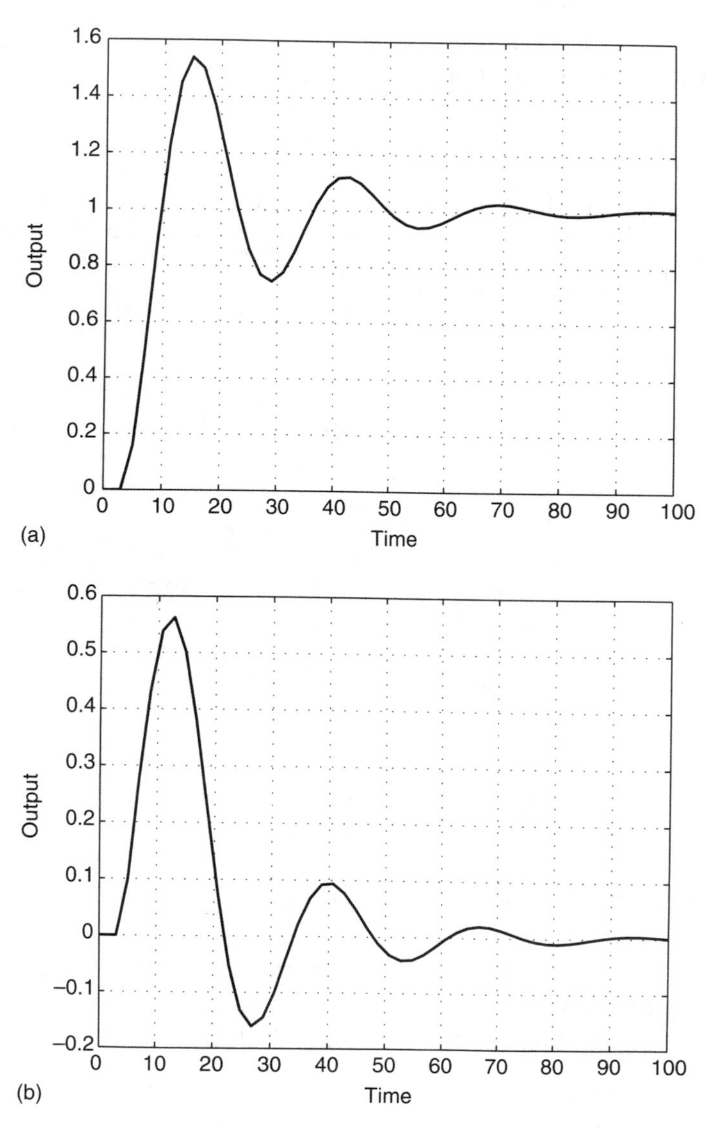


**FIGURE 12.4** Closed-loop responses for PI controller with $k_c = 1.5$ and $\tau_I = 8.16$, (a) Set-point response (b) disturbance response.

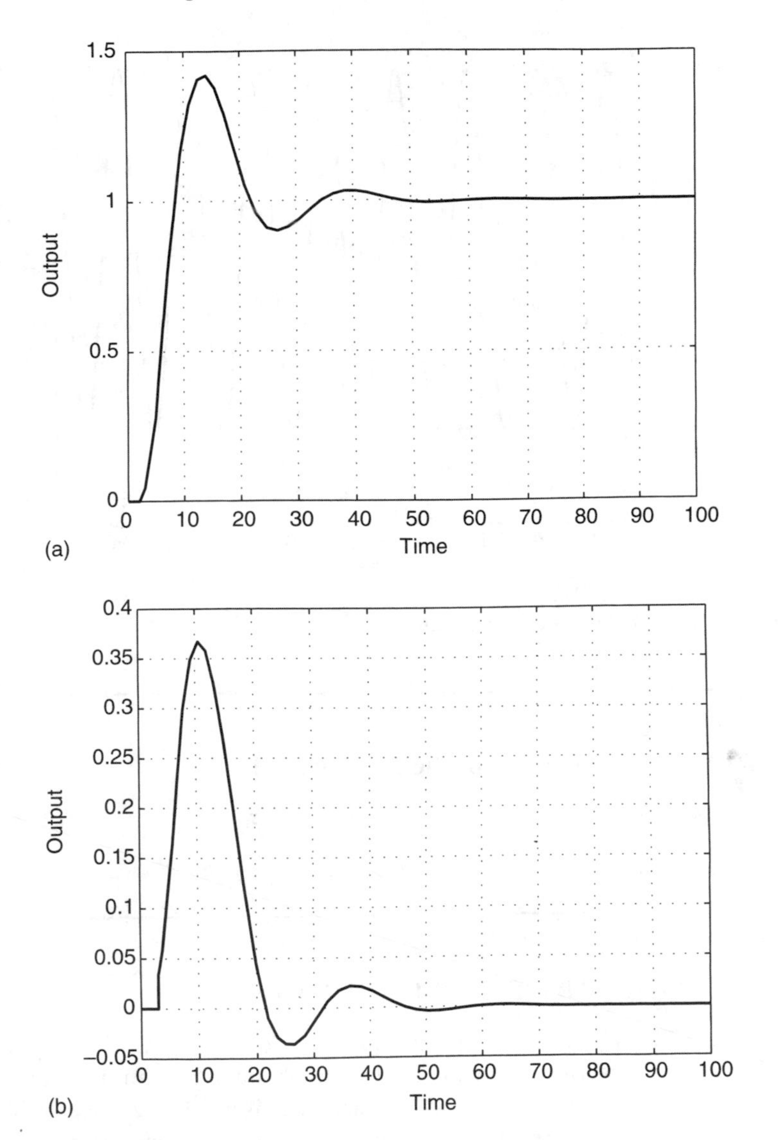


**FIGURE 12.5** Closed-loop responses for PID controller with $k_c = 2.29$, $\tau_I = 8.75$, and $\tau_D = 1.37$, (a) Set-point response (b) disturbance response.

Using the correlations in Table 12.3, the controller parameters given in Table 12.5 are obtained for PI and PID controllers.

Figure 12.7 and Figure 12.8 illustrate the closed-loop responses of each controller for both set-point changes and disturbances. A characteristic of the Z–N technique is that it usually gives rise to overly aggressive control actions. This manifests itself in highly oscillatory closed-loop responses as can also be observed in these figures. Thus, the controller settings obtained by the Z–N method (and often others as well) are typically *detuned* to yield less underdamped responses.

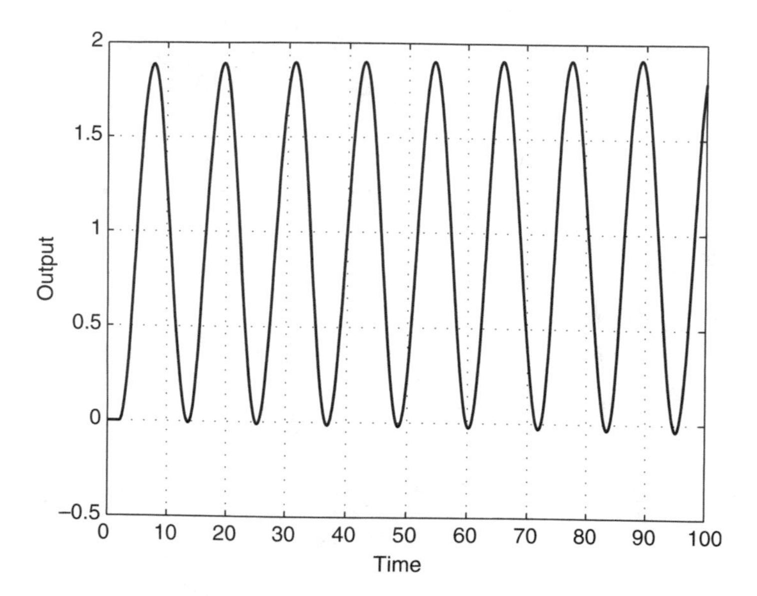


**FIGURE 12.6** Sustained oscillations with P control only.

**TABLE 12.5**
**Parameter Settings for the Z–N Method**

| Controller | $k_c$ | $\tau_I$ | $\tau_D$ |
|---|---|---|---|
| PI | 3.63 | 10 | — |
| PID | 4.70 | 6 | 1.5 |

## 12.4 AUTOTUNING (RELAY) CONTROLLER

In Sections 12.2 and 12.3, two typical approaches, open- and closed-loop tuning techniques, were described allowing the user to obtain a good initial estimate of the controller parameters. The closed-loop approach is appealing since during the procedure the systems remain under closed-loop conditions. However, this approach, as discussed above, may be objectionable, since the process is pushed to the stability limit. Aström and Hägglund[3] developed an alternative approach, which can be used to determine the required parameters in Table 12.3 ($k_{cu}$ and $p_u$) so that a PID controller can be designed from rather simple and limited perturbation experiments.

The basic concept behind the relay controller approach is illustrated in Figure 12.9. The process is operated under a PID controller and a switching strategy allows the user to switch to a relay controller to perform the experiments and then return to the normal PID settings. The relay controller generates a sequence of step changes (a square wave) to perturb the process in closed-loop.

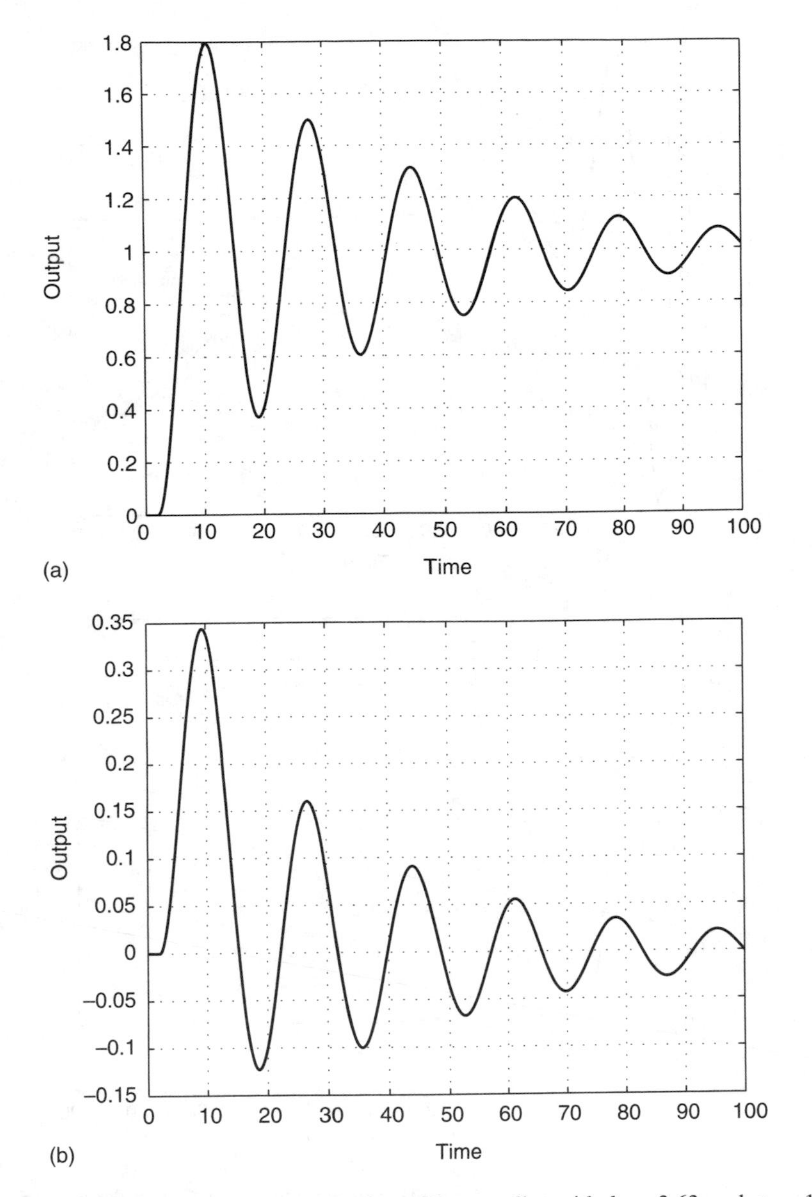


**FIGURE 12.7** Closed-loop responses for a PI controller with $k_c = 3.63$ and $\tau_I = 10$, (a) Set-point response (b) disturbance response.

During the experiment, the process is forced by the relay controller to oscillate with a controlled amplitude $a$ by adjusting the amplitude of the input variation. In general, a single experiment is needed to obtain the basic stability features of the process necessary for the controller tuning. The relay controller has a specified amplitude $d$ and a dead zone to induce oscillations of the process output.

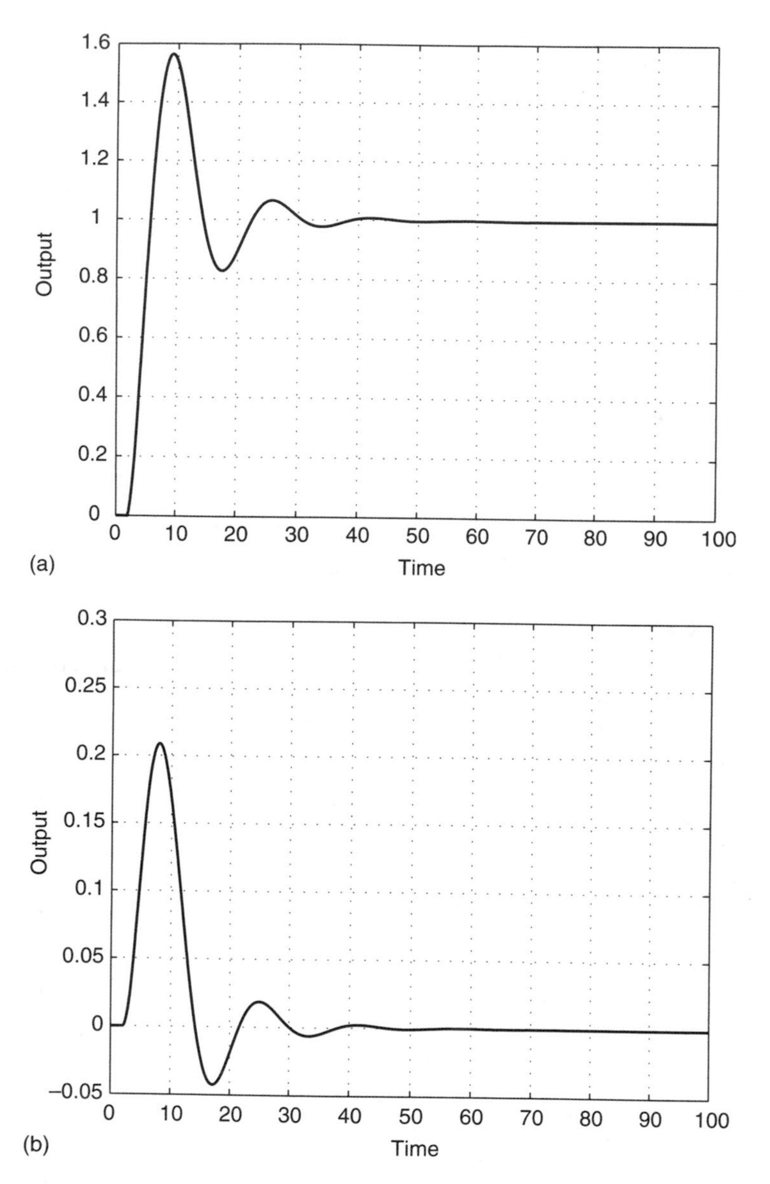


**FIGURE 12.8** Closed-loop responses for PID controller with $k_c = 4.7$, $\tau_I = 6$, and $\tau_D = 1.5$, (a) Set-point response (b) disturbance response.

Following Aström and Hägglund,[3] a good approximation of the ultimate proportional gain ($k_{cu}$) is given by

$$k_{cu} \approx \frac{4\,d}{\pi\,a} \tag{12.9}$$

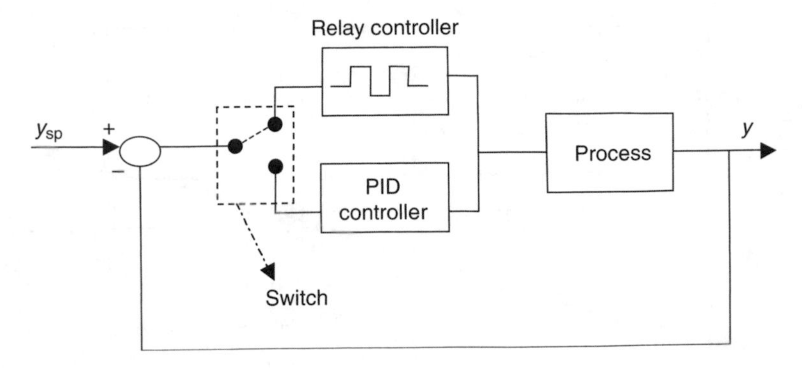


**FIGURE 12.9** Schematic representation of feedback with relay controller.

The ultimate period $p_u$ is given by the period of the process output oscillation. The advantages of the approach can be summarized as:

- Output amplitude can be controlled by selecting $d$
- Procedure is safe for stable systems
- The square wave input is considered to be *near-optimal* for model determination[3]

Most modern computer control systems have built-in autotuning functionalities, which are based along the lines of the structure described above.

### Example 12.3

In this example, we use the built-in capabilities of HYSYS simulation software to design a PI controller to operate a distillation column to recover propylene glycol. The model of the process is available in the HYSYS environment and a copy of the model is provided in the book website.

The controller to be tuned (PI) is used to control the column tray temperature (tray 9) by manipulating the reboiler heat duty and is already implemented in the model. The steps performed within HYSYS are as follows:

1. Run the model until the process steady-state is reached
2. Switch to autotuner option
3. Start autotuner and stop after the process is back to normal
4. From the strip chart provided calculate $k_{cu}$ and $p_u$
5. Calculate the PI parameters using Table 12.3

Figure 12.10 illustrates the results of implementing the autotuning strategy in HYSYS. The control actions forced by the relay controller (% of valve opening) and the resulting oscillating output (tray 9 temperature) are shown in the figure.

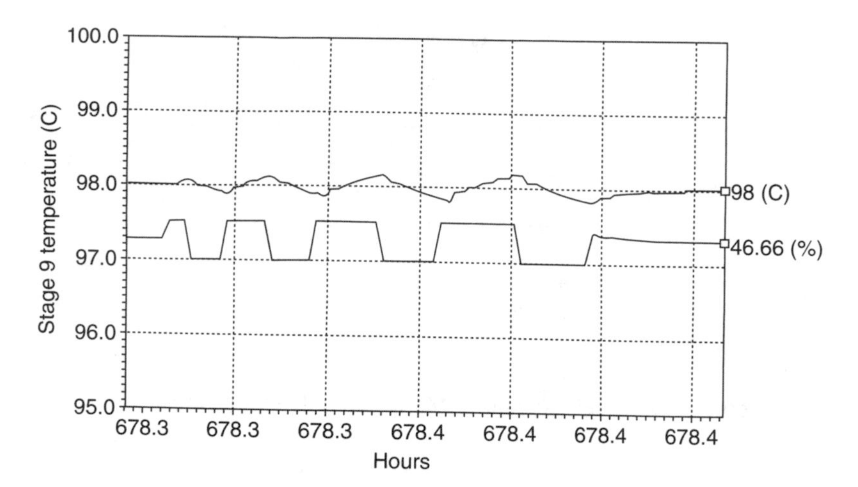


**FIGURE 12.10** Process response under relay controller.

In Figure 12.10, the values of $d$, $a$, and $p_u$ are observed to be 5, 0.20, and 1.4, respectively. Using Eq. (12.9), the ultimate gain is calculated as

$$k_{cu} \approx \frac{4 \times 5}{\pi 0.16} = 39.8$$

Figure 12.11 and Figure 12.12 illustrate the performance of the plant under the new controller settings, for both set-point and disturbance (feed flow rate) changes. We can observe how tray 9 temperature is responding as well as the control valve opening associated with the reboiler rate.

## 12.5 PRACTICAL ISSUES IN PID DESIGN

As shown previously, during the implementation of a PID controller, there are a number of decisions to be made that involve the choice of PID modes to be used (P, PI, or PID) as well as the main PID parameters $k_c$, $\tau_I$ and $\tau_D$. There are, however, other issues that need to be considered before the controller is implemented on an actual process.

In industrial control systems, control vendors usually assign modes to the algorithm that indicates the source of the set-point and the source of the controller output, i.e., the manipulated variable. These modes are classified as follows:

1. *Manual mode* The operator sets the manipulated variable and the controller does not produce any action.
2. *Automatic mode* The controller generates a control action and defines the manipulated variable, and the operator can change the set-point of the controller.
3. *Cascade mode* The set-point is received from another controller (or a computer) and the controller defines the manipulated variable (see Chapter 13).

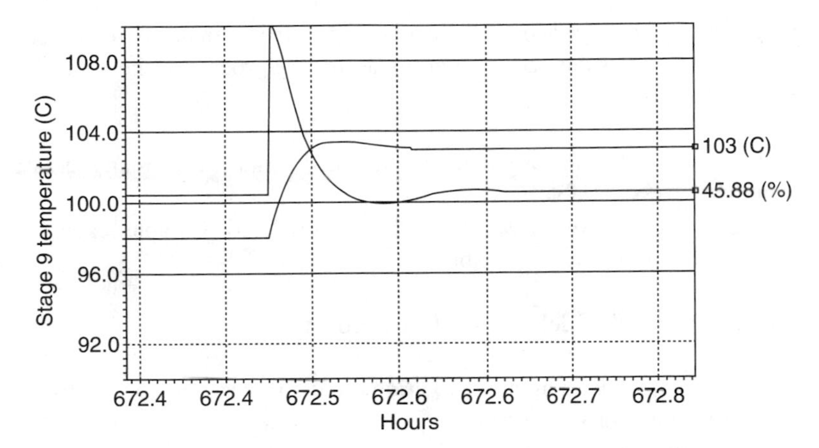


**FIGURE 12.11** Process response for a set-point change under new control settings ($k_c = 19.4$ and $\tau_I = 1.16$ min) using the autotuner.

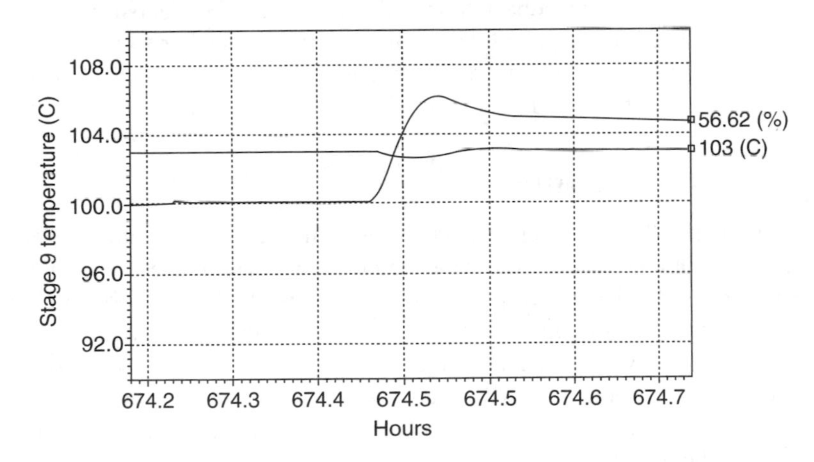


**FIGURE 12.12** Process response for a disturbance under new control setting ($k_c = 19.4$ and $\tau_I = 1.16$ min) using auto-tuner.

4. *Computer mode* The manipulated variable is supplied by a computer where the control law is executed.

In general, these modes indicate who (or what) has control of the set-point and the manipulated variable.

### 12.5.1 Direct and Reverse Action

Direct and reverse actions are related to the sign relationship of the input (manipulated variable) to the output (process variable) and specifically refer to the sign of the gain of either the process or the controller.

When setting the direct or reverse parameter, the user needs to investigate whether the vendor has defined this parameter for the process or the controller. If defined for the process:

1. *Direct acting* An increase in the manipulated variable causes an increase in the process variable.
2. *Reverse acting* An increase in the manipulated variable causes a decrease in the process variable.

On the other hand, if defined for the controller:

1. *Direct acting* The manipulated variable must increase to correct for an increase in the process variable.
2. *Reverse acting* The manipulated variable must decrease to correct for an increase in the process variable.

To match the signs in the feedback loop and to avoid positive feedback, a reverse-acting process needs to be paired with a direct-acting controller and *vice versa*.

The tracking error for a direct-acting process is defined as, $e = y_{sp} - y$, while for a reverse-acting process, it would be given as $e = y - y_{sp}$.

### 12.5.2 Bumpless Transfer

When the controller is switched from Manual mode to Automatic mode, the tracking error often forces the controller to produce a manipulated variable value that may be substantially different from its previous value. This sudden change can damage process equipment or upset downstream processes. With bumpless transfer, when the switch is executed, the new manipulated variable value is either filtered (smoothed) or it is slowly increased (or decreased) to the new value. Often, bumpless transfer is part of the control algorithm supplied by the vendor and may not be disabled.

### 12.5.3 PID Equation Forms

In general, there are several forms of the PID equation for practical purposes, and depending on the vendor, one may or may not have the ability to choose a specific form.

#### 12.5.3.1 Interacting and Noninteracting

This relates to the way the PID equation is expressed mathematically as a function of its parameters. In the noninteracting form, the manipulated variable, $u(t)$, is calculated as:

$$u(t) = k_c e(t) + \frac{1}{\tau_I}\int e(t) + \tau_D \frac{d}{dt} e(t) + \text{Bias} \tag{12.10}$$

On the other hand, in the interacting form, the value of the proportional gain, $k_c$ modifies the integral and derivative terms.

$$u(t) = k_c\left(e(t) + \frac{1}{\tau_I}\int e(t) + \tau_D\frac{d}{dt}e(t)\right) + \text{Bias} \tag{12.11}$$

### 12.5.3.2 Derivative on Error or Derivative on Measurement

This classification depends whether or not the derivative action is calculated based on the error term or on the actual value of the process variable, $y(t)$. The corresponding equations are

$$u(t) = k_c e(t) + \frac{1}{\tau_I}\int e(t) + \tau_D\frac{d}{dt}e(t) + \text{Bias} \tag{12.12}$$

and

$$u(t) = k_c e(t) + \frac{1}{\tau_I}\int e(t) + \tau_D\frac{d}{dt}y(t) + \text{Bias} \tag{12.13}$$

for the noninteracting version of the PID equation. Similar expressions can be derived for the interactive form of the PID equation.

Derivative action on the measurement is actually preferred in most industrial situations, because set-point changes do not cause a large change in the manipulated variable.

### 12.5.3.3 Positional or Velocity Form

In the positional form of the PID algorithm (see Section 10.2), the value of the manipulated variable is calculated and used directly. In the velocity form of the PID, on the other hand, we compute and use the *change* in the manipulated variable.

---

1. Only derivative on error vs. derivative on measurement may be a critical choice for the control engineer.
2. The choice of positional vs. velocity forms of equations has usually been made by the vendor. The particular choice will have an impact on such issues as initialization, bumpless transfer, etc.
3. The choice between interacting vs. non-interacting affects the tuning equations. For example, if the tuning rules calculate the parameters of an interacting PID, one needs to divide $\tau_I$ by $k_c$ and multiply $\tau_D$ by $k_c$ to obtain the noninteracting parameters.
4. If the user needs derivative action, then derivative on measurement is the preferred choice.

---

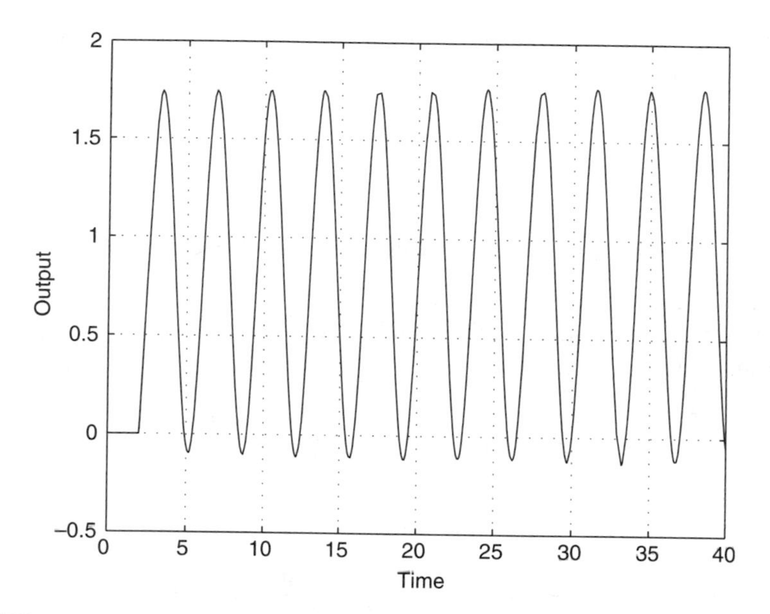


**FIGURE 12.13** Closed-loop response for set-point change with P controller with $k_c = 45$.

**TABLE 12.6**
**Z–N Tuning Parameters for the P, PI and PID Controllers**

| Controller | $k_c$ | $\tau_I$ | $\tau_D$ |
|---|---|---|---|
| P | 22.5 | — | — |
| PI | 20.25 | 3.34 | — |
| PID | 27 | 2 | 0.5 |

## 12.6 SUMMARY

This chapter is devoted to stating the design objectives in terms of the performance criteria and consequently to develop standard measures of closed-loop performance which lead finally to alternative controller design techniques. Specifically, the topics included (1) the definition of the design objectives in terms of a control problem formulation, (2) different performance criteria to develop quantitative measures of the closed-loop performance allowing us to compare alternative controller designs, (3) and design techniques based on the open-loop response, and the closed-loop response of the system. Finally, some practical issues regarding PID controller implementation in an industrial environment are discussed.

## CONTINUING PROBLEM

For the blending process, and with the conditions defined in Chapter 10

1. Design a P, PI, and a PID controller using the Z–N method.
2. Compare and discuss the performance of the resulting controllers.

### SOLUTION

The closed-loop tuning method (Z–N) is used to tune controllers, which could then be implemented in the blending process. The ultimate gain, $k_{cu}$, and the ultimate period, $p_u$, are found to be 45 and 4 respectively (see Figure 12.13). The controller settings are then calculated based on $k_{cu}$ and $p_u$ using the Z–N tuning relations. The tuning parameters are shown in Table 12.6.

Figure 12.14 and Figure 12.15 show the closed-loop response for a PI and a PID controller, respectively, for a set-point change. Inserting the values of $k_c$, $\tau_I$, and $\tau_D$ in the controller field in the APC-Tool interface and then introducing a step change produced these responses. The control action is typical of PI and PID controllers and is shown to achieve the desired set-point and stabilize the system within approximately 15 and 10 seconds, respectively.

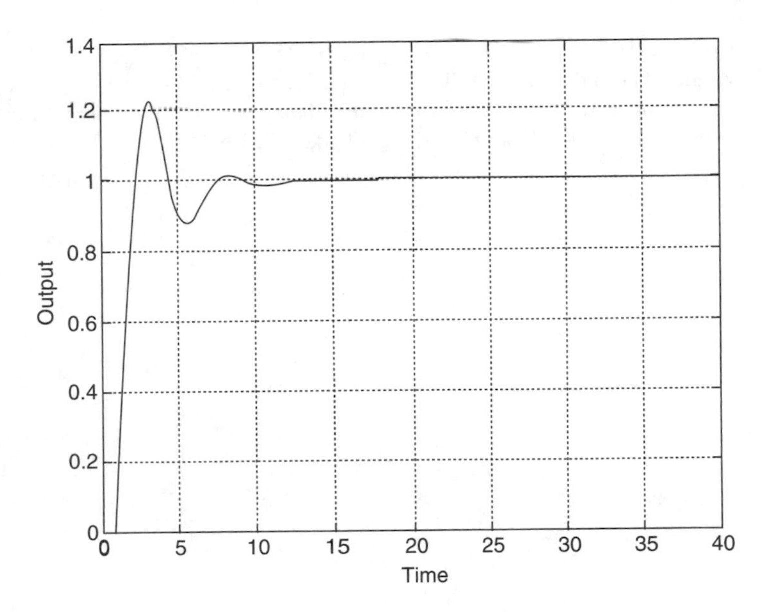


**FIGURE 12.14** Closed-loop response for set-point change with a PI controller with $k_c = 20.25$ and $\tau_I = 3.34$.

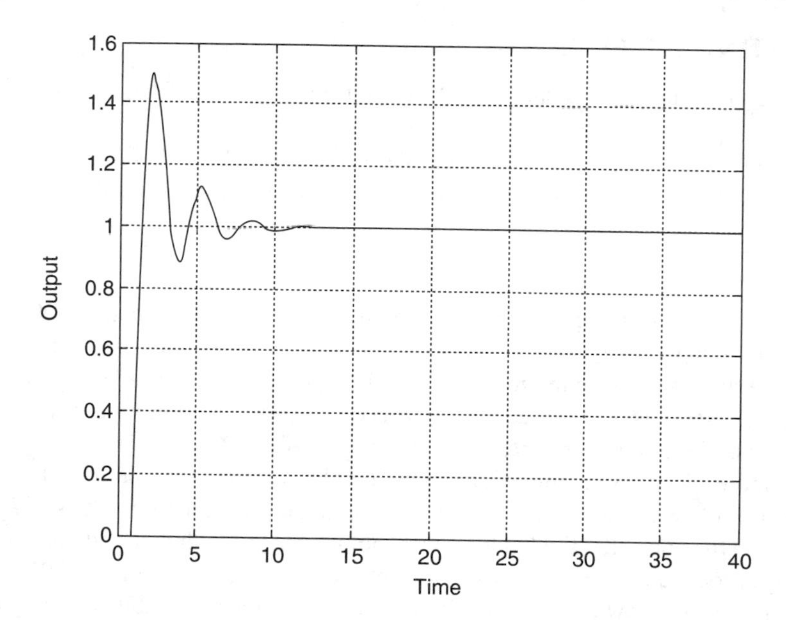


**FIGURE 12.15** Closed-loop response for set-point change with a PID controller with $k_c = 27$, $\tau_I = 2$, and $\tau_D = 0.5$.

## REFERENCES

1. Cohen, G.H. and Coon, G.A., *Trans. ASME*, 75, 827, 1953.
2. Ziegler, J.G. and Nichols, N.B., *Trans. ASME*, 64, 759, 1942.
3. Aström, K. and Hägglund, T., *Automatic Tuning of PID Controllers*, Instrument Society of America, Research Triangle Park, NC, 1998.

# Section IV Additional Reading

The early applications of feedback control systems in the chemical industry have been discussed by a number of texts that focus on the fundamentals of PID control as well as how the advent of digital computers impacted the developing technologies:

Buckley, P.S., *Techniques of Process Control*, Wiley, New York, 1964.
Eckman, D.P., *Principles of Industrial Process Control*, Wiley, New York, 1945.
Eckman, D.P., *Automatic Process Control*, Wiley, New York, 1958.
Smith, C.L., *Digital Computer Process Control*, Intext Educational Publishers, Scranton, PA, 1972.

In the late 1970s and early 1980s, the control of distillation columns presented itself as a major challenge in the chemical and petrochemical industries due to potential savings in energy costs that can be derived from better control:

Buckley, P.S., Luyben, W.L., and Shunta, J.P., *Design of Distillation Column Control Systems*, Instrument Society of America, Research Triangle Park, NC, 1985.
Shinskey, F.G., *Distillation Control: For Productivity and Energy Conservation*, McGraw-Hill, New York, 1977.

The relationship between modeling processes from plant data and its implications for designing PID controllers is explored in the following book.

Wang, L. and Cluett, W.R., *From Plant Data to Process Control: Ideas for Process Identification and PID Design*, Taylor & Francis, London, 2000.

There are a number of reference books that provide a detailed study of the hardware elements associated with feedback control systems.

*Considine's Process/Industrial Instruments and Controls Handbook*, McMillan, G.K.(editor-in-chief), Considine, D.M.(late editor-in-chief), 5th ed., McGraw-Hill, New York, 1999.
DeSá, D.O.J., *Instrumentation Fundamentals for Process Control*, Taylor & Francis, New York, 2001.
Fitzgerald, B., *Control Valves for the Chemical Process Industries*, McGraw-Hill, New York, 1995.
Fraser, R.F., *Process Measurement and Control: Introduction to Sensors, Communication, Adjustment, and Control*, Prentice-Hall, Upper Saddle River, NJ, 2001.
Johnson, C.D., *Process Control Instrumentation Technology*, Wiley, New York, 1988.
Rijnsdorp, J.E., *Integrated Process Control and Automation*, Elsevier, Amsterdam, 1991.

A more in-depth study of practical controller tuning techniques can be found in the books published by the Instrument Society of America (ISA):

Aström, K.J. and Hägglund, T., *Automatic Tuning of PID Controllers*, Instrument Society of America, Research Triangle Park, NC, 1988.

Corripio, A.B., *Tuning of Industrial Control Systems*, Instrument Society of America, Research Triangle Park, NC, 2001.

McMillan, G.K., *Tuning and Control Loop Performance: A Practitioner's Guide*, Instrument Society of America, Research Triangle Park, NC, 1990.

As discussed in Chapter 12, one of the drawbacks of the Cohen–Coon approach was its open-loop nature. This was addressed in the following article:

Yuwana, M. and Seborg, D.E., A new method for on-line controller tuning, *AIChE J.*, 28, 434, 1982.

Skogestad presents a comprehensive study of PID tuning in which he develops a number of analytic rules that yield a satisfactory closed-loop performance:

Skogestad, S., Simple analytic rules for model reduction and PID controller tuning, *J. Process Control*, 13, 291–309, 2003.

Skogestad, S., Erratum to 'simple analytic rules for model reduction and PID controller tuning', *J. Process Control*, 14, 465, 2004.

A book devoted to relay–feedback tuning of PID controllers is given below:

Yu, C-C., *Autotuning of PID Controllers: Relay Feedback Approach*, Springer, New York, 1999.

There are a number of recent studies that aim to improve on the existing PID tuning techniques and a representative sample is given below.

Cluett, W.R. and Wang, L., New tuning rules for PID control, *Pulp Paper-Canada*, 98, 52–55, 1997.

Lee Y., Park, S., Lee, M., and Brosilow, C., PID controller tuning for desired closed-loop responses for SI/SO systems, *AIChE J.*, 44, 106–115, 1998.

Luyben, W.L., Tuning proportional-integral-derivative controllers for integrator/deadtime processes, *Ind. Eng. Chem. Res.,* 35, 3480–3483, 1996.

Natarajan, K. and Gilbert, A.F., On direct PID controller tuning based on finite number of frequency response data, *ISA Trans.,* 36, 139–149, 1997.

Grassi, E., Tsakalis, K.S., Dash, S., Gaikwad, S.V., MacArthur, W., and Stein, G., Integrated system identification and PID controller tuning by frequency loop-shaping, *IEEE Trans. Control Systems Technol.,* 9, 285–294, 2001.

# Section IV Exercises

**IV.1.** For the bioreactor introduced in Exercise II.7, we will control the biomass concentration ($x_2$) using the dilution rate ($D$).

1. Sketch the feedback control loop.
2. Find the closed-loop transfer function between the process output and the set-point for a P controllers.
3. Find the characteristic equation.

**IV.2.** For the oil stream heated as it passes through two well-mixed tanks in series in Exercise II.9, we will develop a control scheme that maintains the temperature in the second reactor ($T_2$) at its set-point, using the heat input ($Q$).

1. Draw the schematic representation of the feedback control loop including the input disturbances.
2. Find the closed-loop transfer function between the output temperature and the set-point and disturbance (inlet flow) for P and PI controllers.

**IV.3.** A process is described by the following state-space model (Exercise II.12):

$$x = Ax + Bu \qquad A = \begin{bmatrix} -2 & 1 & 0 \\ -3 & 0 & 3 \\ -1 & 0 & -3 \end{bmatrix}, \quad B = \begin{bmatrix} 0 \\ 3 \\ 1 \end{bmatrix}, \quad C = [1 \quad 0 \quad 0]$$
$$y = Cx$$

Assume that the sensor has dynamics represented by

$$g_{\mathrm{m}} = \frac{1}{0.5s + 1}$$

1. Find the characteristic equation for the closed-loop system under a P controller.
2. Analyze closed-loop stability using Routh's Criterion and Root-Locus.

**IV.4.** Consider the process in Exercise II.13, represented by the following state-space model:

$$\begin{bmatrix} \dot{x}_1 \\ \dot{x}_2 \end{bmatrix} = \begin{bmatrix} -2.405 & 0 \\ 0.833 & -2.238 \end{bmatrix} \begin{bmatrix} x_1 \\ x_2 \end{bmatrix} + \begin{bmatrix} 7 \\ -1.117 \end{bmatrix} u$$

$$y = [0 \quad 1] \begin{bmatrix} x_1 \\ x_2 \end{bmatrix}$$

1. Draw the schematic representation of the feedback control loop
2. Find the characteristic equation for the closed-loop system with a P controller
3. Analyze the stability of the system using Root-Locus with P control
4. Analyze the stability of the system with a PI controller with $\tau_I = 0.1$
5. Analyze the stability of the system for different values of $\tau_I$

**IV.5.** For the blending process (see Continuing Problem), and using the transfer functions developed in Chapter 5, we will implement a feedback control scheme to control the level of the tank using $F_1$ as manipulated input. $F_2$, in this case, is a disturbance:

1. Write the characteristic equation.
2. Determine the range of controller gains that makes the closed-loop system stable using Routh's Criterion and Root-Locus.

**IV.6.** For the chemical reactor in Exercise II.14, we are interested in developing a control scheme to control the outlet stream composition by manipulating the feed flow rate ($F_{in}$):

1. Design a PI controller using the Cohen–Coon method.
2. Design a PI controller using the Z–N method.
3. Analyze and compare the closed-loop process response for both set-point and disturbance changes.

**IV.7.** For the shell-and-tube heat exchanger in Exercise II.15, we will implement a feedback control system using a PID controller by controlling the temperature with steam as the manipulated variable.

1. Sketch the closed-loop block diagram
2. Find the characteristic equation
3. Find the ultimate controller gain ($k_{cu}$)
4. Design a PID controller using the Z–N tuning method

**IV.8.** Consider the bioreactor problem discussed in Exercise IV.1:

1. Analyze the stability of the closed-loop system using Root-Locus.
2. Design a PI controller using Z–N tuning rules. (**Hint:** introduce a small delay, compared to the process time constant, in the transfer function.)
3. Analyze the set-point response.

**IV.9.** For the process described in Exercise IV.3,

1. Find the ultimate value of the controller gain ($k_{cu}$)
2. Design a PID controller using Z–N tuning method
3. Analyze the closed-loop behavior

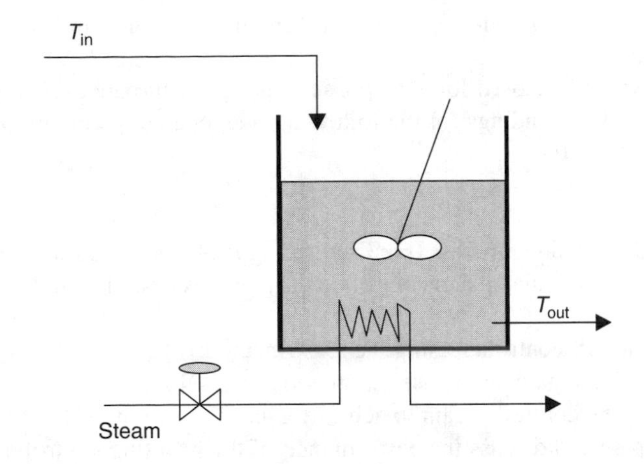


**FIGURE IV. 1** The schematic representation of the heater.

**IV.10.** Consider the continuous-stirred-tank heater given in Figure IV.1. The following information about the process is given:

$$g_p(s) = \frac{1.2e^{-10s}}{(60s + 1)(5s + 1)}, \quad g_d(s) = \frac{0.9e^{-10s}}{60s + 1}$$

where $g_p(s)$ and $g_d(s)$ are the transfer functions between the steam and the inlet temperature and the reactor temperature, respectively. For this process:

1. Sketch the feedback control loop on the process diagram
2. Draw the feedback loop using a block diagram
3. Based on theoretical frequency response equations, design a Z–N controller using:
   - The open-loop method, and find $g_{OL}$ and calculate $k_{cu}$ and $p_u$.
   - The closed-loop method, and find the characteristic equation and use direct substitution method. (**Hint:** Express the delay using the Padé approximation.)

**IV.11.** For the problem defined in Exercise IV.10:

1. Validate the results using the graphical approach.
2. Investigate the performance of the closed-loop system under both set-point and disturbances changes.

**IV.12.** For the heating of the oil stream described in Exercise IV.2 and considering the same operating conditions:

1. Design a PI controller using Z–N tuning rules. (**Hint:** Incorporate a small delay in the transfer function to find the ultimate gain.)

2. Analyze the closed-loop behavior for both set-point and disturbance changes.
3. Analyze the closed-loop response against disturbance changes and explain your findings for the following new operating conditions:
   - $T_{2,s} = 40$
   - $T_{2,s} = 50$

**IV.13.** For the blending process (see Continuing Problem), and following the definitions of the manipulated and controlled variables as described in Exercise IV.5,

1. Design a PI controller using the Z–N method
2. Analyze the gain and phase margins using Z–N settings
3. Adjust the controller gain to achieve a gain margin of 2.5
4. Compare and discuss the performance of the resulting controllers

**IV.14.** For the chemical reactor in Exercise IV.6:

1. Using the frequency analysis in APC-Tool, investigate the gain and phase margins under the settings obtained in Exercise IV.6
2. Obtain new settings to achieve a gain margin equal to 2
3. Compare the closed-loop system responses

**IV.15.** Consider the distillation column model in HYSYS provided on the book website. The column is used to recover propylene oxide from a mixture of several components that come from the reactor upstream. Two control loops are already in place to control the inventories (condenser and reboiler levels). We will implement an additional control loop to control the temperature in Plate 10 by manipulating the reboiler heat duty. There is already a controller in place with some initial settings:

1. Using the relay autotuning functionality available in HYSYS, find the controller settings for the temperature loop at two different operating conditions (98 and 104°C set-points for temperature in Plate 10).
2. Compare and discuss the performance of the controller under the new settings at different operating conditions.

# Section V

## Multivariable Control

# 13 Multivariable Systems: Special Cases

In this chapter, we introduce control systems that may have more than just one manipulated variable or a measured variable. The control systems considered up to this point were single-variable controllers because they had the ultimate objective of maintaining a single variable near its set-point using a single manipulated variable. By contrast, in its most general form, the control of multivariable processes involves the objective of maintaining several controlled variables at independent set-points using multiple manipulated variables.

Before we present the general concepts associated with multivariable systems, we shall study, first, a number of control structures that merit a special discussion. These are:

- Cascade control
- Ratio control
- Override control

## 13.1 CASCADE CONTROL

To motivate the use of cascade control, consider the level control system for a mixing tank depicted in Figure 13.1. The feedback control strategy makes use of one of the feed flow rates as the manipulated variable to maintain the desired level in the tank. There is a possibility that the feed flow rates may be subject to variations due to changing upstream conditions. In summary, we have

- Control Objective: maintain $h$ at a target value $h_{sp}$
- Manipulated Variable: feed flow rate $F_1$
- Possible Disturbances: $F_1$, $F_2$

The control valve is calibrated so that a certain valve opening corresponds to a certain flow rate through it under design conditions. The calibration, of course, assumes that all other resistances in the pipe section (as well as the upstream pressure) are constant by design. Thus, to compensate for changes in the tank level, the controller sends a signal to the control valve to open a specific amount and deliver the corresponding flow rate. This strategy works well when there are disturbances in $F_2$ only, but if $F_1$ is also subject to variations, the results can be unsatisfactory. The control valve would not be able to deliver the expected flow

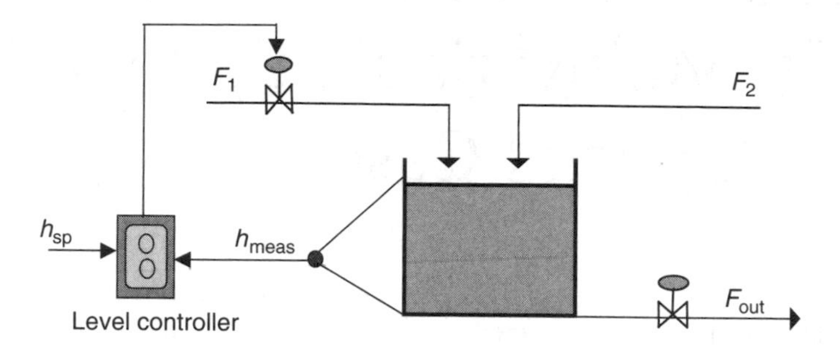


**FIGURE 13.1** Mixing tank with SISO feedback control of liquid level.

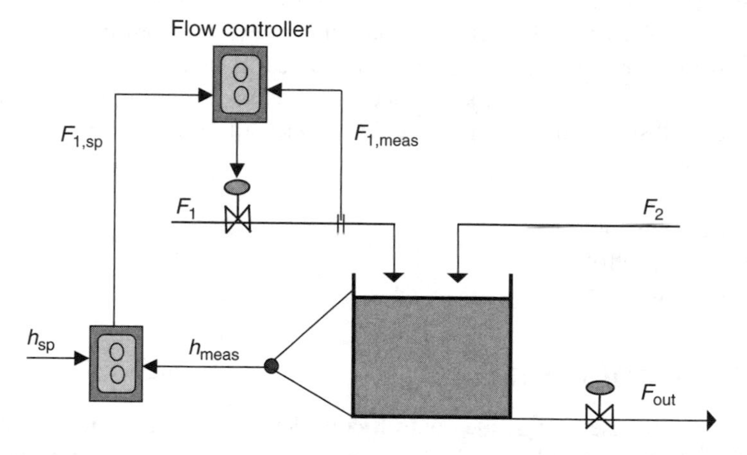


**FIGURE 13.2** Mixing tank with cascade control.

rate and the performance of the level control system would be adversely affected.

---

QUESTION: Is it possible to improve the performance of the feedback system in the presence of disturbances in $F_1$?

ANSWER: Yes, by implementing a *Cascade Control Configuration.*

---

The cascade control configuration is depicted in Figure 13.2. It uses (1) Two measurements, e.g., $F_1$ and $h$, and (2) One manipulated variable, e.g., $F_1$.

As a net result, the cascade configuration is expected to improve the closed-loop system response to changes in $F_1$ due to the direct measurement of possible disturbances in this variable. Effectively, any disturbance arising within this loop would be corrected immediately before it could affect the primary controlled output, the tank level.

The loop that measures the primary controlled variable (e.g., $h$) is called the *primary* control loop. The primary loop uses the set-point provided by the process operator.

Accordingly, the loop that measures the disturbance (e.g., $F_1$) is called the *secondary* control loop. The secondary loop uses the output of the primary loop as its set-point.

---

In chemical processes, flow control loops are almost always cascaded with other control loops so that major control objectives are unaffected by inevitable fluctuations in stream flows.

---

The closed-loop block diagram for the cascade control configuration is depicted in Figure 13.3. For the tank level control example, note that $d_1$ corresponds to disturbances in $F_2$ and $d_2$ accounts for unexpected variations in $F_1$. Thus, the primary loop is the level control loop and the secondary loop is the flow control loop.

To develop the block diagram for the cascade configuration and the associated transfer functions, let us start from open-loop transfer function for the *secondary* system:

$$u_1 = g_{p_2} u_2 + g_{d_2} d_2 \tag{13.1}$$

The secondary feedback loop yields the following transfer function (for simplicity, we shall assume that the transfer functions for measuring elements and actuators are equal to 1),

$$u_1 = \frac{g_{c_2} g_{p_2}}{1 + g_{c_2} g_{p_2}} u_{1sp} + \frac{g_{d_2}}{1 + g_{c_2} g_{p_2}} d_2 \tag{13.2}$$

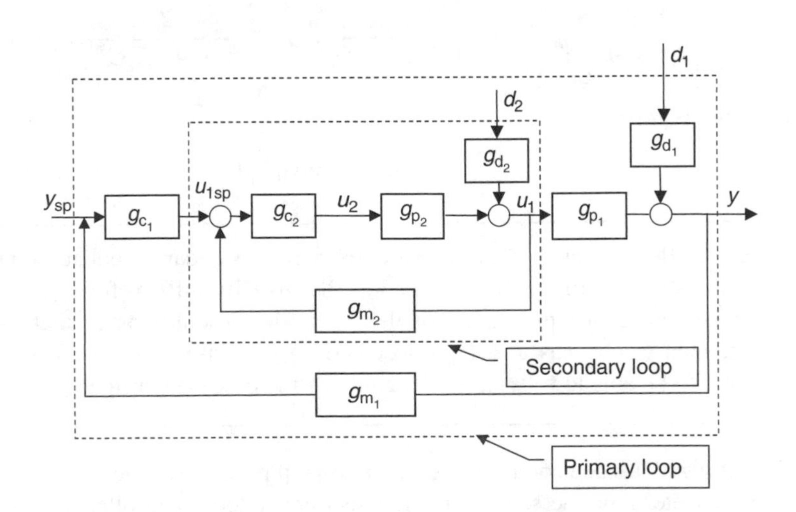


**FIGURE 13.3** Block diagram of cascade control.

Or, in compact form,

$$u_1 = G_{CL_2}^{sp} u_{1sp} + G_{CL_2}^{d_2} d_2 \tag{13.3}$$

As mentioned earlier, output of the primary controller is the set-point for the secondary control loop. In other words, the secondary loop, as shown in Eq. (13.3), ensures that the control action demanded by the primary controller is indeed delivered to the primary process. Now, we can express the open-loop process for the *primary* system as

$$y = g_{p_1} u_1 + g_{d_1} d_1 \tag{13.4}$$

Using Eq. (13.3), we can rewrite the open-loop expression as

$$\begin{aligned} y &= g_{p_1}\left[G_{CL_2}^{sp} u_{1sp} + G_{CL_2}^{d_2} d_2\right] + g_{d_1} d_1 \\ y &= g_{p_1} G_{CL_2}^{sp} u_{1sp} + g_{p_1} G_{CL_2}^{d_2} d_2 + g_{d_1} d_1 \end{aligned} \tag{13.5}$$

The key for the success of a cascade control configuration is the speed of the secondary loop. It is expected that the closed-loop time constant for the secondary loop be fast so that the disturbance $d_2$ is rejected quickly ($G_{CL_2}^{d} \to 0$) and the set-point for $u_1$ is reached immediately ($G_{CL_2}^{sp} \to 1$), thereby resulting in the expected output $y$. The output $y$, however, will not necessarily be at its desired value due to the presence of the disturbance $d_1$. To reject that disturbance, we now need to close the primary loop.

The closed-loop transfer function for the primary control loop becomes

$$y = \frac{g_{c_1} g_{p_1} G_{CL_2}^{sp}}{1 + g_{c_1} g_{p_1} G_{CL_2}^{sp}} y_{sp} + \frac{g_{p_1} G_{CL_2}^{d2}}{1 + g_{c_1} g_{p_1} G_{CL_2}^{sp}} d_2 + \frac{g_{d_1}}{1 + g_{c_1} g_{p_1} G_{CL_2}^{sp}} d_1 \tag{13.6}$$

Or, in compact form,

$$y = G_{CL_1}^{sp} y_{sp} + G_{CL_1}^{d2} d_2 + G_{CL_1}^{d1} d_1 \tag{13.7}$$

---

- Controllers $g_{c_1}$ and $g_{c_2}$ are often chosen from standard feedback controllers (i.e., P, PI, or PID). A P controller usually suffices for $g_{c_2}$.
- As stated earlier, the design of the controllers should ensure that the secondary loop has a faster (closed-loop) response than the primary loop to be able to fully take advantage of the cascade configuration.

---

For tuning a cascade control system, a two-step procedure is used:

***Step 1***: Determine the settings for the secondary loop controller, $g_{c_2}$, using any tuning technique.

***Step 2***: Using the settings above for $g_{c_2}$, place the secondary loop in automatic and determine the settings for the primary loop controller, $g_{c_1}$, using again any tuning technique.

## 13.2 RATIO CONTROL

Ratio control is a typical control strategy used to control the ratio of the flow rates of two or more streams. A typical application is the blending of two liquid streams in a given proportion (or ratio) as shown in Figure 13.4. The blend property of interest can be the stream composition or the stream temperature.

Both stream flow rates are assumed to be measured but only one stream is controlled. This uncontrolled stream is referred to as the "wild stream" or the flow rate as the "wild flow".[1] In the ratio configuration, the flow of the other stream is set on a given ratio with the "wild flow". There are at least two possible alternatives to implement the ratio scheme:

- Ratio computation
- Set-point computation

### 13.2.1 Ratio Computation

Consider the configuration depicted in Figure 13.5. In this case, the ratio is calculated from the measured values of both flow rates and is supplied to the controller. The set-point of the controller is the desired blend ratio and the flow rate of the manipulated stream is adjusted accordingly.

### 13.2.2 Set-Point Computation

Consider now the alternative implementation shown in Figure 13.6. In this case, the wild stream flow rate is multiplied by the desired ratio to generate the set-point of the controller for the manipulated stream.

Suppose that our control objective is to control a property of the blended stream, such as temperature. In this case, the previous configurations need to be

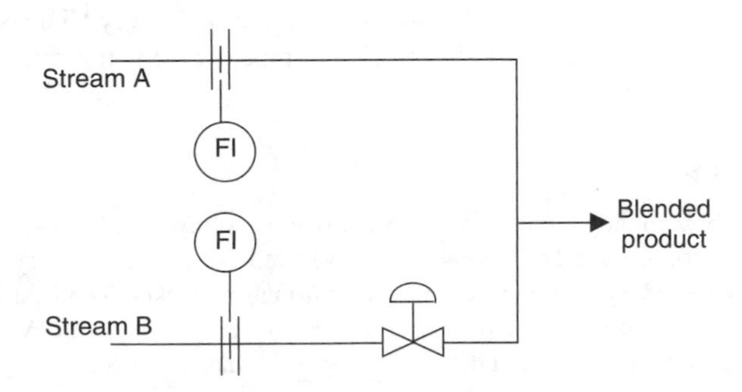


**FIGURE 13.4** Blending of two streams.

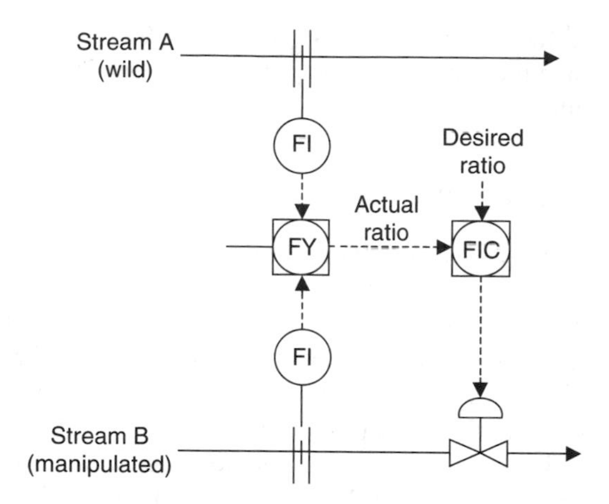


**FIGURE 13.5** Ratio computation.

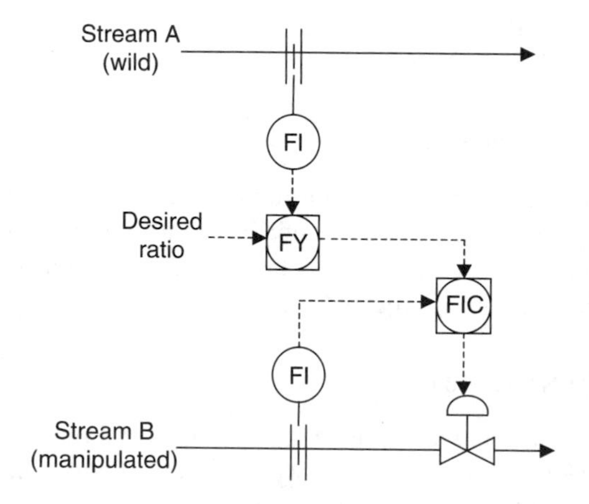


**FIGURE 13.6** Set-point computation.

modified. A possible implementation for such purposes is illustrated in Figure 13.7, which is a modification of the ratio computation strategy.[1] The output of the blended property controller (in this case, the temperature) is the ratio set-point to the feedback flow controller.

### Example 13.1

Consider the schematic representation of a process cooling circuit for a crystallization unit as discussed in Example 4.4 and given in Figure 13.8. The crystallizer is heated and cooled by an oil stream flowing through its jacket. The heat-exchange oil is circulated through hot and/or cold heat exchangers by a pump. A boiler and chiller are used to heat and cool the oil at the heat exchangers, respectively. Cooling is used to generate the required supersaturation. As the solution is cooled, the solute will crystallize in accordance with the saturation concentration at that temperature.

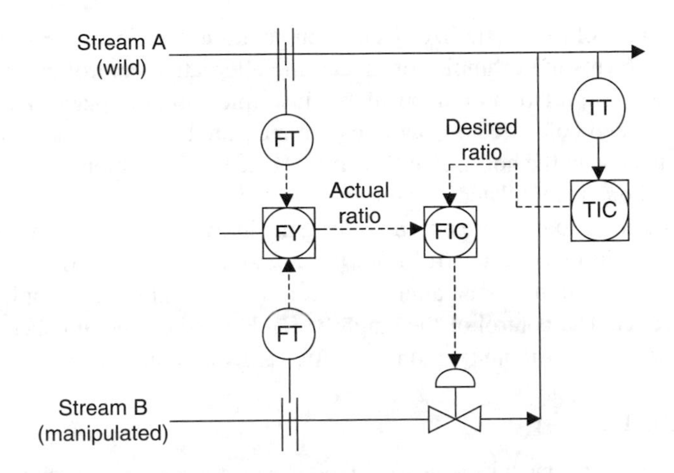


**FIGURE 13.7** Modified set-point computation.

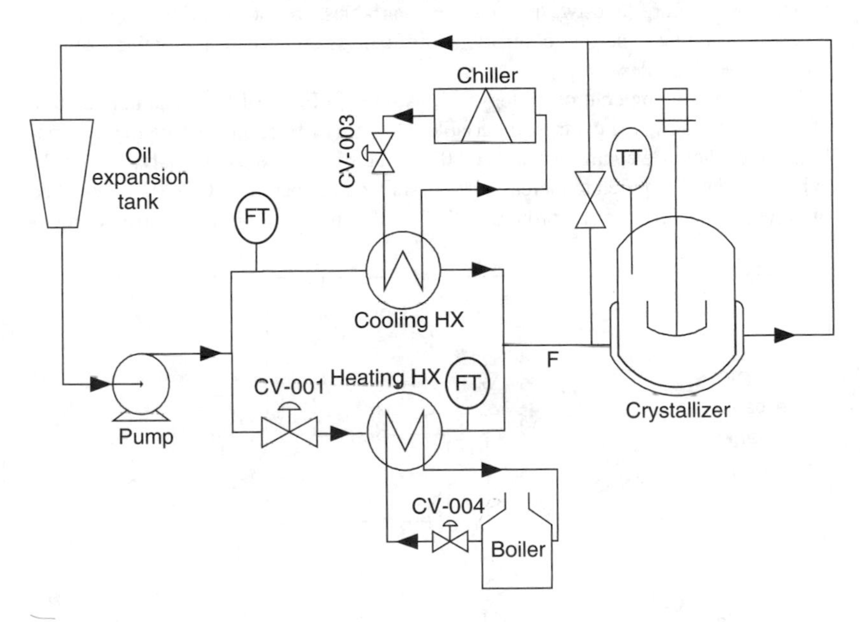


**FIGURE 13.8** Schematic diagram of the oil circuit to control the temperature of a crystallization unit.

This configuration permits the mixing of the hot and cold splits under the action of a ratio controller. It is assumed that both hot and cold streams are measured but only the hot oil stream can be manipulated. The temperatures just downstream of the heat exchangers would remain fixed by controllers acting on the steam and chiller flow rates. The temperature of the mixed stream and the temperature of the stream entering the crystallizer jacket are regulated by adjusting the ratio of the flow rates of the hot and cold splits. This example constitutes a typical application of the modified ratio control illustrated in Figure 13.7.

The control of the crystallizer jacket temperature described in Example 13.1 presents an interesting example for discussing alternative control configurations for a given unit in relation to the available instrumentation in place. We assumed that both hot and cold stream flow rates are measured and there is a single control valve to control the hot stream flow rate. Hence, we had more than one measurement and one manipulated variable.

Let us assume that we are not measuring the individual flows but we have two control valves in place to control both the hot and cold stream flow rates. This provides an opportunity for an alternative control configuration to implement the ratio controller. The control of the single variable is now accomplished by coordinating the action of more than one manipulated variable.

### Example 13.2

Consider again the modified schematic representation of the crystallization unit in Figure 13.9. There is now, in the modified version, only one measurement, the crystallizer temperature, and two manipulated variables, the hot oil flow rate and the cold oil flow rate. The one controller output must coordinate the actions of two manipulated variables.

The corresponding control strategy is illustrated in Figure 13.10. The temperature of the crystallizer is the controlled variable. Degrees of freedom include the manipulation of the hot side control valve (CV-001) and the cold side control valve (CV-002). When heating is required to increase the crystallizer temperature, CV-002 is switched off forcing the oil to pass through CV-001. The reverse applies when cooling is

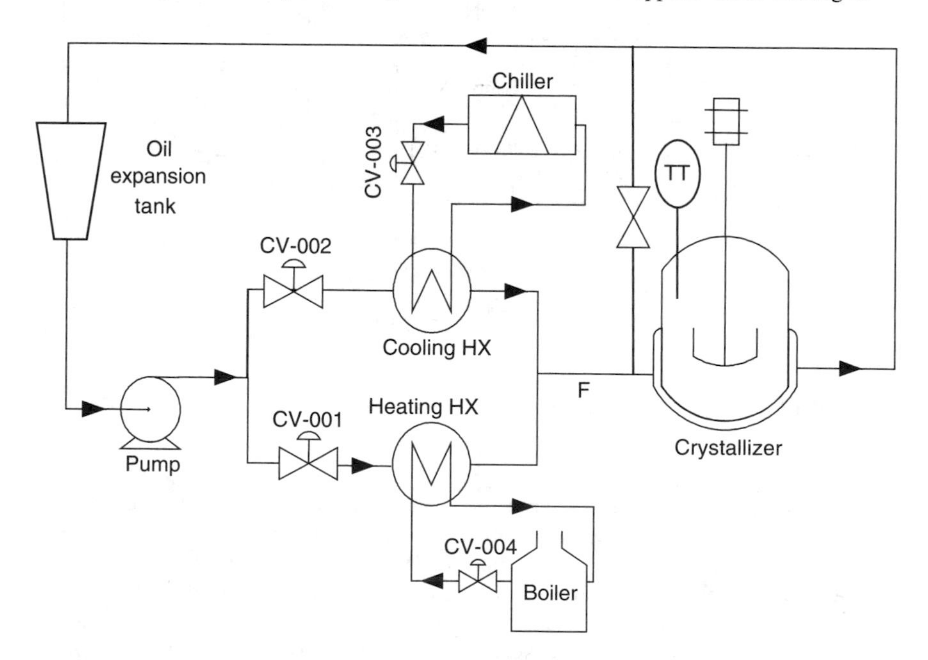


**FIGURE 13.9** Modified schematic diagram of the oil circuit to control the temperature of a crystallization unit.

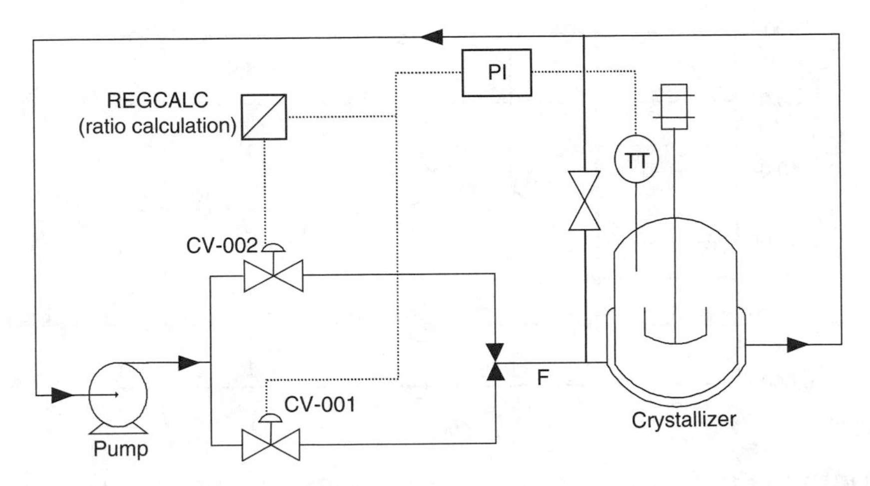


**FIGURE 13.10** Crystallizer control configuration.

required, i.e., CV-001 is shut while CV-002 is open. A ratio controller was implemented to control the crystallizer temperature by manipulating CV-001 and CV-002.

In this specific implementation, the position of the hot stream control valve, CV-001, is determined by a PI controller block measuring the crystallizer temperature and minimizing the error between this measurement and a given set-point. The position of the cold stream control valve, CV-002, is set by the relationship

$$\text{CV-002} = 100 - \text{CV-001}$$

As described in Chapter 4, a hybrid simulation environment is developed for this unit using HYSYS (to simulate the cooling circuit) and gPROMs (to simulate the crystallizer). In the following, the HYSYS model of the cooling circuit is used to illustrate the performance of the ratio control strategy. For simplicity the crystallizer jacket dynamics and the crystallizer temperature are approximated by first-order lags so that the whole simulation can be performed within the HYSYS environment. Figures 13.11–13.13 illustrate the behavior of the process under the proposed control configuration for a number of set-point changes in the temperature of the crystallizer. A full batch cycle is shown in Figure 13.13, excluding the feeding time but including heating, soaking, and cooling stages. Also shown in the figures are the actions of CV-001 and CV-002.

The implementation of this control strategy within an industrial DCS environment for a particular pilot-scale crystallization unit is fully described in Chapter 21.

## 13.3 OVERRIDE CONTROL

Override control is a protective control strategy principally used to maintain process variables within safe operational limits to ensure safety of plant personnel

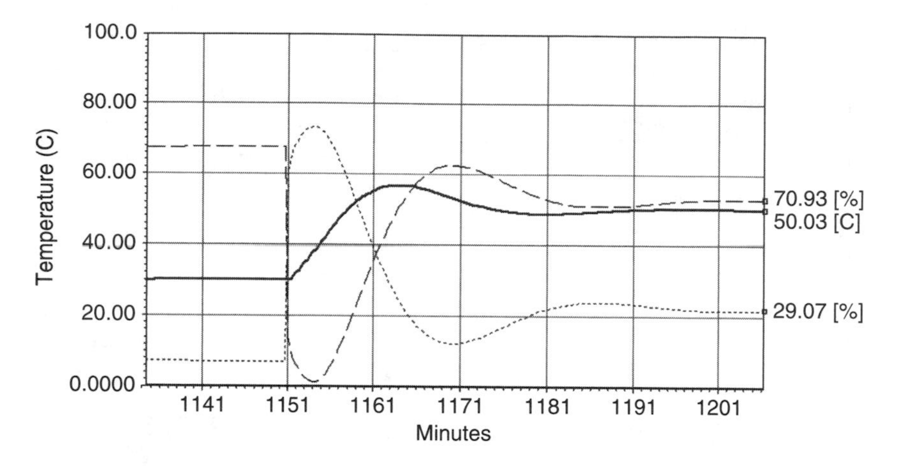


**FIGURE 13.11** The response for a set-point change. CV-001 (OP%) (dashed), CV-002 (OP%) (dotted), and TI-001 (PV) (solid).

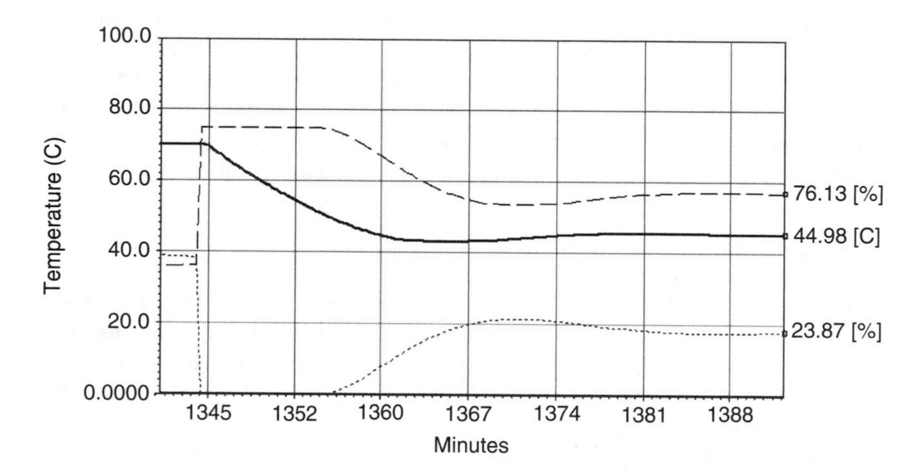


**FIGURE 13.12** The response for a set-point change. CV-001 (OP%), CV-002 (OP%), TI-001 (PV).

and process equipment as well as product quality. Special types of control switches are used to implement an override control strategy. Low-selector switch (LSS) is used to select the lowest of two or more signals and equivalently a high-selector switch (HSS) is used to select the highest of two or more signals.

Consider the simple example of the level control in a reactor where the liquid feed is being pumped from another section of the plant using a constant speed pump. The simplified schematic is given in Figure 13.14 and represents a subsection of a pilot-scale plant built at The University of Sydney running under an industrial computer control system. During start-up and normal operating conditions (inlet flow rate is less or equal to the outlet flow rate) the level of the tank,

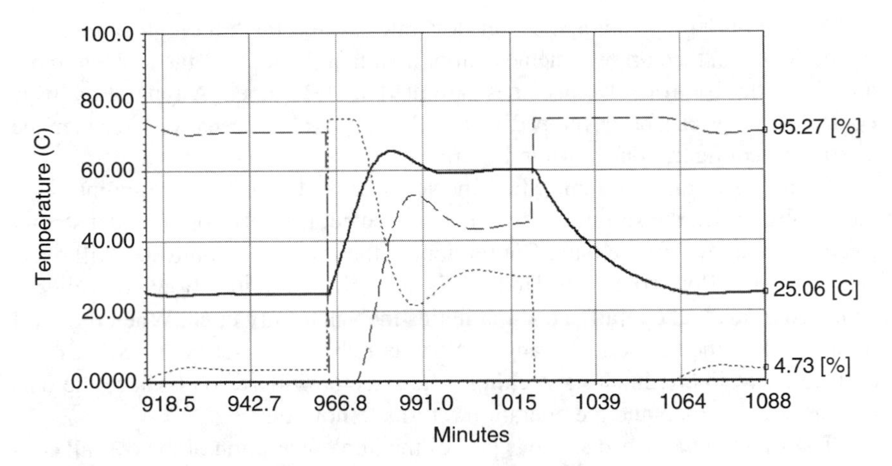


**FIGURE 13.13** The response for a complete operating cycle. CV-001 (OP%), CV-002 (OP%), and TI-001 (PV).

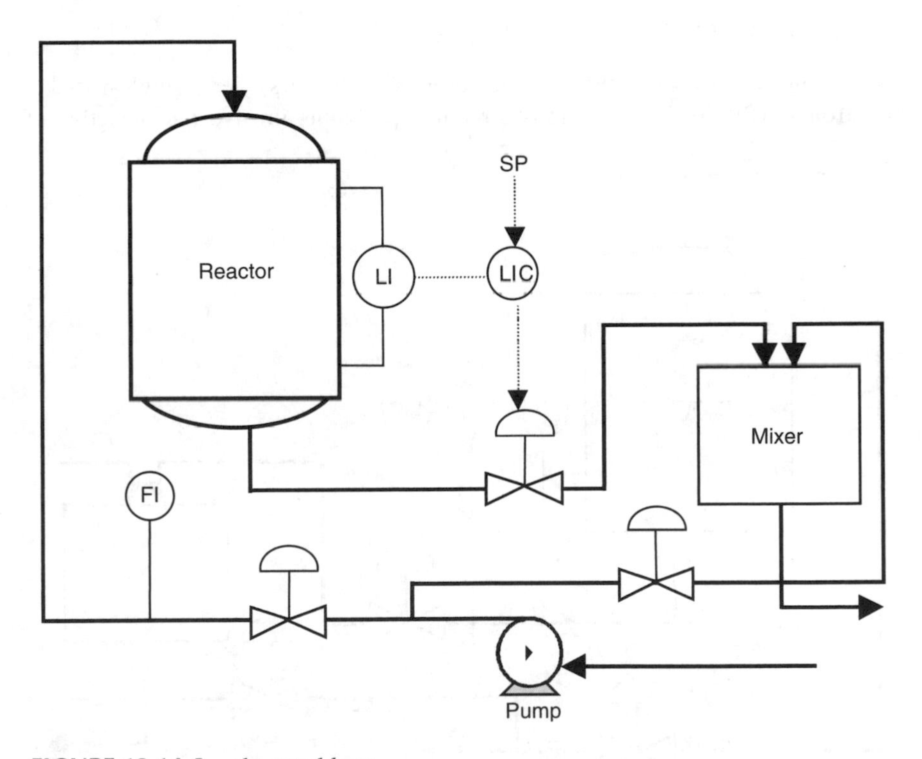


**FIGURE 13.14** Level control loop.

at a given condition, can always be maintained by the current control scheme. The outlet flow rate is determined by gravity and consequently limited by the static pressure of the liquid in the tank.

Now, if during operation, the inlet flow rate exceeds the limit high value of the outlet flow rate, the current scheme cannot maintain the level of the tank anymore and the outlet control valve becomes saturated at 100% open. A type of override control strategy can be developed to avoid this possible scenario. Consider the alternative configuration shown in Figure 13.15.

In the new configuration, when the level exceeds the level set-point (with valve fully open), the selector will switch and select the valve on the inlet stream to control the level of the tank. Consequently, the inlet flow controller will override the outlet flow controller. Under this scenario, the outlet flow controller is switched to manual operation and maintains the valve fully open. When the level returns below the set-point (a range below and above the set-point (SP $\pm$ $\delta$) is defined to avoid continuous switching), the selector will switch back to the outlet flow control loop and the operation returns to normal.

The strategy described above is part of the implementation of the overall control strategy of the flexible pilot-scale plant and will be fully described in detail in Chapter 21.

## 13.4 SUMMARY

The control systems considered in previous chapters were single-variable controllers (SISO); however, most practical problems involve the objective of

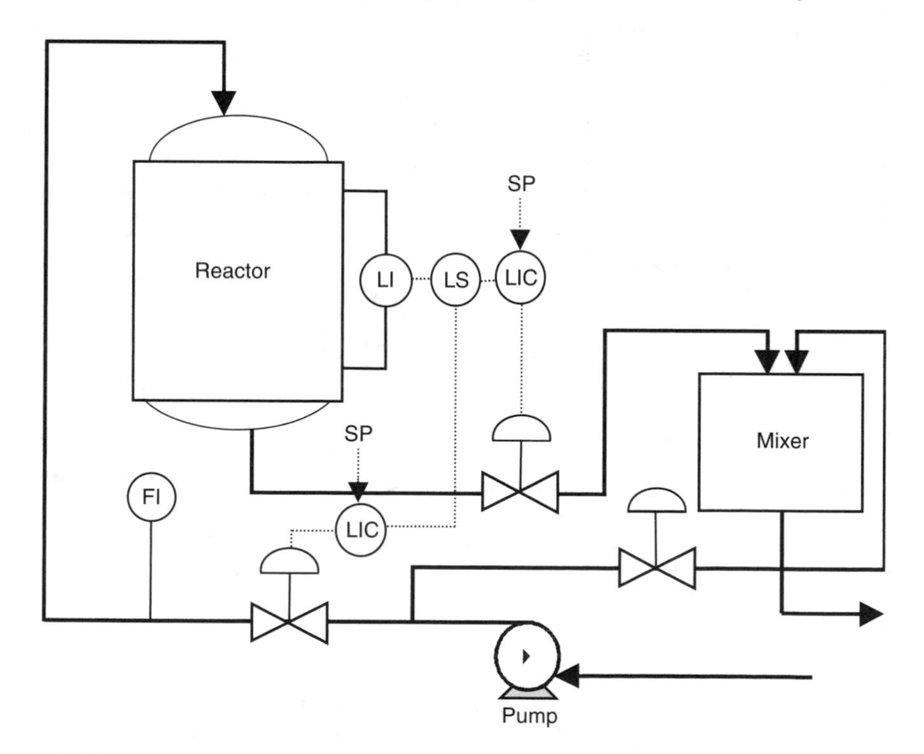


**FIGURE 13.15** Alternative "override" flow control loop.

maintaining several controlled variables at independent set-points using multiple manipulated variables. In this chapter, we have introduced more complex control architectures that may have more than just one manipulated variable and a measured variable, posing additional challenges.

First, the cascade control configuration was discussed involving a single manipulated variable but more than one measured variable. Its advantages over conventional feedback control, implementation issues, as well as controller tuning of such systems were also addressed.

The concept of ratio control was then introduced to provide a flexible strategy for processes with an excess of measured variables. A practical application to a crystallization unit was used as a vehicle to illustrate this strategy.

Finally, the concept of override control was briefly described. This is primarily a protective control strategy principally used to maintain process variables within safe operational limits to ensure safety of plant personnel and process equipment as well as product quality.

## CONTINUING PROBLEM

For the blending process, we wish to control the outlet composition by manipulating the feed flow rate, $F_2$, and the composition loop will be cascaded on the flow loop (Figure 13.16).

### SOLUTION

The process transfer function is given as

$$x(s) = \frac{0.1e^{-s}}{2.5s + 1} F_2(s)$$

The flow process is described by the transfer function

$$F_2(s) = \frac{1}{s + 1} F_{2,c}(s)$$

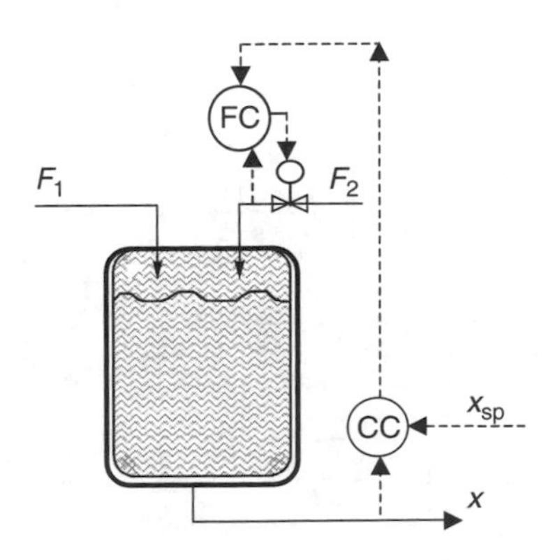


**FIGURE 13.16** Blending process with cascade control configuration.

where $F_{2,c}(s)$ represents the control valve opening. The disturbance transfer functions are given as

$$x(s) = g_{d_1}(s)F_1(s) = \frac{-0.1e^{-s}}{2.5s + 1}F_1(s)$$

$$F_{2,c}(s) = g_{d_2}(s)F_{2,d}(s) = F_{2,d}(s)$$

The variable $F_{2,d}(s)$ denotes the effect of flow variations on the flow rate of stream 2, and is the disturbance for the secondary loop. For simplicity, we shall assume that $g_{d_2}(s) = 1$.

The corresponding block diagram under cascade control for this process is shown in Figure 13.17. For the secondary loop, a PI controller with the settings $k_c = 1$ and $\tau_I = 1$ will be used. It should be noted that since the secondary process is just a first-order process, it would not have a crossover frequency; consequently, the only limitations for the secondary controller settings are based on practical considerations. Thus, a P-only controller with a large gain would be sufficient.

We tune a PI controller for the primary loop using the Z–N technique (Chapter 12), and compare the performance of the closed-loop system with respect to a unit-step change in the disturbance in the feed flow rate with and without the cascade configuration.

The first step is to evaluate the ultimate gain and ultimate period for the primary controller under proportional-only control action and with the secondary controller on. Figures 13.18 and 13.19 show the time and frequency (Nyquist plot) responses obtained from this analysis.

From Figures 13.18 and 13.19, the ultimate gain and the ultimate period for the primary controller can be estimated as 41 and 5, respectively. Following the Z–N tuning rules, the controller gain and the integral time constant for a PI controller would be 18.45 and 4.17, respectively.

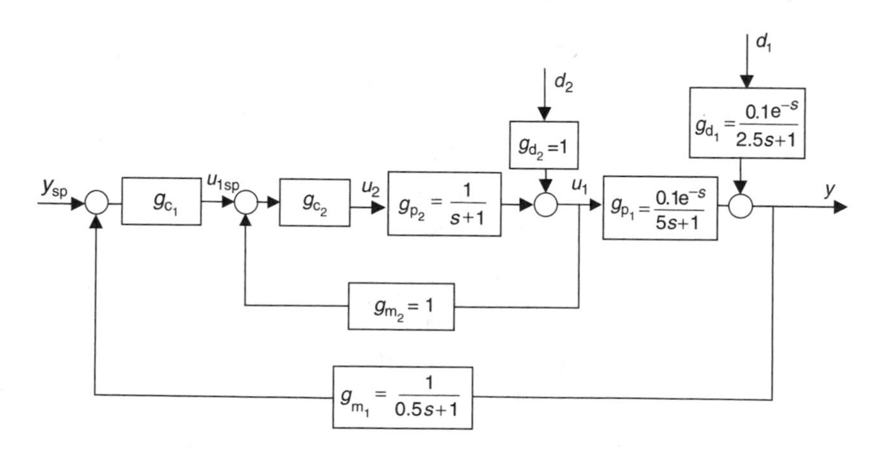


**FIGURE 13.17** Block diagram for blending process with cascade control configuration.

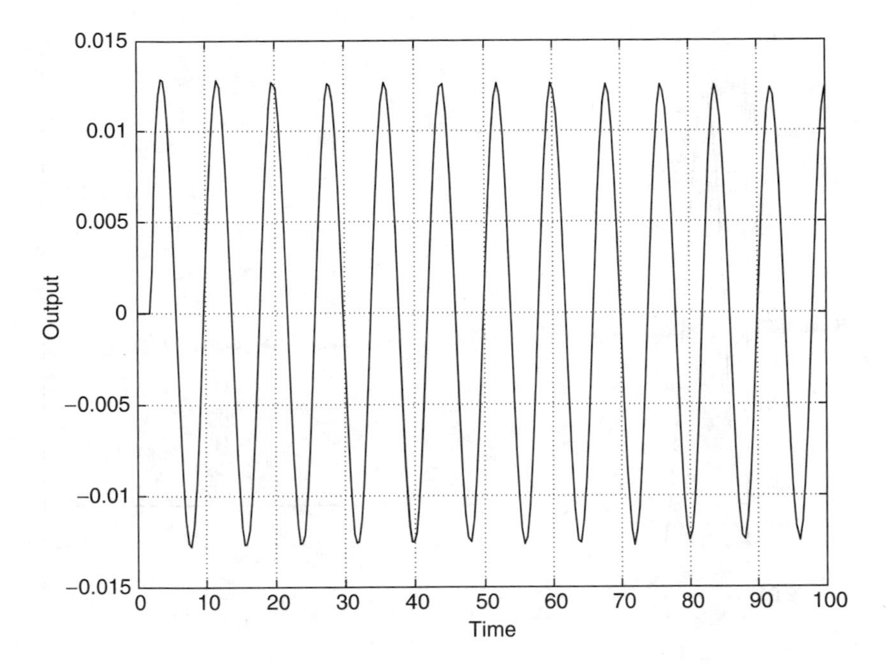


**FIGURE 13.18** Process response for primary controller with proportional action, $k_c$=41.

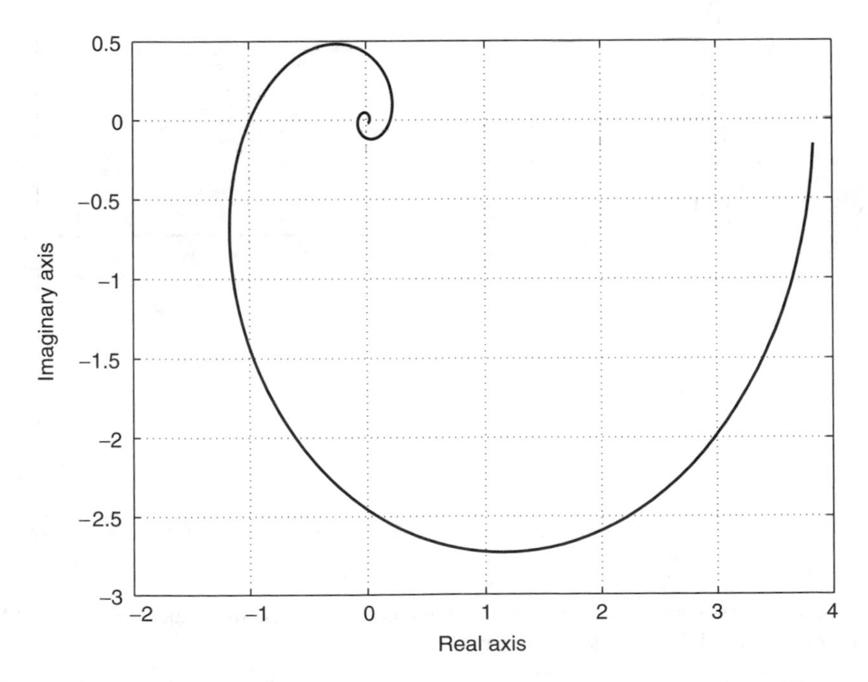


**FIGURE 13.19** Nyquist plot for blending process under cascade control, $k_c = 41$.

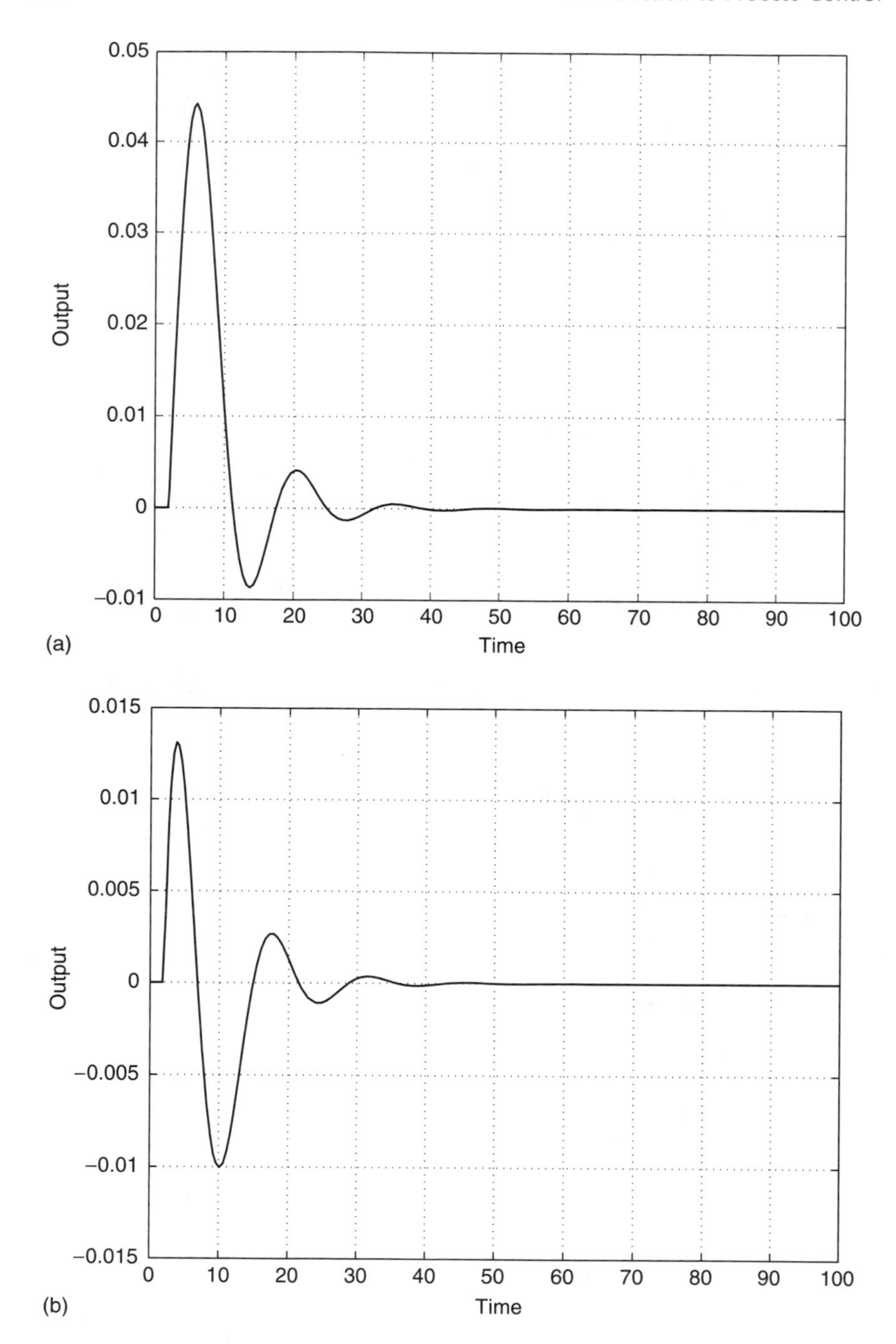


**FIGURE 13.20** Closed-loop response for the blending process using the conventional PI controller (a) and the cascade controller (b).

The closed-loop response to a disturbance for the conventional PI control strategy that measures the composition and manipulates the flow rate directly (with $k_c = 18.45$ and $\tau_I = 4.17$) is shown in Figure 13.20a. Figure 13.20b illustrates the closed-loop responses for the cascade control configuration with the primary controller settings obtained using the Z–N technique ($k_c = 18.45$ and $\tau_I = 4.17$ for the primary controller and $k_c = 2$ and $\tau_I = 1$ for the secondary controller). The improvement obtained with the new configuration regarding the disturbance rejection capabilities for changes in the feed stream is clearly demonstrated.

## REFERENCES

1. Erickson, K.T. and Hedrick, J.L., *Plantwide Process Control*, Wiley, New York, 1999.

# 14 Multivariable Systems: General Concepts

In general, multivariable systems involve multiple inputs (manipulated variables) and multiple outputs (measured variables) to deal with multiple control objectives associated with a process unit or a plant. The presence of multiple control objectives complicates the task of the control engineer, yet it may also offer opportunities for advanced control applications.

## 14.1 CHARACTERISTICS OF MULTIVARIABLE PROCESSES

Consider the MIMO process representation depicted in Figure 14.1. One of the key decisions in multivariable systems involves the structure of the control system, i.e., the pairing of the measured variables with the manipulated variables. It may be possible to assign a single, independent manipulated variable to each and every control objective (measured variable), and the resulting control structure would be referred to as a *multi-loop* controller (i.e., multiple SISO controllers). However, the control engineer may also choose to consider all or a portion of the manipulated variables to satisfy all or a portion of the control objectives simultaneously. Such flexibility is responsible for one of the most important challenges in multivariable control design as it gives rise to a large number of alternative control structures, ranging from multiloop to fully multivariable control.

---

The *selection of the most appropriate* control configuration is the central and the critical task.

---

The control of multivariable systems requires more effort in analysis than that of single-variable systems. Multivariable systems have a number of unique characteristics due to interactions among the variables that demand careful consideration. These will be discussed in detail in Chapter 15.

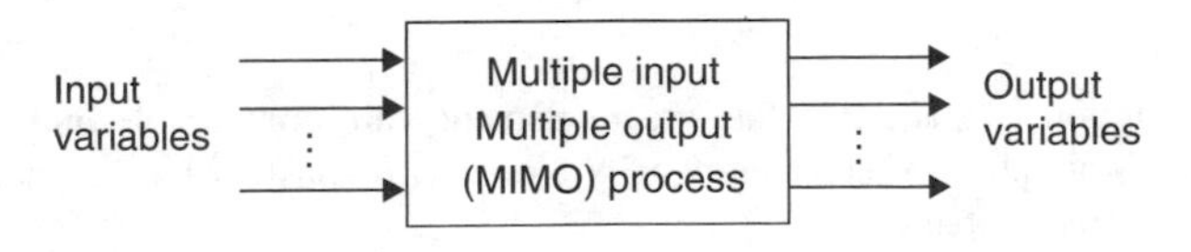


**FIGURE 14.1** Process with multiple inputs and multiple outputs.

**Characteristics Unique to Multivariable Systems**

1. Interactions among variables affect stability and performance of the controlled system.
2. Feasibility of the controlled system depends on the overall process.
3. The pairing of measured and manipulated variables via control loops is a design decision.
4. Some processes may have unequal number of measured and manipulated variables.
5. Some multivariable control designs may be very sensitive to modeling errors.

## 14.2 MODELING OF MULTIVARIABLE PROCESSES

In Chapter 5, we have introduced the state-space model for a given process. Similar arguments follow for multivariable processes. In general, the state-space model is expressed as

$$\begin{aligned} \dot{x} &= \frac{\mathrm{d}x}{\mathrm{d}t} = Ax + Bu \\ y &= Cx \end{aligned} \tag{14.1}$$

We note here that the variables $x$, $y$, and $u$ in Eq. (14.1) are now all vectors with appropriate dimensions. The following example illustrates the construction of a state-space model.

### Example 14.1

Consider the thermal-mixing tank in Figure 14.2. A cold and a hot stream are mixed in a continuously stirred tank with constant cross-sectional area, $A$. The height of the liquid in the tank is denoted as $h$. We will develop the mass and energy balances for this process.

We start with the overall mass balance

$$\begin{aligned} A\rho \frac{\mathrm{d}h}{\mathrm{d}t} &= \rho F_{\mathrm{C}} + \rho F_{\mathrm{H}} - \rho F \\ A \frac{\mathrm{d}h}{\mathrm{d}t} &= F_{\mathrm{C}} + F_{\mathrm{H}} - F \end{aligned} \tag{14.2}$$

The assumption is made that the stream densities are comparable and constant. Using the assumption of constant density along with constant heat capacities, we arrive at the energy balance

$$A \frac{\mathrm{d}(Th)}{\mathrm{d}t} = F_{\mathrm{C}}(T_{\mathrm{C}} - T_{\mathrm{r}}) + F_{\mathrm{H}}(T_{\mathrm{H}} - T_{\mathrm{r}}) - F(T - T_{\mathrm{r}}) \tag{14.3}$$

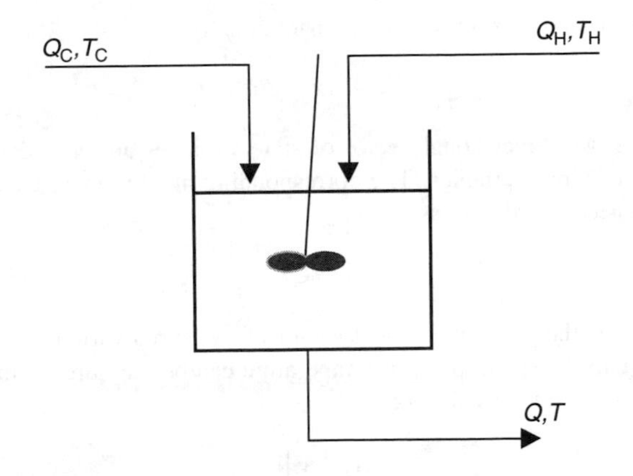


**FIGURE 14.2** Thermal mixing tank.

where $T_r$ represents a reference temperature. We know that

$$\frac{d(Th)}{dt} = T\frac{dh}{dt} + h\frac{dT}{dt}$$

thus,

$$Ah\frac{dT}{dt} = T(F_C + F_H - F) + F_C(T_C - T_r) + F_H(T_H - T_r) - F(T - T_r) \tag{14.4}$$

At steady-state, we can show that

$$0 = F_C + F_H - F \tag{14.5}$$

$$0 = T(F_C + F_H - F) + F_C(T_C - T_r) + F_H(T_H - T_r) - F(T - T_r) \tag{14.6}$$

The following definitions are introduced,

$$F = \beta\sqrt{2gh}, \quad F_s = \beta\sqrt{2gh_s}, \quad x_1 = \frac{h - h_s}{h_s}, \quad x_2 = \frac{T - T_s}{T_s}$$
$$u_1 = \frac{F_C - F_{C_s}}{F_{C_s}}, \quad u_2 = \frac{F_H - F_{H_s}}{F_{H_s}}, \quad \tau = \frac{F_s}{Ah_s}t \tag{14.7}$$

After linearization, we arrive at the following model,

$$\begin{bmatrix} \dfrac{dx_1}{dt} \\ \dfrac{dx_2}{dt} \end{bmatrix} = \begin{bmatrix} -0.5 & 0 \\ 0 & -1 \end{bmatrix} \begin{bmatrix} x_1 \\ x_2 \end{bmatrix} + \begin{bmatrix} 1 & 1 \\ \dfrac{T_{C_s} - T_s}{T_s} & \dfrac{T_{H_s} - T_s}{T_s} \end{bmatrix} \begin{bmatrix} u_1 \\ u_2 \end{bmatrix} \tag{14.8}$$

(14.8) can be expressed in compact notation

$$\dot{x} = Ax + Bu$$

where $x$ is the two-dimensional vector of state variables and $u$ is the two-dimensional vector of input variables. The corresponding measurement equation for the system is defined accordingly as

$$y = Cx$$

As expected, $y$ is the two-dimensional vector of measured variables as we assume that both the tank level and the exit temperature can be measured. This makes the output matrix $C$ the identity matrix

$$y = \begin{bmatrix} 1 & 0 \\ 0 & 1 \end{bmatrix} x = Ix$$

## 14.3 TRANSFER FUNCTIONS OF MULTIVARIABLE PROCESSES

In this section, we extend the concept of transfer functions to systems with multiple inputs and multiple outputs.

---

In an MIMO system, we have a vector of inputs and a vector of outputs. The matrix that relates the Laplace transform of the output vector to that of the input vector is called the transfer function matrix (TFM).

---

Let us consider the MIMO system shown in Figure 14.3. This system has two inputs and two outputs.

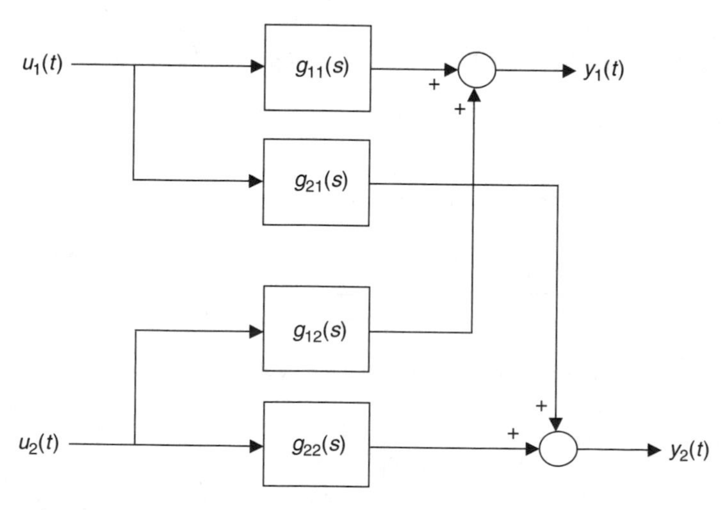


**FIGURE 14.3** Block diagram of a two-input–two-output process.

Based on Figure 14.3, the relationships between the inputs and the outputs are given by

$$
\begin{aligned}
y_1 &= g_{11}(s)u_1(s) + g_{12}(s)u_2(s) \\
y_2 &= g_{21}(s)u_1(s) + g_{22}(s)u_2(s)
\end{aligned} \tag{14.9}
$$

Using matrix notation, we can rewrite Eq. (14.9) as

$$
Y(s) = \begin{bmatrix} y_1(s) \\ y_2(s) \end{bmatrix} = \begin{bmatrix} g_{11}(s) & g_{12}(s) \\ g_{21}(s) & g_{22}(s) \end{bmatrix} \begin{bmatrix} u_1(s) \\ u_2(s) \end{bmatrix} \tag{14.10}
$$

or

$$
Y(s) = G(s)U(s)
$$

where $G(s)$ stands for the TFM of the MIMO system. To generalize, if a system has $m$ inputs and $n$ outputs and the transfer function between the $i$th output and the $j$th input is given by $g_{ij}(s)$, then the $i$th output can be expressed as a function of all the inputs as follows:

$$
y_i(s) = g_{i1}(s)u_1(s) + g_{i2}(s)u_2(s) + \cdots + g_{im}(s)u_m(s) \tag{14.11}
$$

with $i = 1, 2, \ldots, n$. In matrix notation, the $n \times m$ TFM can be expressed as

$$
Y(s) = \begin{bmatrix} y_1(s) \\ \vdots \\ y_n(s) \end{bmatrix} \begin{bmatrix} g_{11}(s) & \ldots & g_{1m}(s) \\ \vdots & & \vdots \\ g_{n1}(s) & \ldots & g_{nm}(s) \end{bmatrix} \begin{bmatrix} u_1(s) \\ \vdots \\ u_m(s) \end{bmatrix} = G(s)U(s) \tag{14.12}
$$

Each element of $G(s)$ represents the individual transfer function between the input $u_j(s)$ and the output $y_i(s)$, i.e.,

$$
g_{ij}(s) = \frac{y_i(s)}{u_j(s)} \tag{14.13}
$$

We note that if $g_{ij}(s) = 0$, then the input $u_j(s)$ does not affect the output $y_i(s)$.

For systems represented by a state-space model such as in Eq. (14.1), the TFM can be obtained using the Laplace transform and matrix algebra as follows:

$$
G(s) = C(sI - A)^{-1}B \tag{14.14}
$$

If one knows the matrices in the state-space model, the TFM can be determined directly using this expression.

**Example 14.2**

Consider again the model of the mixing tank process in Example 14.1.

$$\begin{bmatrix} \dfrac{dx_1}{dt} \\ \dfrac{dx_2}{dt} \end{bmatrix} = \begin{bmatrix} -0.5 & 0 \\ 0 & -1 \end{bmatrix} \begin{bmatrix} x_1 \\ x_2 \end{bmatrix} + \begin{bmatrix} 1 & 1 \\ \Delta T_C & \Delta T_H \end{bmatrix} \begin{bmatrix} u_1 \\ u_2 \end{bmatrix}$$

Here, we defined the following new constants,

$$\Delta T_C = \frac{T_{C_s} - T_s}{T_s} \quad \text{and} \quad \Delta T_H = \frac{T_{H_s} - T_s}{T_s}$$

Then, using Eq. (14.14), we can write

$$Y(s) = [C(sI - A)^{-1}B]U(s) = G(s)U(s)$$

$$= \begin{bmatrix} 1 & 0 \\ 0 & 1 \end{bmatrix} \begin{bmatrix} s + 0.5 & 0 \\ 0 & s + 1 \end{bmatrix}^{-1} \begin{bmatrix} 1 & 1 \\ \Delta T_C & \Delta T_H \end{bmatrix} U(s)$$

$$= \begin{bmatrix} \dfrac{1}{s + 0.5} & 0 \\ 0 & \dfrac{1}{s + 1} \end{bmatrix} \begin{bmatrix} 1 & 1 \\ \Delta T_C & \Delta T_H \end{bmatrix} U(s)$$

$$Y(s) = \begin{bmatrix} \dfrac{1}{s + 0.5} & \dfrac{1}{s + 0.5} \\ \dfrac{\Delta T_C}{s + 1} & \dfrac{\Delta T_H}{s + 1} \end{bmatrix} U(s)$$

Thus, the TFM is given as,

$$G(s) = \begin{bmatrix} \dfrac{1}{s + 0.5} & \dfrac{1}{s + 0.5} \\ \dfrac{\Delta T_C}{s + 1} & \dfrac{\Delta T_H}{s + 1} \end{bmatrix} \tag{14.15}$$

### 14.3.1 Poles and Zeros of MIMO Systems

---

The poles and zeros of a square TFM $G(s)$ are the poles and zeros of the determinant of $G(s)$.

---

The expression that results from the determinant operation is a rational transfer function whose poles and zeros can be identified easily as before (see Chapter 5).

$$\det(G(s)) = \frac{z(s)}{p(s)} \tag{14.16}$$

where $z(s)$ and $p(s)$ (as before) are the zero and the pole polynomials of the TFM, respectively. Consequently, the zeros and poles are given by the roots of $z(s)$ and $p(s)$, respectively.

If the TFM is not square, i.e., the number of output variables is not equal to the number of input variables, $n \neq m$, the poles can still be identified as the collection of the poles of the individual TFM elements. On the other hand, to determine the zeros, we look for values of the Laplace variable $s$ that diminish the rank of the TFM. However, these concepts are not trivial for the general case. Example 14.4 provides a new set of definitions and more insight into the methodology for such computations.

### Example 14.3

Consider again the thermal mixing tank process. First, we note that for a $2 \times 2$ matrix $H$,

$$H = \begin{bmatrix} h_{11} & h_{12} \\ h_{21} & h_{22} \end{bmatrix}$$

the determinant is given by

$$\det H = h_{11}h_{22} - h_{12}h_{21}$$

Therefore, the determinant of the TFM in Eq. (14.15) can be expressed as

$$\begin{aligned} \det G(s) &= \frac{\Delta T_{\mathrm{H}}}{(s+0.5)(s+1)} - \frac{\Delta T_{\mathrm{C}}}{(s+0.5)(s+1)} \\ &= \frac{\Delta T_{\mathrm{H}} - \Delta T_{\mathrm{C}}}{(s+0.5)(s+1)} \end{aligned} \tag{14.17}$$

Thus, the process has (1) Two poles located at $-0.5$ and $-1.0$, and (2) No finite zeros, only two zeros at infinity.

### Example 14.4

The following $3 \times 2$ transfer function is given for a process as

$$G(s) = \begin{bmatrix} \dfrac{s+1}{(s+3)^2} & 0 \\ 0 & \dfrac{s+2}{(s+4)^2} \\ \dfrac{s+2}{(s+5)^2} & \dfrac{s+1}{(s+5)^2} \end{bmatrix} \tag{14.18}$$

The TFM can be rewritten by factoring out the poles as follows:

$$G(s) = \frac{\begin{bmatrix} (s+1)(s+4)^2(s+5)^2 & 0 \\ 0 & (s+2)(s+3)^2(s+5)^2 \\ (s+2)(s+3)^2(s+4)^2 & (s+1)(s+3)^2(s+4)^2 \end{bmatrix}}{(s+3)^2(s+4)^2(s+5)^2} \tag{14.19}$$

We can easily observe that the denominator polynomial contains the poles of $G(s)$ as the TFM becomes unbounded for those values of $s$:

$$p_{1,2} = -3, \quad p_{3,4} = -4, \quad p_{5,6} = -5$$

Next, the zeros can be found by determining the values of $s$ that cause $G(s)$ to lose rank. Note that the normal rank of $G(s)$ is defined as $\text{Rank}[G(s)] = \min[n, m]$ where $n$ is the number of outputs and $m$ is the number of inputs of TFM. Thus, for this example, we have $\text{Rank}[G(s)] = 2$. Since the calculation of the rank needs to be carried out for all values of $s$, we run into problems when $s$ takes on the values of the poles. Thus, for the rank operation, the pole locations are ignored. This strategy presents a subtle (yet important) problem when the multivariable system zeros coincide with the multivariable system poles. Therefore, the identification of zeros has to be performed in a more rigorous setting. Let us briefly illustrate how one should calculate the zeros of a TFM in general. First, we need to define right and left coprime factorizations of a rational $n \times m$ TFM:[1]

$$G(s) = N_R(s)D_R^{-1}(s) = D_L^{-1}(s)N_L(s) \tag{14.20}$$

where, for instance, $N_L(s)$ is a $n \times m$ polynomial matrix and $D_L(s)$ is a $n \times n$ polynomial matrix. The key property of these matrices is that they are coprime, in other words, there is no cancellation in the fraction (no nontrivial common divisors).[2] There are algorithms (e.g., in MATLAB) that can perform this factorization. Once these factors are found, the determination of the system zeros (as well as the system poles) is greatly facilitated.

---

For a $n \times m$ rational TFM $G(s)$, a complex number $s_0$ is called a (transmission) zero of $G(s)$ if $\text{Rank}[N_L(s_0)] < \min[n, m]$.

---

Such a factorization can also be used in the determination of the system poles.

---

For a $n \times m$ rational TFM $G(s)$, a complex number $s_0$ is called a pole of $G(s)$ if $\det[D_L(s_0)] = 0$.

---

Now, let us apply these definitions to our example. The right coprime factorization for $G(s)$ is given as:

$$G(s) = \begin{bmatrix} \frac{s+1}{(s+3)^2} & 0 \\ 0 & \frac{s+2}{(s+4)^2} \\ \frac{s+2}{(s+5)^2} & \frac{s+1}{(s+5)^2} \end{bmatrix}$$

$$= \begin{bmatrix} (s+3)^2 & 0 & 0 \\ 0 & (s+4)^2 & 0 \\ 0 & 0 & (s+5)^2 \end{bmatrix}^{-1} \begin{bmatrix} s+1 & 0 \\ 0 & s+2 \\ s+2 & s+1 \end{bmatrix} \quad (14.21)$$

By applying the previous definitions, we can easily confirm the poles as determined before as the values of $s$ that make $D_L(s)$ zero. Similarly, by observing $N_L(s)$, we can confirm that there are no values of $s$ that would cause the matrix to lose rank, thus indicating that there are no MIMO zeros.

In general, there is no relationship between the zeros of $G(s)$ and the zeros of the individual transfer function elements of $G(s)$ as we have seen in Example 14.4 where the individual elements had zeros but not the MIMO system. Furthermore, the poles of the individual transfer function elements do appear as the poles of $G(s)$, but the multiplicity of these poles may differ from what they may be with the individual elements of the TFM.

In the sequel, we will strictly deal with square matrices, thus avoiding the issues discussed in Example 14.4.

## 14.4 MULTIVARIABLE FEEDBACK CONTROL STRUCTURE

To illustrate the construction of the feedback control structure for a multivariable system, consider the 2 × 2 process in Figure 14.4.

By combining the individual transfer functions in TFMs, and defining the following matrices (note that we omitted the Laplace variable $s$ for convenience),

$$G_c = \begin{bmatrix} g_{c_1} & 0 \\ 0 & g_{c_2} \end{bmatrix}, \quad G = \begin{bmatrix} g_{11} & g_{12} \\ g_{21} & g_{22} \end{bmatrix} \quad (14.22)$$

with the vectors, $D = [d_1 \;\; d_2]^T$, $Y = [y_1 \;\; y_2]^T$ and $Y_{sp} = [y_{1,sp} \;\; y_{2,sp}]^T$, we can redraw the block diagram in Figure 14.4 in a more compact form as in Figure 14.5.

With the above notation, the representation can be easily generalized. Thus, for a multivariable feedback system, the following relationship can be developed:

$$Y = GG_c(Y - Y_{sp}) + D \quad (14.23)$$

By collecting the terms, we get

$$(I + GG_c)Y = GG_cY_{sp} + D \quad (14.24)$$

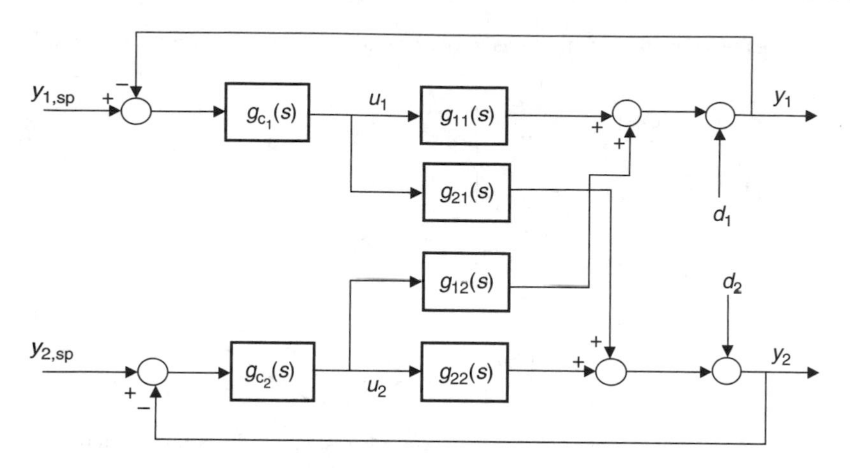


**FIGURE 14.4** A two-input–two-output feedback system.

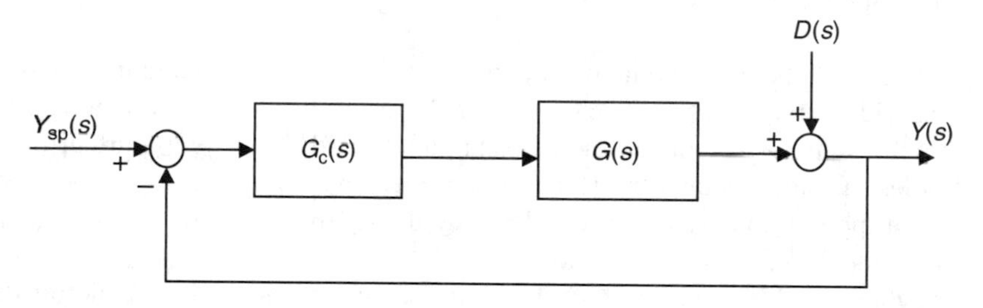


**FIGURE 14.5** Compact representation of the two-input–two-output feedback system.

or, finally, the closed-loop equation becomes

$$Y = (I + GG_c)^{-1}GG_cY_{sp} + (I + GG_c)^{-1}D \tag{14.25}$$

We stress here again that we are dealing with matrices and vectors. In general, the matrix product is not commutative, thus, the order of multiplication of the matrices in the closed-loop equation is important.

### 14.4.1 Closed-Loop Poles and Zeros

For the multivariable feedback system in Figure 14.5, the corresponding open-loop transfer function is

$$G_{OL}(s) = G(s)G_c(s) \tag{14.26}$$

The open-loop characteristic function (*OLCF*) can then be defined as

$$OLCF = \det(GG_c) = \det(G)\det(G_c) = \frac{z_{OLCF}(s)}{p_{OLCF}(s)} \tag{14.27}$$

where $z_{OLCF}(s)$ and $p_{OLCF}(s)$ are the zero and the pole polynomials of the *OLCF*, respectively.

For the closed-loop system shown in Figure 14.5, the closed-loop zeros can be found from the following expression of the closed-loop characteristic function (*CLCF*):

$$CLCF = \det((I + GG_c)^{-1}GG_c) = \det(GG_c)\det(I + GG_c)^{-1} = \frac{\det(GG_c)}{\det(I + GG_c)} = \frac{z_{CLCF}(s)}{p_{CLCF}(s)} \tag{14.28}$$

**Example 14.5**

There is a special relationship between the open- and the closed-loop zeros. To uncover it, we need to use the factorization introduced in Example 14.4. Define

$$G(s) = N_R(s)D_R^{-1};\ G_c(s) = D_L^{-1}(s)N_L(s) \tag{14.29}$$

We arbitrarily used the right coprime factorization for the process and the left coprime factorization for the controller. To illustrate, the closed-loop transfer function for the set-point tracking problem is given as ($s$ dependence is omitted for brevity of expression)

$$G_{sp} = GG_c(I + GG_c)^{-1} \tag{14.30}$$

Using matrix operations and the factorizations, Eq. (14.30) can be rephrased as

$$G_{sp} = G(I + G_cG)^{-1}G_c = N_RD_R^{-1}(I + D_L^{-1}N_LN_RD_R^{-1})^{-1}D_L^{-1}N_L = N_R(D_LD_R + N_LN_R)^{-1}N_L \tag{14.31}$$

If we assume that we are working with stable transfer functions, the argument of the inverse operation would be nonsingular. Then, the only way in which the closed-loop transfer function $G_{sp}(s)$ can lose rank is at the values of $s$ that cause either $N_R(s)$ or $N_L(s)$ to lose rank. This prompts the following observation:

---

The closed-loop zeros are equal to the open-loop zeros, i.e., the process zeros are invariant (do not change) under feedback.

---

### 14.4.2 Stability of MIMO Closed-Loop Systems

---

The closed-loop system is stable if and only if all the poles of the closed-loop TFM are in the LHP.

---

The characteristic equation for MIMO systems is defined as

$$\det(I + GG_c) = 0 \tag{14.32}$$

In general, *if the open-loop system is stable*, the stability of the closed-loop system can be determined from the poles of the characteristic equation, $\det(I + GG_c)$. In the control literature, the TFM $(I + GG_c)$ is referred to as the *return difference operator*.

Following similar arguments to the SISO case, we can generalize the Nyquist stability criterion (Chapter 11) to MIMO systems. Consider the constant-gain feedback control system as shown in Figure 14.6, and assume that $G(s)$ is stable. Then, we have the following definition:

---

The closed-loop system is stable if and only if the number of net clockwise encirclements of the origin $(-1, 0)$ by the plot of $\det(I + k\,(G(s))$ is zero as $s$ goes from $-\infty$ to $+\infty$.

---

### Example 14.6

Consider the following $2 \times 2$ process,

$$\begin{aligned} y_1 &= \frac{1}{0.1s+1}u_1 + \frac{5}{s+1}u_2 \\ y_2 &= \frac{1}{0.5s+1}u_1 + \frac{2}{0.4s+1}u_2 \end{aligned} \tag{14.33}$$

The TFM for this system is given by

$$G(s) = \begin{bmatrix} \dfrac{1}{0.1s+1} & \dfrac{5}{s+1} \\ \dfrac{1}{0.5s+1} & \dfrac{2}{0.4s+1} \end{bmatrix} \tag{14.34}$$

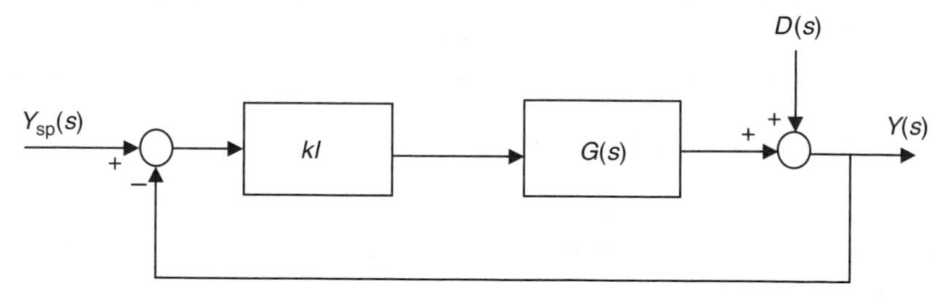


**FIGURE 14.6** Multivariable system with a constant-gain controller.

We shall choose two proportional controllers for this multiloop feedback control system.

$$g_{c_1} = k_{c_1} \; ; \; g_{c_2} = k_{c_2}$$
$$G_c = \begin{bmatrix} k_{c_1} & 0 \\ 0 & k_{c_2} \end{bmatrix} \tag{14.35}$$

Now, we can construct the return difference operator:

$$I + GG_c = \begin{bmatrix} 1 & 0 \\ 0 & 1 \end{bmatrix} + \begin{bmatrix} \dfrac{1}{0.1s+1} & \dfrac{5}{s+1} \\ \dfrac{1}{0.5s+1} & \dfrac{2}{0.4s+1} \end{bmatrix} \begin{bmatrix} k_{c_1} & 0 \\ 0 & k_{c_2} \end{bmatrix}$$
$$= \begin{bmatrix} 1 & 0 \\ 0 & 1 \end{bmatrix} + \begin{bmatrix} \dfrac{k_{c_1}}{0.1s+1} & \dfrac{5k_{c_2}}{s+1} \\ \dfrac{k_{c_1}}{0.5s+1} & \dfrac{2k_{c_2}}{0.4s+1} \end{bmatrix} \tag{14.36}$$
$$= \begin{bmatrix} 1 + \dfrac{k_{c_1}}{0.1s+1} & \dfrac{5k_{c_2}}{s+1} \\ \dfrac{k_{c_1}}{0.5s+1} & 1 + \dfrac{2k_{c_2}}{0.4s+1} \end{bmatrix}$$

The characteristic equation can be expressed as

$$\det(I + GG_c) = \left(1 + \frac{k_{c_1}}{0.1s+1}\right)\left(1 + \frac{2k_{c_2}}{0.4s+1}\right) - \left(\frac{5k_{c_2}}{s+1}\right)\left(\frac{k_{c_1}}{0.5s+1}\right) = 0 \tag{14.37}$$

By eliminating the denominator of the resulting rational polynomial, we can express the characteristic equation in polynomial form as

$$0.02s^4 + 0.1(3.1 + 2k_{c_1} + k_{c_2})s^3 + (1.29 + 1.3k_{c_2} + k_{c_1} + 0.8k_{c_1}k_{c_2})s^2 + (2 + 3.2k_{c_2} + 1.9k_{c_1} + 0.5k_{c_2}k_{c_1})s + (1 + 2k_{c_1} + k_{c_1} - 3k_{c_2}k_{c_1}) = 0 \tag{14.38}$$

We are now ready to apply, for example, the Routh's Criterion (Chapter 11) for stability to find those values of the gains $k_{c_1}$ and $k_{c_2}$ that have the roots in the LHP. This is left as an exercise for the reader.

## 14.5 SUMMARY

In this chapter, we have introduced the multivariable control problem. It is clear that when there are more than one set of controlled and manipulated variables, the selection and design of a control system becomes equally complex. To understand the way multivariable systems behave, we have reintroduced some of the concepts that we discussed earlier in the context of SISO systems. The idea of the MIMO transfer function and the associated poles and zeros are presented along with the MIMO block diagrams and the closed-loop transfer functions. The stability analysis of the closed-loop system is rephrased in the MIMO case by exploiting the motion of the newly defined MIMO closed-loop poles across the imaginary axis with the help of the Nyquist theorem. These now set the stage for analysis and design of MIMO controllers.

## REFERENCES

1. Kailath, T., *Linear Systems*, Prentice-Hall, Englewood Cliffs, NJ, 1980.
2. Rugh, W.J., *Linear System Theory*, Prentice-Hall, Englewood Cliffs, NJ, 1996.

# 15 Design of Multivariable Controllers

In this chapter, we focus on the analysis and design of the control system for multivariable processes introduced in Chapter 14. Specifically, we discuss aspects related to the quantification of the degree of interaction in a multiloop system through the concept of relative gain array (RGA). This allows us to choose the manipulated variable–controlled variable pairings that best suit the control problem.

## 15.1 MIMO FEEDBACK ANALYSIS

We pointed out before that one of the unique features of MIMO systems is the presence of interactions among the control variables (objectives). To propose an effective control strategy for MIMO systems, the nature and the impact of such interactions need to be explored in detail.

### Example 15.1

Consider a distillation column (Figure 15.1) that separates a binary mixture of ethanol and water.[1]

The control objective of this process unit is as follows:

---

To maintain the compositions of the distillate product ($x_D$) and the bottoms product ($x_B$) at their target values (set-points) using the reflux rate $F_R$ and the reboiler duty $Q_r$ as manipulated variables.

---

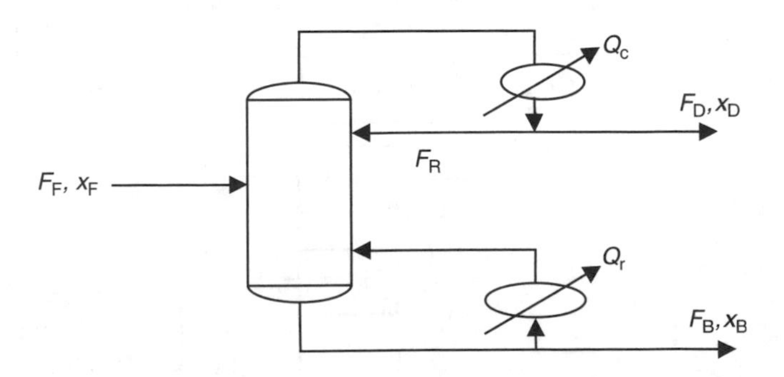


**FIGURE 15.1** Schematic representation of a distillation column.

The following MIMO TFM is determined experimentally:

$$\begin{bmatrix} x_D \\ x_B \end{bmatrix} = \begin{bmatrix} \dfrac{12.8\,e^{-s}}{1+16.7s} & \dfrac{-18.9\,e^{-3s}}{1+21s} \\ \dfrac{6.6\,e^{-7s}}{1+10.9s} & \dfrac{-19.4\,e^{-3s}}{1+14.4s} \end{bmatrix} \begin{bmatrix} F_R \\ Q_r \end{bmatrix} + \begin{bmatrix} \dfrac{3.8\,e^{-8.1s}}{14.9s+1} \\ \dfrac{4.9\,e^{-3.4s}}{13.2s+1} \end{bmatrix} F_F \tag{15.1}$$

In designing the multivariable control system for this process, two questions arise:

**Question 1:** Should we use $F_R$ for controlling $x_D$ and $Q_r$ for $x_B$ or vice versa?

**Question 2:** Should we design the controllers separately or have two single loops?

In this example, both off-diagonal elements of the plant TFM are nonzero. This means that changing one of the manipulated variables will affect both compositions, thus indicating the source of the interaction between two control loops (Figure 15.2).

For demonstration purposes, let us design two SISO controllers with the following loop assignments:

$$F_R \Rightarrow x_D$$

$$Q_r \Rightarrow x_B$$

This selection makes intuitive sense, as the reflux rate appears to affect the distillate product quality more directly and quickly just as the reboiler duty does for the bottoms product quality. We shall use the Z–N tuning technique (Chapter 12) for tuning two PID controllers.

### *Loop* $F_R - x_D$

We first consider the $F_R - x_D$ loop, and ignore the other loop (placed in manual). Figure 15.3 shows the closed-loop response for a P controller for this loop with the

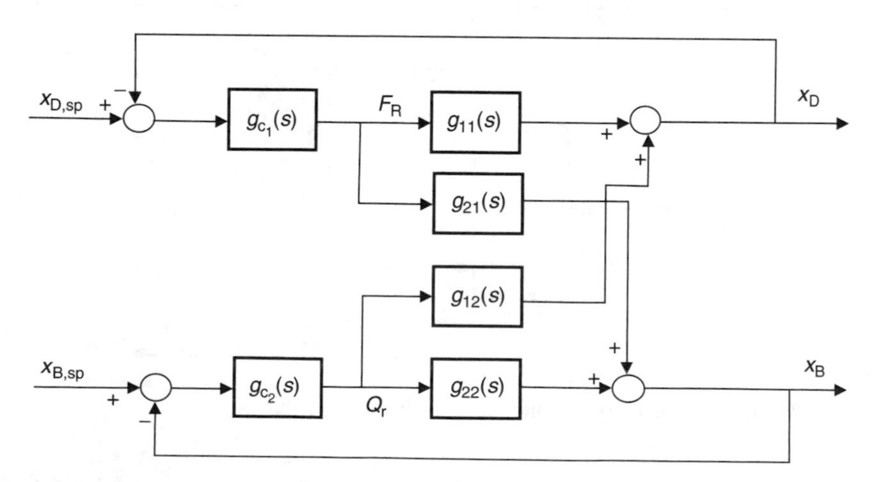


**FIGURE 15.2** Interactions in the distillation control system.

gain $k_c = 2$. Hence, the ultimate gain is $k_{cu} = 2$, and the ultimate period can be determined as $p_u = 4$.

Figure 15.4 illustrates the performance of the PI controller with $k_c = 0.9$ and $\tau_I = 3.33$ for a unit-step set-point change. The performance, slightly underdamped, appears to be satisfactory.

### *Loop* $Q_r - x_B$

Following a similar procedure, we design a PI controller for the $Q_r - x_B$ loop, while keeping the other loop in manual. In this case, we get $k_{cu} = -0.42$ and $p_u = 11$.

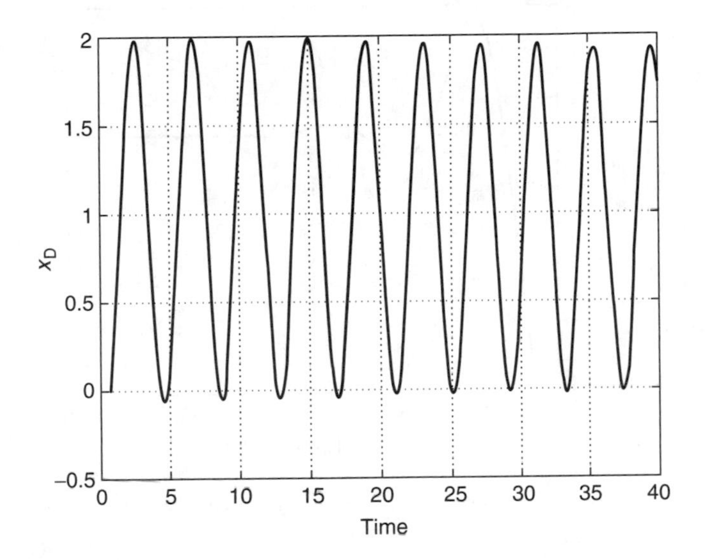


**FIGURE 15.3** Closed-loop response for $x_D$ at the ultimate gain.

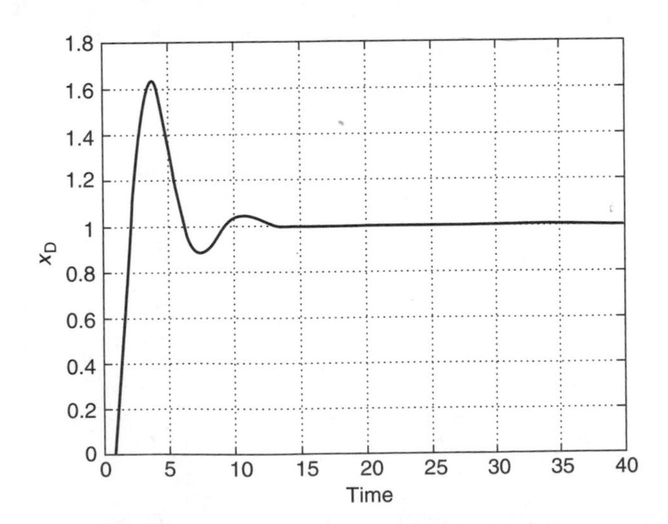


**FIGURE 15.4** Closed-loop response of $x_D$ to a set-point change.

Note that the gain is negative due to the negative steady-state gain of the corresponding transfer function element (Eq. [15.1]). Figure 15.5 displays the closed-loop response to a unit-step set-point change for a PI controller with $k_c = -0.189$ and $\tau_I = 0.16$. Again, the performance appears to be acceptable.

Now, let us test the performance of the control system. Figures 15.6–15.8 show the performance of each controller. We can clearly see that both controllers perform again very well independently (when the other loop is in manual). Notably, in Figure 15.6 and Figure 15.7, the set-point is reached very quickly for the automatic

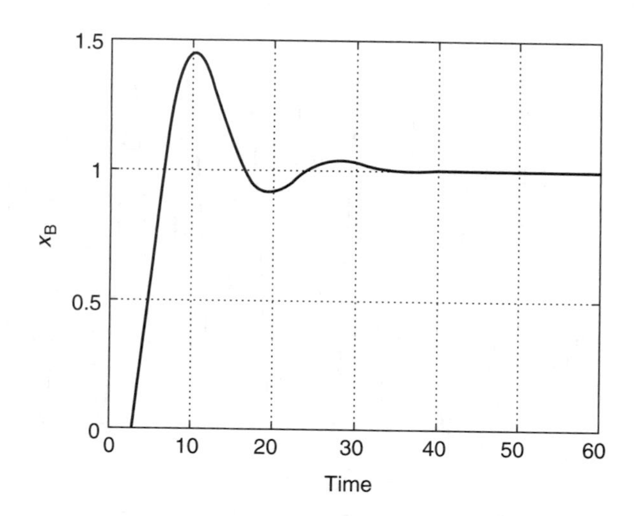


**FIGURE 15.5** Closed-loop response of $x_B$ to a set-point change.

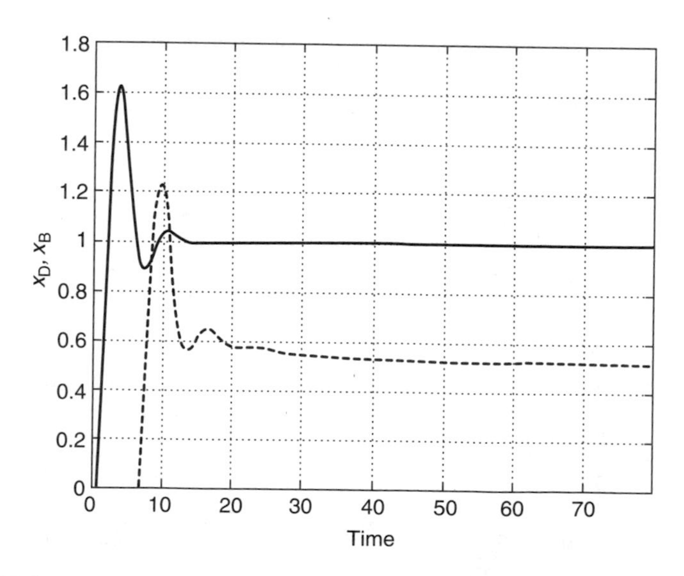


**FIGURE 15.6** $F_R - x_D$ (solid) loop closed and $Q_r - x_B$ loop (dashed) open.

loop and the other variable deviates from its desired value as it is not under feedback control. When a higher purity distillate product is desired, which causes the reflux rate to increase as this, in turn, makes the bottoms composition higher (Figure 15.6). To increase the bottoms composition, the boilup rate is decreased by the controller and this causes the distillate product also to become purer (Figure 15.7).

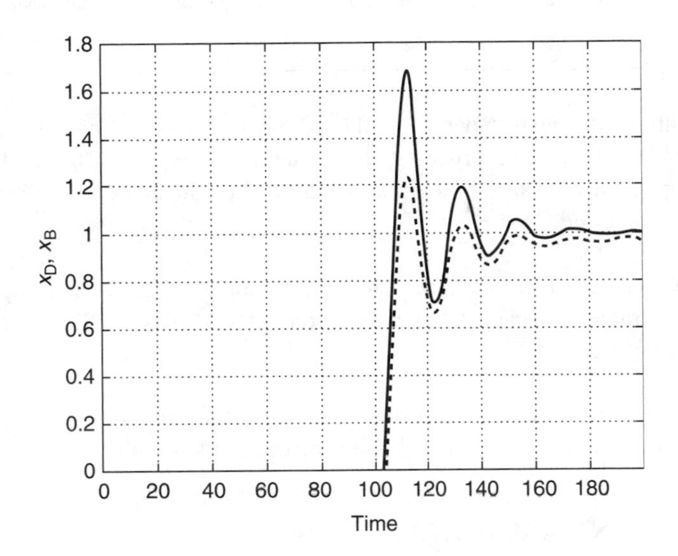


**FIGURE 15.7** $F_R - x_D$ loop (solid) open and $Q_r - x_B$ loop (dashed) closed.

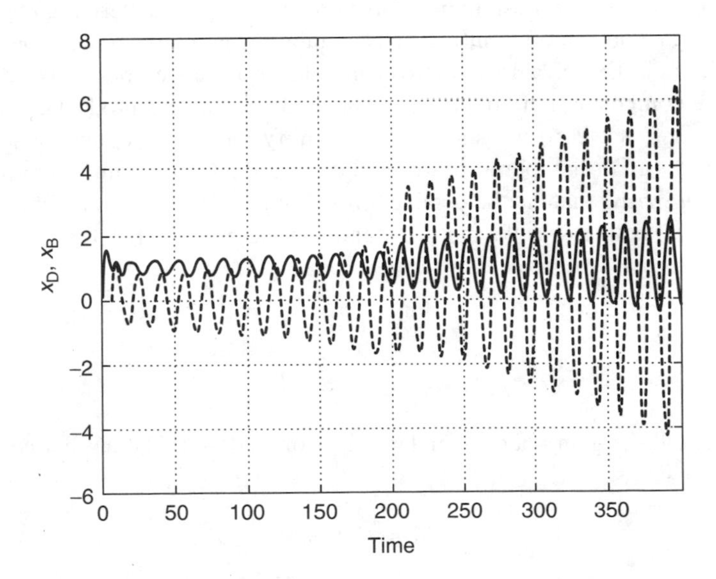


**FIGURE 15.8** Set-point response for $x_D$ when both $F_R - x_D$ loop and $Q_r - x_B$ loop are closed.

When both loops are placed in automatic (Figure 15.8) and a set-point change is made in the distillate product, the closed-loop performance degrades significantly, and becomes unstable.

---

If the feedback controllers of each individual loop are tuned independently, we cannot guarantee stability of the multivariable system when both loops are closed.

---

The above example underscores the need for understanding the interactions among control loops in a process before attempting to design the multivariable (multiloop) control system. To guide the control engineer in this effort, the following are required:

- A method to measure the extent of interaction among the loops, the RGA.
- A strategy to cancel the interaction effects between the loops, the decoupling strategy.

In other words, we need to establish the means to account for and counteract the presence of interactions in designing the control system.

## 15.2 RGA INTERACTION MEASURE

For an MIMO system to be noninteracting, the TFM $G(s)$ should be diagonal, but this is not the case for most multivariable processes as the underlying physical or chemical phenomena often influence each other. A good example is the distillation column in Example 15.1 where changing, for instance, the reflux rate influences the composition of both the distillate and the bottoms products.

The first step is to measure and to quantify the extent of interactions in a given process. For that purpose, we shall use the RGA (sometimes referred to as the *Bristol Array*) that can be obtained from the gains of the plant TFM.[2]

We shall define $K$ as the matrix of gains of the process transfer function. For example, for a $n \times n$ process, we would have

$$K = \begin{bmatrix} k_{11} & \dots & k_{1n} \\ \vdots & \dots & \vdots \\ k_{n1} & \dots & k_{nn} \end{bmatrix} \tag{15.2}$$

As expected, the gain matrix is independent of $s$ (static). Having defined $K$, the RGA is given as:

$$\Lambda = K^{-1} * K^{\mathrm{T}} \tag{15.3}$$

The operator $*$ is referred to as Hadamard product and indicates element-by-element multiplication of two matrices.

---

The $i$th loop of a process is considered to be *interacting* with the $j$th loop if the $(i,j)$ element of $\Lambda$ is nonzero.

---

**Example 15.2**

Let the process gain matrix for a process be given as

$$K = \begin{bmatrix} 1 & 2 \\ 3 & 4 \end{bmatrix}$$

Then, its inverse can be calculated as

$$K^{-1} = \frac{1}{4-6}\begin{bmatrix} 4 & -2 \\ -3 & 1 \end{bmatrix} = \begin{bmatrix} -2 & 1 \\ 1.5 & -0.5 \end{bmatrix}$$

The transpose of the gain matrix is

$$K^{\mathrm{T}} = \begin{bmatrix} 1 & 3 \\ 2 & 4 \end{bmatrix}$$

Consequently, the RGA is given as

$$\Lambda = \begin{bmatrix} -2 & 3 \\ 3 & -2 \end{bmatrix}$$

All elements of RGA are different from zero, thus indicating the presence of interactions between both loops. The sign and the magnitude of the RGA elements also have some significance as will be explained next.

The relative gain between a controlled variable $y_i$ and a manipulated variable $u_j$ is defined by

$$\lambda_{ij} = \frac{(\Delta y_i/\Delta u_j)_u}{(\Delta y_i/\Delta u_j)_y} \tag{15.4}$$

The subscripts on the parentheses have the following meaning: $u$ denotes constant values for all manipulated variables except $u_j$ (i.e., all loops open) as in Figure 15.9a; $y$ indicates that all outputs except $y_i$ are kept constant by the control loops (i.e., all loops are closed) as in Figure 15.9b.

Thus, for a process with $n$ inputs and $n$ outputs, the RGA is expressed as

$$\Lambda = \begin{bmatrix} \lambda_{11} & \cdots & \lambda_{1n} \\ \vdots & \cdots & \vdots \\ \lambda_{n1} & \cdots & \lambda_{nn} \end{bmatrix} \tag{15.5}$$

For the special case of a $2 \times 2$ process, the RGA reduces to

$$\Lambda = \begin{bmatrix} \lambda & 1-\lambda \\ 1-\lambda & \lambda \end{bmatrix} \tag{15.6}$$

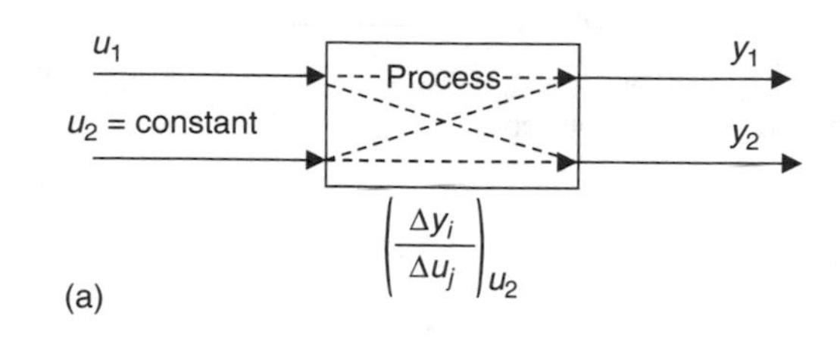


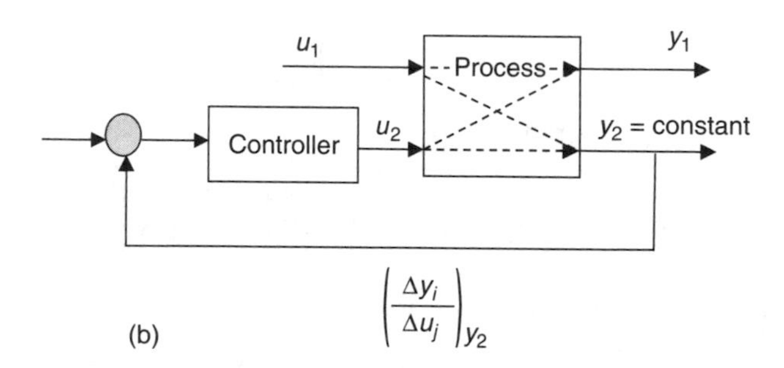


**FIGURE 15.9** Schematic representation for the calculation of the relative gain.

The RGA element is given by

$$\lambda = \frac{1}{1 - \xi} \tag{15.7}$$

where the gain ratio is calculated as

$$\xi = \frac{k_{12}k_{21}}{k_{11}k_{22}} \tag{15.8}$$

from the gain matrix elements.

### 15.2.1 Selection of Loops

The RGA helps the control designer by indicating which controlled variable (output) should be paired with which manipulated variable (input) for the application of a multiloop control strategy. Intuitively, the designer selects the pairing that has the largest relative gain. Ideally, as can be seen in Eq. (15.6), if $\lambda$ is close to 1, then RGA would indicate the obvious pairing between the first output and the first input and the second output and the second input.

There are three important properties of RGA:

1. The rows and columns of RGA add up to 1.
2. If an element of the gain matrix is zero, $k_{ij} = 0$, then the corresponding relative gain is also zero, $\lambda_{ij} = 0$.
3. If a loop is paired on a negative relative gain, the result is an unstable system or a system with inverse response.

**Example 15.3**

Consider the TFM of the distillation column given in Eq. (15.1). The gain matrix is given as

$$K = \begin{bmatrix} 12.85 & -18.89 \\ 6.62 & -19.4 \end{bmatrix} \tag{15.9}$$

Then, we can calculate the following:

$$\xi = \frac{(-18.89)6.62}{12.85x(-19.4)} = 0.50$$

$$\lambda = \frac{1}{1 - 0.5} = 2 \tag{15.10}$$

Finally, the RGA is given as

$$\Lambda = \begin{bmatrix} 2 & -1 \\ -1 & 2 \end{bmatrix} \tag{15.11}$$

Thus, RGA indicates that $x_D$ should be controlled by $F_R$, and $x_B$ should be paired with $Q_r$. We also note that the off-diagonal terms are large, indicating the presence of significant interactions.

For a process with $n$ outputs and $n$ inputs, there are $n!$ different ways to form the control loops, i.e., $n!$ *control configurations*. The RGA offers the following rule that aims at pairings that result in minimum loop interactions:

---

Select the control loop pairing between the outputs $y_i$ with the inputs $u_j$ in such a way that the relative gains $\lambda_{ij}$ are positive and as close as possible to unity.

---

We note that:

1. The RGA method provides a measure of interaction based on steady-state considerations only.
2. It does not guarantee that the dynamic interaction between loops will also be minimal. Dynamic extensions of RGA have also been developed.[2]

**Example 15.4**

Let us consider the thermal-mixing tank described in Example 15.2. In this case, the gain matrix is given by

$$K = \begin{bmatrix} 1 & 1 \\ \Delta T_C & \Delta T_H \end{bmatrix} \tag{15.12}$$

The gain ratio can be expressed as

$$\xi = \frac{\Delta T_C}{\Delta T_H}$$

Thus, $\lambda$ is given by

$$\lambda = \frac{1}{1 - \dfrac{\Delta T_{\mathrm{C}}}{\Delta T_{\mathrm{H}}}}$$

Therefore, we observe that the RGA depends on the process operating conditions. Note that from the definitions of $\Delta T_{\mathrm{C}}$ and $\Delta T_{\mathrm{H}}$ we have,

$$\Delta T_{\mathrm{C}} = \frac{T_{\mathrm{C_s}} - T_{\mathrm{s}}}{T_{\mathrm{s}}} \quad \text{and} \quad \Delta T_{\mathrm{H}} = \frac{T_{\mathrm{H_s}} - T_{\mathrm{s}}}{T_{\mathrm{s}}}$$

But $T_{\mathrm{C_s}} < T_{\mathrm{s}} < T_{\mathrm{H_s}}$, consequently $\Delta T_{\mathrm{C}}$ is always negative. Then, $\lambda$ is also always positive and less than 1, and we can have the following scenarios:

- If $\Delta T_{\mathrm{C}} \ll \Delta T_{\mathrm{H}}$, then $\lambda$ is close to 1.
- If $\Delta T_{\mathrm{C}} \gg \Delta T_{\mathrm{H}}$, then $\lambda$ is close to 0.

Thus, the following conclusions can be drawn from the RGA analysis:

- If $\Delta T_{\mathrm{C}} \ll \Delta T_{\mathrm{H}}$, then $u_1$ should be used to control the level of the tank and $u_2$ should be used to control the temperature of the tank.
- If $\Delta T_{\mathrm{C}} \gg \Delta T_{\mathrm{H}}$, then $u_1$ should be used to control the temperature of the tank and $u_2$ should be used to control the level of the tank.

## 15.3 MULTILOOP CONTROLLER DESIGN

The design of multiple single-loop controllers for MIMO systems relies on two key factors:

- Judicious pairing of loops
- Effective tuning of controllers for each loop

It should not be surprising that when the RGA is close to ideal ($\lambda_{ii}$ very close to 1), the multiloop controllers are likely to function very well. When the RGA indicates strong interactions for the chosen loop pairing, multiloop controllers are *not* likely to perform well even for the best possible tuning.

One of two possible approaches may be taken to address this problem. First, controller tuning for multiloop systems can be performed as

- Tune the first loop using a conventional tuning technique
- Tune the second loop with the first loop under closed-loop

The key to this approach is the initial independent tuning of each loop. Then, the controllers are detuned using the following iterative procedure:

- Keep other loops in manual, tune each controller independently so that satisfactory performance is obtained for that loop
- Restore all controllers under automatic control and readjust the tuning parameters until overall performance becomes satisfactory

In almost all cases, the controllers will need to be made more conservative, i.e., the controller gains reduced and the integral action increased in comparison to when operating alone. Naturally, these steps require a substantial amount of simulation and testing.

**Example 15.5**

We shall illustrate a method that relies on the detuning strategy based on the following:[1]

- Use any of the single loop tuning techniques (e.g., Z–N tuning technique) to obtain starting values for individual controllers
- These gains should be reduced using the following expressions that depend on the relative gain $\lambda$

$$k_{ci} = \left(\lambda - \sqrt{\lambda^2 - \lambda}\right)k'_{ci} \quad \text{when} \quad \lambda > 1 \tag{15.13}$$

$$k_{ci} = \left|\lambda + \sqrt{\lambda^2 - \lambda}\right| k'_{ci} \quad \text{when} \quad \lambda < 1 \tag{15.14}$$

where $k'_{ci}$ is the initial proportional gain for the $i$th loop. We note that in Eq. (15.14), the argument of the absolute value operation would be a complex number, thus it would also have implications for the phase. A more in-depth discussion of these concepts can be found elsewhere.[1]

Consider the distillation column example (Example 15.1) with the original PI-tuning parameters for both loops:

$$\text{Top Loop: } k_{c_1} = 0.9 \text{ and } \tau_I = 3.33.$$

$$\text{Bottom Loop: } k_{c_2} = -0.189 \text{ and } \tau_I = 9.16.$$

Based on Eq. (15.13), we will detune the controllers as follows:

$$k_{c,\text{new}} = 2 - \sqrt{4 - 2}k_{c,\text{old}}$$

Thus, the new controller gains are $k_{c_1} = 0.526$ and $k_{c_2} = -0.11$. Figure 15.10 shows the performance improvement for the closed-loop system with both controllers in automatic, with the new settings, compared with Figure 15.8.

## 15.4 DESIGN OF NONINTERACTING CONTROL LOOPS: DECOUPLERS

The RGA analysis indicates how the inputs should be paired with the outputs to form control loops with the smallest possible amount of interaction. However, the remaining interactions may be as small as possible yet may not be small enough. As can be seen in Example 15.5 (Figure 15.10), the bottoms composition does react substantially to a change in the set-point of the distillate composition,

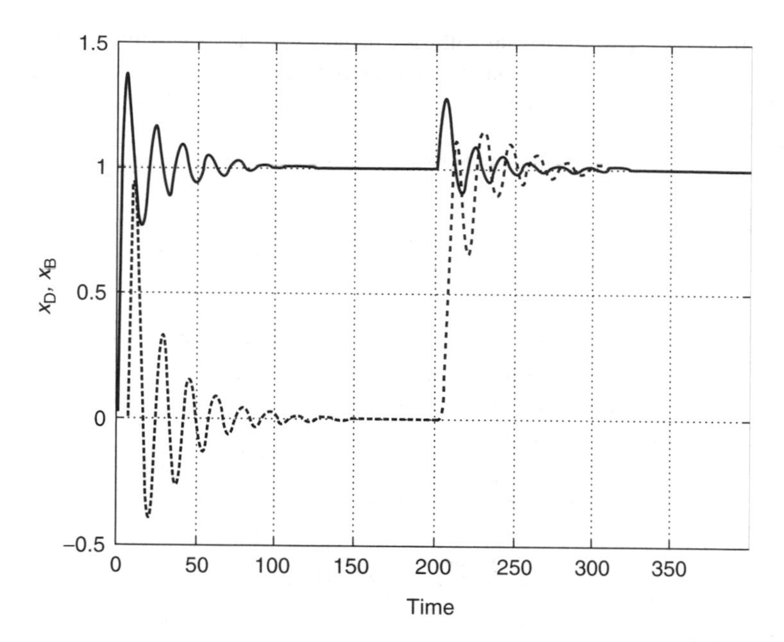


**FIGURE 15.10** Closed-loop response for a set-point change in the distillate composition after the detuning procedure.

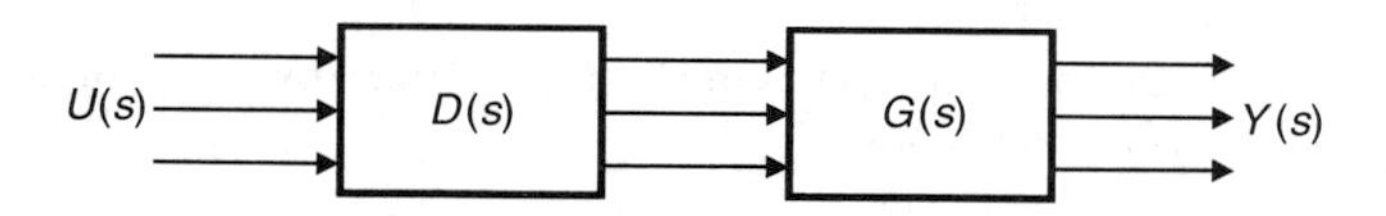


**FIGURE 15.11** Process with a decoupler.

indicating that the two loops are indeed interacting, and this performance may not be acceptable from the viewpoint of product quality objectives.

A possible strategy to achieve a noninteracting closed-loop system is *decoupling*. Consider the process with a *decoupler* as depicted in Figure 15.11. The objective is to find a matrix $D(s)$ such that

$$Y(s) = G(s)D(s)U(s) = Q(s)U(s) \tag{15.15}$$

where the matrix $Q(s)$ has some desirable properties. Depending on the extent of decoupling desired, the control engineer specifies $Q(s)$. To illustrate, we will use $2 \times 2$ system, described by the following process TFM:

$$G(s) = \begin{bmatrix} g_{11}(s) & g_{12}(s) \\ g_{21}(s) & g_{22}(s) \end{bmatrix} \tag{15.16}$$

For perfect (*ideal*) decoupling, we can specify the matrix $Q(s)$ to be

$$Q(s) = \begin{bmatrix} g_{11}(s) & 0 \\ 0 & g_{22}(s) \end{bmatrix} \tag{15.17}$$

This eliminates the off-diagonal (interaction) terms completely and results in the following structure of the decoupling matrix:

$$D(s) = \begin{bmatrix} \dfrac{g_{11}(s)g_{22}(s)}{g_{11}(s)g_{22}(s) - g_{12}(s)g_{21}(s)} & \dfrac{-g_{12}(s)g_{22}(s)}{g_{11}(s)g_{22}(s) - g_{12}(s)g_{21}(s)} \\ \dfrac{-g_{21}(s)g_{11}(s)}{g_{11}(s)g_{22}(s) - g_{12}(s)g_{21}(s)} & \dfrac{g_{11}(s)g_{22}(s)}{g_{11}(s)g_{22}(s) - g_{12}(s)g_{21}(s)} \end{bmatrix} \tag{15.18}$$

This, of course, is a complex (possibly high order) TFM, the price being paid for *perfect* decoupling. One can also specify the decoupler elements to obtain a *simplified* decoupling structure:

$$D(s) = \begin{bmatrix} 1 & d_{12}(s) \\ d_{21}(s) & 1 \end{bmatrix} = \begin{bmatrix} 1 & \dfrac{-g_{12}(s)}{g_{11}(s)} \\ \dfrac{-g_{21}(s)}{g_{22}(s)} & 1 \end{bmatrix} \tag{15.19}$$

This results in the decoupled system given as

$$Q(s) = \begin{bmatrix} g_{11}(s)\left(1 - \dfrac{g_{12}(s)g_{21}(s)}{g_{11}(s)g_{22}(s)}\right) & 0 \\ 0 & g_{22}(s)\left(1 - \dfrac{g_{12}(s)g_{21}(s)}{g_{11}(s)g_{22}(s)}\right) \end{bmatrix} \tag{15.20}$$

Note that the system is again decoupled but the diagonal elements are not the same as the original process. The closed-loop system with a simplified decoupling strategy is shown in Figure 15.12. We should note that the decoupler elements are considered to be part of the control structure, and the new controller

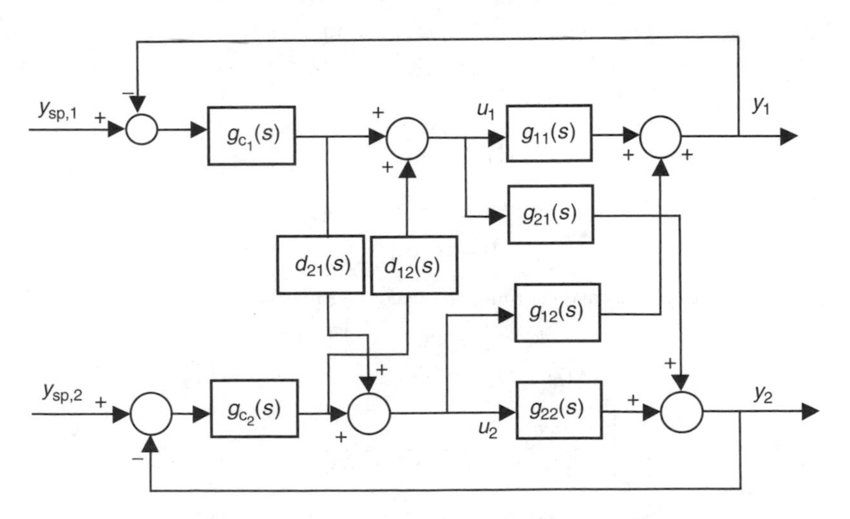


**FIGURE 15.12** Closed-loop representation of the simplified decoupling scheme.

matrix is no longer diagonal in structure due to the decoupler elements aimed at compensating for the loop interactions:

$$G_{c,new}(s) = G_c(s)D(s) = \begin{bmatrix} g_{c_1} & 0 \\ 0 & g_{c_2} \end{bmatrix} \begin{bmatrix} 1 & d_{12} \\ d_{21} & 1 \end{bmatrix} = \begin{bmatrix} g_{c_1} & g_{c_1}d_{12} \\ g_{c_2}d_{21} & g_{c_2} \end{bmatrix} \quad (15.21)$$

In general, for any order system, with the desired decoupling, $Q(s) = G(s)D(s)$, we can compute $D(s) = G(s)^{-1}Q(s)$.

---

1. The decoupling compensators are dynamic decouplers that will decouple the loops perfectly.
2. The decoupling technique is heavily model dependent.
3. A dynamic decoupler is physically realizable if both $d_{12}(s)$ and $d_{21}(s)$ are realizable.
4. If the decoupler is designed based on the steady-state models (steady-state gains), then the decoupler is referred to as static.

---

In summary, different approaches to decoupling are:

- *Static decoupling* uses only static part of the transfer function model (gains)
- *Dynamic decoupling* uses the whole transfer function model
- *One-way decoupling* uses decoupler for only one variable
- *Two-way* (*simplified* and *ideal*) *decoupling* uses decouplers for all variables

**Example 15.6**

Consider again the thermal-mixing tank process, whose TFM is given by

$$G(s) = \begin{bmatrix} \frac{1}{s+0.5} & \frac{1}{s+0.5} \\ \frac{\Delta T_C}{s+1} & \frac{\Delta T_H}{s+1} \end{bmatrix}$$

We shall design a steady-state decoupler, given as

$$D = [G(0)]^{-1}Q$$

Thus, in this case, we have the following decoupler matrix

$$D = \frac{1}{2(\Delta T_H - \Delta T_C)} \begin{bmatrix} \Delta T_H & 2 \\ -\Delta T_C & 2 \end{bmatrix} \begin{bmatrix} 2 & 0 \\ 0 & \Delta T_H \end{bmatrix}$$

$$= \begin{bmatrix} \frac{\Delta T_H}{\Delta T_H - \Delta T_C} & -\frac{\Delta T_H}{\Delta T_H - \Delta T_C} \\ \frac{\Delta T_C}{\Delta T_H - \Delta T_C} & \frac{\Delta T_H}{\Delta T_H - \Delta T_C} \end{bmatrix} \quad (15.23)$$

The following are the decoupled process results:

$$Q(s) = G(s)D(s)$$

$$= \begin{bmatrix} \dfrac{1}{s+0.5} & \dfrac{1}{s+0.5} \\ \dfrac{\Delta T_C}{s+1} & \dfrac{\Delta T_H}{s+1} \end{bmatrix} \begin{bmatrix} \dfrac{\Delta T_H}{\Delta T_H - \Delta T_C} & -\dfrac{\Delta T_H}{\Delta T_H - \Delta T_C} \\ \dfrac{\Delta T_C}{\Delta T_H - \Delta T_C} & \dfrac{\Delta T_H}{\Delta T_H - \Delta T_C} \end{bmatrix} \tag{15.24}$$

$$= \begin{bmatrix} \dfrac{1}{s+0.5} & 0 \\ 0 & \dfrac{\Delta T_H}{s+1} \end{bmatrix}$$

For this specific case, the steady-state decoupler appears to have decoupled the system dynamically as well.

The reliance of the decoupling schemes on explicit process model elements can be problematic if the model is subject to uncertainty. In other words, the model transfer function elements may not accurately describe the process dynamics, and the resulting decouplers may not produce the expected performance. Arkun et al.[4] studied a number of decoupling strategies and their performance in the presence of model uncertainty. The issue of model uncertainty will be discussed in detail in Chapter 17.

## 15.5 SUMMARY

The presence of interactions among loops is a unique feature that distinguishes MIMO systems from SISO systems. In this chapter, we have discussed the implications of loop interactions and defined RGA as a means of quantifying them. The RGA helps the control designer by pointing out the loop pairings that would result in the least amount of interaction when a multiloop controller is designed for the MIMO system. To address the presence of interactions more effectively (even though one chooses the pairings that would minimize them), we have introduced an additional compensator, the decoupler. Depending on the type and extent of interactions that the control designer is willing to allow, the decoupler matrix takes a variety of forms.

## CONTINUING PROBLEM

Consider the blending process shown in Figure 15.13 with two control loops that are intended to maintain level and product composition (mole fraction) at their set-points.

In this process, the controlled and manipulated variables and possible disturbance variables are given below.

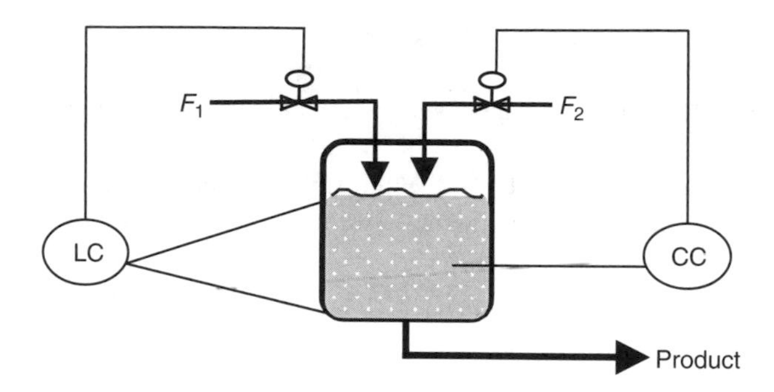


**FIGURE 15.13** Blending process with control system.

## Controlled

- Level $h$ of tank
- Product mole fraction $x$

## Manipulated

- The flow rate $F_1$ of stream 1
- The flow rate $F_2$ of stream 2

## Disturbances

- The mole fraction $x_2$ of stream 2

The transfer functions for this process are given as

$$\frac{x(s)}{F_2(s)} = \frac{0.1\mathrm{e}^{-s}}{2.5s + 1}, \qquad \frac{h(s)}{F_2(s)} = \frac{\mathrm{e}^{-0.1s}}{5s + 1}$$
$$\frac{x(s)}{F_1(s)} = \frac{-0.1\mathrm{e}^{-s}}{2.5s + 1}, \qquad \frac{h(s)}{F_1(s)} = \frac{\mathrm{e}^{-0.1s}}{5s + 1} \tag{15.25}$$

Using the standard matrix notation, the plant TFM becomes

$$G(s) = \begin{bmatrix} \dfrac{\mathrm{e}^{-0.1s}}{5s + 1} & \dfrac{\mathrm{e}^{-0.1s}}{5s + 1} \\ \dfrac{-0.1\mathrm{e}^{-s}}{2.5s + 1} & \dfrac{0.1\mathrm{e}^{-s}}{2.5s + 1} \end{bmatrix} \tag{15.26}$$

The multiloop PI controller parameters based on Z–N tuning are given as follows:

$$G_c = \begin{bmatrix} 22.95\left(1 + \dfrac{1}{0.5s}\right) & 0 \\ 0 & 20.25\left(1 + \dfrac{1}{3.33s}\right) \end{bmatrix} \tag{15.27}$$

We shall

- Perform an interaction analysis using RGA and detune the controllers if necessary.
- Design and implement a decoupling controller for the process and compare the performance with conventional multiloop control.

### Solution

For a 2 × 2 system, the RGA is given by

$$\Lambda = \begin{bmatrix} \lambda & 1-\lambda \\ 1-\lambda & \lambda \end{bmatrix}$$

$$\lambda = \frac{1}{1-\xi} \quad \text{and} \quad \xi = \frac{k_{12}k_{21}}{k_{11}k_{22}}$$

In this case,

$$\xi = \frac{1(-0.1)}{1\,0.1} = -1 \quad \text{and} \quad \lambda = \frac{1}{1-(-1)} = \frac{1}{2} = 0.5$$

And we have,

$$\Lambda = \begin{bmatrix} 0.5 & 0.5 \\ 0.5 & 0.5 \end{bmatrix} \tag{15.28}$$

Thus, RGA indicates that either $F_1$ or $F_2$ could be used to control both the output variables, and the system has strong interactions at steady state, thus, independently designed multiloop controllers would not be expected to perform well when both loops are closed simultaneously.

Figure 15.14 illustrates the closed-loop responses for the tank level and product mole fraction under conventional feedback multiloop controllers using Z–N settings (independent design) and Z–N settings obtained by tuning the level controller with the concentration controller closed (sequential design). It can be seen that:

- The multiloop controller using independent tuning becomes unstable when both loops are closed simultaneously.
- The *sequential* tuning by first adjusting the composition controller and with this loop closed, tuning the level controller yields a stable response but shows a large degree of interaction.

Since the loops show a significant amount of interaction, a decoupler needs to be designed for this process. In this case, the decoupler parameters are obtained as follows:

$$D(s) = \begin{bmatrix} 1 & d_{12}(s) \\ d_{21}(s) & 1 \end{bmatrix} = \begin{bmatrix} 1 & \dfrac{-g_{12}(s)}{g_{11}(s)} \\ \dfrac{-g_{21}(s)}{g_{22}(s)} & 1 \end{bmatrix}$$

$$
= \begin{bmatrix} 1 & \dfrac{-(1)e^{-0.1s}(5s+1)}{(5s+1)(1)e^{-0.1s}} \\ \dfrac{-(0.1)e^{-s}(2.5s+1)}{(2.5s+1)(-0.1)e^{-s}} & 1 \end{bmatrix} \tag{15.29}
$$

$$
= \begin{bmatrix} 1 & -1 \\ 1 & 1 \end{bmatrix}
$$

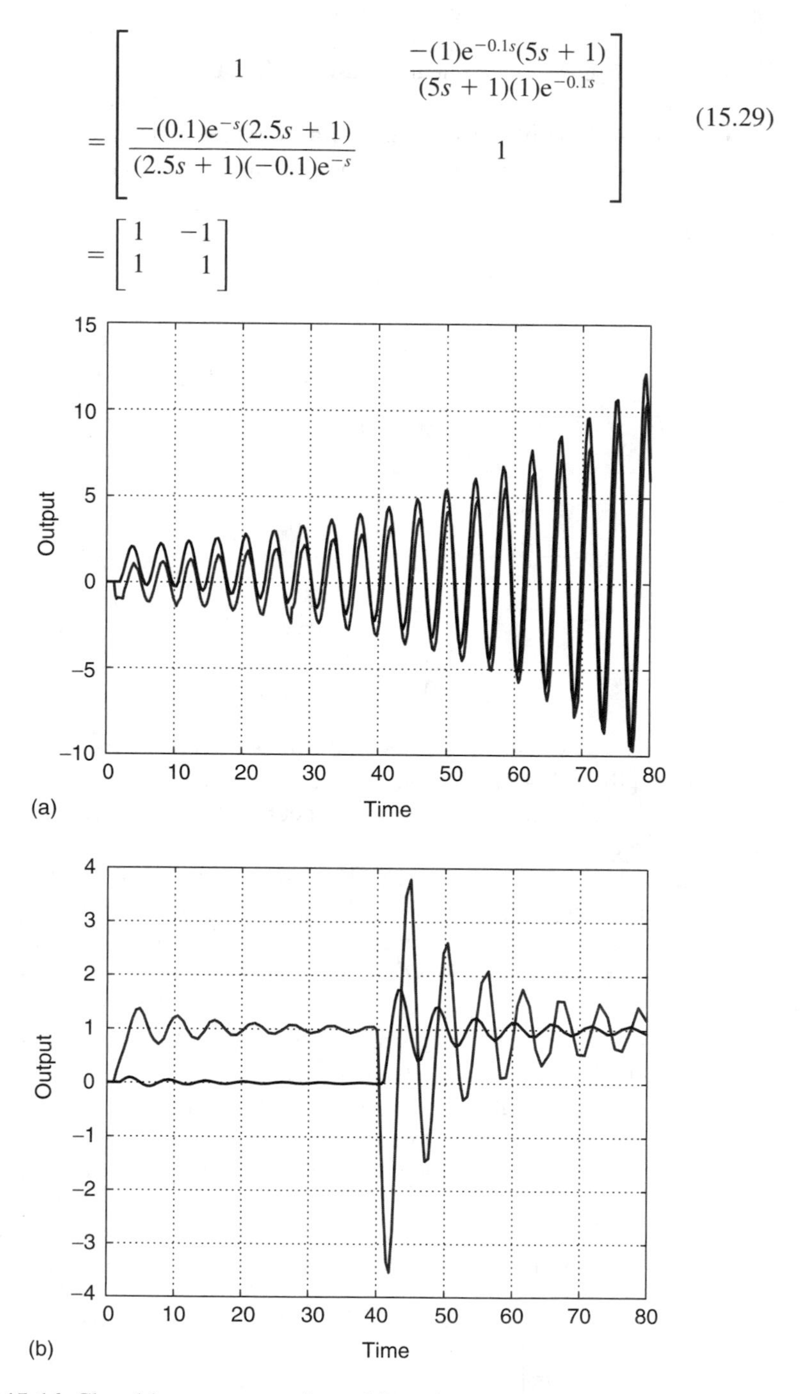


**FIGURE 15.14** Closed-loop responses for multiloop independent Z–N design (a) and for multiloop sequential Z–N design (b).

Figure 15.15 illustrates the closed-loop performance of the process using one- and two-way (simplified) decoupling schemes with perfect model. As can be seen from the responses, the decoupling strategy effectively eliminates the interactions in the system.

However, the closed response may not be acceptable due to significant oscillatory response of the composition. Detuning of the composition loop only (using Eq. [15.14]) results in the closed-loop responses in Figure 15.16.

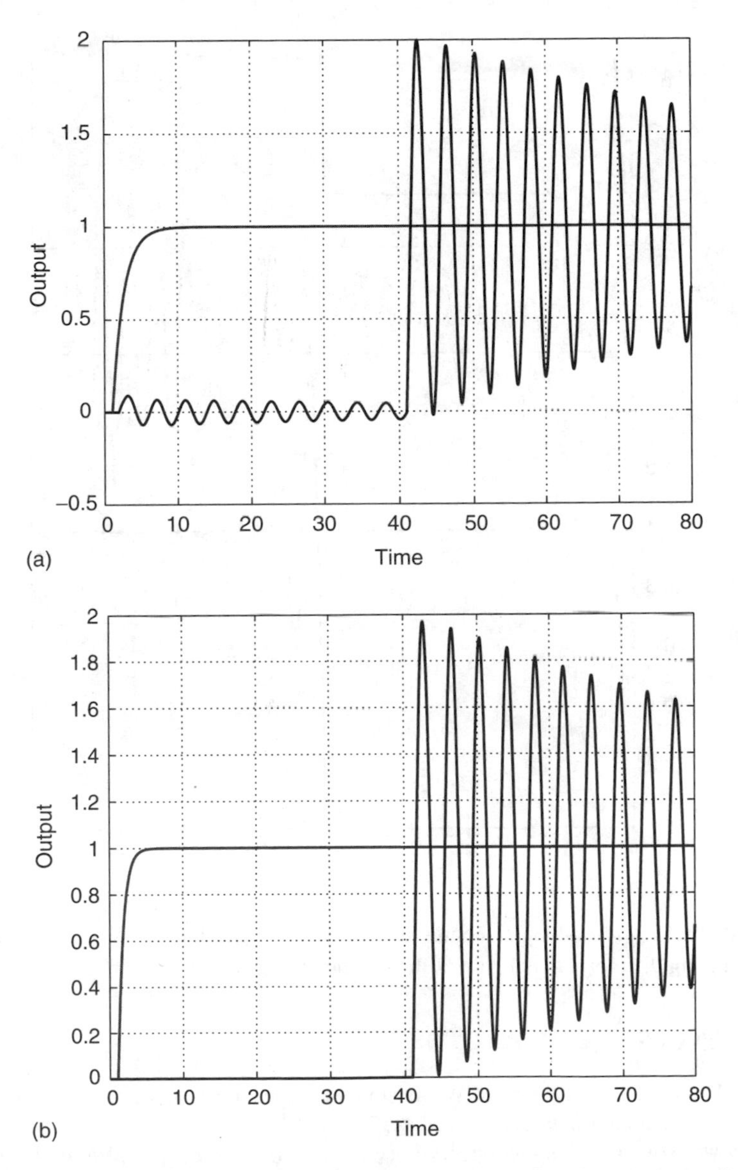


**FIGURE 15.15** Closed-loop responses for one-way (a) and for two-way decoupling (b).

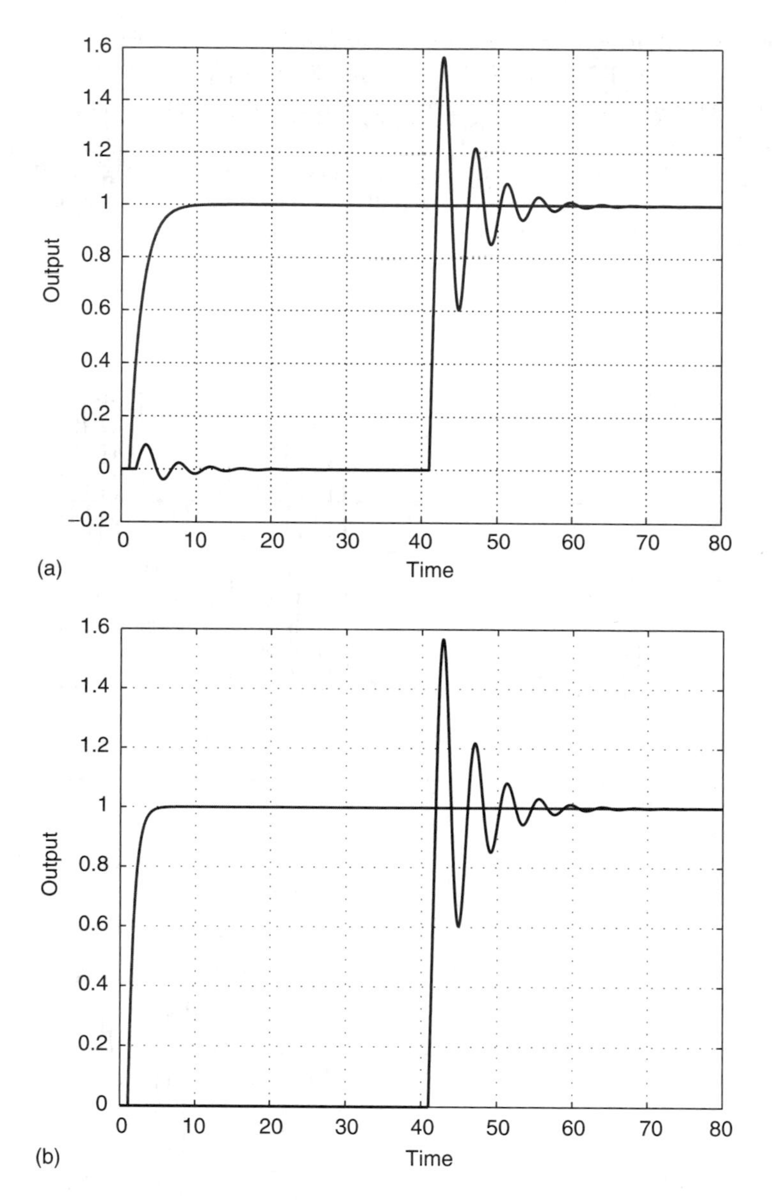


**FIGURE 15.16** Closed-loop responses for one-way (a) and for two-way decoupling (b) after detuning the proportional gain of the composition controller ($k_c = 15$).

## REFERENCES


1. Wood, R.K. and Berry, M.W., Terminal composition control of a binary distillation column, *Chem. Eng. Sci.*, 28, 1707–1717, 1973.
2. McAvoy, T.J., *Interaction Analysis — Principles and Applications*, Instrument Society of America, Research Triangle Park, NC, 1983.

3. McAvoy, T.J., Arkun, Y., Chen, R., Robinson, D., and Schnelle, P.D., A new approach to defining a dynamic relative gain, *Control Engineering Practice*, 11, 907–914, 2003.
4. Arkun, Y., Manousiouthakis, V., and Palazoglu, A., Robustness analysis of process control systems — A case study of decoupling control in distillation, *Ind. Eng. Chem. Process Des. Dev.*, 23, 93–100, 1984.

# Section V Additional Reading

With the increasing complexity of chemical plants and the demands on better controllability, multivariable control of chemical processes became one of the key challenges in the 1970s. The first contribution that explored some of the advanced multivariable control topics in the context of chemical processes was the book by Ray:

Ray, W.H., *Advanced Process Control*, McGraw-Hill, New York, 1981.

For more detailed information on special multivariable control structures such as cascade, ratio, etc., and the application of multivariable control concepts to process systems, the reader is directed to the following books:

Deshpande, P.B. (Ed.), *Multivariable Process Control*, Instrument Society of America, Research Triangle Park, NC, 1989.

Shinskey, F.G., *Controlling Multivariable Processes*, Instrument Society of America, Research Triangle Park, NC, 1981.

A double-effect evaporator process has been studied by Newell and Lee to demonstrate a number of multivariable control analysis and design techniques:

Newell, R.B. and Lee, P.L., *Applied Process Control: A Case Study*, Prentice-Hall, New York, 1989.

We have pointed out in Chapter 14 that multivariable systems exhibit some unique dynamic properties. The analysis of the pole-zero structure of multivariable systems and the special extensions of the theories of stability for linear multivariable systems can be found in the following books:

Maciejowski, J.M., *Multivariable Feedback Design*, Addison-Wesley, Reading, MA, 1989.

Skogestad, S. and Postlethwaite, I., *Multivariable Feedback Control – Analysis and Design*, Wiley, New York, 1990.

A detailed exposition on the theory of decoupling control for multivariable systems is by Hui.

Hui, L.C., *General Decoupling Theory of Multivariable Process Control Systems*, Springer, New York, 1983.

# Section V Exercises

**V.1.** Food safety is a primary concern during operation of the high-temperature short time (HTST) pasteurization of milk.[1] High temperatures result in increased shelf life, and pasteurization destroys enzymatic systems to safeguard product quality. Milk is supplied into a plate heat exchanger (PHE), where it is heated to the pasteurization temperature and then sent to a holding tube (section) where the temperature is held constant. The sketch in Figure V.1 illustrates the process.

A cascade control system is suggested to insure constant pasteurization temperature. Draw the block diagram of this feedback control system, clearly identifying all process elements (blocks) and the variables involved.

**V.2.** Consider the closed-loop diagram in Figure V.2.

For this system:

1. Define a cascade control configuration exploiting the dynamic characteristics of the process
2. Design and implement a cascade controller to improve the performance of the control system under the effect of the secondary
3. Compare the closed-loop results with those using the conventional feedback configuration shown in Figure V.2

**V.3.** Consider the process shown in Figure V.3 that adjusts the concentration of a calcium chloride salt solution. As can be seen, the flow rate of calcium chloride solution can be adjusted with the available control valve, while the water flow rate does not have any control valve:

1. Show where the flow meters would be properly placed if a ratio controller is to be implemented

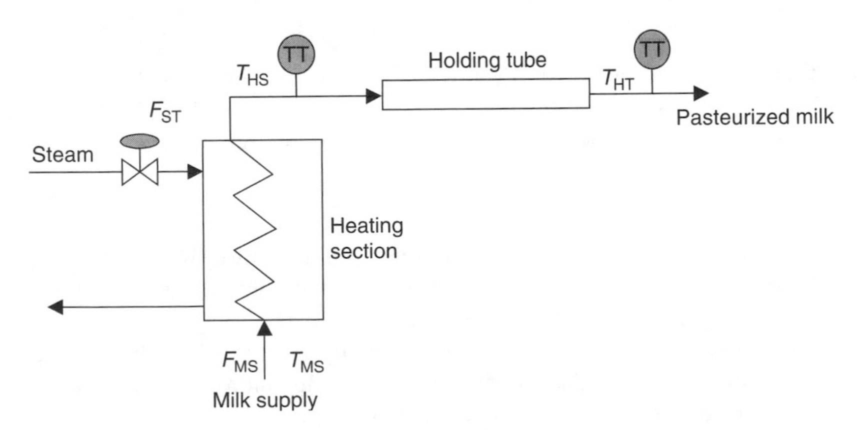


**FIGURE V.1** The sketch of high temperature short time (HTST) pasteurization of milk.

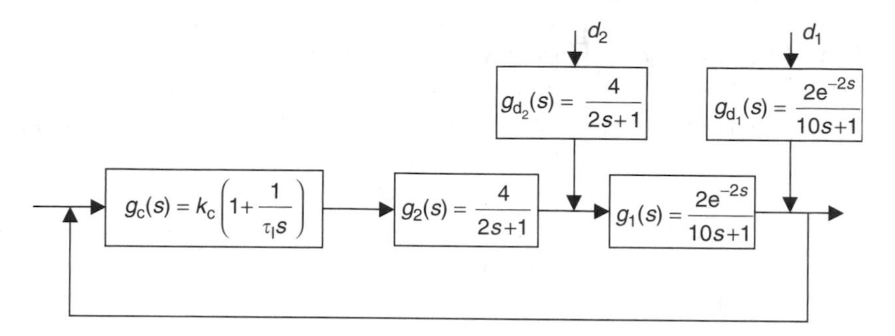


**FIGURE V.2** Closed-loop block diagram for Exercise V.2.

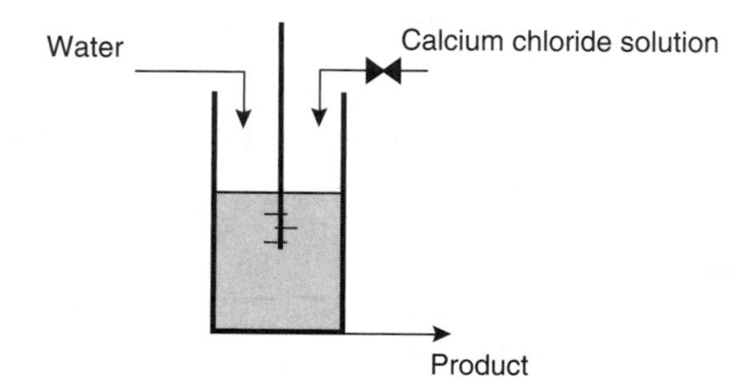


**FIGURE V.3** The schematic representation of the calcium chloride mixing process.

2. Propose two control configurations that would maintain a uniform outlet salt concentration by maintaining a constant ratio of the water and the calcium chloride solution flow rates
3. Discuss and compare each configuration

**V.4.** Consider the oil cooling and heating circuit of a crystallization plant described in Example 13.1. A model for this process has been developed in HYSYS and is available on the book website. A ratio control scheme has been implemented in the model. Using the model provided:

1. Investigate the effect of changing operating conditions on the heating side of the circuit. Change the set-point of the controller up and down with respect to the nominal operating conditions
2. Investigate the effect of changing operating conditions on the cooling side of the circuit. Change the set-point of the controller up and down with respect to the nominal operating conditions
3. Discuss your observations

**V.5.** For the furnace problem in Exercise I.9, we mentioned the safety implications associated with the temperature of the tubes inside the furnace. In addition to the feedback and feedforward controllers discussed in Exercise I.9, the safety problem needs to be tackled using a different control configuration. For this problem, discuss and formulate an override control scheme that would address the safety issue assuming that the surface temperature of the tubes is measured.

**V.6.** For the system introduced in Exercise II.11 and described by the following dynamic equations:

$$2\frac{dx_1}{dt} = -2x_1 + \exp(-x_1) - 3u_1x_2$$

$$\frac{dx_2}{dt} = -x_2 + \frac{2x_1}{1 + x_2} + 4u_2$$

1. Analyze the process interaction using the RGA approach
2. Define the best pairing of inputs and outputs with the least amount of interaction
3. Design a multiloop control scheme
4. Design a one- and a two-way decoupler
5. Analyze the closed-loop behavior in terms of the dynamic response

**V.7.** For the blending process (see Continuing Problem), and the transfer functions developed in Chapter 5, consider the following two cases:

1. Assume that the steady-state value of the inlet composition $x_{1,s}$ is now 0.1
2. Assume that the steady-state value of the inlet composition $x_{2,s}$ is now 0.7

For these two cases:

1. Perform an interaction analysis using RGA
2. Choose the optimal pair of control and manipulated variables to minimize interactions and compare with the base case discussed in Chapter 15

**V.8.** For the blending process (see Continuing Problem), and using the conditions defined for Case 1, in Exercise V.7,

1. Design and implement a multiloop PI controller, with parameters based on Z–N tuning for both level and composition loops. Take into account that the level controller is the same as designed before since the corresponding transfer functions have not changed
2. Design and implement a decoupling controller for the process and compare the performance with a conventional multiloop control

**V.9.** Following transfer function matrices were developed from experimental data obtained from a bench-scale distillation column:

$$\begin{bmatrix} T_1 \\ T_8 \end{bmatrix} = G_1(s)\begin{bmatrix} F_{\text{reflux}} \\ F_{\text{reboiler}} \end{bmatrix} = \begin{bmatrix} \dfrac{-0.007e^{-9.73s}}{23.97s+1} & \dfrac{0.029e^{-3.67s}}{3.82s+1} \\ \dfrac{-0.025e^{-65.5s}}{18.34s+1} & \dfrac{0.369}{79.42s+1} \end{bmatrix}\begin{bmatrix} F_{\text{reflux}} \\ F_{\text{reboiler}} \end{bmatrix}$$

$$\begin{bmatrix} T_1 \\ T_8 \end{bmatrix} = G_2(s)\begin{bmatrix} F_{\text{reflux}} \\ F_{\text{reboiler}} \end{bmatrix} = \begin{bmatrix} \dfrac{-0.006e^{-9.73s}}{59.5s+1} & \dfrac{0.016e^{-7.03s}}{37.12s+1} \\ \dfrac{-0.034e^{-23s}}{81.45s+1} & \dfrac{0.185e^{-11.81s}}{18.62s+1} \end{bmatrix}\begin{bmatrix} F_{\text{reflux}} \\ F_{\text{reboiler}} \end{bmatrix}$$

The transfer function matrix $G_1(s)$ was obtained by changing the reflux and the boilup rates in the positive direction and observing the response of two tray temperatures. The transfer function matrix $G_2(s)$, on the other hand, was obtained by changing the reflux and the boilup rates in the negative direction.

1. For each model, design a multiloop PI controller and analyze the amount of interactions using RGA
2. Test each multiloop controller on both models and discuss the results with respect to closed-performance
3. Explore the use of decouplers if necessary

# REFERENCES

1. Negiz, A., Ramanauskas, P., and Cinar, A., Modeling, monitoring and control strategies for high temperature short time (HTST) pasteurization systems, *Food Control*, 9, 1–47, 1998.

# *Section VI*

## *Model-Based Control*

# 16 Model-Based Control

The performance of the closed-loop system can be better shaped if the model of the process is directly incorporated into the controller. This concept leads to the paradigm of model-based control that advocates the use of the process model explicitly in the formulation of the control law. This naturally allows the designer to take advantage of any information provided by the model, thus creating a more intelligent control system. We have seen a hint of this in Chapter 15 where decouplers that used the transfer function elements improved the performance of the closed-loop system by minimizing the effect of loop interactions.

In this chapter, we shall start with two special cases of model-based control, namely the feedforward control and the delay compensation. Both strategies make use of the process model information directly in obtaining the control law. Then, we shall introduce the internal model control (IMC) concept that offers a general framework for designing model-based controllers.

## 16.1 FEEDFORWARD CONTROL

Unlike the feedback system, a feedforward control configuration measures the disturbance (load) directly and takes a preemptive control action to eliminate its impact on the process (see Chapter 2).

To design the feedforward compensation for measured disturbances, we shall start by considering the block diagram of the process given in Figure 16.1. For this process, we have the following open-loop transfer functions:

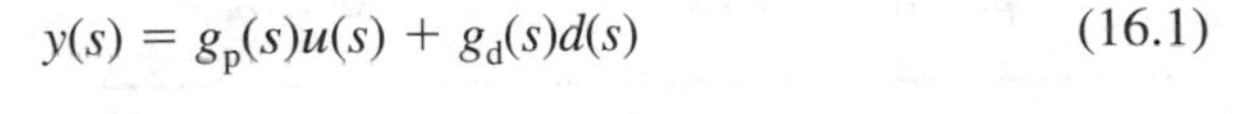

$$y(s) = g_p(s)u(s) + g_d(s)d(s) \tag{16.1}$$

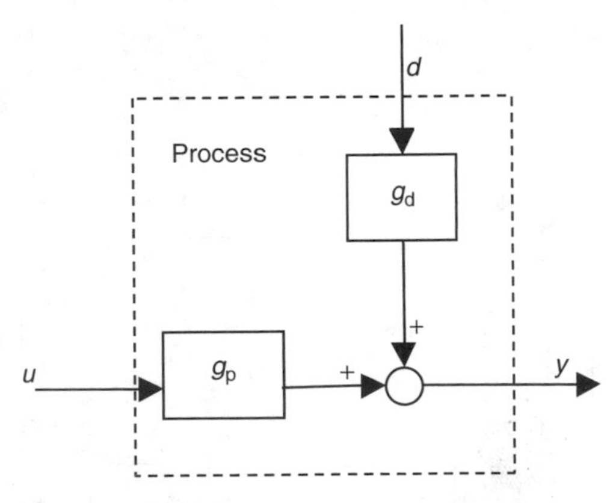


**FIGURE 16.1** Process block diagram.

The control objective dictates that we have $y(s) = y_{sp}(s)$, thus strictly enforcing that specification, we obtain

$$y(s) = y_{sp}(s) = g_p(s)u(s) + g_d(s)d(s) \tag{16.2}$$

We can solve for the necessary control action from Eq. (16.2) as follows

$$u(s) = \left[\frac{1}{g_d(s)}y_{sp}(s) - d(s)\right]\frac{g_d(s)}{g_p(s)} \tag{16.3}$$

Since the set-point $y_{sp}(s)$ is specified and the disturbance $d(s)$ is measured, the control action to achieve perfect set-point tracking can be evaluated using the known transfer function models $g_p(s)$ and $g_d(s)$. The disturbance is measured through the sensor transfer function $g_{md}(s)$, and the resulting feedforward schematic representation is depicted in Figure 16.2.

From Figure 16.2, the following transfer functions associated with the feedforward control action can be identified as

$$g_{ff}^1(s) = \frac{g_{md}(s)}{g_d(s)}, \quad g_{ff}^2(s) = \frac{g_d(s)}{g_p(s)g_f(s)g_{md}(s)} \tag{16.4}$$

---

1. The feedforward controller cannot be a conventional controller (P, PI or PID).
2. Feedforward control depends heavily on the knowledge of the process transfer function models, $g_p(s)$ and $g_d(s)$.
3. It can be developed for more than one disturbance and easily extended to systems with multiple controlled variables.
4. With the exception of the controller, all hardware elements in the loop are the same as for feedback control.

---

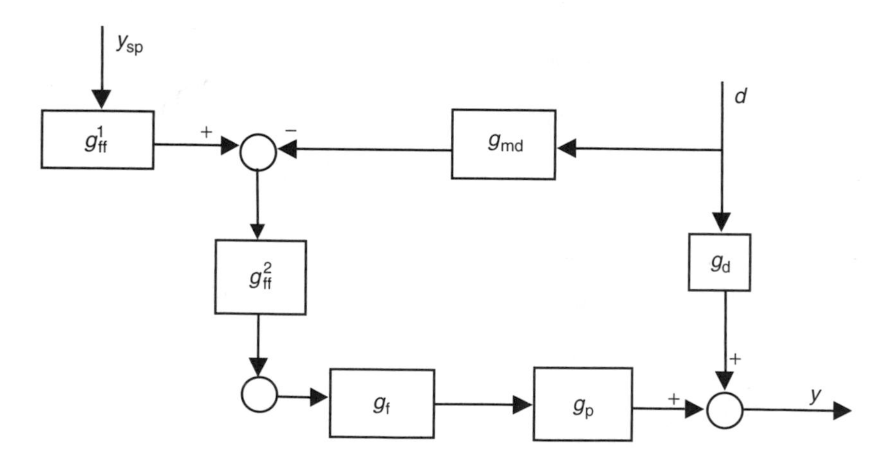


**FIGURE 16.2** Feedforward control configuration.

### 16.1.1 Feedforward–Feedback Control Strategy

Feedforward control has the potential for perfect control ($y(s) = y_{sp}(s)$), but one must consider the following with respect to the feedback control to understand the opportunities better.

#### Feedforward Control

| Advantages | Disadvantages |
|---|---|
| Acts before the disturbance hits the process | Must identify and measure all disturbances |
| Cannot cause instability | Fails for unmeasured disturbances |
| Good for slow process dynamics | Relies on availability of process models |
| | Fails if process behavior varies |
| | No indication of control quality |

#### Feedback Control

| Advantages | Disadvantages |
|---|---|
| No disturbance measurements needed | Will always have some transient error |
| No process model necessary | Poor for slow process dynamics and interactions |
| Can cope with changes within process | Instability is possible |

Therefore, one can combine the advantages of both configurations and design a feedforward–feedback controller as shown in Figure 16.3.

For this system, the closed-loop transfer function can be developed as follows (details of the derivation is left as an exercise):

$$y(s) = \frac{g_p g_f \left[ g_{ff}^2 g_{ff}^1 + g_{c_1} \right]}{1 + g_p g_f g_{c_1} g_m} y_{sp}(s) + \frac{\left[ g_d - g_p g_f g_{ff}^2 g_{md} \right]}{1 + g_p g_f g_{c_1} g_m} d(s) \tag{16.5}$$

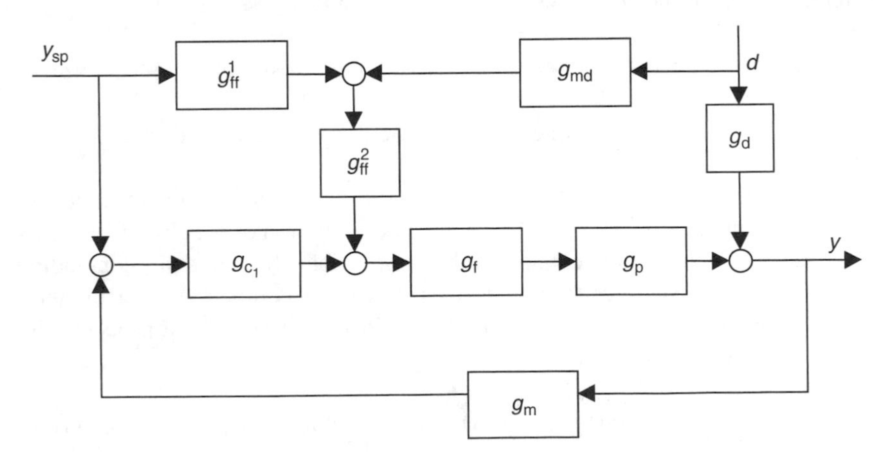


**FIGURE 16.3** Feedback–feedforward control configuration.

The characteristic equation for this closed-loop system is

$$1 + g_p g_f g_{c_1} g_m = 0$$

Therefore, the stability characteristics of the feedback control system will not change with the addition of a feedforward loop.

The method will be illustrated by the Continuing Example at the end of the chapter.

## 16.2 DELAY COMPENSATION (SMITH PREDICTOR)

For processes with large dead-times (time delays), conventional controllers (P, PI, and PID) may not be sufficient, requiring more sophisticated control strategies. For a typical closed-loop configuration (e.g., Figure 10.4), each dynamic component of the feedback loop may exhibit significant time delays in their response. Thus,

- A disturbance entering the process will not be detected until after some time period has passed.
- The control action based on the delayed information will be inadequate.
- The control action may take some time to make its effect felt by the process.

Hence, we can conclude that dead-time has a detrimental effect on the performance of the control system. Consider, for example, a process represented by the following open-loop transfer function:

$$g_{OL}(s) = \frac{k_c e^{-\tau_D s}}{0.5s + 1}$$

From the Bode plot showing the ultimate gain and the crossover frequency, for different values of the dead-time (Figure 16.4), we can observe that

- As the process dead-time increases, the crossover frequency $\omega_c$ decreases.
- As the process dead-time increases, the ultimate gain $k_{cu}$ decreases.

Consequently, we must reduce the gain of the controller. This results in reducing the amount of feedback, possibly causing deterioration of the closed-loop performance. We, thus, need to counteract the unfavorable effect of delays, including the possibility of instability, leading us to explore delay compensation techniques.

Consider the simple feedback loop given in Figure 16.5 for a process with transfer function

$$y(s) = g_p(s)u(s) = (g(s)e^{-t_D s})u(s) \tag{16.6}$$

where the time-delay element is factored out and shown explicitly.

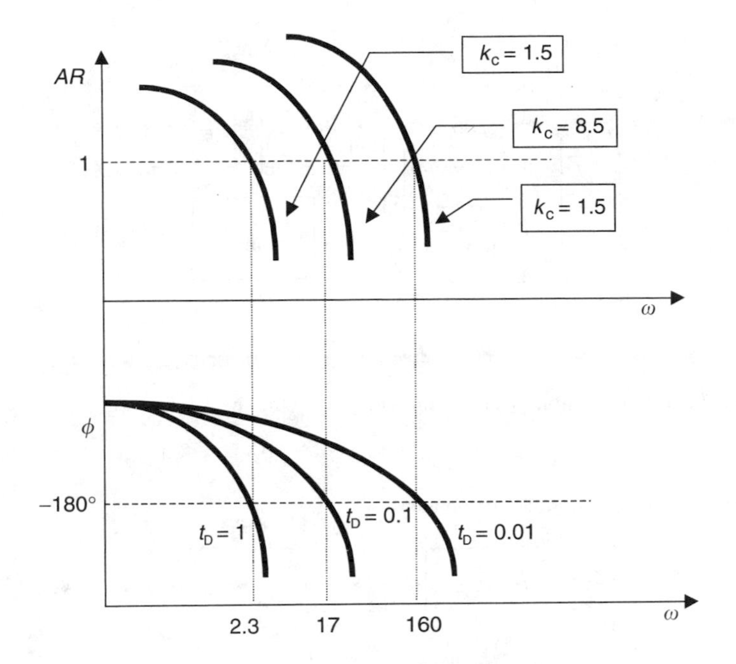


**FIGURE 16.4** Bode plots of process with the transfer function in Eq. 16.5.

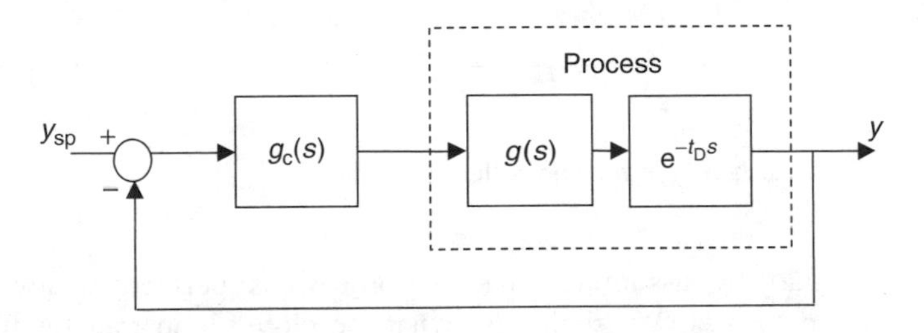


**FIGURE 16.5** Feedback control configuration with the delay shown explicitly.

The delay compensator proposed by Smith[1] aims to eliminate the delay element from the feedback loop. To accomplish this, we start with a model of the process as follows:

$$\tilde{y}(s) = \tilde{g}e^{-\tilde{t}_D s}u(s) \tag{16.7}$$

Here, $\tilde{y}(s)$ denotes the model output, $\tilde{g}(s)$ the delay-free model transfer function and $\tilde{t}_D$ is our estimate of the process delay. We will introduce two internal loops as shown in Figure 16.6. We can move the model transfer function in series with the process around the loop and incorporate it into the loop around the controller as shown in Figure 16.7. This yields the new controller mechanism that represents the delay compensation. In order to understand how Smith originally envisioned the compensation for the delay, we need to perform further block-diagram algebra.

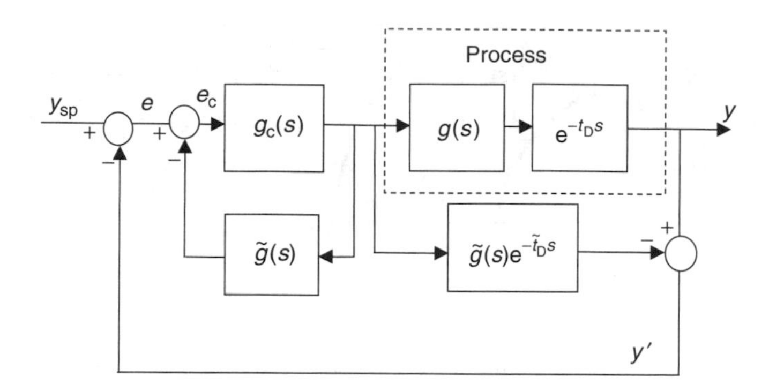


FIGURE 16.6 Introduction of model information in the feedback loop.

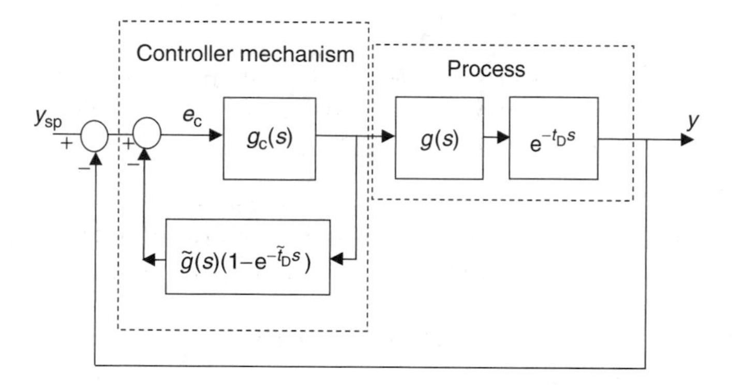


FIGURE 16.7 Construction of the controller mechanism.

Let us start by assuming that the process is perfectly known, i.e., $g(s) = \tilde{g}(s)$, and $t_D = \tilde{t}_D$. We shall find what the closed-loop transfer function looks like:

$$y(s) = \frac{g_c^* \tilde{g}(s) e^{-\tilde{t}_D s}}{1 + g_c^* \tilde{g}(s) e^{-\tilde{t}_D s}} y_{sp}(s) \tag{16.8}$$

Here, $g_c^*(s)$ represents the controller mechanism transfer function and is given by

$$u(s) = g_c^*(s) e(s) = \frac{g_c(s)}{1 + g_c(s)\tilde{g}(s)(1 - e^{-\tilde{t}_D s})} e(s) \tag{16.9}$$

If we incorporate Eq. (16.9) into Eq. (16.8), we obtain

$$y(s) = \frac{\left[g_c(s)/[1 + g_c(s)\tilde{g}(s)(1 - e^{-\tilde{t}_D s})]\right]\tilde{g}(s) e^{-\tilde{t}_D s}}{1 + \left[g_c(s)/[1 + g_c(s)\tilde{g}(s)(1 - e^{-\tilde{t}_D s})]\right]\tilde{g}(s) e^{-\tilde{t}_D s}} y_{sp}(s) \tag{16.10}$$

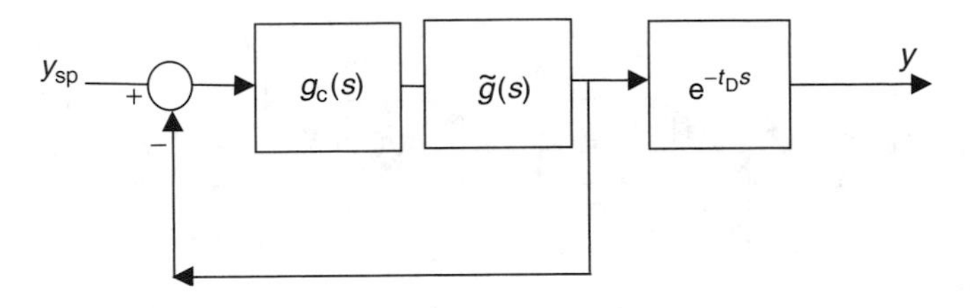


**FIGURE 16.8** Delay-free feedback loop.

By making the obvious cancellations, Eq. (16.10) yields the final closed-loop transfer function

$$y(s) = \frac{g_c\tilde{g}(s)}{1 + g_c\tilde{g}(s)} e^{-\tilde{t}_D s} y_{sp}(s) \tag{16.11}$$

This expression corresponds to the closed-loop block diagram as shown in Figure 16.8. As one can see, the delay element is removed from the feedback loop and thus the characteristic equation is free from delay.

---

- The compensator "predicts" the delayed effect (see Eq. [16.9]) that the manipulated variable will have on the process output (thus the name Smith predictor).
- In most process control problems, the model is only an approximation of the actual process (plant). Therefore, in practice, we are faced with modeling errors that would alter the performance of such delay compensation.

---

## 16.3 INTERNAL MODEL CONTROL (IMC) STRUCTURE

The IMC approach is based on an assumed process model and relates the controller settings to the model parameters in a straightforward manner. Very similar to the Smith predictor discussed in the previous section, the IMC approach explicitly uses the process model in the feedback loop. In this section, we will focus on the SISO implementation of IMC and refer the reader to other publications (see Additional Reading) for further study of the MIMO IMC design.

Consider the classical closed-loop block diagram given in Figure 16.9a and introduce the process model into the feedback loop in two locations as shown in Figure 16.9b. As the model transfer function is added and subtracted from the loop, its net effect on the loop is zero. Thus, the two block diagrams are equivalent. Using block-diagram algebra, the configuration in Figure 16.9b can be expressed as shown in Figure 16.9c. This is the IMC structure, and $c(s)$ is the so-called IMC controller.

Owing to the equivalence of two configurations, there is a direct link between the classical control structure and the IMC structure. This link can be illustrated by

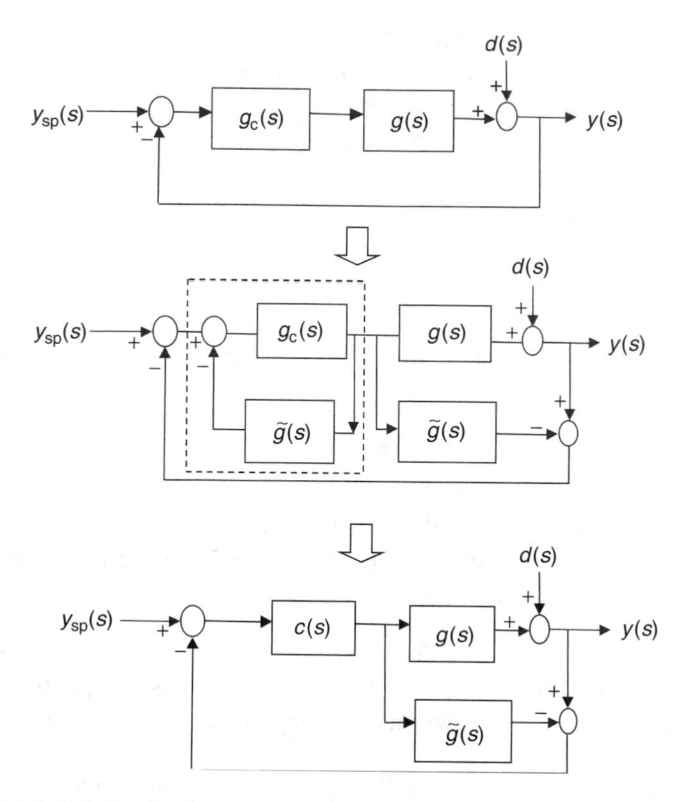


**FIGURE 16.9** Relationship between classical and IMC control configurations.

defining the controller $g_c(s)$ in the classical control structure and the controller $c(s)$ in the IMC structure as follows (the Laplace variable $s$ is omitted for brevity):

$$g_c = \frac{c}{1 - c\tilde{g}}, \quad c = \frac{g_c}{1 + g_c\tilde{g}} \tag{16.12}$$

The closed-loop transfer functions for the IMC configuration can be derived as

$$y(s) = \frac{c\tilde{g}}{1 + c(g - \tilde{g})} y_{sp}(s) + \frac{1 - c\tilde{g}}{1 + c(g - \tilde{g})} d(s) \tag{16.13}$$

---

Considering the characteristic equation, i.e., $1 + c(g - \tilde{g}) = 0$, we can assert that the stability of the closed-loop system is insured if both the process (and its model) and the controller are stable.

---

Let us analyze the case where we have perfect knowledge of the process

$$g(s) = \tilde{g}(s), \quad d(s) = 0 \tag{16.14}$$

In other words, we know the model perfectly and there are no disturbances. In that case, Eq. (16.13) reduces to

$$y(s) = c(s)\tilde{g}(s)y_{sp}(s) \tag{16.15}$$

which is simply a feedforward control action. This result immediately points to the following:

---

We need feedback *only* when we do not have perfect knowledge of the process (perfect model) and there are unmeasured (unknown) disturbances (see our original discussion in Chapter 2).

---

### 16.3.1 Concept of Perfect Control

Now, we will elaborate further on the perfect control concept that we have just introduced. Note that for disturbance rejection only ($y_{sp}(s) = 0$), the closed-loop transfer function is given by

$$y(s) = \frac{1 - c\tilde{g}}{1 + c(g - \tilde{g})} d(s) \tag{16.16}$$

In the ideal situation where the disturbance is rejected perfectly ($y(s) = 0$), we have

$$y(s) = \frac{1 - c\tilde{g}}{1 + c(g - \tilde{g})} d(s) = 0 \tag{16.17}$$

This can *only* be accomplished if the IMC controller is chosen as the exact inverse of the process model:

$$c(s) = \frac{1}{\tilde{g}(s)} \tag{16.18}$$

This establishes the *perfect control* case. If we were to implement the perfect controller (Eq. [16.18]) using the classical control configuration, we would have

$$g_c = \frac{c}{1 - c\tilde{g}} = \frac{1/\tilde{g}}{1 - \tilde{g}/\tilde{g}} = \infty \tag{16.19}$$

Indeed, we would need infinite control action (infinite gain) to achieve perfect control. Thus, this is just an idealization of our performance expectation and not practically possible. Although *perfect control* cannot be achieved, it is of great theoretical and practical interest to determine how closely this ideal can be *approached*. To understand this, we need to review the fundamental limitations to inverting a process model and using it as the controller.

**Limitations to Inverting a Process Model**

1. If the process model contains a time delay, its inverse would have to make a *prediction*. This would require the controller to have access to the future values of the feedback error and that is not possible.
2. If the process model contains an RHP zero, the controller would be unstable. This causes the feedback loop to lose stability and cannot be tolerated.
3. For process transfer functions that are proper (see Chapter 5), the inverse could give rise to an improper transfer function. This implies that the controller output would be extremely sensitive to changes in the error and would not be practically implementable.
4. The process model is always an approximation of the actual process. The feedback system is often very sensitive to the mismatch between the model and the plant (see Chapter 17).

### 16.3.2 IMC Design Procedure

The design procedure first establishes if the model has any time delays or RHP zeros (referred to as nonminimum-phase elements) and decomposes the model in such a way as to create a minimum-phase part, $\tilde{g}_M(s)$, and a nonminimum-phase part, $\tilde{g}_A(s)$:

$$\tilde{g}(s) = \tilde{g}_A(s)\tilde{g}_M(s) \tag{16.20}$$

The nonminimum phase-part is also called the *all-pass* element since, by choice, the *AR* of its Bode plot remains at 1 for all frequencies. Naturally, the minimum-phase element represents the invertable part of the transfer function as far as the fundamental limitations 1 and 2 above are concerned (i.e., it is free of time delays and RHP zeros) (see Example 16.2).

The decomposition in Eq. (16.20) yields an optimal closed-loop response based on the ISE criterion (see Chapter 12) with respect to an input of interest. In other words, assuming that the model is a perfect representation of the process, $g(s) = \tilde{g}(s)$, the closed-loop error is minimized if the *nominal* IMC controller is chosen as

$$\tilde{c}(s) = \frac{1}{\tilde{g}_M(s)} \tag{16.21}$$

Thus, this is as close as we can get to the ideal performance of the *perfect* controller discussed earlier.

**Example 16.1**

The ISE criterion for performance is expressed in the time domain as follows (Eq. [12.1]):

$$\text{ISE} = \int_0^\infty e^2(t)\mathrm{d}t$$

Let us see how we can express this criterion in the frequency domain, specifically for the IMC structure. The output equation (16.13) can be used to define the output error, $e(s) = y_{sp}(s) - y(s)$ assuming that $g(s) = \tilde{g}(s)$:

$$e(s) = (1 - c\tilde{g})[d(s) - y_{sp}(s)] \tag{16.22}$$

In other words, the *nominal* closed-loop performance is based on the error response transfer function

$$\frac{e(s)}{d(s) - y_{sp}(s)} = (1 - c\tilde{g}) \equiv \varepsilon(s) \tag{16.23}$$

which is referred to as the *sensitivity* function. In response to changes in the set-point or disturbances, the performance expectation is to keep the *AR* of the sensitivity function below a limit as specified by the designer. Figure 16.10 depicts how the sensitivity function may be "shaped" by the definition of a bound on its magnitude, $w(s)$. The designer can specify $w(s)$ so that the sensitivity function has the desired (tolerable) attenuation at low frequencies (around steady-state), as well as having a crossover frequency that imposes an acceptable closed-loop speed of response.

The minimization of the ISE criterion is equivalent to the following definition in the frequency domain ($s = j\omega$):[2]

$$J_2 = \min_c \|(1 - c\tilde{g})w\|_2 \tag{16.24}$$

The term $\|\cdot\|_2$ refers to the 2-norm of a transfer function, $q(s)$, as defined by

$$\|q(s)\|_2 = \left(\frac{1}{2\pi}\int_{-\infty}^{\infty} |q(j\omega)|^2 d\omega\right)^{1/2} \tag{16.25}$$

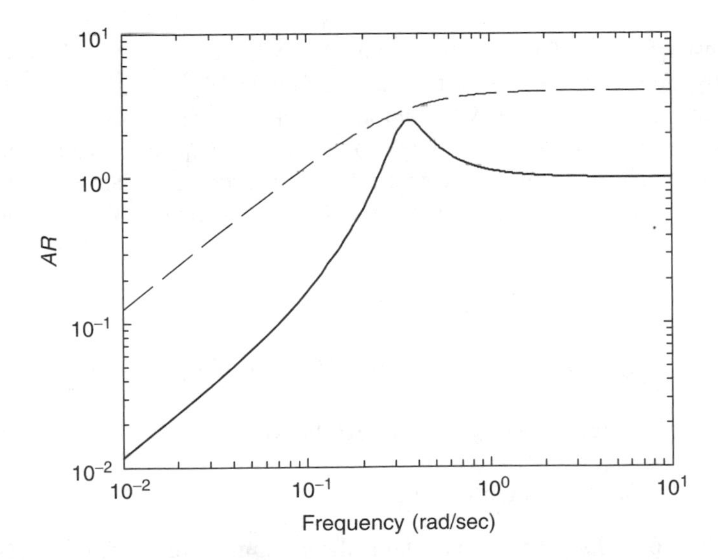


**FIGURE 16.10** The *AR* of the sensitivity function $|\varepsilon(j\omega)|$ (solid line) and its bound $|1/w(j\omega)|$ (dashed line).

Thus, from Eq. (16.24) we can immediately observe that by choosing $c = \tilde{g}^{-1}$, we achieve perfect control, irrespective of $w(s)$. In addition, when we confine ourselves to stable and causal controller transfer functions, thus satisfying restrictions 1 and 2 above, the result would depend on $w(s)$. For step functions, which are the typical inputs a control designer uses for performance testing, the minimum of Eq. (16.24) is reached when[3]

$$\tilde{c} = \tilde{g}_M^{-1}$$

Note that we use $\tilde{c}$ to refer to the *nominal* IMC controller, as it uses only the *nominal* model information. For the ISE to be bounded as shown in Figure 16.10, the error has to vanish as time grows ($t \to \infty$), i.e., at low frequencies. For processes with no integrators (no poles at the origin), the final value theorem (Chapter 5) for the set-point response (Eq. [16.15]) yields

$$\lim_{s \to 0} c\tilde{g} = 1 \tag{16.26}$$

This indicates that the IMC controller delivers offset-free performance for such processes.

To address the possible improperness of the resulting inverse (limitation 3), the IMC controller is modified by another transfer function

$$f(s) = \frac{1}{(\lambda s + 1)^n} \tag{16.27}$$

The resulting IMC controller now becomes

$$c(s) = \tilde{c}(s)f(s) = \tilde{g}_M^{-1}(s)f(s) \tag{16.28}$$

The transfer function $f(s)$ is referred to as the IMC filter and its order $n$ is determined by the relative degree* of the transfer function $\tilde{g}_M^{-1}(s)$, so that the resulting controller is proper. Now the IMC controller is referred to as $c$ as it has been altered from its nominal form. The filter time constant $\lambda$ is an adjustable parameter.

Let us consider, again, the case where there is no model-plant mismatch, i.e., $g(s) = \tilde{g}(s)$. The closed-loop transfer function (Figure 16.11) takes the following form:

$$\begin{aligned} y(s) &= c(s)\tilde{g}(s)y_{sp}(s) + [1 - c(s)\tilde{g}(s)]d(s) \\ &= \tilde{g}_M^{-1}(s)f(s)\tilde{g}_A(s)\tilde{g}_M(s)y_{sp}(s) + [1 - \tilde{g}_M^{-1}(s)f(s)\tilde{g}_A(s)\tilde{g}_M(s)]d(s) \\ &= f(s)\tilde{g}_A(s)y_{sp}(s) + [1 - f(s)\tilde{g}_A(s)]d(s) \end{aligned} \tag{16.29}$$

We can make the following observations:

1. The closed-loop response contains the nonminimum-phase elements. In other words, one cannot eliminate the effect of time delays or RHP zeros on the output response by feedback.

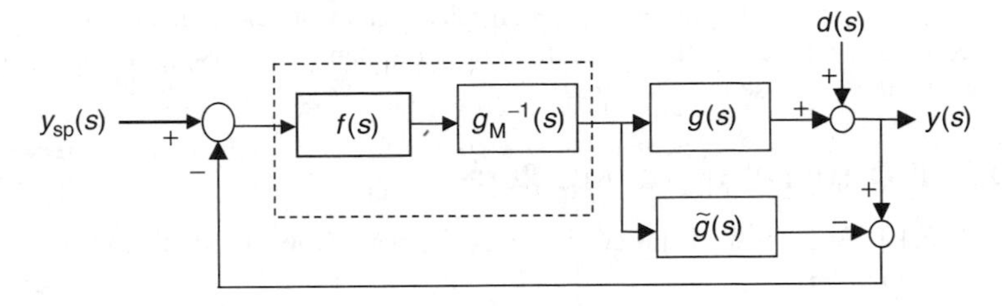


**FIGURE 16.11** Schematic representation of the IMC design.

2. The speed of response of the closed-loop system depends on the filter parameter, $\lambda$. Thus, the key adjustable parameter for IMC has a direct physical meaning, and allows for a very intuitive tuning problem.
3. The presence of the IMC filter and the associated filter constant $\lambda$ addresses the limitation 4 discussed earlier. We shall elaborate on this in Chapter 18.

### Example 16.2

Consider the process represented by the following transfer function:

$$\tilde{g}(s) = \frac{k(-\beta s + 1)e^{-t_D s}}{\tau s + 1} \tag{16.30}$$

The parameters $k$, $\beta$, $\tau$, and $t_D$ are all constants. The IMC design proceeds through the following steps.

***Step 1*:** By simple inspection, factor the model into invertible and noninvertible parts

$$\tilde{g}_A(s) = e^{-t_D s}\frac{-\beta s + 1}{\beta s + 1}; \quad \tilde{g}_M(s) = \frac{k}{\tau s + 1} \tag{16.31}$$

Note that the all-pass part contains the mirror image of the RHP zero in the denominator to make sure that the *AR* of the all-pass transfer function remains at 1 for all frequencies.

***Step 2*:** Design the IMC controller:

$$c(s) = \tilde{g}_M^{-1}(s)f(s) = \left(\frac{\tau s + 1}{k}\right)\frac{1}{\lambda s + 1} \tag{16.32}$$

Note that the order of the IMC filter is chosen as 1 to make the controller proper. The closed-loop response for set-point changes when there is no model-plant mismatch is given as

$$y(s) = \frac{(-\beta s + 1)e^{-t_D s}}{\lambda s + 1}y_{sp}(s) \tag{16.33}$$

One can clearly see that the nonminimum-phase elements are still present in the closed-loop response and the closed-loop time constant is $\lambda$. Also, one can easily show that the response is offset-free.

### 16.3.3 PID Tuning using IMC Rules

In Eq. (16.12), we showed the equivalence between the classical controller and the IMC controller.

$$g_c = \frac{c}{1 - c\tilde{g}} \tag{16.12}$$

This equation points to a rather intuitive fact, asserting that the complexity of the classical controller is determined by the complexity of the model. This is a valuable observation that underscores the point about designing controllers whose complexity commensurates with the process that they are being implemented on. This equivalence can be exploited to derive parameters of PID controllers for a number of specific process models.[3]

**Example 16.3**

Design an IMC controller for the first-order process

$$\tilde{g}(s) = \frac{k}{\tau s + 1} \tag{16.34}$$

In this case, the factorization of the model yields

$$\tilde{g}_A(s) = 1, \quad \tilde{g}_M(s) = \frac{k}{\tau s + 1} \tag{16.35}$$

Consequently the IMC controller becomes

$$c(s) = \frac{\tau s + 1}{k} \frac{1}{\lambda s + 1} = \frac{\tau s + 1}{k(\lambda s + 1)} \tag{16.36}$$

The order of the IMC filter is chosen as $n = 1$. A first-order filter is sufficient to make the controller proper.

---

The controller is a lead-lag function and thus can be implemented with modern microprocessor-based controllers.

---

To obtain the conventional controller, $g_c(s)$, we use Eq. (16.12),

$$g_c(s) = \frac{\dfrac{\tau s + 1}{k(\lambda s + 1)}}{1 - \dfrac{\tau s + 1}{k(\lambda s + 1)} \dfrac{k}{\tau s + 1}} = \frac{\tau s + 1}{k(\lambda s + 1) - k} = \frac{1}{k}\left(\frac{\tau s + 1}{\lambda s}\right) \tag{16.37}$$

The final expression in Eq. (16.37) can be rephrased in the standard PID notation as

$$g_c(s) = \frac{\tau}{k\lambda}\left(1 + \frac{1}{\tau s}\right) \tag{16.38}$$

We can immediately notice that this is a PI controller with $k_c = \tau/k\lambda$ and $\tau_I = \tau$. We can make the following observations:

- To choose the proportional gain larger than the inverse of the process gain ($k_c > 1/k$), we select $\tau > \lambda$. This implies that the closed-loop response is faster than the open-loop response.
- When the proportional gain is smaller than the inverse of the process gain ($k_c < 1/k$), we have $\tau < \lambda$. This implies that the closed-loop response is more sluggish than the open-loop response.

Table 16.1 displays the PID-tuning constants as a function of the model parameters and the IMC filter constant for a variety of process models, designed for step inputs. To maintain a general PID equation that is practically implementable (see Chapter 12), we consider the following PID structure:

$$g_c(s) = k_c\left(1 + \frac{1}{\tau_I s} + \tau_D s\right)\frac{1}{\tau_F s + 1} \tag{16.39}$$

Here, $\tau_F$ brings in an additional lag element that ensures properness of the PID transfer function.

## 16.4 SUMMARY

In this chapter, we have introduced the key concept of model-based control, which implies that the controller design makes explicit use of the available

**TABLE 16.1**
**PID Tuning Parameters using IMC Rules for Some Process Models**

| Model/PID Parameters | $k_c$ | $\tau_I$ | $\tau_D$ | $\tau_F$ |
|---|---|---|---|---|
| $\frac{k}{\tau s+1}$ | $\frac{\tau}{k\lambda}$ | $\tau$ | — | — |
| $\frac{k}{(\tau_1 s+1)(\tau_2 s+1)}$ | $\frac{\tau_1+\tau_2}{k\lambda}$ | $\tau_1+\tau_2$ | $\frac{\tau_1\tau_2}{\tau_1+\tau_2}$ | — |
| $\frac{k(-\beta s+1)}{\tau s+1}$ | $\frac{\tau}{k(2\beta+\lambda)}$ | $\tau$ | — | $\frac{\beta\lambda}{(2\beta+\lambda)}$ |

process model. This can be most easily seen in the formulation of the feedforward control strategy that relies on the availability of the disturbance model (information) to counteract its effect on the control variable. For processes with significant dead-time, the use of the delay compensator (Smith predictor) improves the closed-loop response of the process by incorporating delay information into the controller mechanism. The ultimate model-based control design framework is offered by the IMC paradigm that explicitly uses the process model in computing the control action. One of the key drawbacks of model-based control is its exclusive reliance on process models that may not necessarily reflect the dynamic behavior of the actual process. This uncertainty may impact the closed-loop behavior significantly and the next chapter addresses this issue directly.

## CONTINUING PROBLEM

Consider the blending process first described in Chapter 2 with the transfer functions

$$x(s) = \frac{-0.1e^{-s}}{2.5s+1}F_1 + \frac{0.1e^{-s}}{2.5s+1}F_2 + \frac{0.5e^{-s}}{2.5s+1}x_1 + \frac{0.5e^{-s}}{2.5s+1}x_2$$

Here, we will design a feedforward controller for the blending process and also demonstrate how a delay compensator as well as an IMC controller can be designed.

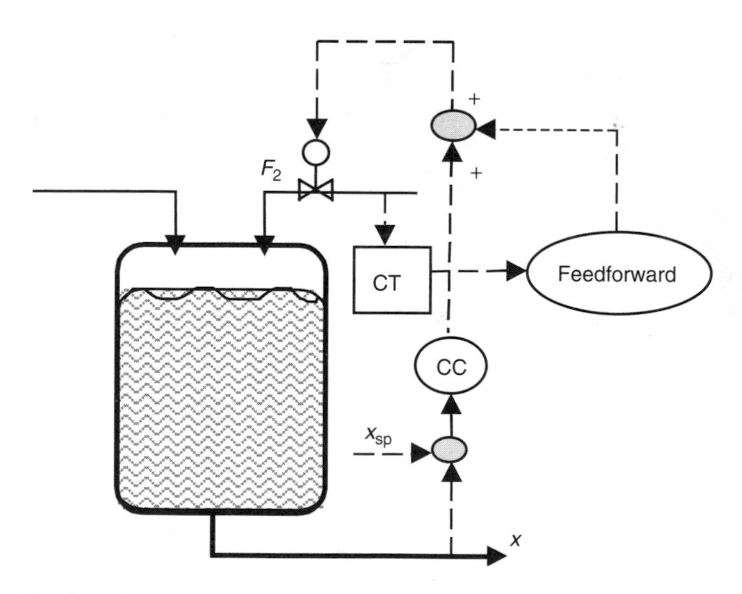


**FIGURE 16.12** Blending process with feedback–feedforward control strategy.

## Feedforward Control

The control objective is to maintain the composition of the mixture in the tank at its desired value by manipulating the feedflow rate $F_2$.

- Design a feedforward–feedback controller for this system to compensate for disturbances in feed composition.
- Evaluate the performance of this system with and without the feedforward controller.

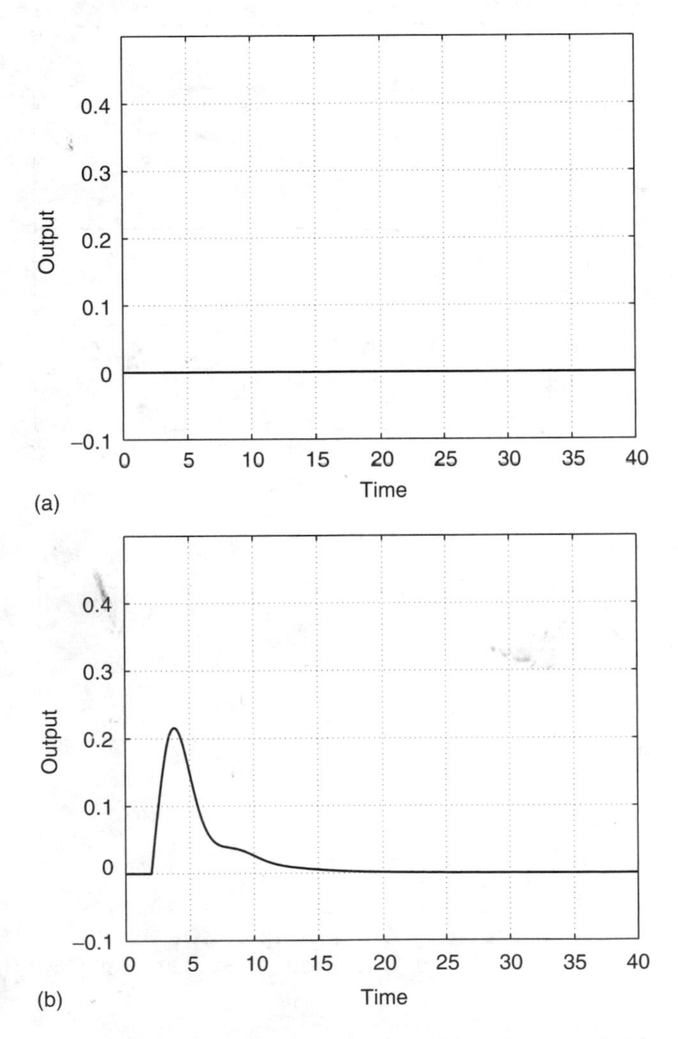


**FIGURE 16.13** (a) Feedforward–feedback scheme with perfect model, (b) conventional PID control performance for a disturbance change.

## Solution

Consider the configuration in Figure 16.12 to control the product composition (mass fraction). In this case, we add the feedforward compensation to the composition loop, assuming we measure the composition of the feed stream (disturbance). The transfer functions are:

$$g_p(s) = \frac{0.1e^{-s}}{2.5s+1}, \quad g_d(s) = \frac{0.5e^{-s}}{2.5s+1}$$

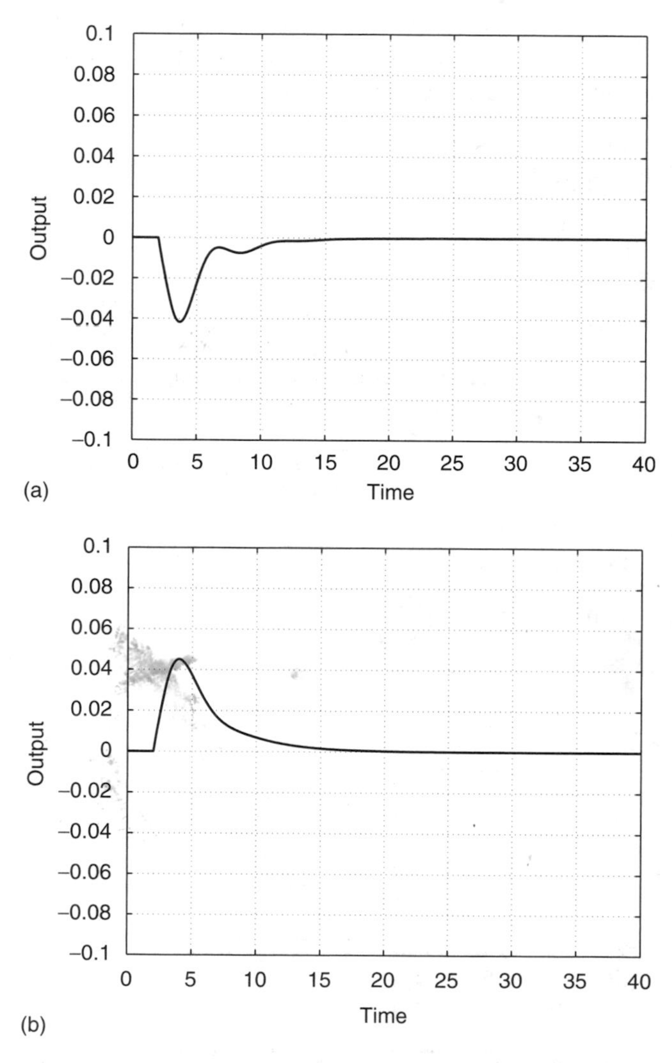


**FIGURE 16.14** Feedforward–feedback scheme with 20% modeling error in the process gain; (a) positive error, (b) negative error.

Consequently, the feedforward compensator is given by

$$g_{\mathrm{ff}}^{2}(s) = -\frac{0.5e^{-s}(2.5s+1)}{(2.5s+1)\,0.1e^{-s}} = -\frac{0.5}{0.1} = -5.0$$

Figure 16.13 illustrates the process behavior for disturbance changes in the feed composition for both conventional and feedforward controls with perfect model. It is evident from the figures that the feedforward controller provides perfect compensation for the disturbance when the model represents the plant exactly. Figure 16.14 illustrates the behaviour of the feedforward controller with 20% modeling error in the process gain. In both cases, the improvement achieved with the feedforward compensation is clear.

## Delay Compensator

In this section, our goals are

- To design a delay compensator for this system.
- To compare the performance of this system with and without delay compensator.
- To perform tests on the sensitivity of the compensator to modeling errors.

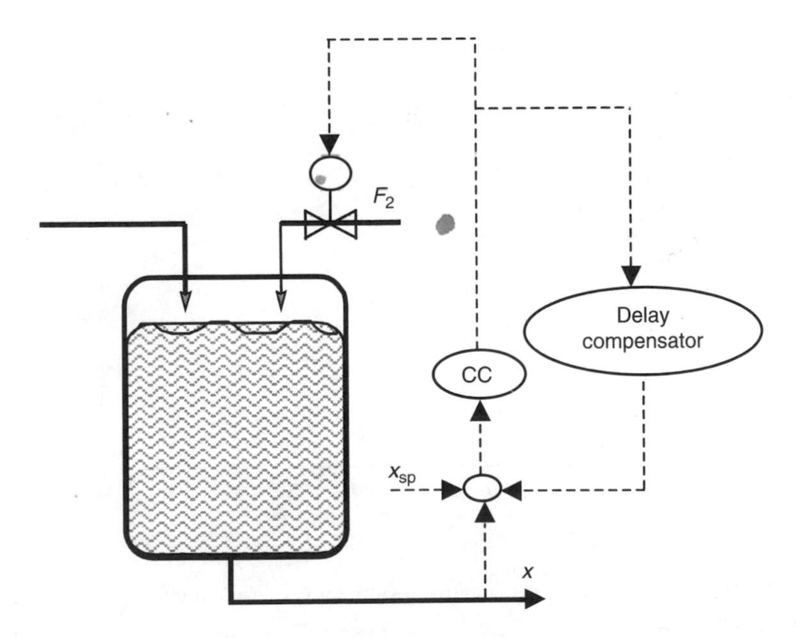


**FIGURE 16.15** Delay compensation strategy for blending process.

## Solution

Consider the configuration in Figure 16.13 to control the outlet composition. In this case, we add delay compensation to the composition loop (Figure 16.15). The closed-loop responses for the conventional PI control strategy (for set-point and disturbance changes) are shown in Figure 16.16. Figure 16.17 illustrates the

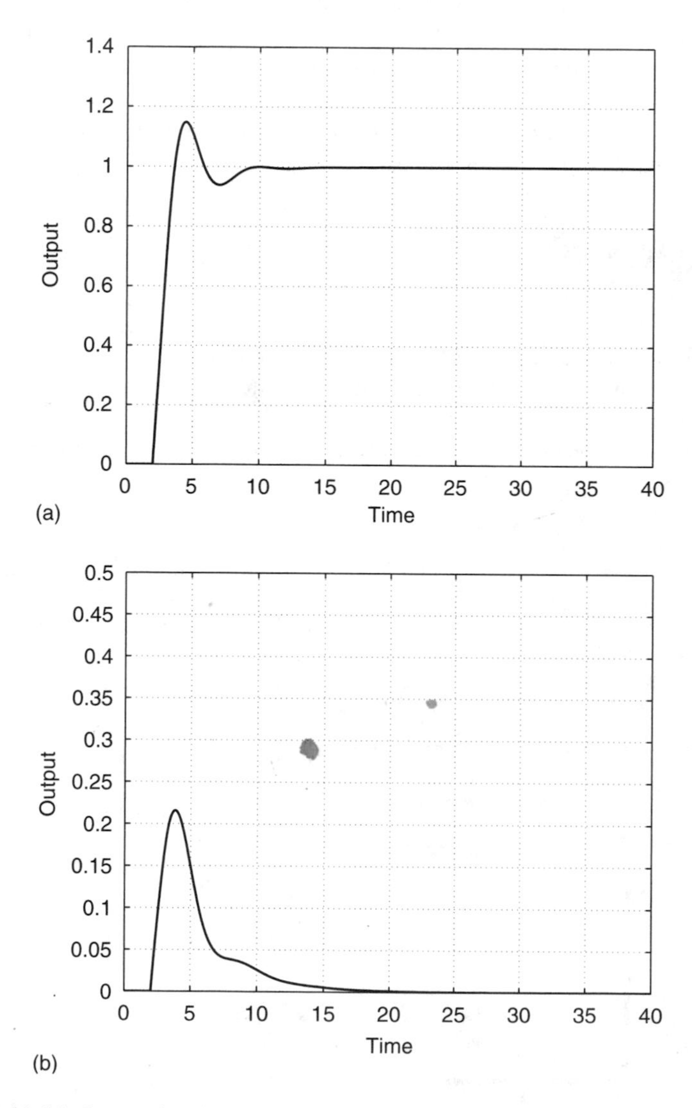


**FIGURE 16.16** Conventional PID control results for (a) set-point change and (b) disturbance change.

closed-loop responses for the conventional feedback controller with delay compensation with PI controller settings $k_c = 20.25$, $\tau_I = 3.34$, and for the case of perfect model.

Figure 16.18 illustrates the closed-loop responses for the conventional feedback controller with delay compensation with PI controller settings and with the previous settings when there is 25% error in the prediction of the delay term in the model for both set-point and disturbance changes.

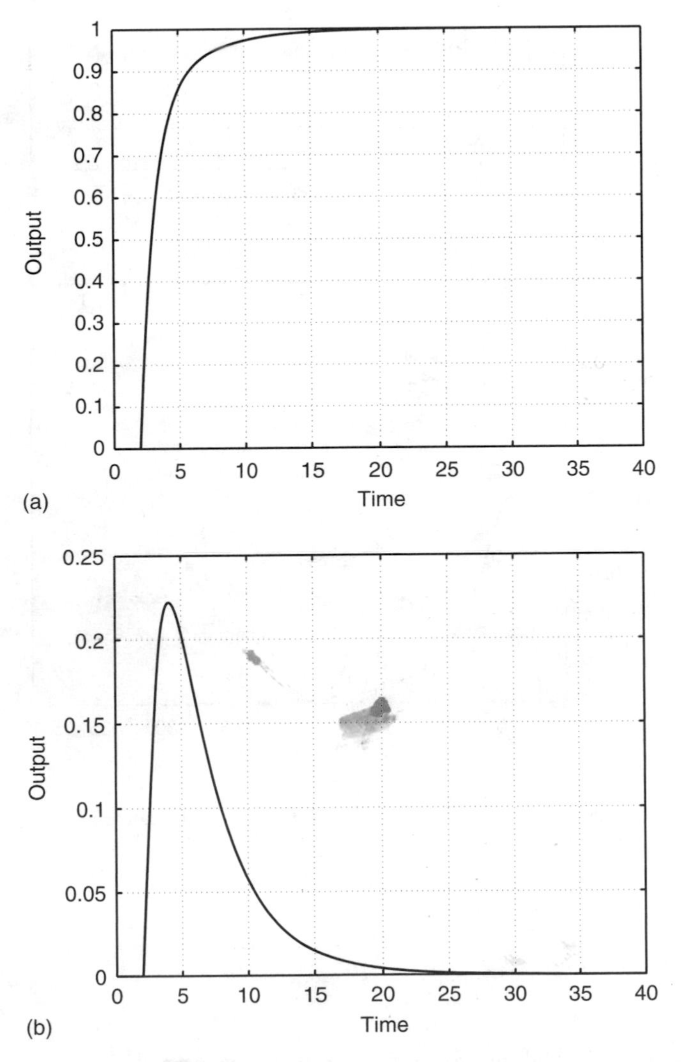


**FIGURE 16.17** Delay compensation results for (a) set-point change and (b) disturbance change.

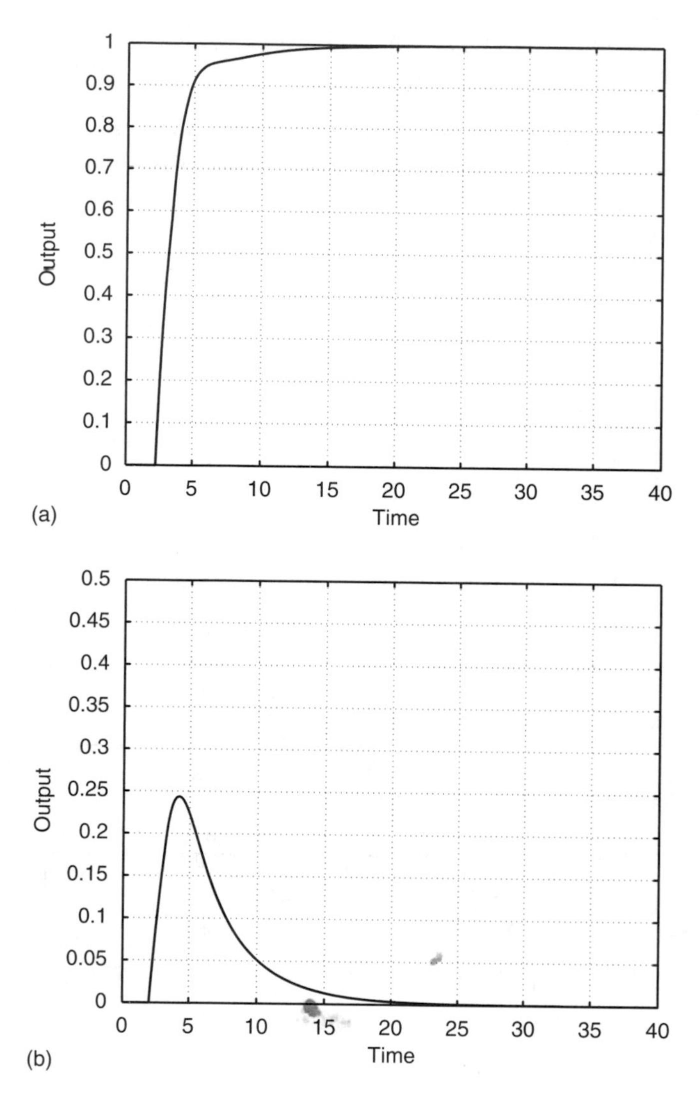


**FIGURE 16.18** Delay compensation results for (a) set-point change and (b) disturbance change with 25% error in the delay term.

## IMC Controller

Finally, we shall

- Design an IMC controller for the blending process and analyze the effect of the IMC filter constant on the closed-loop performance.
- Assume 20 and 25% modeling error in the time delay of the model and simulate the set-point response in the presence of uncertainty.

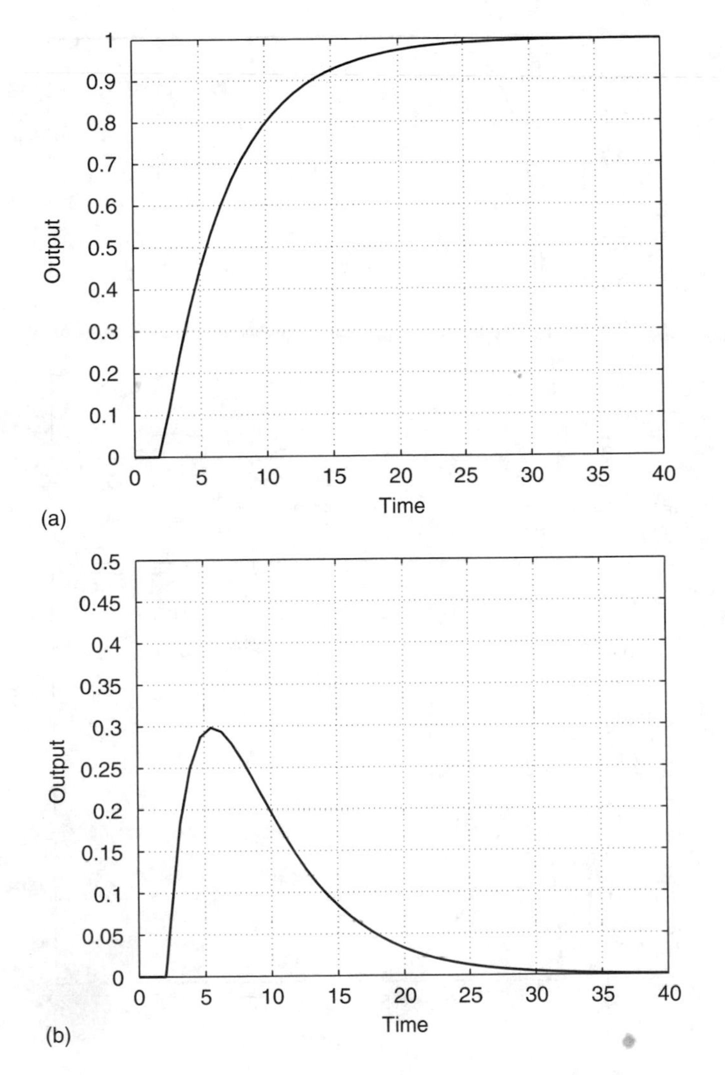


**FIGURE 16.19** An IMC with perfect model; (a) set-point change and (b) disturbance change, with ($\lambda = 5$).

## Solution

The IMC controller can be shown to be

$$c(s) = \frac{(2.5s+1)}{0.1(1+\lambda s)} = \frac{25s+10}{(1+\lambda s)}$$

Figure 16.19 displays the set-point and disturbance responses for the closed loop with the parameter $\lambda = 5$. When the filter constant is decreased to $\lambda = 2$,

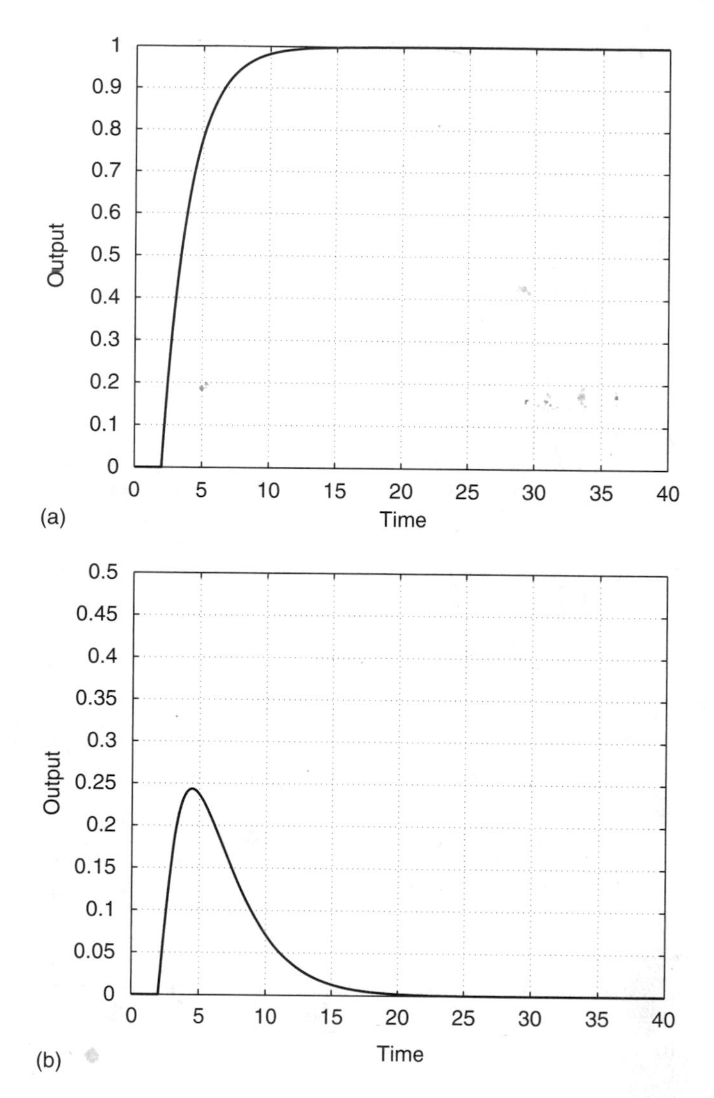


**FIGURE 16.20** Effect of filter time constant $\lambda = 2$; (a) set-point change and (b) disturbance change.

indicating faster closed-loop response, we get the performance depicted in Figure 16.20. The set-point response with two different values of the time delay is shown in Figure 16.21. The performance degrades slightly as the uncertainty increases.

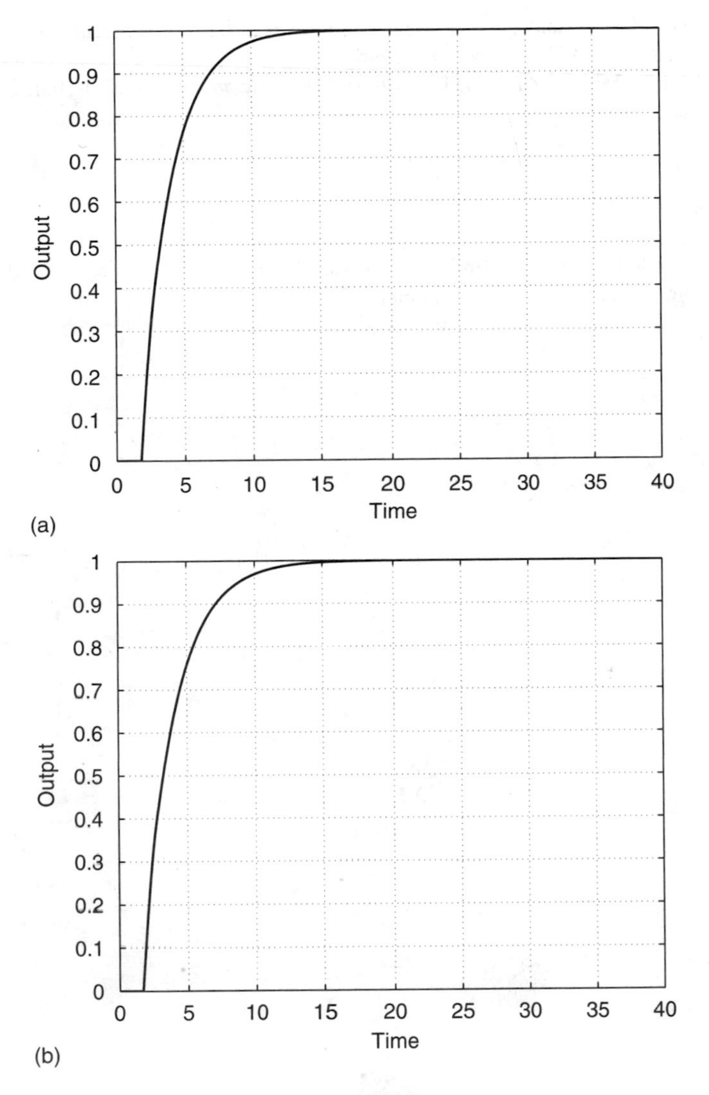


**FIGURE 16.21** An IMC with (a) 20% model–plant mismatch in the delay term, (b) 25% model–plant mismatch in the delay term. ($\lambda = 2$).

## REFERENCES


1. Smith, O.J.M., Close control of loops with dead time, *Chem. Eng. Prog.*, 53, 217, 1957; and Smith, O.J.M., A controller to overcome dead time, *ISA Journal*, 6, 28, 1959.

2. Skogestad, S. and Postlethwaite, I., *Multivariable Feedback Control — Analysis and Design*, Wiley, New York, 1990.
3. Morari, M. and Zafiriou, E., *Robust Process Control*, Prentice-Hall, New York, 1989.

## NOTES

* "Relative order" is simply defined as the difference between the order of the denominator polynomial and the order of the numerator polynomial of a transfer function.

# 17 Model Uncertainty and Robustness

When a controller is designed based on an assumed model and implemented on the actual plant, its closed-loop performance may be arbitrarily poor depending on the extent of the mismatch between the model and the plant. This simply means that the controller tuning parameters, based on a specific model of the plant, may not be the best choices for a plant that behaves differently from its model. Thus, we need to study model uncertainty (model–plant mismatch) more carefully, and quantify its impact on the expected performance of the control system.

The sources of model uncertainty can be varied:

- Nonlinear effects when a linear model is used
- High-order dynamics when model neglects such phenomena
- Slow-varying parameters such as heat-transfer coefficients (during fouling) and kinetic parameters (during catalyst decay)
- Unknown phenomena (unmeasured disturbances)

Up to this point, we have used mainly two objectives in designing a control system: stability and performance. Now, we will add a third objective, *robustness*. Ensuring stability and a certain level of performance based on a nominal model of the process is not sufficient. We shall demand that stability and performance requirements for the closed-loop system also be maintained for all possible models that may represent the actual plant dynamics.

---

A closed-loop system is *robust* with respect to a property (such as stability and performance), if it maintains that property in the presence of model uncertainty.

---

In this chapter, we discuss how one can quantify the extent of uncertainty associated with a process model and incorporate that information into the design of the IMC controller.

## 17.1 IMC STRUCTURE WITH MODEL UNCERTAINTY

The dynamic behavior of a plant can never be described exactly, but may be assumed to lie in some *neighborhood* of a *nominal* (reference) model. Consequently, the

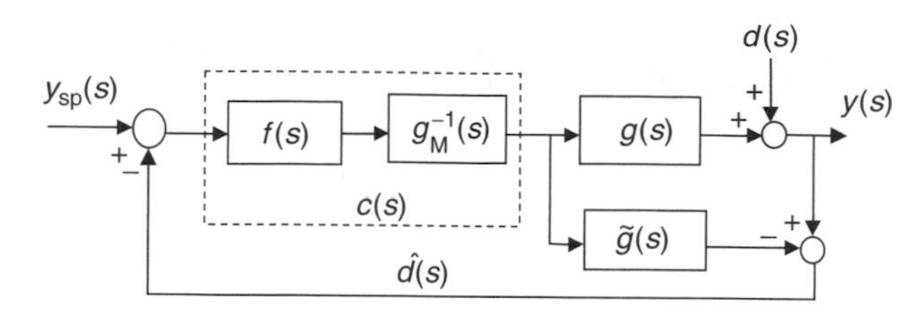


**FIGURE 17.1** An IMC block diagram.

challenge is to design a controller that guarantees performance and stability for all possible plant realizations in the neighborhood of that model.

Consider the IMC block diagram in Figure 17.1. The following relationships can be observed

$$\begin{aligned} \hat{d}(s) &= d(s) + [g(s) - \tilde{g}(s)]u(s) \\ e(s) &= y_{sp}(s) - \hat{d}(s) = y_{sp}(s) - d(s) + [g(s) - \tilde{g}(s)]u(s) \end{aligned} \tag{17.1}$$

The presence of model uncertainty (and disturbances) leads to the feedback of the control action $u(s)$ and, thus, may be a source of instability in the loop.

The IMC filter $f(s)$ not only needs to be designed for a particular input, $y_{sp}(s) - d(s)$, but also for the expected model uncertainty.

## 17.2 DESCRIPTION OF MODEL UNCERTAINTY

To be able to incorporate the model uncertainty information into the control design, the model uncertainty needs to be quantified. There are a number of ways that this can be achieved. Mostly, the uncertainty description relies on the definition of bounds on the available *nominal* model information. For example, we can describe the model uncertainty as:

- Bounds on the parameters of a linear *nominal* model
- Bounds on the nonlinearities
- Bounds on the frequency response of a nominal model

The theory of robustness is well defined for bounds established in the frequency domain. Other types of uncertainties can also be expressed in the frequency domain but there may be some limitations. Here, we will focus on two possible representations of the uncertainty.

### 17.2.1 Additive Uncertainty

In this case, the actual process dynamics will be expressed in the additive perturbation form (Figure 17.2)

$$g(s) = \tilde{g}(s) + l_a(s) \tag{17.2a}$$

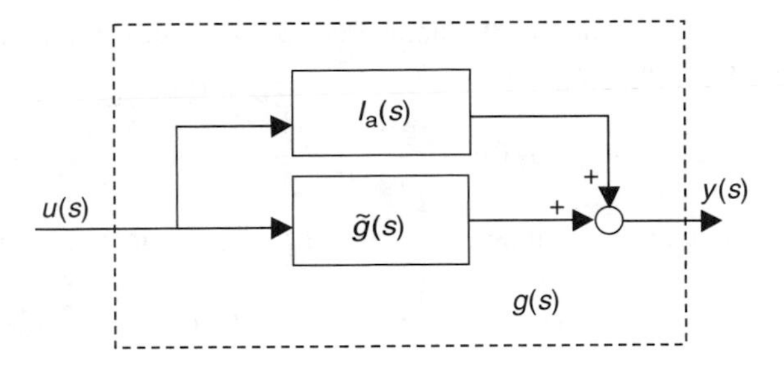


**FIGURE 17.2** Schematic representation of additive model uncertainty.

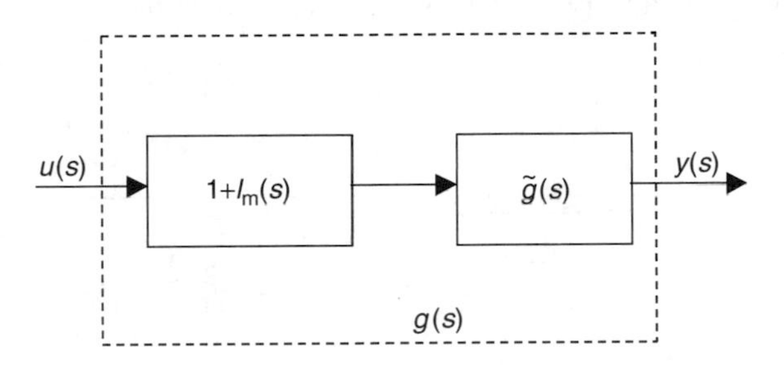


**FIGURE 17.3** Schematic representation of multiplicative model uncertainty.

where $l_a(s)$ is an unknown transfer function representing the uncertainty. This form would be appropriate for uncertainty associated with the process directly.

### 17.2.2 Multiplicative Uncertainty

In this case, the actual process dynamics will be expressed in the multiplicative perturbation form (Figure 17.3)

$$g(s) = \tilde{g}(s)[1 + l_m(s)] \tag{17.2b}$$

Again, $l_m(s)$ is an unknown transfer function. This form can describe uncertainties associated with the sensors or dynamics that affect the output response. In SISO systems, since the transfer functions can commute, any uncertainty associated with the actuators or the process input can also be expressed using Eq. (17.2b). In MIMO systems, we have to differentiate between *input* and *output* multiplicative uncertainties.

Multiplicative uncertainty is the form most often used due to its ability to capture a wide range of uncertain dynamics.

### 17.2.3 Estimation of Uncertainty Bounds

For both of the above uncertainty descriptions, sometimes the only information one may have about the perturbation $l_m(s)$ (or $l_a(s)$) is some bound on its magnitude expressed in the frequency domain. By rewriting Eq. (17.2b), we can express the perturbation as a function of the unknown actual process model and a known nominal process model:

$$l_{\mathrm{m}}(s) = \frac{g(s) - \tilde{g}(s)}{\tilde{g}(s)} \tag{17.3}$$

Then, an upper bound on the magnitude of the perturbation can be proposed as

$$|l_{\mathrm{m}}(s)| = \left| \frac{g(s) - \tilde{g}(s)}{\tilde{g}(s)} \right| < \bar{l}_{\mathrm{m}}(\omega)$$

Here, we shall assume that $\bar{l}_{\mathrm{m}}(\omega)$ is a known (or can be estimated) function of frequency. Graphically, such a bound can be expressed using either Bode or Nyquist plots as shown in Figure 17.4. The solid lines represent the plot for the nominal

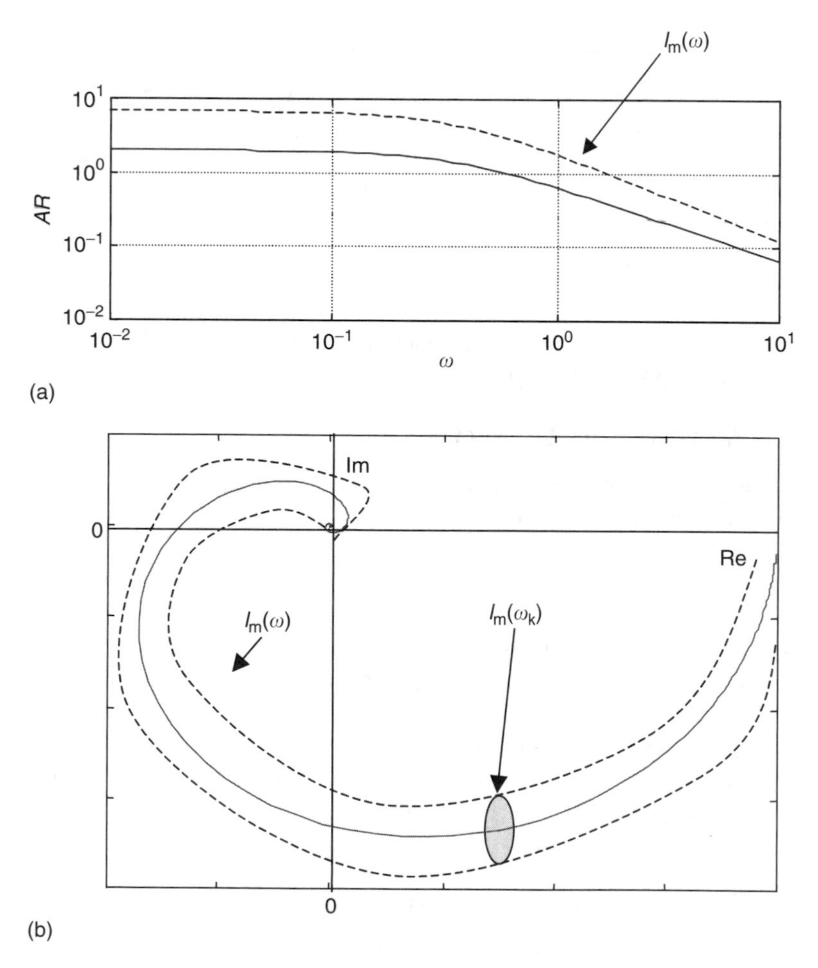


**FIGURE 17.4** Frequency domain representation of uncertainty. (a) Bode magnitude plot, (b) Nyquist plot with the disk uncertainty shown at a specific frequency $\omega_k$.

model. Note that we obtain a "fuzzy" Nyquist plot that contains, within the tube-like uncertainty, all Nyquist plots representing all possible plant realizations. At a given frequency, the bound manifests itself as a disk around the nominal model.

---

These uncertainty descriptions capture the variance associated with the magnitude of the frequency response only, and the phase is considered to be arbitrary. This assumption does lead to some conservatism in the estimation of the uncertainty but more complex uncertainty descriptions can be used that can also capture known variance of the phase, and decrease the conservatism.

---

**Example 17.1**

Consider a process described by the transfer function

$$g(s) = \frac{as + 1}{a^2 s^2 + 3s + 1}$$

where the unknown parameter $a$ is expected to vary in the range $1.5 \leq a \leq 2.5$. The nominal model is given by $a = 2$:

$$\tilde{g}(s) = \frac{2s + 1}{4s^2 + 3s + 1}$$

Figure 17.5 shows the magnitudes of the uncertainty functions for a family of plants that result from the possible realizations of the unknown parameter in the range $1.5 \leq a \leq 2.5$. From these figures, it would be possible to estimate an upper bound on the perturbations.

## 17.3 IMC DESIGN UNDER MODEL UNCERTAINTY

Consider the IMC structure, including a process model with a multiplicative uncertainty description (Figure 17.6).

The closed-loop relationships for this structure can be expressed as (see Eq. [16.13])

$$\begin{aligned} y(s) &= \frac{c\tilde{g}}{1 + c(g - \tilde{g})} y_{sp}(s) + \frac{1 - c\tilde{g}}{1 + c(g - \tilde{g})} d(s) \\ &= \frac{c\tilde{g}}{1 + c(g - \tilde{g})} (y_{sp}(s) - d(s)) + d(s) \end{aligned} \tag{17.4}$$

Substituting for the controller as well as for the uncertain model, we obtain

$$y(s) = \frac{\tilde{g}_A(s) f(s)[1 + l_m(s)]}{1 + \tilde{g}_A(s) f(s) l_m(s)} (y_{sp}(s) - d(s)) + d(s) \tag{17.5}$$

As one can observe, the uncertainty directly affects the stability of the closed-loop system (through the characteristic equation) as well as the closed-loop performance to set-point changes and disturbances.

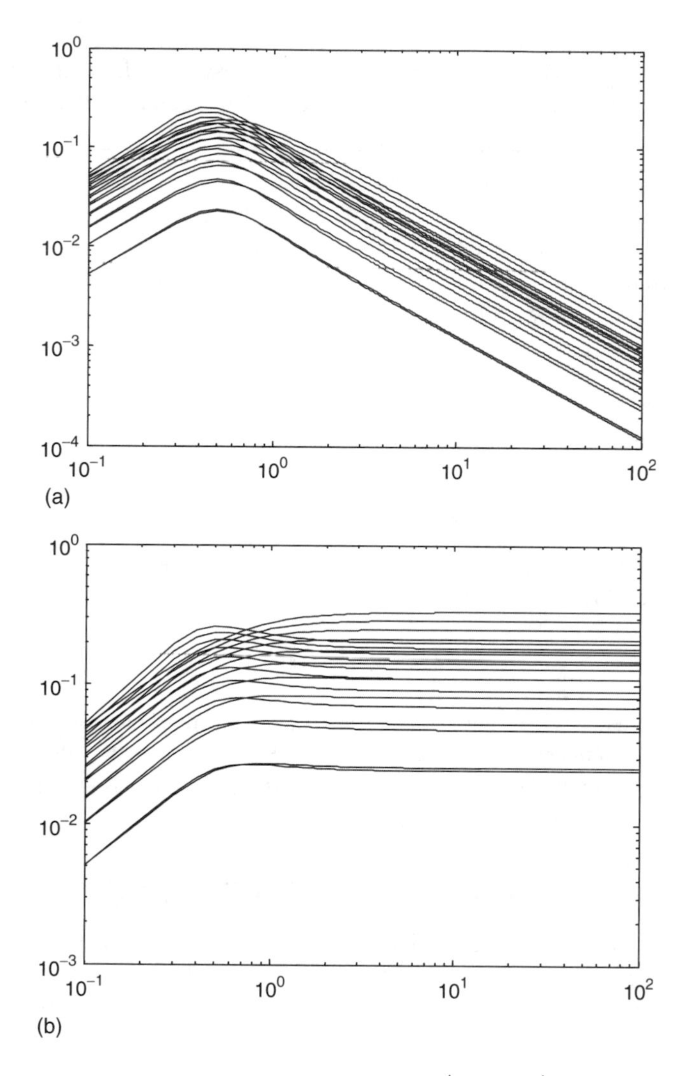


**FIGURE 17.5** Uncertainty realizations: (a) additive, $\left|l_{\mathrm{a}} = (\mathrm{j}\omega)\right|$ and (b) multiplicative, $\left|l_{\mathrm{m}} = (\mathrm{j}\omega)\right|$.

Next, we will establish conditions under which we can maintain stability and performance characteristics in the presence of model uncertainty. These conditions will then guide the selection of the IMC filter parameter, $\lambda$.

### 17.3.1 Robust Stability

To maintain stability in the presence of uncertainty, the characteristic equation of the closed-loop system (17.5) should satisfy the following condition:

$$\det\left[1 + \tilde{g}_{\mathrm{A}}(s)f(s)l_{\mathrm{m}}(s)\right] \neq 0 \tag{17.6}$$

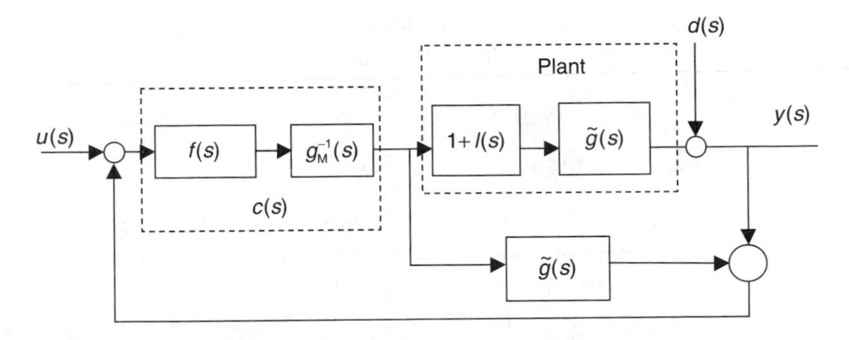


**FIGURE 17.6** An IMC structure with multiplicative uncertainty.

However, since the perturbation transfer function is not known, except for a bound on its magnitude, this condition is not practical. Instead, a more practical but a more conservative condition can be obtained. Equation (17.6) is satisfied if the following condition holds:

$$\left|1 + \tilde{g}_A(s)f(s)l_m(s)\right| < 0 \tag{17.7}$$

for all frequencies, $s = j\omega$. Equation (17.7) is the Nyquist stability criterion rephrased for the case of model uncertainty. It can be rearranged to yield the following practical stability condition:

$$\left|\tilde{g}_A(s)f(s)l_m(s)\right| < 1 \tag{17.8}$$

On further rearrangement, and noting that $\left|\tilde{g}_A(s)\right| = 1$ and $\left|l_m(s)\right| < \bar{l}_m(\omega)$, we get the condition on the IMC filter:

$$\left|f(s)l_m(s)\right| < 1 \Rightarrow \left|f\right| < \frac{1}{\bar{l}_m(\omega)} \tag{17.9}$$

When this condition is satisfied, robust stability of the closed-loop system is guaranteed.

The implication of this condition is that in frequency regions where the uncertainty is large (typically at high frequencies), the filter magnitude needs to be attenuated. Thus, the filter time constant $\lambda$ needs to be made large (sluggish response), and the closed-loop bandwidth would be limited. In other words, the presence of model uncertainty limits the speed of response so that closed-loop stability can be maintained.

### 17.3.2 Robust Performance

To establish the robust performance condition, we need to start from a performance objective. In other words, we need to state what our performance expectation is for the nominal model and then check how that can be maintained for the actual plant. In Chapter 16, we defined the *sensitivity* function (Eq. [16.23]) and showed how it can be used to assess closed-loop performance. Let us rephrase the nominal closed-loop equation for the IMC structure:

$$\begin{aligned} y(s) &= c(s)\tilde{g}(s)y_{sp}(s) + (1-c(s)\tilde{g}(s))d(s) \\ &\equiv \eta(s)y_{sp}(s) + \varepsilon(s)d(s) \end{aligned} \tag{17.10}$$

Here, we also introduce the *complementary sensitivity* function, $\eta(s)$. Note that $\varepsilon(s) + \eta(s) = 1$ by definition. Good performance dictates suppression of disturbances ($|\varepsilon| \to 0$) and set-point tracking ($|\eta| \to 1$) for all frequencies.

Normally (see Figure 17.7), the *AR* of the sensitivity function diminishes at low frequencies (perfect disturbance rejection near and at steady-state), and the *AR* of the complementary sensitivity function approaches unity at low frequencies also (perfect set-point tracking near and at steady-state). However, since both the *AR*s deviate from this objective at high frequencies, we need to articulate well, how wide the bandwidth can be set so that high frequency phenomena (such as noise and, more importantly, unmodeled dynamics) have little influence on performance. This is a key trade-off in *robust* control design.

### Example 17.2

In Example 16.1, we discussed the frequency domain expression of the ISE criterion of closed-loop performance. Here, we will introduce another performance

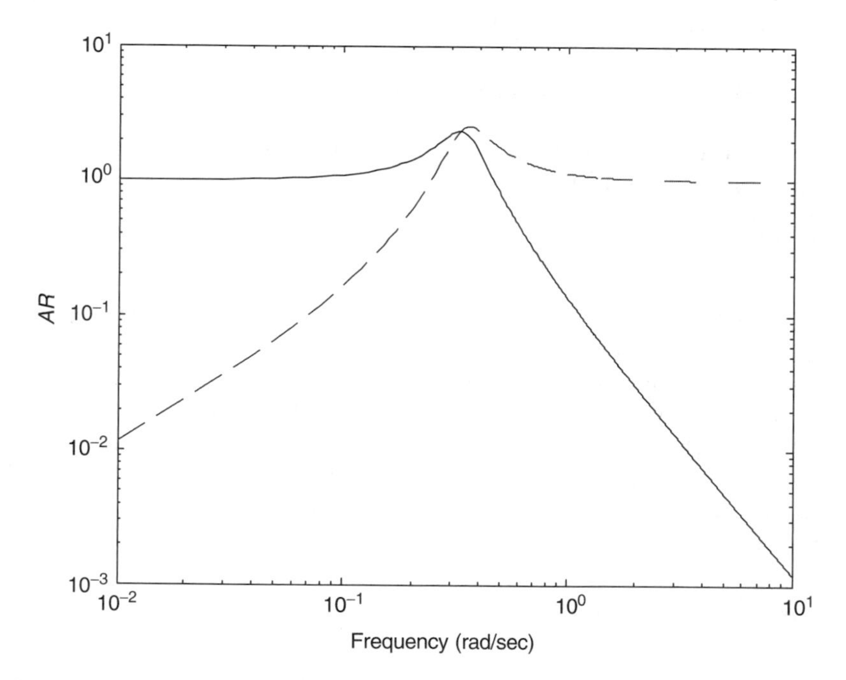


**FIGURE 17.7** *AR* of typical sensitivity (dashed line) and complementary sensitivity (solid line) functions.

objective, again expressed in the frequency domain, in terms of the sensitivity function. This objective is formulated through the ∞-norm, as opposed to the 2-norm defined in Eq. (16.25):

$$J_{\infty} = \min_{c} \|\varepsilon w\|_{\infty} \tag{17.11}$$

The ∞-norm of a transfer function is defined as

$$\|q(\mathrm{j}\omega)\|_{\infty} = \sup_{\omega} |q(\mathrm{j}\omega)| \tag{17.12}$$

where the *supremum* operation simply finds the peak value of the *AR* of the transfer function in question. From a time-domain perspective, the ∞-norm can be interpreted as the worst-case (maximum possible) steady-state gain (amplification) for sinusoidal input signals at any frequency.

Finally, Eq. (17.11) can be rephrased as

$$J_{\infty} = \min_{c} \|\varepsilon w\|_{\infty} = \min_{c} \sup_{\omega} |\varepsilon w| \tag{17.13}$$

This objective function is aimed at finding the controller that minimizes the peak value of the *AR* of the *weighted* sensitivity function. The practicality of this performance criterion stems from the fact that the designer can simply specify the maximum disturbance attenuation (and the bandwidth) through the choice of $w(s)$ and solve the control design problem. More importantly, as we shall see next, this objective facilitates the formulation of the robust performance criterion.

Given the multiplicative uncertainty (Eq. [17.2b]) and a bound on its magnitude (Eq. [17.3]), the closed-loop system will be *robust with respect to performance* (Eq. [17.11]) if the following condition is satisfied:

$$|\tilde{\eta} l_{\mathrm{m}}| + |\tilde{\varepsilon} w| < 1, \quad \forall \omega \tag{17.14}$$

As we can see, if the robust performance condition is satisfied, this automatically implies robust stability (Eq. [17.9]). In other words, we need to satisfy the robust stability condition with a sufficient margin to accommodate robust performance.

Equation (17.14) also points to a fundamental trade-off. The dependence of the robust performance condition on both the complementary sensitivity function and the sensitivity function also adds to the challenge to meet this criterion due to their interdependence. If the designer wishes to improve nominal performance by increasing $|\tilde{\varepsilon} w|$, this limits $|\tilde{\eta} l_{\mathrm{m}}|$, thereby leading the closed-loop system to potential instability for a plant realization allowed by the uncertainty description.

**Example 17.3**

A process is described by the transfer function

$$g(s) = \frac{\mathrm{e}^{-2.1s}}{(5s+1)(0.5s+1)(s+1)} \tag{17.15}$$

This dynamic behavior will be assumed unknown and we have an estimated model for this process as follows:

$$\tilde{g}(s) = \frac{e^{-2s}}{(4s+1)} \tag{17.16}$$

For the purpose of this example, the multiplicative error can be estimated as in Eq. (17.2b), by simply computing the magnitude of the error (since the uncertainty is specific and represented by a single realization):

$$\bar{l}_m(\omega) = \left| \frac{g(s) - \tilde{g}(s)}{\tilde{g}(s)} \right| = \left| \frac{\dfrac{e^{-2.1s}}{(5s+1)(0.5s+1)(s+1)} - \dfrac{e^{-2s}}{(4s+1)}}{\dfrac{e^{-2s}}{(4s+1)}} \right|$$

The uncertainty bound is given in Figure 17.8. As expected, the bound increases at high frequencies, indicating the presence of unmodeled dynamics. The bound also tends to zero at low frequencies, which is an indication that the model is capturing at least the steady-state behavior well.

We can design the IMC controller based on the following decomposition of the model:

$$\tilde{g}(s) = \tilde{g}_A(s)\tilde{g}_M(s) = (e^{-2s})\left(\frac{1}{(4s+1)}\right) \tag{17.17}$$

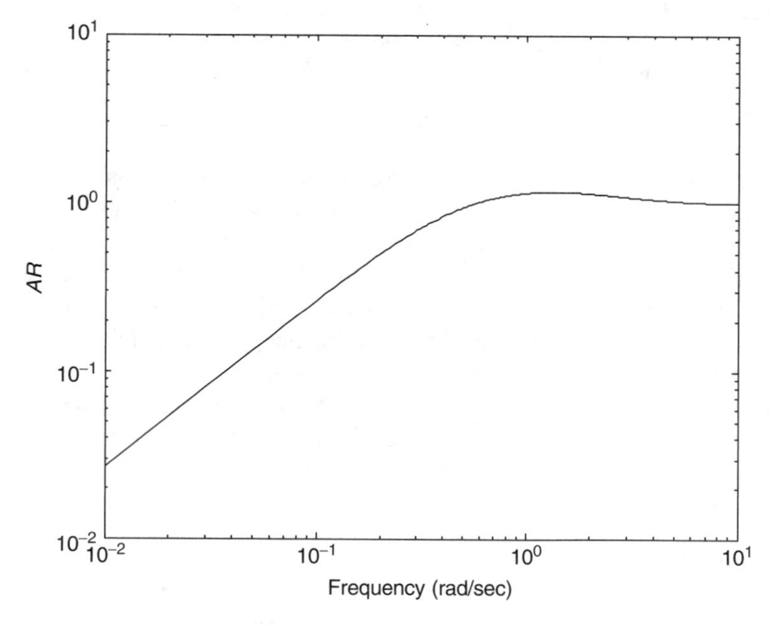


**FIGURE 17.8** The uncertainty bound for Example 17.3.

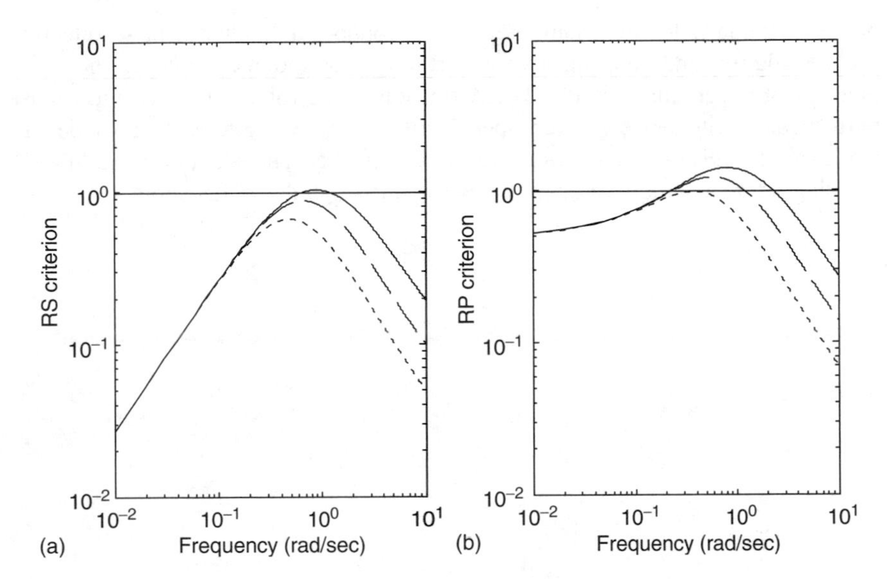


**FIGURE 17.9** The robust stability (a) and robust performance criteria (b) for three values of the filter time constant, $\lambda = 0.5$ (solid), $\lambda = 1$ (dashed), and $\lambda = 2$ (dotted).

The IMC controller becomes

$$c(s) = (4s + 1)\frac{1}{\lambda s + 1} \tag{17.18}$$

For performance, we will simply demand that the highest peak of the *AR* of the nominal complementary sensitivity function remain below 2.0, i.e.,

$$\sup_{\omega} \left|\tilde{\eta}\right| < 2.0 \tag{17.19}$$

We can easily show that

$$\begin{aligned} \tilde{\varepsilon} &= 1 - c\tilde{g} = 1 - \left(\frac{4s + 1}{\lambda s + 1}\right)\left(\frac{e^{-2s}}{4s + 1}\right) = \frac{\lambda s + 1 - e^{-2s}}{\lambda s + 1} \\ \tilde{\eta} &= c\tilde{g} = \left(\frac{4s + 1}{\lambda s + 1}\right)\left(\frac{e^{-2s}}{4s + 1}\right) = \frac{e^{-2s}}{\lambda s + 1} \end{aligned} \tag{17.20}$$

Figure 17.9 depicts the robust stability and robust performance (Eqs. [17.9] and [17.14]) criteria for a number of filter time constants. We can see that as the filter time constant is decreased (more aggressive control), the bounds are violated. For this process, the choice of $\lambda = 2$ appears to be acceptable.

## 17.4 SUMMARY

In this chapter, we have discussed the impact of model uncertainty on closed-loop performance in a more rigorous setting. The discussion was placed in the context of the IMC controller design and we have introduced a number of new concepts, such as additive and multiplicative uncertainty descriptions, including the computation of uncertainty bounds. The definition of the robust stability and robust performance criteria were given specifically for SISO systems to provide the reader a relatively simple exposure to these concepts. The section on Additional Reading offers a number of alternatives for further study of this subject.

# 18 Model Predictive Control (MPC)

Model predictive control (MPC) is a control strategy that has received wide acceptance in the process industries in recent years as an effective means of implementing multivariable model-based control in the presence of process constraints. An extensive survey conducted by Qin and Badgwell[1] discusses in detail the history, the variety of formulations and algorithms, as well as a number of reported industrial applications.

In essence, MPC belongs to the class of model-based controllers discussed in Chapter 16, as it explicitly uses the process model as part of the control design. The unique manner with which it exploits the model makes it a suitable alternative for a wide spectrum of control problems. The main strength of MPC is demonstrated when applied to problems with:

- a large number of manipulated and controlled variables
- constraints imposed on both the manipulated and controlled variables (this is perhaps the most valuable feature of MPC)
- changing control objectives and equipment (sensor and actuator) failures
- time delays
- strong interactions among variables, where the use of multiloop control may be problematic
- many unmeasured disturbances (if disturbances can be measured, MPC has built-in feedforward capabilities)

In this chapter, we will highlight the key principles behind the MPC strategy and introduce the traditional dynamic matrix control (DMC) implementation as a means to demonstrate why MPC has been such an appealing technology for the industry.

## 18.1 GENERAL PRINCIPLES

Figure 18.1 depicts the conceptual implementation of the MPC technique. The key element is the embedded process model that allows the prediction of output variables. Currently, the most comprehensive MPC algorithms are those based on optimization of an objective function involving the error between the predicted future outputs and their set-points. These algorithms may often be computationally intensive, but can still be readily implemented on a process control computer.

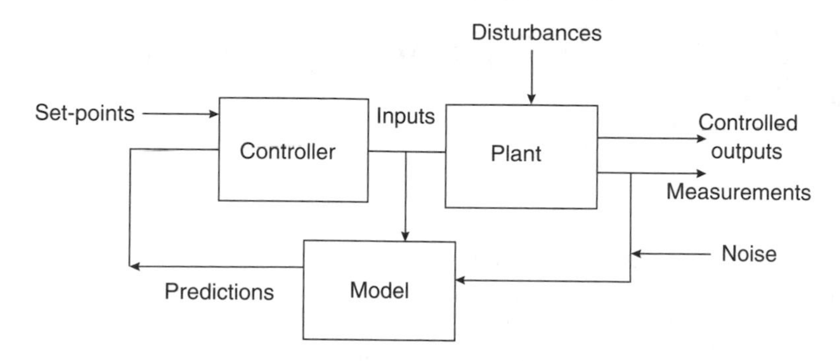


**FIGURE 18.1** Conceptual schematic representation of MPC.

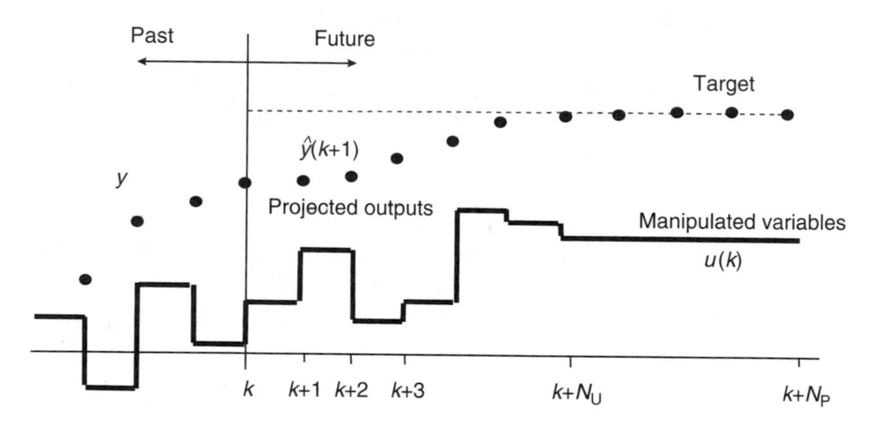


**FIGURE 18.2** Implementation of the MPC strategy.

Many control system vendors offer comprehensive software packages and services for a variety of industries.

There are three elements that characterize a typical MPC implementation. Figure 18.2 illustrates how these elements collectively define the MPC strategy.

- *Reference trajectory*: This is associated with the definition of a desired target (set-point) trajectory for the process output.
- *Process output prediction*: A model is used to predict the process output over a predetermined prediction horizon (current time is used as the prediction origin). In Figure 18.2, the prediction is made up to $N_P$ steps into the future.
- *Control action computation*: The process model is also used to calculate the sequence of control moves to drive the specified target according to a certain optimal objective. In Figure 18.2, the first $N_U$ moves are calculated and the last move is kept constant until the end of the output prediction horizon.

The key to the MPC strategy is that the future control action (move) sequence is computed at the current time $k$, but only the next move (at time $k+1$) is implemented. The computation of the control moves is repeated at time $k+1$ and the algorithm proceeds forward in time. This strategy is often referred to as the *receding horizon* strategy and mimics an expert chess player who, after surveying the board, develops a strategy for the next multiple moves, implements the next move, and then makes another assessment of the situation after observing how his or her opponent has moved.

It should be noted that since no process model is perfect, plant measurements are used to compare with the model predictions to compute the prediction error that is used to update future predictions. This is the key reason for the *receding* horizon strategy that incorporates some sense of a feedback mechanism to an otherwise feedforward implementation.

### 18.1.1 Model Forms

For linear MPC applications, most algorithms use one of the three model forms: impulse response models, step-response models, and state-space models. Here, we will give a brief overview of each.

*Impulse-response models*: The finite impulse response (FIR) model can be expressed as

$$y(k) = \sum_{i=1}^{N} h_i u(k-i) \tag{18.1}$$

The output at a discrete time step $k$ is expressed as a function of the values of the input at previous time steps. The model is an $N$-dimensional FIR model as it has a *memory* of $N$ time steps into the past. $h_i$ are referred to as the impulse response coefficients. The early version of the IDCOM algorithm[2] used this model form for prediction.

*Step-response (convolution) models*: These models are obtained when the process is subjected to a step input. One of the key assumptions is that the process is stable. The model is expressed as

$$y(k) = \sum_{i=1}^{N} a_i \Delta u(k-i) + a_{ss} u(k-N-1) \tag{18.2}$$

where $a_i$ is the $i$th step-response coefficient and the last term on the right hand side represents the steady-state bias. The model horizon $N$ defines the memory of the model where $\Delta u(k) = u(k) - u(k-1)$. $a_i$'s are found by using the unit step response for the process at sampling periods $\Delta t$ (see Figure 18.3). We define $a_i = 0$ for $i < 0$ and $N\Delta t$ can be taken as the settling time of the process. We also note the following relationship between the impulse- and the step-response coefficients, as the step-response can be defined as the integral of the impulse-response:

$$h_i = a_i - a_{i-1} \tag{18.3}$$

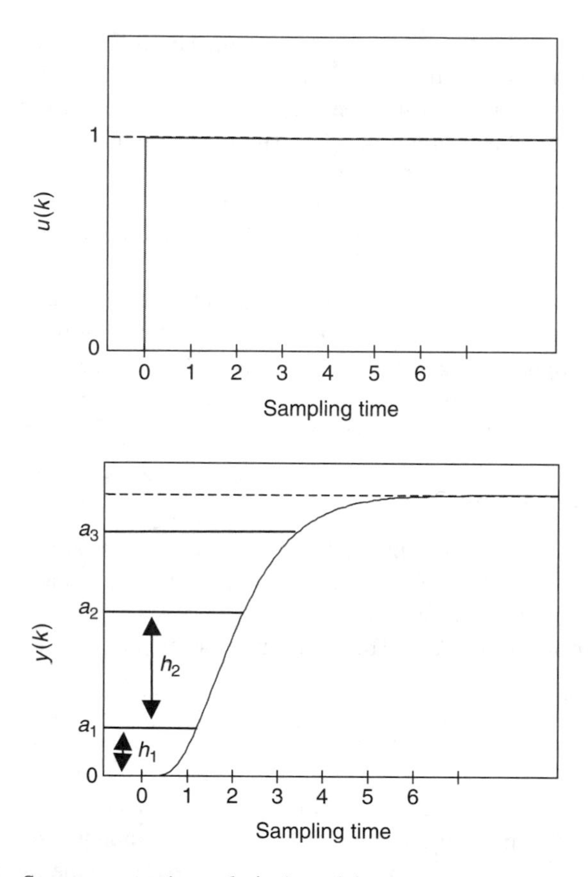


**FIGURE 18.3** Step-response (convolution) model.

The step-response models are used in the DMC algorithm originally proposed by Cutler and Ramaker.[3]

*State-space models*: In discrete time steps, a state-space model (see Eq. [5.18]) can be expressed as follows:

$$
\begin{aligned}
\mathbf{x}(k+1) &= A\mathbf{x}(k) + B\mathbf{u}(k) \\
\mathbf{y}(k+1) &= C\mathbf{x}(k+1)
\end{aligned}
\tag{18.4}
$$

This model form is more general and can also represent unstable plants. Most of the recent MPC algorithms utilize this form.[4]

---

- It is important to remember that it is possible to convert from one model form to another, although some conversions are more difficult than others are.

- We reiterate that finite impulse and step-response models are intuitive but they cannot represent unstable plants, and may require a large number of terms. On the other hand, state-space models are general with significant theory behind them, and can address unstable systems.

---

In the following section, we shall illustrate the MPC approach through its application using step-response models.

## 18.2 DYNAMIC MATRIX CONTROL

Previously, we have used transfer function models to describe the dynamic behavior of a process where we generally used first-order models with time delay. Transfer function models are parametric models that also require the order to be specified. One alternative approach is to employ a "discrete response (*convolution*) model." The advantage of these models is that the model coefficients can be obtained directly from the experimental step response as shown in Figure 18.3.

Let us define the control horizon $N_U$ as the number of control actions (or control moves) that are calculated to affect the outputs predicted using the process model over the model prediction horizon $N_P$.

We shall make the following statements:

- $a = [a_1, a_2, \ldots, a_N]^T$ represents the vector of step-response coefficients of the convolution model of a process (see Figure 18.3).
- The current time instant is $k$, and a control action has been performed at time $k - 1$.

Then, the predicted output of the process over the prediction horizon, whose origin is the current time instant $k$, will be represented by

$$y(k) = \sum_{i=1}^{N} a_i \Delta u(k - i) + a_{ss} u(k - N - 1) + d(k) \tag{18.5}$$

where $d(k)$ represents the effect of disturbances. Normally, the disturbance in Eq. (18.5) is unknown. However, one can use the same equation to get an *estimate* of the disturbance based on the plant measurements. In other words, we can rephrase Eq. (18.5) as follows:

$$\hat{d}(k) = y_{\text{measured}}(k) - \sum_{i=1}^{N} a_i \Delta u(k - i) + a_{ss} u(k - N - 1) \tag{18.6}$$

where $\hat{d}(k)$ becomes the disturbance estimate. For a one-step-ahead prediction, we can write Eq. (18.5) as

$$\begin{aligned} y(k + 1) = {} & a_1 \Delta u(k) + a_2 \Delta u(k - 2) + \cdots \\ & + a_N \Delta u(k - N + 1) + a_{ss} u(k - N + 1) + d(k) \end{aligned} \tag{18.7}$$

But often, we have a prediction horizon greater than 1 ($N_P > 1$); in that case, we can generalize Eq. (18.7) as

$$\begin{bmatrix} y(k+1) \\ y(k+2) \\ \vdots \\ y(k+N_P) \end{bmatrix} = \begin{bmatrix} a_{ss}u(k-N) \\ a_{ss}u(k-N+1) \\ \vdots \\ a_{ss}u(k-N+N_P-1) \end{bmatrix} + \begin{bmatrix} a_2 & a_3 & \ldots & a_N \\ a_3 & a_4 & \ldots & 0 \\ a_4 & a_5 & \ldots & 0 \\ a_{N_P+1} & \cdots & \ldots & 0 \end{bmatrix} \begin{bmatrix} \Delta u(k-1) \\ \Delta u(k-2) \\ \vdots \\ \Delta u(k-N+1) \end{bmatrix}$$

$$+ \begin{bmatrix} a_1 & 0 & \ldots & 0 \\ a_2 & a_1 & \ldots & 0 \\ \vdots & \vdots & \vdots & \vdots \\ a_{N_P} & a_{N_P-1} & \cdots & a_1 \end{bmatrix} \begin{bmatrix} \Delta u(k) \\ \Delta u(k+1) \\ \vdots \\ \Delta u(k+N_P-1) \end{bmatrix} + \begin{bmatrix} d(k+1) \\ d(k+2) \\ \vdots \\ d(k+N_P) \end{bmatrix} \tag{18.8}$$

We will make an assumption that the future values of the disturbance are equal to its present estimated value,

$$d(k+l) = \hat{d}(k), \quad l = 1, 2, \ldots, N_P \tag{18.9}$$

Also, by incorporating the control horizon, we can rephrase Eq. (18.8) as follows:

$$\begin{bmatrix} y(k+1) \\ y(k+2) \\ y(k+3) \\ \vdots \\ y(k+N_U) \\ \vdots \\ y(k+N_P) \end{bmatrix} = \begin{bmatrix} a_{ss}u(k-N) \\ a_{ss}u(k-N+1) \\ a_{ss}u(k-N+2) \\ \vdots \\ a_{ss}u(k-N+N_U-1) \\ \vdots \\ a_{ss}u(k-N+N_P-1) \end{bmatrix} + \begin{bmatrix} a_2 & a_3 & \cdots & \cdots & a_N \\ a_3 & a_4 & \cdots & a_N & 0 \\ a_4 & a_5 & \cdots & 0 & 0 \\ \vdots & \vdots & \vdots & \vdots & \vdots \\ a_{N_U+1} & \cdots & \cdots & 0 & 0 \\ \vdots & \vdots & \vdots & \vdots & \vdots \\ a_{N_P+1} & \cdots & \cdots & 0 & 0 \end{bmatrix} \begin{bmatrix} \Delta u(k-1) \\ \Delta u(k-2) \\ \vdots \\ \Delta u(k-N+1) \end{bmatrix}$$

$$+ \begin{bmatrix} a_1 & 0 & 0 & \ldots & 0 \\ a_2 & a_1 & 0 & \ldots & 0 \\ a_3 & a_2 & a_1 & \ldots & 0 \\ \vdots & \vdots & \vdots & \vdots & \vdots \\ a_{N_U} & a_{N_U-1} & \cdots & a_1 & 0 \\ \vdots & \vdots & \vdots & \vdots & \vdots \\ a_{N_P} & a_{N_P-1} & \cdots & \cdots & a_{N_P-N_U+1} \end{bmatrix} \begin{bmatrix} \Delta u(k) \\ \Delta u(k+1) \\ \vdots \\ \Delta u(k+N_U-1) \end{bmatrix} + \begin{bmatrix} d(k) \\ d(k) \\ \vdots \\ d(k) \end{bmatrix} \tag{18.10}$$

This equation can be compactly written using matrices as

$$Y(k+N_P) = Y^{\text{past}} + A\Delta U(k) + D \tag{18.11}$$

In this equation, $Y^{\text{past}}$ represents the past known information,

$$Y^{\text{past}} = \begin{bmatrix} a_{ss}u(k-N) \\ a_{ss}u(k-N+1) \\ a_{ss}u(k-N+2) \\ \vdots \\ a_{ss}u(k-N+N_U-1) \\ \vdots \\ a_{ss}u(k-N+N_P-1) \end{bmatrix} + \begin{bmatrix} a_2 & a_3 & \cdots & \cdots & a_N \\ a_3 & a_4 & \cdots & a_N & 0 \\ a_4 & a_5 & \cdots & 0 & 0 \\ \vdots & \vdots & \vdots & \vdots & \vdots \\ a_{N_U+1} & \cdots & \cdots & 0 & 0 \\ \vdots & \vdots & \vdots & \vdots & \vdots \\ a_{N_P+1} & \cdots & \cdots & 0 & 0 \end{bmatrix} \begin{bmatrix} \Delta u(k-1) \\ \Delta u(k-2) \\ \vdots \\ \Delta u(k-N+1) \end{bmatrix} \tag{18.12}$$

Now, we are ready to discuss various formulations of MPC.

### 18.2.1 SISO Unconstrained DMC Problem

Let us consider the desired output trajectory as,

$$Y^{sp} = \begin{bmatrix} y^{sp}(k+1) \\ y^{sp}(k+2) \\ y^{sp}(k+3) \\ \vdots \\ y^{sp}(k+N_U) \\ \vdots \\ y^{sp}(k+N_P) \end{bmatrix} \quad (18.13)$$

And require that this trajectory be followed perfectly up to the end of the prediction horizon:

$$Y^{sp}(k+l) = Y(k+l), \quad l = 1, 2, \ldots, N_P \quad (18.14)$$

---

The control problem is to choose and implement a control sequence $\Delta u(k)$ such that the predicted output follows and remains at the desired trajectory.

---

Equations (18.11) and (18.14) allow us to solve for the control moves that would accomplish this objective:

$$\Delta U = A^{-1}(Y^{sp} - Y^{past} - D) \quad (18.15)$$

The solution of Eq. (18.15) demands further discussion. If $N_U = N_P$, the solution would be exact as the dynamic matrix $A$ would become square and could be inverted in a straightforward manner. Often, this results in what is referred to as the deadbeat control that takes the output to the desired trajectory in one time step. As this requires excessive control action, one often chooses $N_U < N_P$. Then, the optimal solution for this overdetermined problem is the least-squares solution given by,

$$\Delta U = (A^T A)^{-1} A^T (Y^{sp} - Y^{past} - D) \quad (18.16)$$

Indeed, Eq. (18.16) is the solution of the optimization problem:

$$\min_{\Delta U} \quad J = E^T E \quad (18.17)$$

where we have defined the error matrix as $E = Y^{sp} - Y$. In practice, a penalty for excessive control action is also incorporated into the objective function

$$\min_{\Delta U} \quad J = E^T W_1 E + \Delta U^T W_2 \Delta U \quad (18.18)$$

The matrices $W_1$ and $W_2$ determine the trade-off between the amount of control action and the tracking error. The control law that takes into account the control action penalty is, then, given by

$$\begin{aligned}\Delta U &= (A^T W_1^T W_1 A + W_2^T W_2)^{-1} A^T W_1^T W_1 (Y^{sp} - Y^{past} - D) \\ &= K_{DMC}(Y^{sp} - Y^{past} - D)\end{aligned} \tag{18.19}$$

The above equation illustrates that the DMC is implemented through a constant gain matrix $K_{DMC}$ that can be calculated off-line using the model and weighting matrices.

## 18.2.2 Controller Tuning

The DMC implementation described above includes a number of design parameters, which can be adjusted to give the desired response as well as an appropriate amount of controller effort. The following guidelines are commonly used:

1. The *model horizon* $N$ should be selected so that $N\Delta t \geq$ open-loop settling time. Values of $N$ between 20 and 70 are typical in the literature.
2. The *prediction horizon* $N_P$ determines how far into the future the control objective reaches.
   - Increasing $N_P$ results in a more conservative control action but increases the computational effort.
   - Most practitioners recommend using $N_P = N + N_U$.
3. The *control horizon* $N_U$ determines the number of control actions calculated into the future.
   - Too large a value of $N_U$ results in excessive control action.
   - A smaller value of $N_U$ yields a controller relatively insensitive to model errors.
4. The *weighting matrices* $W_1$ and $W_2$ contain potentially a large number of parameters.
   - It is usually sufficient to select $W_1 = I$ and $W_2 = \rho I$, with $\rho$ being a constant.
   - Larger values of $\rho$ penalize the magnitude of $\Delta U$ more and yield less vigorous control actions.

**Example 18.1**

Consider a process with the following process and disturbance transfer functions:

$$\tilde{g}(s) = \frac{1}{5s^3 + 15.5s^2 + 11.5s + 1}, \quad g_d(s) = \frac{1.52}{50s + 1} \tag{18.20}$$

We shall implement an unconstrained DMC with the horizon parameters set as $N_P = 15$, $N_U = 5$ and the following set of weights: (1) $W_1 = 1$ and $W_2 = 1$; (2) $W_1 = 5$ and $W_2 = 1$; (3) $W_1 = 1$ and $W_2 = 5$.

Figure 18.4 shows the closed-loop response of the system for a set-point change for case (1) Figure 18.5 illustrates the effect of changing the ratio between the weights; case (2) We can observe that by increasing the weight on the error in the objective function, the system responds faster at the expense of more aggressive control. Figure 18.6, on the other hand, shows the effect of increasing $W_2$ with respect to $W_1$; case (3) In this case, since the penalty on the control action is increased, the response of the system becomes more sluggish.

## 18.3 TREATMENT OF PROCESS CONSTRAINTS

A key feature of many practical control problems is the presence of inequality constraints on both controlled variables and manipulated variables. Such

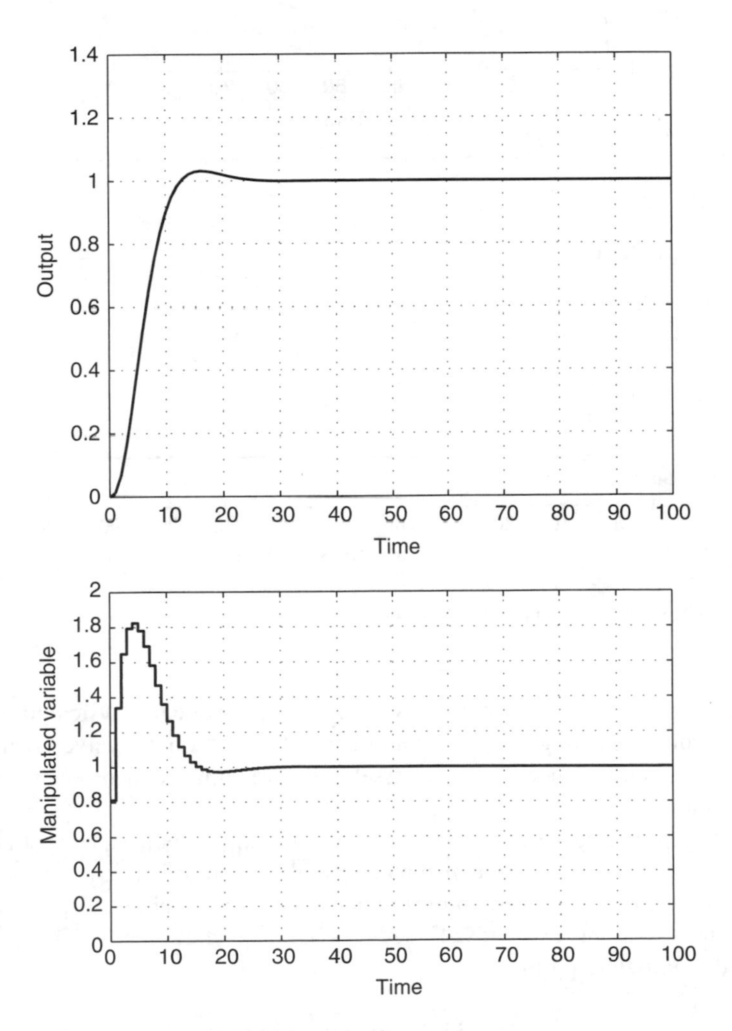


**FIGURE 18.4** Set-point response for case (1).

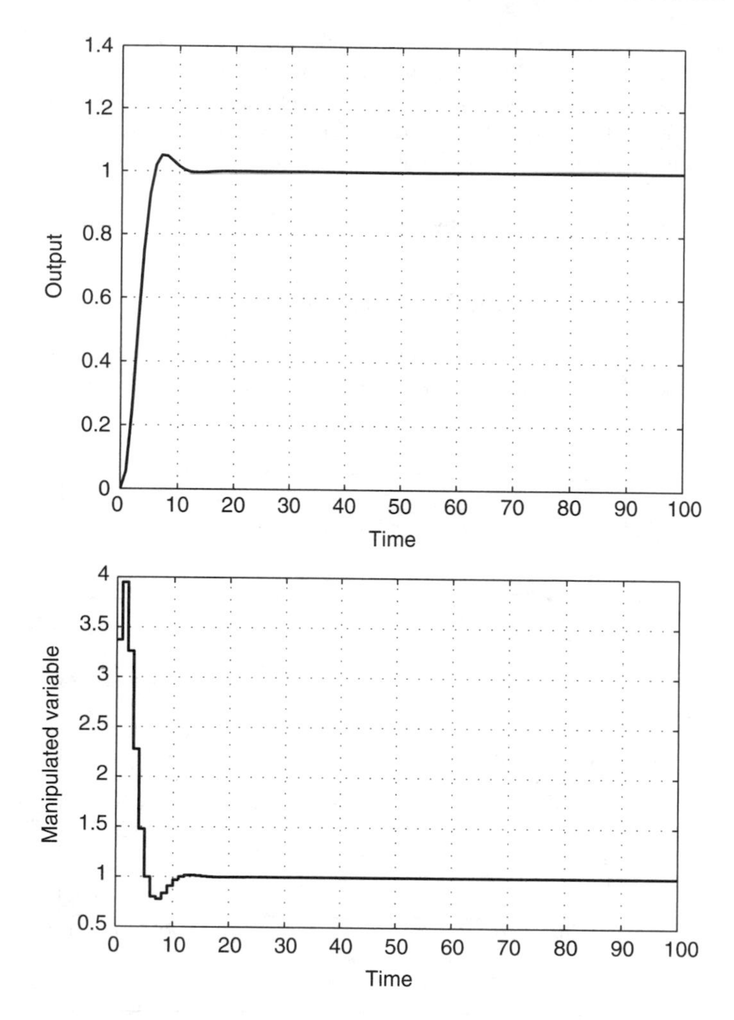


**FIGURE 18.5** Set-point response for case (2).

constraints arise commonly in process control problem due to physical limitations of plant equipment (e.g., limited pump capacity and control valve sizing) and safety and quality concerns (e.g., vessel pressure limits and product impurity specifications).

MPC approach allows us to design a controller that satisfies the constraints on current and future values of the manipulated and the controlled variables. We need to rephrase the unconstrained optimization problem solved above (Eq. [18.18]) by adding the constraints explicitly. This results in the quadratic DMC (QDMC)[5] that solves the following problem:

$$\min_{\Delta U} \quad J = E^{\mathrm{T}} W_1 E + \Delta U^{\mathrm{T}} W_2 \Delta U \tag{18.21}$$

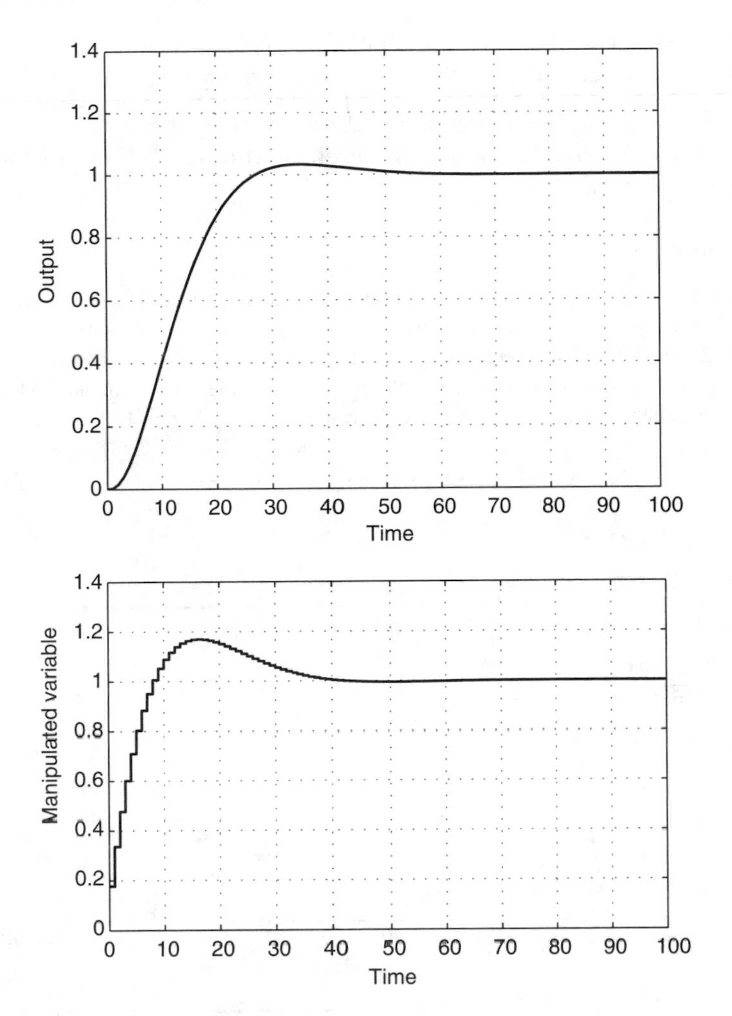


**FIGURE 18.6** Set-point response for case (3).

subject to

$$Y(k + N_{\mathrm{P}}) = Y^{\mathrm{past}} + A\Delta U(k) + D$$

$$U_{\min} \leq U \leq U_{\max}$$

$$\Delta U_{\min} \leq \Delta U \leq \Delta U_{\max}$$

$$Y_{\min} \leq Y \leq Y_{\max}$$

This approach produces a mathematical programming problem, specifically a quadratic programming (QP) problem because

- The objective function is quadratic.
- The process model and the inequality constraints are linear.

There are many commercial algorithms, some involving shortcut procedures, that can solve the constrained optimization problem (Eq. [18.21]) in real time with reasonable efficiency.[6]

### Example 18.2

Consider the SISO process in the previous example and let us assume that we have the following constraints on the control inputs: (1) $W_1 = 1$ and $W_2 = 1$; $U_{max} \leq 1.5$, and (2) $W_1 = 1$ and $W_2 = 1$; $\Delta U_{max} \leq 0.1$.

Figure 18.7 shows the behavior of the process with the original default values of the controller parameters, but for the constrained case (1). Figure 18.8 illustrates the

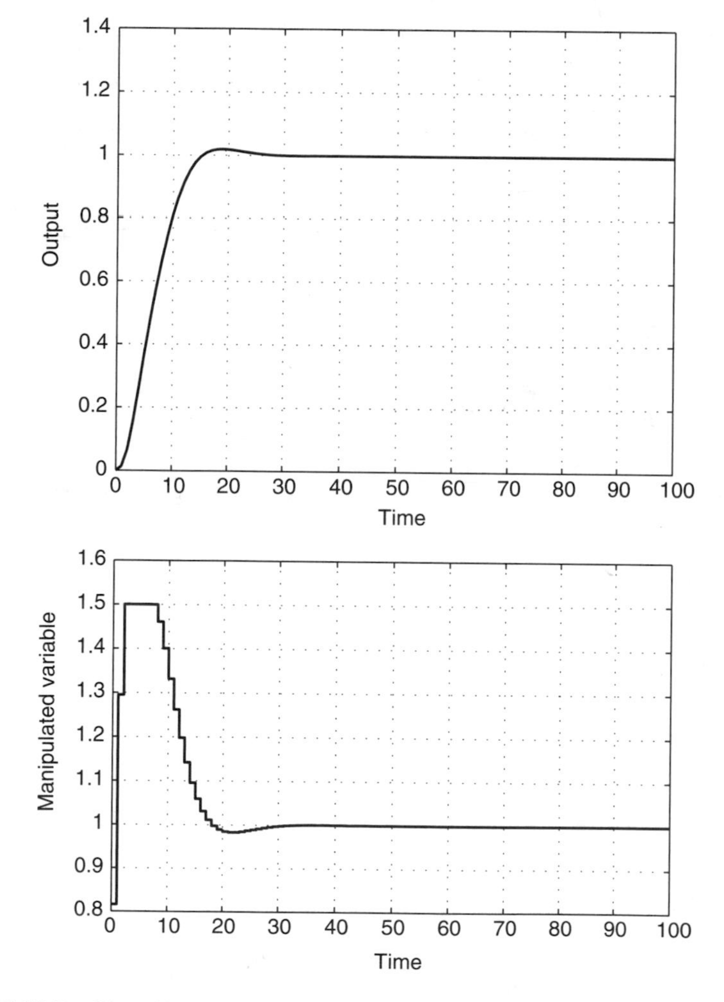


**FIGURE 18.7** Closed-loop response for case (1).

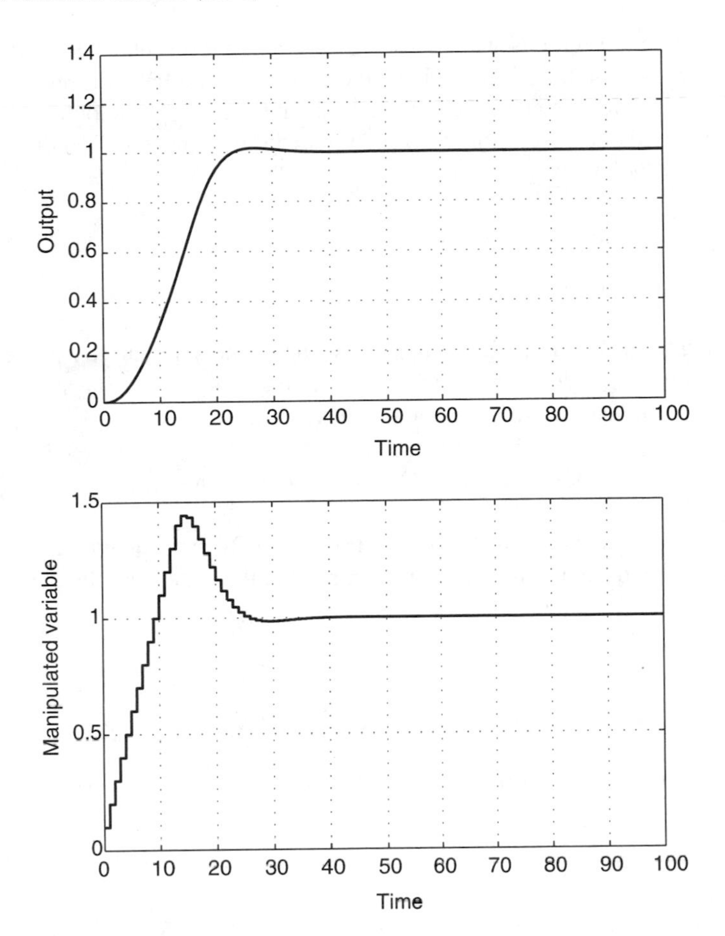


**FIGURE 18.8** Closed-loop response for case (2).

effect of incorporating a constraint in terms of the rate of change of the input in case (2). In both cases, as can be expected, the response of the process is more sluggish than for the system without constraints (Example 18.1).

## 18.4 STATE-SPACE FORMULATION OF MPC

It should be noted that the step–response models are a special realization of the state-space models. However, the state-space models are much more general and allow for a more elegant formulation of the problem that can also accommodate MIMO systems in a straightforward manner.

Consider the state-space model (e.g., Eq. [18.4]) given by

$$\begin{aligned} \mathbf{x}(k+1) &= A\boldsymbol{x}(k) + B\boldsymbol{u}(k) + \boldsymbol{d}(k) \\ \mathbf{y}(k+1) &= C\boldsymbol{x}(k+1) + \boldsymbol{v}(k+1) \end{aligned} \tag{18.22}$$

The disturbance vector is given by $\boldsymbol{d}(k)$, and $\boldsymbol{v}(k)$ represents the measurement noise (error) vector. Considering that the disturbance and the measurement noise may be stochastic in nature, the prediction of the states and the outputs depends on the state and output information from the previous time step.[7] Equation (18.22) can be rephrased as

$$\begin{aligned} \mathbf{x}(k|k-1) &= A\boldsymbol{x}(k-1|k-1) + B\boldsymbol{u}(k-1|k-1) \\ \mathbf{y}(k|k-1) &= C\boldsymbol{x}(k-1|k-1) \end{aligned} \quad (18.23)$$

The notation $\mathbf{x}(k|k-1)$ indicates that the value of the state at the sample time $k$ is based only on information available at sampling time $k-1$. The states need to be corrected (updated) using the current measurement:

$$\mathbf{x}(k|k) = \boldsymbol{x}(k|k-1) + K(\boldsymbol{y}_{\mathrm{m}}(k) - \boldsymbol{y}(k|k-1)) \quad (18.24)$$

The gain matrix $K$ is obtained as the solution of a Riccati equation and referred to as the Kalman filter gain.[7] Now, we can predict the states and the output using the model

$$\begin{aligned} \mathbf{x}(k+1|k) &= A\boldsymbol{x}(k|k) + B\boldsymbol{u}(k|) \\ \mathbf{y}(k+1|k) &= C\boldsymbol{x}(k+1|k) \end{aligned} \quad (18.25)$$

The objective function for the SISO-MPC problem can be rephrased as,

$$\min_{u(k),u(k+1),\ldots,u(k+N_{\mathrm{U}}-1)} \sum_{i=1}^{N_{\mathrm{P}}} \boldsymbol{x}(k+i|k)^{\mathrm{T}} Q_i \boldsymbol{x}(k+i|k) + \sum_{i=1}^{N_{\mathrm{U}}} u(k+i-1)^{\mathrm{T}} R_i u(k+i-1) \quad (18.26)$$

The positive matrices $Q_i$ and $R_i$ (i.e., $\lambda_j(Q_i)$ and $\lambda_j(R_i) > 0$ for all $j$) offer a means to consider a range of objectives that can emphasize (weigh more) the states versus the inputs or some states toward the beginning, or the end of the prediction horizon. As we have done before, one can solve the optimization problem in Eq. (18.26) subject to constraints on the states, the outputs, the inputs, and the rate of change of the inputs.

The DMC problem would be obtained, as a special case, if we assume constant disturbance effects, $\boldsymbol{d}(k+1) = \boldsymbol{d}(k)$, and no measurement noise, with no state updates ($K = 0$).

### 18.4.1 Infinite Horizon Problem

In Eq. (18.26), if we set the prediction and control horizons to infinity, i.e., $N_{\mathrm{P}} = N_{\mathrm{U}} = \infty$, and do not consider any constraints, we obtain the well-known optimal control problem referred to as the linear quadratic Gaussian (LQG).[7] This problem has a closed-form solution for the control action and, in fact, it is a constant gain controller that acts on the system states. A very important property of this controller is that the closed-loop stability is guaranteed, which is very difficult to prove with the DMC formulation, or the finite horizon MPC.

If we assume that there are no disturbances or measurement noise, and that all the states are measurable, the MPC formulation can be rephrased as,[8]

$$\min_{u(k),u(k+1),\ldots,u(k+N_U-1)} \sum_{i=1}^{\infty} \boldsymbol{x}(i)^{\mathrm{T}} Q\boldsymbol{x}(i) + u(i)^{\mathrm{T}} Ru(i) \tag{18.27}$$

subject to the state and input constraints that can be compactly represented as

$$Y\boldsymbol{x}(i) \le \zeta, \quad i = k_1, k_1 + 1, \ldots$$

$$\Psi u(i) \le \psi, \quad i = k, k + 1, \ldots, k + N_U - 1$$

where $Y$ and $\Psi$ are matrices with appropriate dimensions, with $\zeta$ and $\psi$ being vectors. We also note that the state and sampling rates are different, and often, we select $k_1 > k$.

This formulation also has an elegant stability result that has significant implications for linear feedback systems subject to constraints. If the constrained MPC problem in Eq. (18.27) is feasible, it has been shown[8] that $\boldsymbol{x} = 0$ is an asymptotically stable solution. This is an important result and one can argue that the stability can be assured by relaxing the constraints, and reshaping the region of feasibility.

---

- There are a number of issues that still remain within the realm of research challenges. For instance, while some stability results have been proven for infinite horizon formulations, the impact of model uncertainty (robustness) on the performance of MPC is still largely unclear. Practical approaches to this problem involve extensive experimentation (simulations) with varying tuning parameters and constraint bounds, and may offer some physical understanding toward the trends involved in making the MPC performance insensitive to model uncertainty. This is, of course, far from a rigorous solution to the problem.
- Another challenge lies in the development of models (model identification) and the search for practically meaningful means to determine models that exhibit dynamic behavior most relevant from a closed-loop performance standpoint. For algorithms that rely on state information directly, one needs to develop state estimation algorithms as well, and these can also be phrased within the context of receding horizon formulation.

---

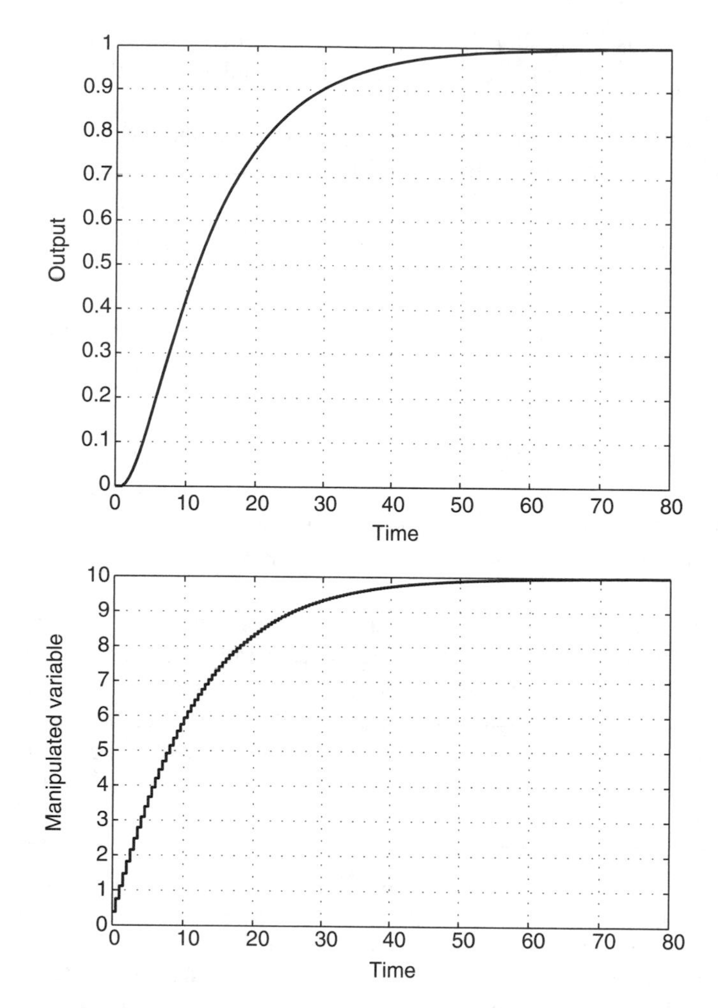


**FIGURE 18.9** Closed-loop response for set-point change with $W_1 = 2$, $W_2 = 1$, and $N_P = 5$, $N_U = 1$.

## 18.5 SUMMARY

The MPC formulation opened up a new era in process control, making it feasible to develop effective control strategies for processes with constraints. In this chapter, we have touched upon some of the key concepts and illustrated the use of MPC in practical applications.

More recent studies on MPC focus on the extensions to nonlinear processes and there are a number of approaches one can take to introduce some degree of nonlinearity in the MPC formulation.

We will discuss some of the recent advances in MPC in more detail in the Additional Reading section.

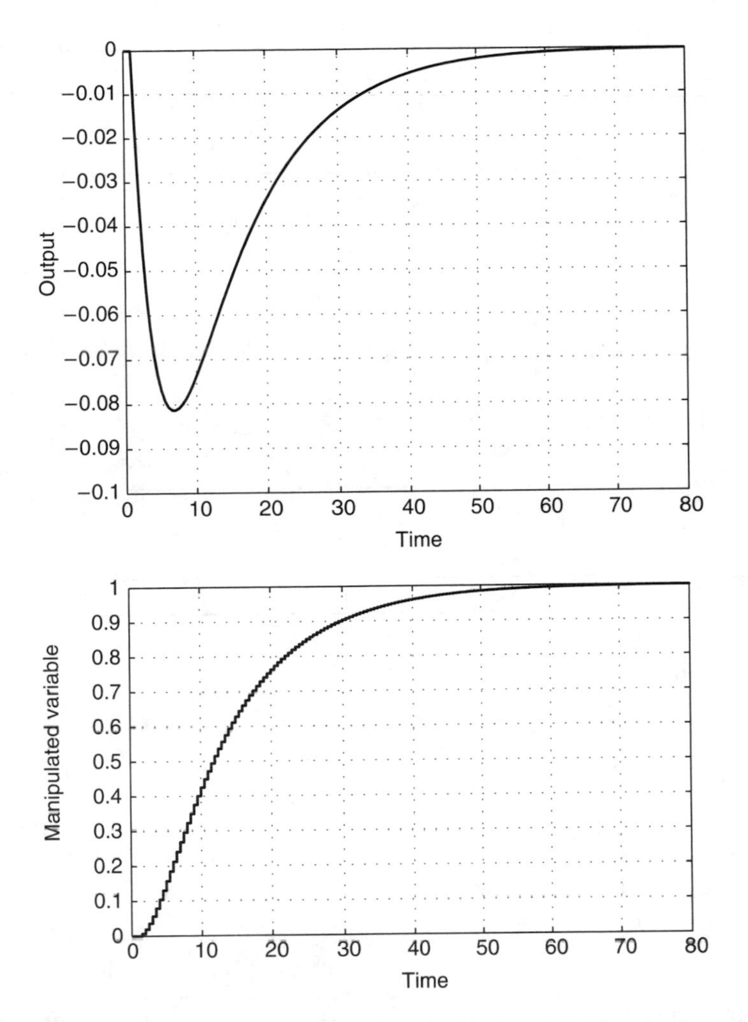


**FIGURE 18.10** Closed-loop response for disturbance with $W_1 = 2$, $W_2 = 1$, and $N_P = 5$, $N_U = 1$.

## CONTINUING PROBLEM

For the blending process, implement the following control strategies using the MPC-Tool package:

1. An SISO-MPC controller to control the composition by manipulating the flow rate $F_2$
   - Analyze the behavior for $W_1 = 2$, $W_2 = 1$ and $N_P = 5$, $N_U = 1$
   - Consider $W_1 = 2$, $W_2 = 1$ and $N_P = 15$, $N_U = 5$
2. Using $W_1 = 10$, $W_2 = 1$, and $N_P = 15$, $N_U = 5$
   - Incorporate an upper bound in the manipulated variable, $F_{2,max} \leq 11$

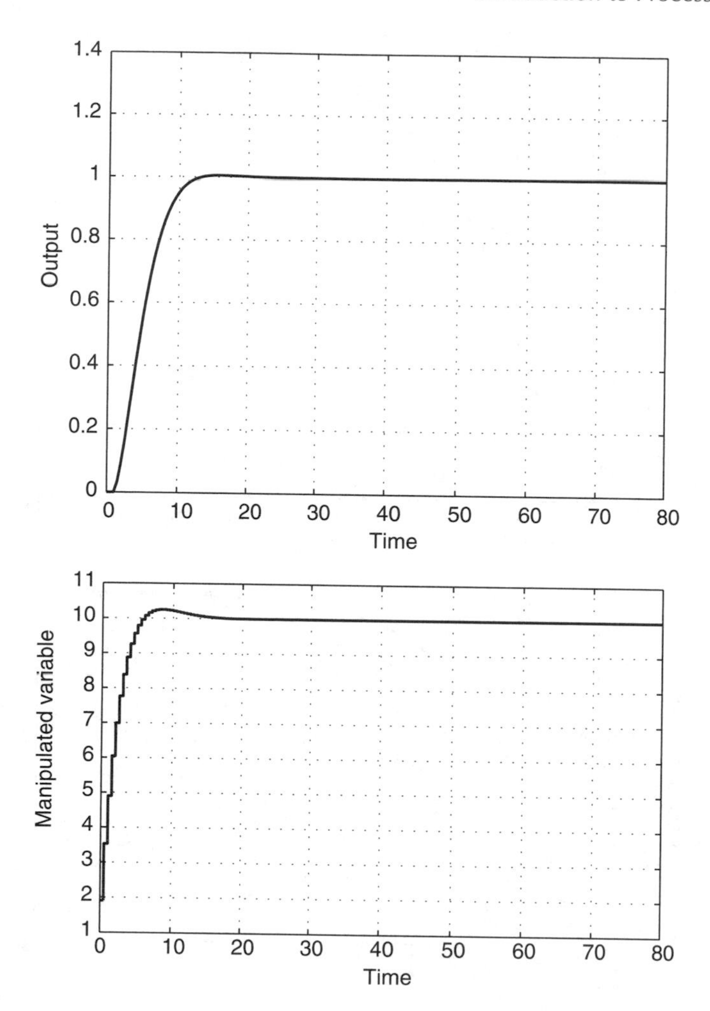


**FIGURE 18.11** Closed-loop response for set-point change with $W_1 = 2$, $W_2 = 1$, and $N_P = 15$, $N_U = 5$.

- Add an upper bound in the rate of change of the controlled variable $\Delta F_{2,\max} \leq 1$

## Solution

Figure 18.9 and Figure 18.10 show the results of the first case when we have a short prediction horizon and a one-step ahead control horizon. The response appears to be sluggish. We then extend both the prediction and the control horizons. Figure 18.11 shows that the response is now much faster. It also appears that more control effort is required for this performance.

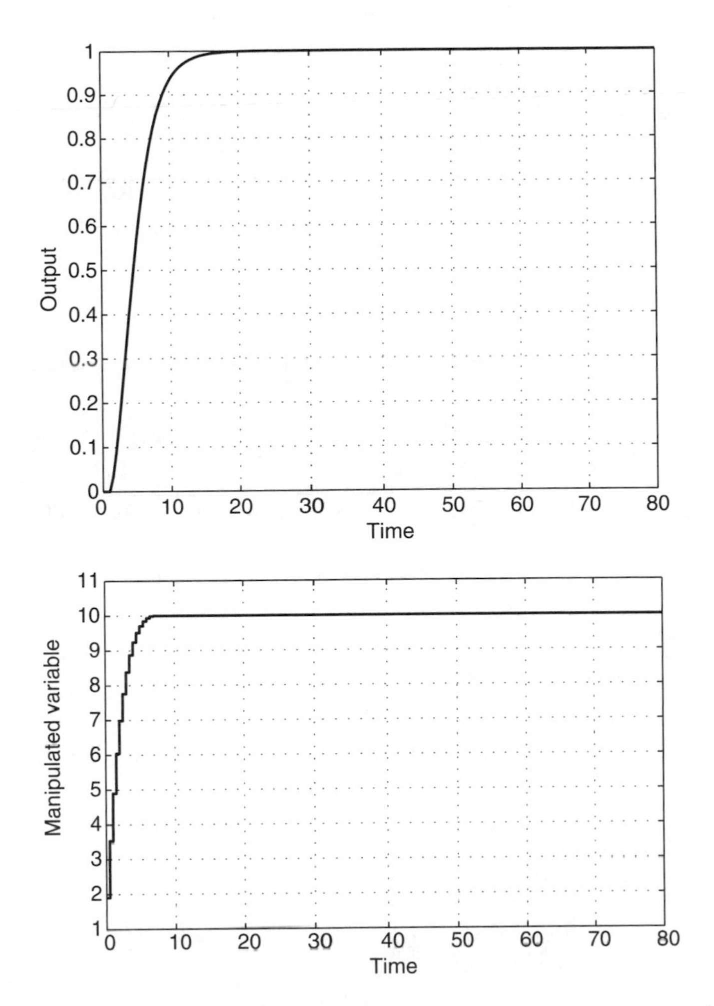


**FIGURE 18.12** Closed-loop response for set-point change with $W_1 = 2$, $W_2 = 1$, $N_P = 15$, $N_U = 5$ and $F_{2,max} \leq 11$.

Next, we test the performance of the closed-loop system when constraints are imposed. Figure 18.12 displays the case where the control action is bounded. It appears that the control action saturates early at its upper bound, which causes the set-point response to be more sluggish but still acceptable.

In Figure 18.12, we also observe that the manipulated variable moves very fast in reaching its upper bound. There may be cases where the control hardware may not allow such rapid changes. Figure 18.13 depicts the case where we impose a bound on the rate of change of the manipulated variable. As expected, the manipulated variable changes less rapidly and the output response is more cautious.

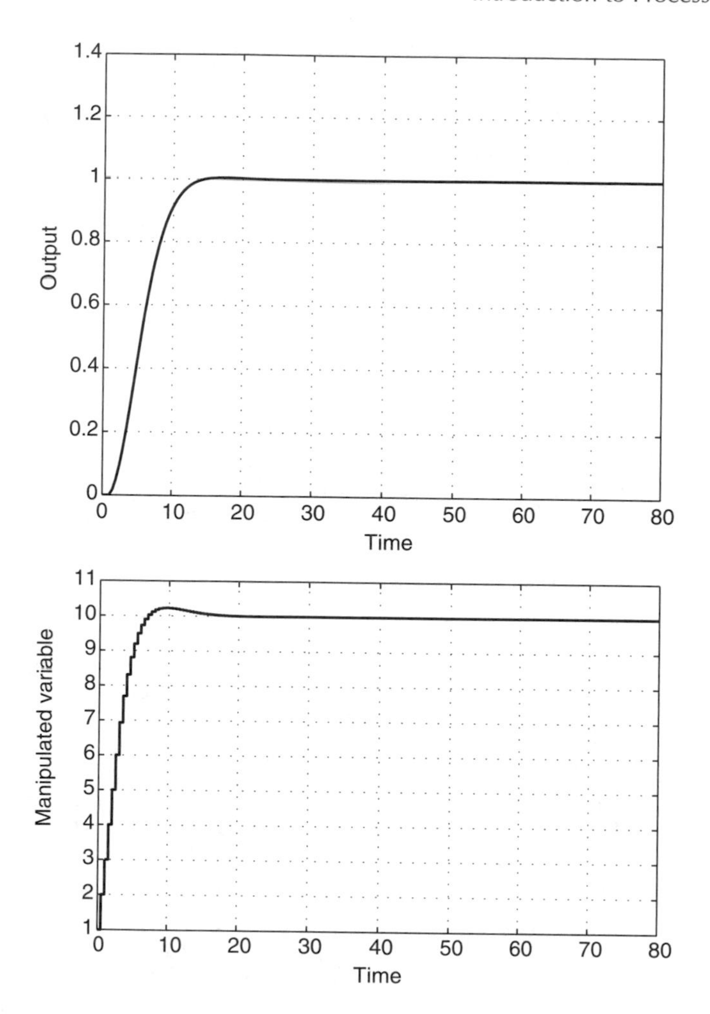


**FIGURE 18.13** Closed-loop response for set-point change with $W_1 = 2$, $W_2 = 1$, $N_P = 15$, $N_U = 5$ and $\Delta F_{2,max} \leq 1$.

We leave the implementation of a MIMO-MPC to the reader. The control system is a $2 \times 2$ system when controlling both composition and the level of the process.

## REFERENCES

1. Qin, S.J. and Badgwell, T.A., A survey of industrial model predictive control technology, *Control Engineering Practice*, 11, 733–764, 2003.
2. Richalet, J., Rault, A., Testud, J.L., and Papon, J., Model predictive heuristic control: Applications to industrial processes, *Automatica*, 14, 413–428, 1978.

3. Cutler, C.R. and Ramaker, B.L., Dynamic matrix control — A computer control algorithm, *Proceedings of the Joint Automatic Control Conference*, WP5-B/6, 1980.
4. Lee, J.H., Morari, M., and Garcia, C.E., State-space interpretation of model predictive control, *Automatica*, 30, 707–717, 1994.
5. Garcia, C.E. and Morshedi, A.M., Quadratic programming solution of dynamic matrix control (QDMC), *Chem. Eng. Comm.*, 46, 73–87, 1986.
6. Richalet, J., Industrial applications of model based predictive control, *Automatica*, 29, 1251–1274, 1993.
7. Astrom, K.J. and Wittenmark, B., *Computer Controlled Systems: Theory and Design*, 2nd Ed., Prentice-Hall, Englewood Cliffs, NJ, 1990.
8. Rawlings, J. and Muske, K., The stability of receding horizon control, *IEEE Trans. Automatic Control*, 38, 1512–1516, 1993.

# 19 Practical Control of Nonlinear Processes

Linear control strategies provide powerful control design tools under general hypotheses as we have seen in the previous chapters. The limitation of the linear control design is that a nominal model, expressed around a single operating point, may not be able to represent the nonlinear plant in the whole operating range. Therefore, the effectiveness of linear controllers may be compromised when the operation strays away from the nominal operating point. In process control, the traditional approach to dealing with nonlinear processes has been to determine a mechanism to modify the proportional gain of the PID controller as the process conditions vary.[1] Referred to as *gain-scheduling*, this strategy has been reasonably effective in controlling nonlinear processes such as pH neutralization. *Adaptive control* is yet another approach to deal with varying process conditions, where the controller parameters are adjusted according to a model, continuously determined online from process input–output data.[2] More recently, motivated by the model-based control paradigm, we have witnessed the emergence of multi-linear control design strategies that rely on determining multiple linear models that, together, can capture the dynamic behavior of a nonlinear process within the whole operating regime.[3]

In this chapter, we introduce first the operating regime modeling approach as the basis for developing multiple (linear) models. This will be followed by an introduction to the concept of gain-scheduling that makes special use of multiple process models. The rest of the chapter is devoted to the study of multi-linear model-based control, a rigorous generalization of the gain-scheduling approach, for dealing with nonlinear processes.

## 19.1 OPERATING REGIME MODELING APPROACH

For a nonlinear process, it is possible to express its dynamic model as

$$\text{NL:}\begin{cases} \dot{x} = f(x, u) \\ y = h(x) \end{cases} \tag{19.1}$$

where $f(x, u)$ and $h(x)$ are smooth and bounded functions of the state vector $x$ and the input $u$. To keep the discussion simple, we will focus on SISO systems. The main idea behind the multi-linear model approach is to decompose the complex nonlinear system domain into multiple regions, i.e., $\Omega = \bigcup_{p=1}^{M} \Omega_p$, where $M$ is the number of regions selected, such that the uncertain regions represented by

$\partial\Omega_p$ are small enough, i.e., $\partial\Omega_p \ll \Omega_p$ (Figure 19.1). The ideal case is obtained when $M \to \infty$, which implies $\partial\Omega_p \to 0$.

As we have seen in Chapter 5, a nonlinear function can be expanded in the neighborhood of a point by a Taylor series and a linear approximation is obtained when second- and higher-order terms are neglected. Thus, a nonlinear function can be locally approximated around $x_{p,s}$ as

$$h(x) \cong h(x_{p,s}) + \frac{\partial h(x_{p,s})}{\partial x}(x - x_{p,s}) \tag{19.2}$$

The nonlinear function is defined in the domain $x \in \Omega$, where the local approximation is carried out around $x_{p,s} \in \Omega_p$, where $\Omega_p$ represents the $p$th subregion. The global approximation suggests that it may be possible to span this function space as a linear combination of local approximations:

$$h(x) \cong h(x_{p,s}) + \sum_{j=1}^{\infty} \delta_j^p \frac{\partial h(x_{j,s})}{\partial x}(x - x_{j,s}) \tag{19.3}$$

where

$$\delta_j^p = \begin{cases} 1 & \text{if } j = p \\ 0 & \text{otherwise} \end{cases} \tag{19.4}$$

Then, we pose the following question: "How does one know if there are enough terms in the approximation?" In other words, how many subregions expressed by a local linear approximation would be sufficient to describe the global region? In practice, our knowledge about the behavior of the process can provide some insight. While not mathematically rigorous in all cases, standard practice is to terminate when the next term does not have a significant contribution to the magnitude of the residual error. This illustrates the trade-off between model complexity and accuracy.

The region identification is an *off-line* procedure where prior knowledge of the system is essential. In fact, this information often comes from the input–output map where it may be possible to identify regions by means of visual and analytical

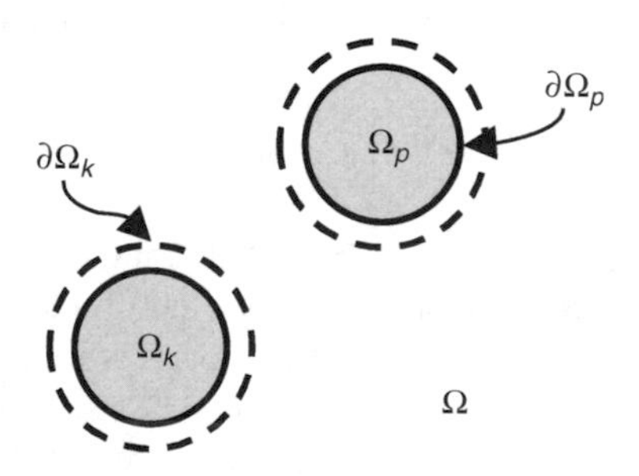


**FIGURE 19.1** Venn diagram where Ω represents the universe of models.

procedures. After region identification, the next step is to choose the local linear models, i.e., the linearization of the system (19.1) around the $p$th operating regime. The first-order approximation in subregion $p$ is given by

$$\text{L:}\begin{cases}\dot{\bar{x}} = A_p\bar{x} + B_p\bar{u} \\ \bar{y} = C_p\bar{x}\end{cases} \tag{19.5}$$

where, as before, we have

$$A_p = \frac{\partial f(x_{p,s}, u_{p,s})}{\partial x}, \quad B_p = \frac{\partial f(x_{p,s}, u_{p,s})}{\partial u}, \quad C_p = \frac{\partial h(x_{p,s}, u_{p,s})}{\partial x}$$

The deviation variables for the states, inputs, and outputs are defined as $\bar{x} = x - x_{p,s}$, $\bar{u} = u - u_{p,s}$, and $\bar{y} = y - y_{p,s}$, respectively. Hence, the nonlinear system can be approximated by the following linear combination of $M$ linear systems:

$$f(x, u) \cong f(x_{p,s}, u_{p,s}) + \left(\sum_{j=1}^{M} \delta_j^p A_j\right)\bar{x} + \left(\sum_{j=1}^{M} \delta_j^p B_j\right)\bar{u} \tag{19.6}$$

$$h(x) \cong h(x_{p,s}) + \left(\sum_{j=1}^{M} \delta_j^p C_j\right)\bar{x} \tag{19.7}$$

We note that the modeling errors associated with the truncation step can be accounted for during control design by considering the model uncertainty descriptions discussed in Chapter 17.

### 19.1.1 Global Model Structure

The operating regimes can either have different local model structures (heterogeneous) or same local model structures (homogeneous). It is assumed that the full operating range of the process operation is completely covered by these local models. At some operating regions, there may be an overlap between regimes where several models may be valid. To address this issue, a variety of approaches are available. One is to assign a membership (weighting) function to each local model, and, then, the nonlinear process would be expressed by a weighted combination of the local models.[3]

Let us consider a slightly more general system and assume that the system operating range is decomposed into $M$ regimes. Then, the following interpolation strategy provides the global model structure:

$$\dot{x} = \sum_{p=1}^{M} f_p(x, u)\ \bar{\rho}_p(y; \pi_p) \tag{19.8}$$

$$y = \sum_{p=1}^{M} h_p(x)\ \bar{\rho}_p(y; \pi_p) \tag{19.9}$$

The interpolation functions, $\bar{\rho}_p$, are obtained using $\rho_p$, a Gaussian (see Appendix D) local model validity function,[3]

$$\bar{\rho}_p(y;\pi_p) = \frac{\rho_p(y;\pi_p)}{\sum_{i=1}^{M}\rho_i(y;\pi_i)} \tag{19.10}$$

where $\pi_p$ is a parameter vector, representing the mean and the standard deviation for the membership function associated with the $p$th model. The interpolation function has the property $\bar{\rho}_p$: $(y;\pi_p) \rightarrow [0, 1]$ and is the normalization of the model validity function $\rho_p$. The interpolation function also has the property that $\sum_{p=1}^{M}\bar{\rho}_p(y;\pi_p) = 1$ for all set of operating points.

### Example 19.1

The batch fermentation of gluconic acid is studied. The bioconversion of glucose into gluconic acid is a simple oxidation of the aldehyde group of the sugar to a carboxyl group. This transformation can be achieved by a microorganism, *Pseudomonas Putido* (*Ovalis*). Diagrammatic description of the fermenter assembly for batch culture of *P. Putido* is shown in Figure 19.2. The experimental and operational details can be found elsewhere.[4]

During batch fermentation, the microorganisms pass through several phases and, thus, the physiological characteristics of the culture are not constant but rather variable. Two important state variables for the fermentation system that could be used to characterize the operating regimes are dissolved oxygen (DO) concentration and

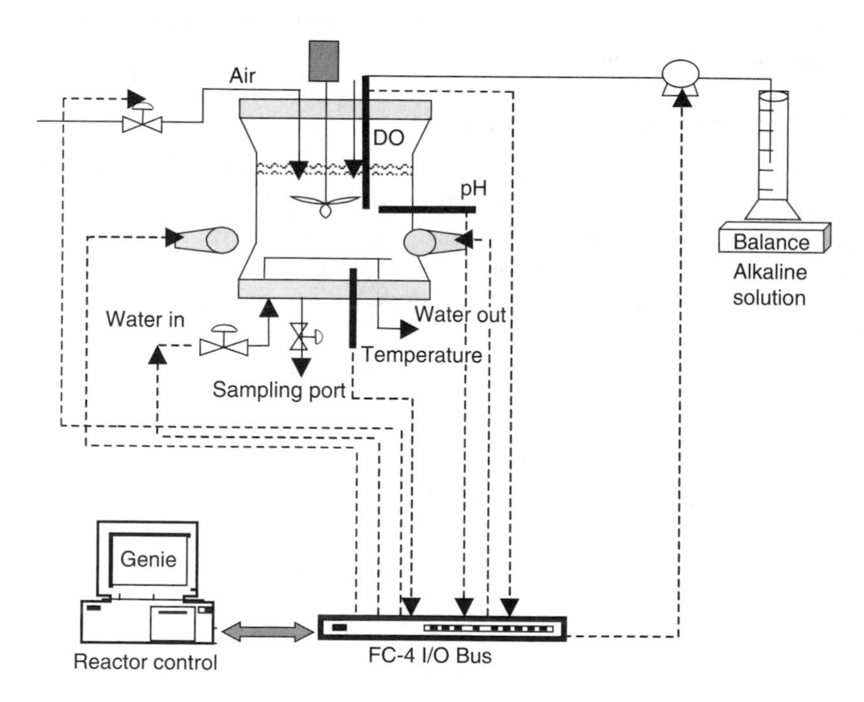


**FIGURE 19.2** Diagrammatic description of the fermenter assembly.

glucose concentration. The operating regimes are characterized using these two rate-limiting substrates based on the assumed set of phenomena that is relevant within each operating regime. The fermentation system is decomposed into five regimes. These phases are shown using the DO concentration profile based on online measurements with an oxygen probe for one of the batch runs (Figure 19.3).

The model structure in each local operating regime is assumed to be a linear state-space model and the following model is defined:

$$\begin{bmatrix} \dot{x}(t) \\ \dot{s}(t) \\ \dot{c}(t) \\ \dot{l}(t) \\ \dot{p}(t) \end{bmatrix} = \begin{bmatrix} a_{ix} & b_{ix} & c_{ix} & 0 & 0 \\ a_{is} & b_{is} & c_{is} & 0 & 0 \\ a_{ic} & b_{ic} & c_{ic} & 0 & 0 \\ a_{il} & b_{il} & 0 & d_{il} & 0 \\ 0 & 0 & 0 & d_{ip} & 0 \end{bmatrix} \begin{bmatrix} x(t) \\ s(t) \\ c(t) \\ l(t) \\ p(t) \end{bmatrix} + \begin{bmatrix} e_{ix} & g_{ix} \\ e_{is} & g_{is} \\ e_{ic} & g_{ic} \\ e_{il} & g_{il} \\ e_{ip} & g_{ip} \end{bmatrix} \begin{bmatrix} \mathrm{pH}(t) \\ T(t) \end{bmatrix}$$

$$y(t) = \begin{bmatrix} 1 & 0 & 0 & 0 & 0 \\ 0 & 1 & 0 & 0 & 0 \\ 0 & 0 & 1 & 0 & 0 \\ 0 & 0 & 0 & 1 & 0 \\ 0 & 0 & 0 & 0 & 1 \end{bmatrix} \begin{bmatrix} x_n(t) \\ s_n(t) \\ c_n(t) \\ l_n(t) \\ p_n(t) \end{bmatrix} + \begin{bmatrix} 0 & 0 \\ 0 & 0 \\ 0 & 0 \\ 0 & 0 \\ 0 & 0 \end{bmatrix} \begin{bmatrix} \mathrm{pH}(t) \\ T(t) \end{bmatrix} \quad (19.11)$$

where $x$, $s$, $c$, $l$, and $p$ denote the biomass concentration, glucose concentration, DO concentration, gluconolactone (intermediate product) concentration, and the gluconic acid (product) concentration, respectively, and the subregion index $i$ goes from 1 to 5. The inputs are the pH and the batch temperature, $T$. Thus, within each operating region, the system is defined by a set of linear differential equations. Note that the zeros in the *state* matrix are structural zeros and are obtained based on our

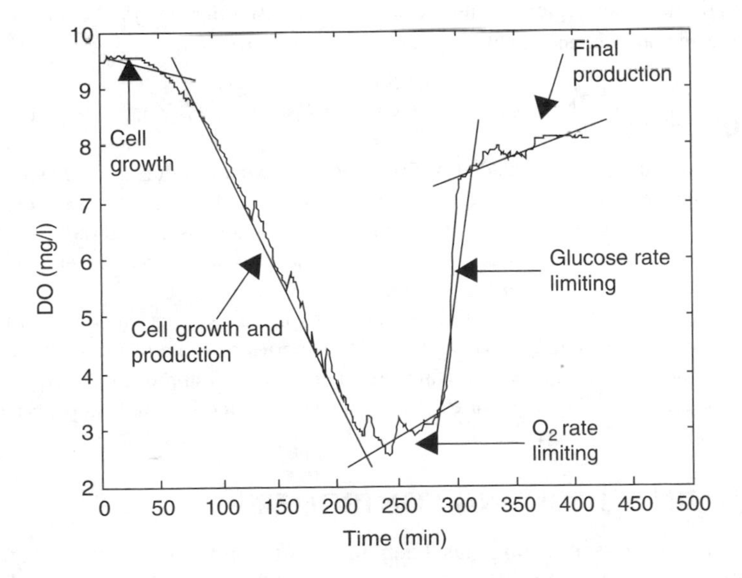


**FIGURE 19.3** Regime decomposition of fermentation process.

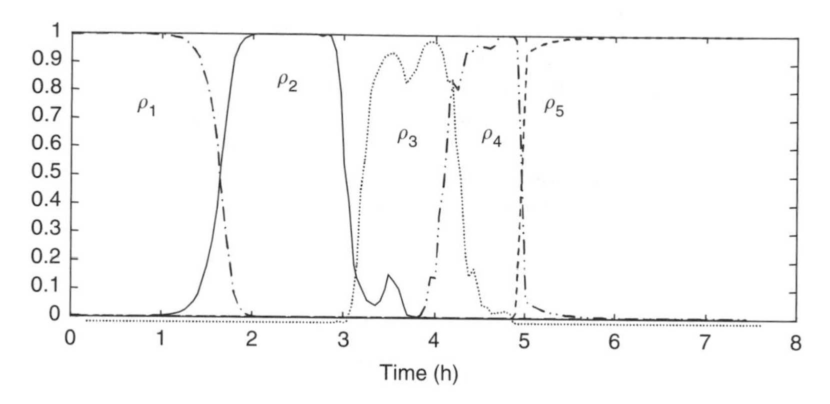


**FIGURE 19.4** The relative weights (validity functions) of the five local models in the interpolation when the fermentation run operated at pH = 6.0.

understanding of the overall reaction mechanism. The local model validity function, *for regime i*, is defined as follows:

$$\rho_i(c, s; \alpha) = \exp\left[-\frac{1}{2}\left(\left(\frac{c_{\text{data}} - \bar{c}_i}{\sigma_{ic}}\right)^2 + \left(\frac{s_{\text{data}} - \bar{s}_i}{\sigma_{is}}\right)^2\right)\right]$$

$$\alpha = [\bar{c}_i, \bar{s}_i, \sigma_{ic}, \sigma_{is}]^{\text{T}}, \quad \beta = [c_{\text{data}}, s_{\text{data}}]^{\text{T}} \tag{19.12}$$

where $c_{\text{data}}$ and $s_{\text{data}}$ are the experimental data values for DO and glucose concentrations, respectively. The parameters associated with the local model validity function are $\bar{c}_i$, $\bar{s}_i$, $\sigma_{ic}$, and $\sigma_{is}$ which correspond to the mean and the standard deviation of the $i$th subregion. Using local model validity functions $\rho_i$, the local models are smoothly connected through a global model structure

$$x(k+1)_{\text{global}} = \sum_{i=1}^{5}(A_i x(k) + B_i u(k)) \quad \bar{\rho}_i(\beta; \alpha) \tag{19.13}$$

This smoothing prevents sudden or discontinuous dynamics during feedback control of the process. The model identification is performed off-line using the experimental data, in two stages. First, local model identification is carried out, followed by the global model identification to estimate the local model parameters and the global model parameters for the weighting functions.[4]

Figure 19.4 illustrates the validity functions for the five local models during a complete batch operated at pH = 6.0. The trajectories of some of the states of the fermentation process generated using the multiple-model approach are compared with the corresponding trajectories of states measured experimentally at pH = 6.0 in Figure 19.5 and Figure 19.6.

## 19.2 GAIN-SCHEDULING CONTROLLER

Traditionally, gain-scheduling has been the most common systematic approach to control nonlinear processes in practice due to its ease of development and

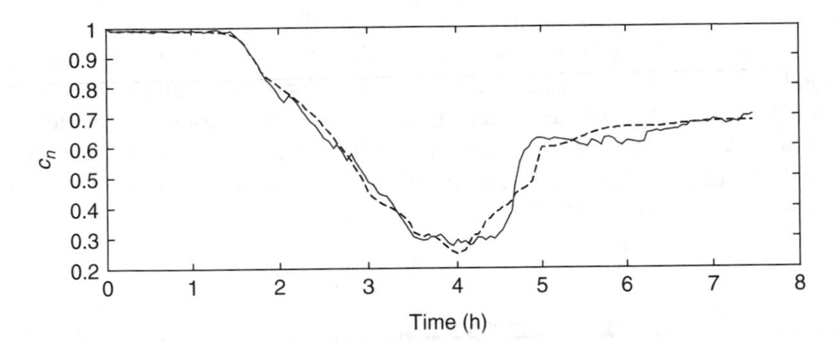


**FIGURE 19.5** Trajectory of DO concentration in normalized form. The dashed trajectory is generated using the multiple-model approach, while the solid trajectory is the experimental data measured at pH = 6.0.

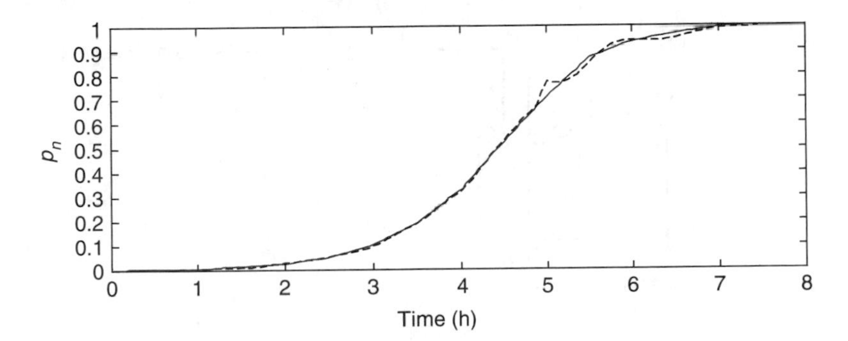


**FIGURE 19.6** Trajectory of gluconic acid concentration in normalized form. The dashed trajectory is generated using the multiple-model approach, while the solid trajectory is the experimental data measured at pH = 6.0.

implementation, as well as intuitive appeal.[5] It is based on the recognition that process characteristics change due to the presence of nonlinearities, as the process traverses all possible operating regimes. It is, then, expected that the controller should also change to compensate for changes in the process dynamics. One parameter that appears to reflect the change of process characteristics most is the *process* gain. And it follows immediately that the *controller* gain should be adjusted in such a way as to follow and compensate for such changes to maintain a relatively constant closed-loop behavior whichever operating regime the process happens to be in.

We should remember that for a linear (or linearized) process, the process gain reflects how much the output variable changes for a change in the input variable, i.e.,

$$k_p = \frac{\Delta y}{\Delta u}$$

Indeed, the process gain is the *local slope* of the process steady-state curve (input–output map), which is the locus of all feasible input–output values that would satisfy the steady-state model equations. For a linear process, this curve would be a straight line, whereas for a nonlinear process, it could have more complex shapes. Consider the titration curve for a pH neutralization reactor (Figure 19.7).[6] The titration curve is the steady-state curve that captures all possible combinations of reagent (base) addition and resultant pH level in the tank.

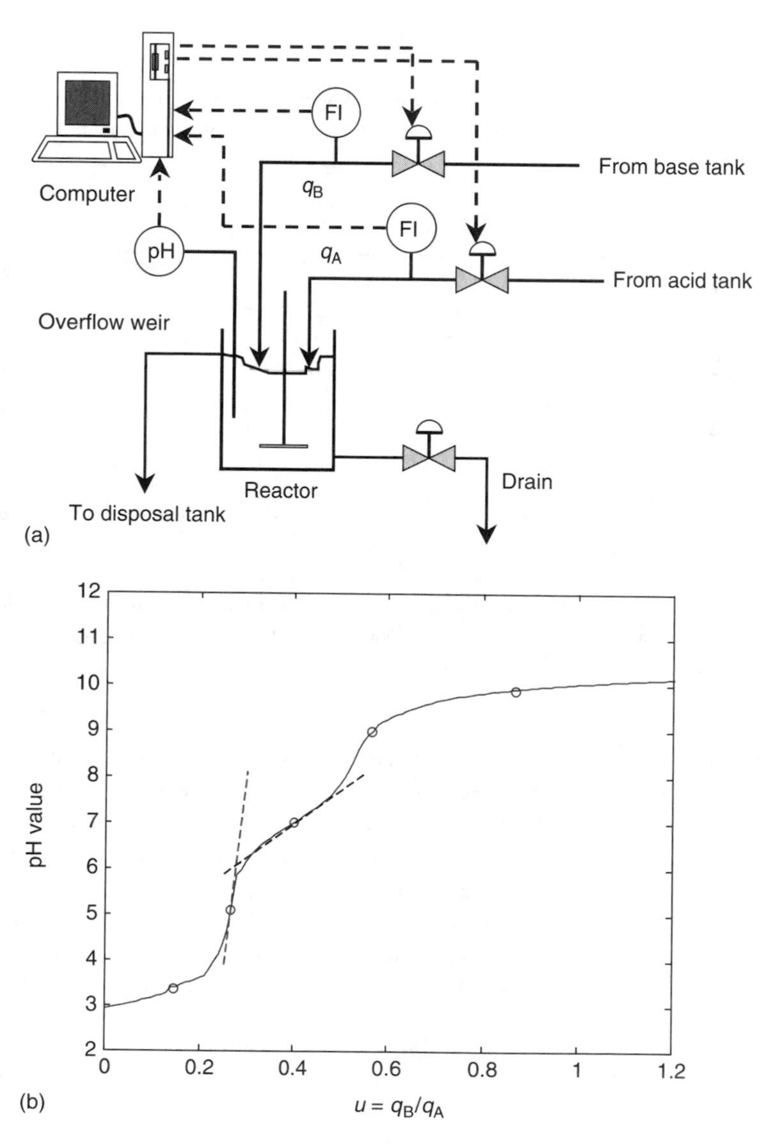


**FIGURE 19.7** (a) pH neutralization process, (b) normalized titration curve.

If the operating point is chosen to be at pH = 7, then the process gain would be reflected by the slope of the tangent line at that point (Figure 19.7b). We can also observe that as the process moves away from this operating point, the slope would change, thus, the process gain estimated at pH = 7 would no longer be accurate (see the slope at pH = 5). This is the major source of closed-loop performance degradation since, in a typical application, the controller (gain) is fixed, based on a linear model of the process around the desired operating point, which for this case would be pH = 7. Naturally, the severity of the performance loss depends on the degree of nonlinearity of the process and also how much the process is expected to move away from the desired operating point due to disturbances and set-point changes.

The gain-scheduling controller relies on the knowledge of how the process changes through the operating regime, and one can envision this in the context of multiregime modeling as having a number of models that represent such process information within the operating regime. However, the problem is much simplified for gain-scheduling as the only model information required is the local process gain along the steady-state curve. In other words, the local model consists solely of the local gain.

### 19.2.1 Determination of Process Gains

The local process gains (or the steady-state curve) can be either determined experimentally or through a nonlinear process model. In essence, we need to determine a sufficient number of gains along the steady-state curve so that we have confidence in the way that the evolution of the process gain is captured. Again, considering the curve in Figure 19.7b, if the process gains are known at the points indicated by the circles, then, we would possibly have sufficient insight toward how the gain is characterized at different pH levels. If we are making the determination experimentally, we have to limit the number of experiments. However, if a nonlinear model is being used, the whole curve can be easily depicted.

### 19.2.2 Gain-Scheduling Implementation

The first step in gain-scheduling is to establish the desired closed-loop performance for the nominal operating regime, using a PID controller. This can be done using any of the PID tuning techniques such as Z–N tuning studied in Chapter 12, or IMC tuning reported in Chapter 16. Let us express this nominal controller by

$$\tilde{g}_c(s) = \tilde{k}_c\left(1 + \frac{1}{\tau_I s} + \tau_D s\right) = \tilde{k}_c\hat{g}_c(s) \tag{19.15}$$

The nominal process model can be expressed as

$$\tilde{g}_p(s) = \tilde{k}_p\hat{g}_p(s) \tag{19.16}$$

with the nominal process gain, $\tilde{k}_p$, shown explicitly, and $\hat{g}_p(s)$ represents the process transfer function after the gain is factored out. The closed-loop transfer function (assuming unity transfer functions for the measuring and the final control elements) for a set-point change can be written as

$$y(s) = \frac{\tilde{g}_c(s)\tilde{g}_p(s)}{1 + \tilde{g}_c(s)\tilde{g}_p(s)} y_{sp}(s) = \frac{\tilde{k}_c\tilde{k}_p\hat{g}_c(s)\hat{g}_p(s)}{1 + \tilde{k}_c\tilde{k}_p\hat{g}_c(s)\hat{g}_p(s)} y(s) \qquad (19.17)$$

Assuming that the process behavior is dominated by the variations in the gain, the goal of gain-scheduling is to keep the closed-loop gain as constant as possible, i.e.,

$$k_c k_p = \tilde{k}_c \tilde{k}_p \qquad (19.18)$$

during process operation so that the nominal closed-loop performance (Eq. [19.17]) is maintained despite nonlinearities manifested by the varying process gain. Thus, when the process exhibits gains different from the nominal gain, the proportional gain can be adjusted using the formula

$$k_c = \frac{\tilde{k}_c \tilde{k}_p}{k_p} \qquad (19.19)$$

The next step, then, is to develop a mechanism to adjust the controller gain to compensate for the evolution of the process gain. Figure 19.8 shows one such strategy based on measurements of the output variable.

The task of the scheduler shown in Figure 19.8 is to determine the status of the process through the measurement of the output variable. In other words, the scheduler determines which regime the process is in, thus assigning the correct local model, or, in this case, the correct local process gain. Once the local process gain is known, the proportional gain for the controller can be adjusted using Eq. (19.19), thus establishing the gain-scheduling strategy.

The scheduler may use one of the two methods to assign the local process gain:

- *Look-up table*. Often, the control engineer has a small number of local process gains that are representative of the process behavior at various operating regimes (corresponding to output variable values). For values of the output variable that fall in between the available set, one can perform a linear interpolation to compute the relevant value of the gain.

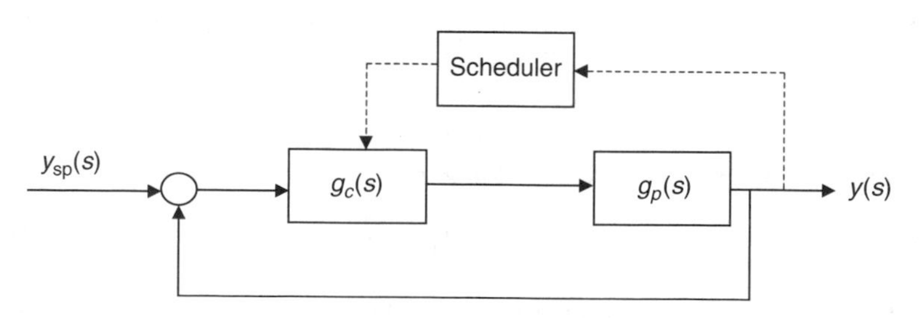


**FIGURE 19.8** The gain scheduler based on output measurements.

- *Steady-state function.* If one can have (or fit) a mathematical expression for the steady-state curve representing how the output varies as a function of the input ($y_{ss} = f(u_{ss})$), then, it may be possible to differentiate that function at desired locations to obtain the local process gain.

### Example 19.2

Consider a pH process depicted in Figure 19.7a where an acidic stream (HCl and NaOH buffer) is titrated by a base stream (NaCl). This bench-scale process is located at UC Davis, and has been studied in detail in various publications.[6,7] The process is described by the following model:

$$\dot{x}_1 = \frac{1}{\theta}(x_{1,i} - x_1) - \frac{1}{\theta}x_1 u$$

$$\dot{x}_2 = -\frac{1}{\theta}x_2 + \frac{1}{\theta}(x_{2,i} - x_2)u \qquad (19.20)$$

$$\dot{x}_3 = -\frac{1}{\theta}x_3 + \frac{1}{\theta}(x_{3,i} - x_3)u$$

The states are the acid ($x_1$), the base ($x_2$), and the buffer ($x_3$) concentrations, respectively. $q_A$ and $q_B$ represent the acid and base stream flow rates, $V$ is the constant vessel volume, and $\theta = V/q_A$. The input is defined as $u = q_B/q_A$. The pH equation is given as

$$\eta(x, y) \equiv \xi + x_2 + x_3 - x_1 - \frac{K_W}{\xi} - \frac{x_3}{1 + K_x\xi/K_W} = 0 \qquad (19.21)$$

where we define $\xi = 10^{-y}$ and the output variable $y$ is the pH measurement. The model parameters for the process are given in Table 19.1. Observing the steady-state curve in Figure 19.7b, five local models can be obtained. A first-order transfer function appears to represent the system dynamics well.

$$g_{p,i}(s) = \frac{k_{p,i}}{\tau_{p,i}s + 1}$$

Table 19.2 depicts the parameters for these models and the steady-state conditions that they correspond to. Note that the gains (and the time constants) show a significant variation.

The control objective is to maintain good set-point tracking of the output pH, $y$, by manipulating the input, $u$, the normalized base flow rate. The input variable is constrained at a maximum of 2 L/min. Using IMC tuning rules (with $\lambda = 10$), a PI controller is designed for Regime 2 that is considered as the nominal regime for illustration purposes:

$$\tilde{g}_c = 0.32\left(1 + \frac{1}{118.55s}\right) \qquad (19.22)$$

We shall design a gain-scheduling controller for this system and compare its set-point tracking performance to the case where only the nominal controller is used.

**TABLE 19.1**
**Model Parameters for the Experimental System**

| Parameter | Value |
|---|---|
| $x_{1,i}$ | 0.0012 mol/L HCl |
| $x_{2,i}$ | 0.002 mol/L NaOH |
| $x_{3,i}$ | 0.0025 mol/L $NaHCO_3$ |
| $K_X$ | $10^{-7}$ mol/L |
| $K_W$ | $10^{-14}$ mol$^2$/L$^2$ |
| $q_A$ | 1 L/min (16.67 mL/sec) |
| $V$ | 2500 mL |

**TABLE 19.2**
**Nominal Operating Conditions and Parameters for the Local Linear Models**

| *Regime i* | $u_{ss}$ | $y_{ss}$ | $k_{p,i}$ | $\tau_{p,I}$ |
|---|---|---|---|---|
| 1 | 0.14 | 3.3 | 3.89 | 131.25 |
| 2 | 0.26 | 4.6 | 101.62 | 118.55 |
| 3 | 0.45 | 7.4 | 5.24 | 103.52 |
| 4 | 0.60 | 8.9 | 25.95 | 94.23 |
| 5 | 0.82 | 10.4 | 1.76 | 82.58 |

The gain scheduler will use the pH measurements to determine which regime the system is in, and determine the value of the corresponding local gain. Then, the proportional gain of the nominal controller will be adjusted by using the expression

$$k_c = \frac{\tilde{k}_c \tilde{k}_p}{k_p} = \frac{(0.32)(101.62)}{k_p} = \frac{32.52}{k_p} \tag{19.23}$$

Figure 19.9 and Figure 19.10 show the results. We can observe that the gain-scheduling has a better overall performance while the nominal PI controller performs well only locally. The gain-scheduling controller is more aggressive in regions where the local gain is slow, thus resulting in better overall set-point tracking.

## 19.3 MULTIMODEL CONTROLLER DESIGN

Multimodel control strategies extend the linear design techniques to nonlinear systems, and generalize the gain-scheduling controllers as presented in the previous section.

The first step is to design local controllers for each operating region as modeled in Section 19.1. We show how a *nonlinear* PID controller can be developed based on the decomposition of the process operating range. Figure 19.11 illustrates the concept.

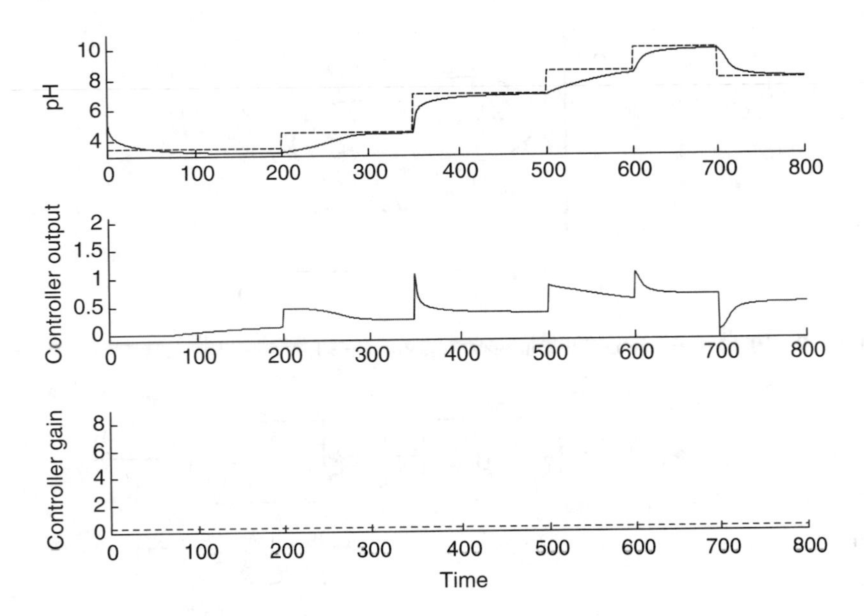


**FIGURE 19.9** Set-point tracking performance using the nominal controller.

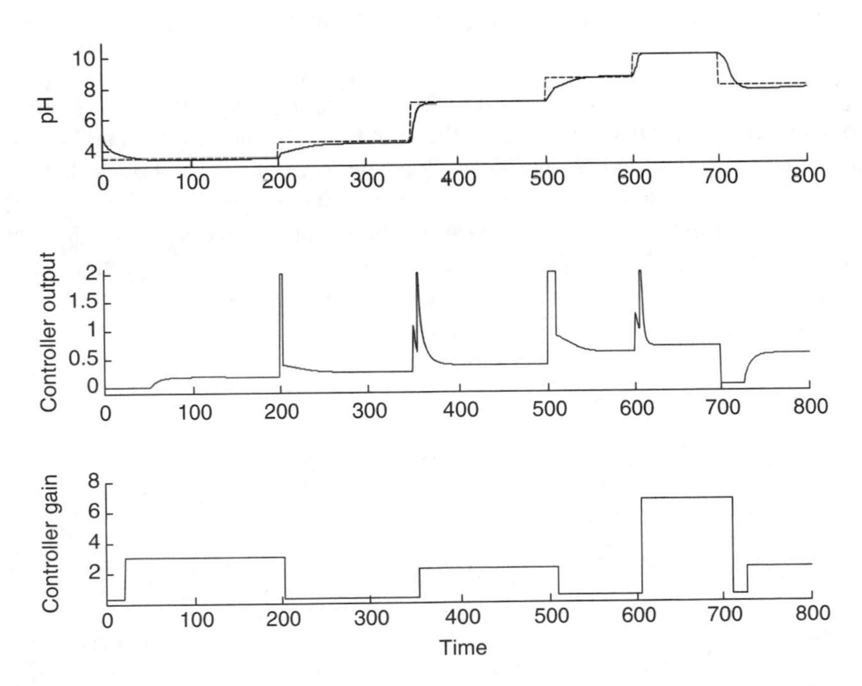


**FIGURE 19.10** Set-point tracking performance using the gain-scheduling controller.

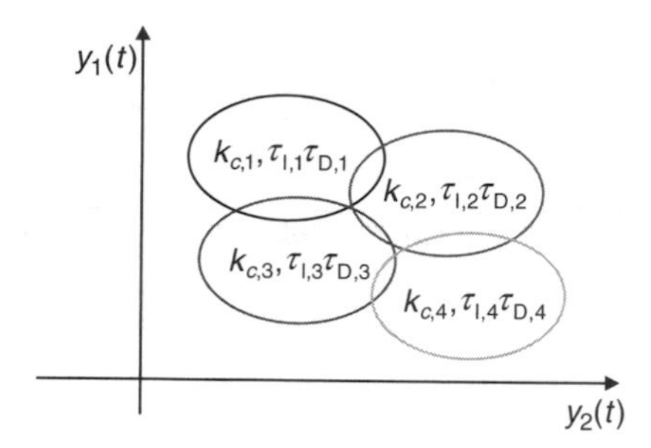


**FIGURE 19.11** Conceptual illustration of a nonlinear PID controller.

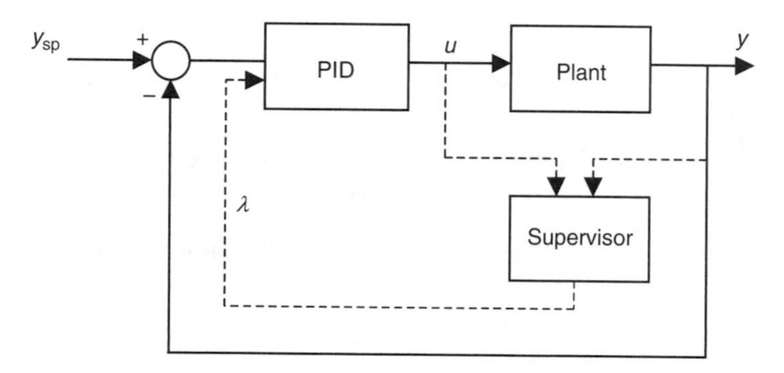


**FIGURE 19.12** Conceptual description of the multi-model control environment.

A PID controller is designed for each operating region of the process. In the regions between the operating points, the controller parameters are to be interpolated. The control designer must select feasible interpolating functions that suit the application requirements. A conceptual schematic is provided in Figure 19.12, where $\lambda$, determined by the *supervisor*, is the parameter that switches the controllers. The task of the supervisor is to coordinate the local controllers. This coordination may result in the selection of a single controller, or some combination of the outputs of a number of local controllers.

### 19.3.1 Multimodel MPC

An MPC can also be extended to take advantage of the multimodel control framework. A typical problem formulation (see Chapter 18) would be given as follows:

$$\min_{u(k),u(k+1),\ldots,u(k+N)} J = \sum_{k=1}^{N_P} W_1(k) \left\| y_{sp}(k) - y^*(k) \right\|_2 + \sum_{k=1}^{N_U} W_2(k) \left\| \Delta u(k) \right\|_2 \quad (19.24)$$

subject to

$$U_{\min}(k) \le U(k) \le U_{\max}(k)$$

where

$$\begin{aligned}
\hat{y}(k) &= \sum_{p=1}^{M} \bar{\rho}_p(y;\ \pi_p)[y_p^*(k) + y_{p,s}] \\
y^*(k) &= \hat{y}(k) + \hat{e}(k) \\
\hat{e}(k) &= \hat{y}(k) - y \\
y^*(N_P) &- y^*(N_P - 1) = 0
\end{aligned} \tag{19.25}$$

Note that Eq. (19.25) specifies an end condition that guarantees the stability of the control system. Since the implementation of this strategy is based on the multimodel strategy, one selects first the membership functions (Eq. [19.10]) and local models to explain the nonlinear plant behavior in the whole operating range. Therefore, the nonlinear constrained optimization problem becomes a linear constrained optimization problem, which is convex, hence, its convergence is guaranteed. A schematic illustration of the multimodel MPC strategy is given in Figure 19.13.

### Example 19.3

A two-layer control strategy is implemented for controlling the fermentation process in Example 19.1. At the lower layer, pH trajectories are controlled by a conventional PID controller. At the upper layer, an MPC strategy is used to determine this trajectory online using as input the desired optimal condition for the production of gluconic acid. The objective function minimizes the batch time.[4] This strategy is implemented using a multimodel strategy. Figure 19.14 illustrates the overall MPC strategy.

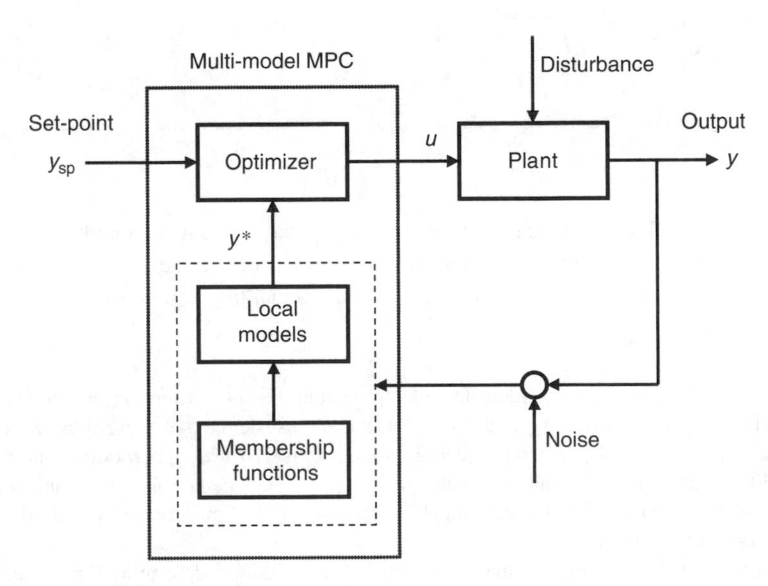


**FIGURE 19.13** Multimodel MPC structure.

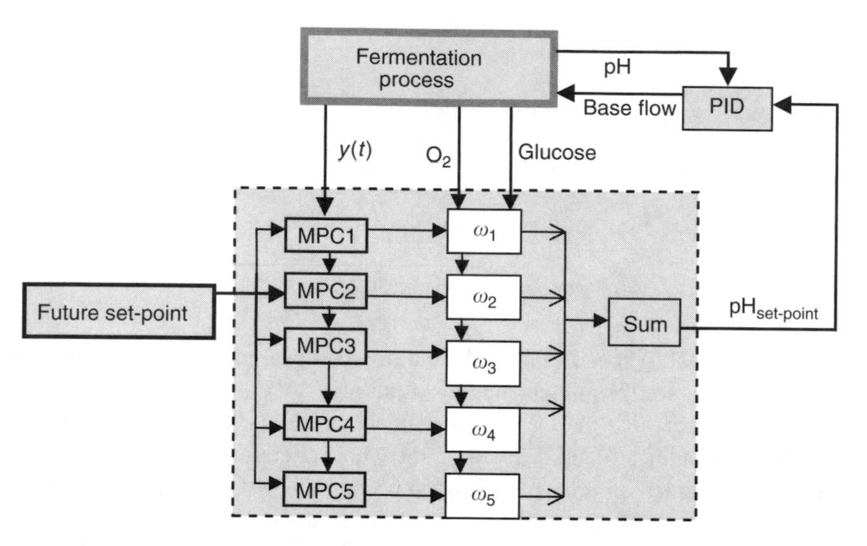


**FIGURE 19.14** Overall (MPC) strategy.

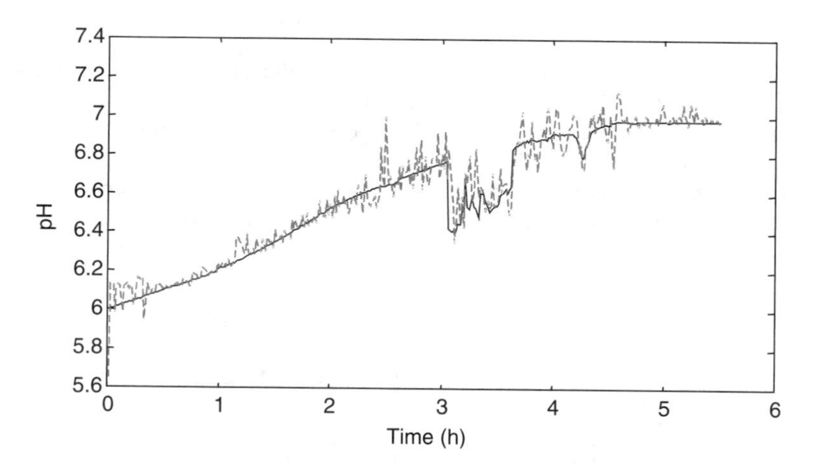


**FIGURE 19.15** The optimum trajectory of pH computed online using multi-model MPC. The solid line is the optimum trajectory (ideal set-point) of pH and $pH_{opt}$, computed online using multi-model MPC. The dashed line is the actual pH of the system regulated by the conventional controller.

The key reason for using multi-model approach in the optimization layer is to use a partitioning of the operating range of the fermentation system to solve the optimization problem. The local linear models identified in Example 19.1 are incorporated into the MPC scheme. The generalized predictive control (GPC) methodology is employed, using each identified local linear model to obtain local ideal trajectory of pH at different operating conditions.[4]

The values of tuning parameters within MPC design are determined by simulation studies using the mechanistic model of the gluconic acid fermentation system.

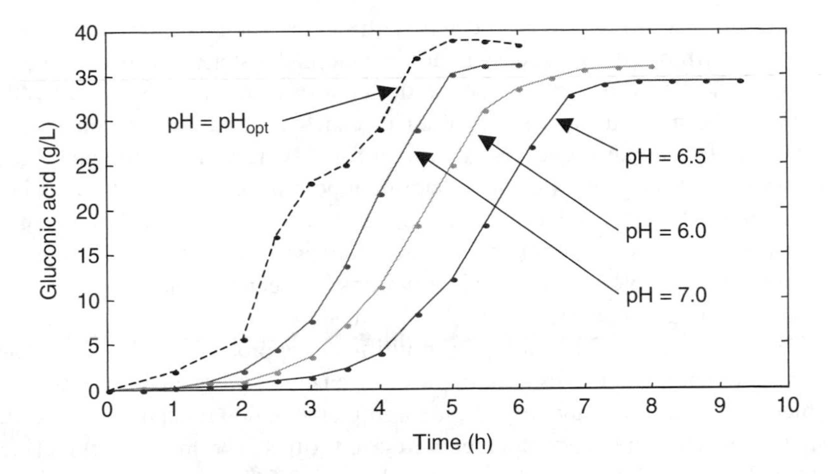


**FIGURE 19.16** The profiles of gluconic acid concentration measured experimentally. The dashed trajectory is found when the system followed the optimum pH trajectory and $pH_{opt}$, computed on-line, compared to a fixed pH.

The ideal set-point for the process output, $y_{sp}(k)$, is considered to be constant and equal to a desired value of interest (e.g., 45 g/L). The sample time is taken as 5 sec.

As linear MPC uses the local linear models to predict the response of the process locally, the actual global set-point signal which is sent to the pH controller within the real-time control layer is averaged by means of interpolation functions. The interpolation functions depend on two important state variables of the system, namely the glucose concentration, and the DO concentration (see Example 19.1). The interpolation functions are updated online based on the current information from the fermentation process. The DO concentration is measured directly while the glucose concentration is estimated from secondary measurements.[4]

Figure 19.15 and Figure 19.16 illustrate the performance of the strategy in controlling the process. Figure 19.15 shows the optimum trajectory of pH computed online using the multimodel MPC scheme and the actual pH of the system regulated by the conventional controller. Figure 19.16 depicts the profiles of gluconic acid concentration measured experimentally, and compares the trajectory obtained when the system followed the on-line computed optimum pH trajectory, $pH_{opt}$, as opposed to a fixed pH operation.

## 19.4 SUMMARY

In this chapter the operating regime approach or multimodel formulation for modeling and controlling a process was introduced to deal with the inherent nonlinear characteristics of chemical processes. Strategies to approximate a nonlinear system using a combination of simple (linear) models were explored and the concepts of local and global models introduced.

It should be noted that the important issue in considering the control problem is not just to develop a sophisticated control theory, as obviously it would be irrelevant in practice, but to consider it with factors such as simplicity, ease of application, effort required in design, and development stage. It is by the consideration of these factors that the core of the control techniques discussed in this chapter has been based on the multimodel approach, a "divide and conquer" strategy, which is a simple yet general approach for solving complex nonlinear control problems. Transparency of the model is important because it simplifies both the model development and its application. The major motivation behind the multimodel approach is its transparency. The rationale is that the development of local models is simpler because the interactions between the relevant phenomena in each operating regime are simpler locally than globally.

The use of an MPC based on the multimodel approach to solve a complex optimization problem divides the complex optimization problem into simple subproblems, which is very important from an implementation point of view because it significantly lowers the computational complexities. The multiple models are identified off-line. Thus the change in the dynamic of the process with the operating conditions is known a priori based on multi-model representation. This means that the controller parameters are obtained depending on the operating conditions in a prespecified manner. The main advantage is the ease of changing the controller parameters as the process dynamic changes.

## REFERENCES


1. Shinskey, F.G., *Process Control Systems, Application, Design and Tuning*, 4th Ed., McGraw-Hill, New York, 1996.
2. Aström, K.J. and Wittenmark, B., *Computer Controlled Systems — Theory and Design*, 2nd Ed., Prentice-Hall, Englewood Cliffs, NJ, 1990.
3. Murray-Smith, R. and Johansen, T.A. (Eds.), *Multiple Model Approaches to Modelling and Control*, Taylor & Francis, London, 1997.
4. Azimzadeh, F., Galan, O., and Romagnoli, J.A., On-line optimal trajectory control for a fermentation process using multi-linear models, *Computers and Chemical Engineering*, 5, 15–26, 2001.
5. Bequette, B.W., Practical approaches to nonlinear control: A review of process applications, in *Nonlinear Model Based Process Control*, Berber, R. and Kravaris, C. (Eds.), NATO ASI Series, vol. 353, Kluwer Academic Publishers, Boston, MA, 1998.
6. Galan, O., Romagnoli, J.A., and Palazoglu, A., Robust $H_\infty$ control of nonlinear plants based on multi-linear models — An application to a bench scale pH neutralization reactor, *Chem. Eng. Sci.,* 55, 4435–4450, 2000.
7. Galan, O., Romagnoli, J.A., and Palazoglu, A., Real-time implementation of multi-linear model-based control strategies — An application to a bench-scale pH neutralization reactor, *J. Process Control*, 14, 571–579, 2004.

# Section VI Additional Reading

The original concept of using a process model in the development of the controller dates back to the formulation of the Smith predictor in 1957. The roots of the IMC structure can be traced back to the book by Frank.

Frank, P.M., *Entwurf von Regelkreisen mit vorgeschriebenem Verhalten*, G. Braun, Karlsruhe, 1974.

The IMC structure is also a special case of the $q$-parametrization that is aimed at defining the class of all stabilizing controllers for a feedback system.

Youla, D.C., Bongiorno, J.J., and Jabr, H.A., Modern Wiener–Hopf design of optimal controllers, Part 1, *IEEE Trans. Automat. Control*, AC-21, 3–13, 1976.

Zames, G., Feedback and optimal sensitivity: Model reference transformations, multiplicative semi-norms and approximate inverses, *IEEE Trans. Automat. Control*, AC-26, 301–320, 1981.

Brosilow's paper that exploits the Smith predictor form to design an inferential controller is also considered to be one of the early forms of the IMC structure.

Brosilow, C.B., The structure and design of Smith predictors from the viewpoint of inferential control, *Proceedings of the Automatic Control Conference*, Denver, CO, 1979.

The IMC concept was finally formulated and unified within the context of process control applications in the following pioneering articles:

Garcia, C.E. and Morari, M., Internal model control – 1. A unifying review and some new results, *Ind. Eng. Chem. Process Des. Dev.*, 21, 308–323, 1982.

Garcia, C.E. and Morari, M., Internal model control – 2. Design procedure for multivariable systems, *Ind. Eng. Chem. Process Des. Dev.*, 24, 472–484, 1985.

A more detailed formulation of IMC and MPC can be found in the book by Prett and Garcia:

Prett, D.M. and Garcia, C.E., *Fundamental Process Control*, Butterworths, Boston, 1988.

A number of articles that cover a large spectrum of model-based control strategies for both linear and nonlinear process have been collected in the proceedings of two NATO workshops.

Berber, R. (Ed.), *Methods of Model Based Process Control*, NATO ASI Series, vol. 293, Kluwer Academic Publishers, London, 1995.

Berber, R. and Kravaris, C. (Eds.), *Nonlinear Model Based Process Control*, NATO ASI Series, vol. 353, Kluwer Academic Publishers, London, 1998.

With the explicit utilization of a model in control design comes the question of the impact of model uncertainty and robustness. This topic is dealt with in detail in the following books.

Doyle, J.C., Francis, B.A., and Tannenbaum, A.R., *Feedback Control Theory*, Macmillan, New York, 1992.

Lunze, J., *Robust Multivariable Feedback Control*, Prentice-Hall, New York, 1989.

Safonov, M.G., *Stability and Robustness of Multivariable Feedback Systems*, MIT Press, Cambridge, MA, 1980.

Zhou, K. with Doyle, J.C., *Essentials of Robust Control*, Prentice-Hall, Upper Saddle River, NJ, 1998.

An excellent tutorial on the MPC can be found in

Rawlings, J.B., Tutorial overview of model predictive control, *IEEE Control Systems Mag.*, 20, 38–52, 2000.

The topic of MPC and its applications in various industries are covered very well in a number of recently published textbooks:

Camacho, E.F. and Bordons, C., *Model Predictive Control in the Process Industry*, Springer, New York, 1995.

Camacho, E.F. and Bordons, C., *Model Predictive Control*, Springer, New York, 2004.

Maciejowski, J.M., *Predictive Control with Constraints*, Prentice-Hall, New York, 2002.

Rossiter, J.A., *Model-Based Predictive Control: A Practical Approach*, CRC Press, Boca Raton, FL, 2003.

Soeterboek, R., *Predictive Control: A Unified Approach*, Prentice-Hall, Englewood Cliffs, NJ, 1992.

For the practical application of nonlinear control strategies, as well as a good overview of some of the theoretical developments, the following books can be consulted:

Shinskey, F.G., *pH and pI on Control in Process and Waste Streams*, Wiley, New York, 1973.

Michael A.H. and Dale E.S. (Eds.), *Nonlinear Process Control*, Prentice-Hall, Upper Saddle River, NJ, 1997.

Modeling and control of nonlinear systems using a special model structure can be found in the following reference:

Doyle III, F.J., Pearson, R.K., and Ogunnaike, B.A., *Identification and Control using Volterra Models*, Springer, New York, 2002.

An excellent reference for gain-scheduling theory and applications is the book by Shamma:

Shamma, J.S., *Linearization and Gain-Scheduling, CRC Controls Handbook*, CRC Press, Boca Raton, FL, 1996.

For processes modeled by partial differential equations, the so-called DPS, Christofides offers an indepth study of nonlinear control design methodologies.

Christofides, P.D., *Nonlinear and Robust Control of PDE Systems: Methods and Applications to Transport-Reaction Processes*, Birkhäuser, Boston, 2001.

# Section VI Exercises

**VI.1.** For the blending process (see Continuing Problem), and using the transfer functions developed in Chapter 5, we will implement a feedforward–feedback control scheme to compensate for variations in the feed flow rate of stream 2 ($F_2$):

1. Design the feedforward compensator
2. Implement the designed feedforward controller within a feedforward–feedback scheme using APC-Tool
3. Analyze the closed-loop dynamics and compare the results with those obtained using conventional feedback scheme
4. Study the performance of the closed-loop system under model uncertainties

**VI.2.** Consider the continuous-stirred-tank heater introduced in Exercises IV.10 and IV.11:

1. Sketch a diagram to show a feedforward or feedback control loop
2. Redraw the loop in (1) in block diagram format
3. Design a realizable feedforward controller
4. Using the new controller (feedforward and feedback together), compute the dynamic response of the system for a unit-step change in disturbance (using SFB-Tool) and compare the responses with those obtained in Exercise IV.11

**VI.3.** For the oil stream being heated as it passes through two well-mixed tanks in series in Exercises IV.2 and IV.12 and considering the same operating conditions as in Case 3a,

1. Redraw the loop in block diagram format for the feedforward or feedback scheme
2. Design a realizable feedforward controller
3. Using the new controller (feedforward and feedback together), compute the dynamic response of the system for a unit-step change in disturbance (using SFB-Tool) and compare the responses with those obtained in Exercise IV.12

**VI.4.** For the shell-and-tube heat exchanger of Exercise IV.7, we will implement a delay compensator in addition to the existing PID scheme, controlling the outlet stream temperature using steam flow as the manipulated variable:

1. Sketch the closed-loop block diagram for the new configuration
2. Design the delay compensator using the process model

3. Implement the new configuration within the APC-Tool environment
4. Compare the performance of the closed-loop system with and without delay compensation
5. Investigate the effect of modeling errors on the performance of the delay compensator

**VI.5.** For the shell-and-tube heat exchanger of Exercise IV.7, we will implement an IMC control configuration, controlling the outlet stream temperature using steam flow as the manipulated variable:

1. Design an IMC controller
2. Implement the designed IMC controller within APC-Tool
3. Analyze the closed-loop behavior under different filter time constants
4. Analyze the closed-loop dynamics under a 30% error in the process gain

**VI.6.** For the blending process (see Continuing Problem), and using the transfer functions developed in Chapter 5, we will implement a feedback scheme that controls the level of the tank using $F_1$ as the manipulated variable. $F_2$, in this case, is considered as a disturbance:

1. Design a PI controller using IMC tuning rules
2. Analyze the closed-loop behavior and compare with the one obtained using Z–N settings
3. Find the IMC filter time constant to achieve a gain margin of 2 for the closed-loop system

**VI.7.** Consider the process represented by the following transfer function:

$$y(s) = \frac{2}{5s + 1} e^{-2s} u(s) + \frac{1}{s + 1} e^{-2s} d(s)$$

Assume that there is a 25% error in the time-delay of the model. Calculate the bound on the IMC filter for robust stability.

**VI.8.** Consider the following plant transfer function:

$$p(s) = \frac{2(\tau_1 s + 1)e^{-\tau_d s}}{(5s + 1)(3s + 1)(\tau_2 s + 1)}$$

where $0 \leq \tau_1 \leq 2$, $0 \leq \tau_2 \leq 1$, $0 \leq \tau_d \leq 0.5$. It is approximated by the following simple model:

$$p(s) = \frac{2}{(5s + 1)(3s + 1)}$$

We use a PI controller with initial settings, $k_c = 1$, and $\tau_I = 2$.

The performance objective is defined as

$$w(s) = \frac{s/2 + 0.01}{s}$$

Explore the robust stability and robust performance of the closed-loop system as a function of the controller parameters (by varying $k_c$ and $\tau_I$). Comment on the implications of the performance objective. You can also use simulations to support your arguments.

**VI.9.** For the shell-and-tube heat exchanger discussed in Exercise IV.7, and using MPC-Tool,

1. Implement an SISO MPC controller to control the temperature using the steam rate.
   - Analyze the behavior for $W_1 = 1$, $W_2 = 1$ and $N_p = 15$, $N_u = 5$
   - Consider $W_1 = 5$, $W_2 = 1$ and $N_p = 15$, $N_u = 5$
2. Using $W_1 = 1$, $W_2 = 1$ and $N_p = 15$, $N_u = 5$:
   - Incorporate an upper bound in the manipulated variable, $Q \leq 0.8$
   - Add an upper bound on the rate of change of the manipulated variable $\Delta Q \leq 0.1$
3. Analyze the performance against modeling errors:
   - Error in the process time-delay ($t_d = 0.7$)
   - Error in the process time constant in addition to the error in the time-delay

**VI.10.** For the blending process (see Continuing Problem), and using the conditions defined for the base case,

1. Implement an MIMO MPC controller using MPC-Tool to control the tank level and the product concentration by manipulating the inlet flow rates:
   - Analyze the behavior for $W_1 = 1$, $W_2 = 1$ and $N_p = 15$, $N_u = 5$
   - Consider $W_1 = 5$, $W_2 = 1$ and $N_p = 15$, $N_u = 5$
2. Using $W_1 = 1$, $W_2 = 5$ and $N_p = 15$, $N_u = 5$, add an upper bound in the rate of change of the manipulated variables $\Delta F_1 \leq 0.8$ and $\Delta F_2 \leq 0.8$
3. Analyze the closed-loop response for the modified problem in Case 1, in Exercise V.4 with:
   - $W_1 = 1$, $W_2 = 1$ and $N_p = 15$, $N_u = 5$
   - $W_1 = 1$, $W_2 = 5$ and $N_p = 15$, $N_u = 5$

**VI.11.** In the distillation column problem discussed in Exercise IV.15, we explored the use of autotuning for adjusting the controller parameters for a range of operating conditions. However, this problem can also be tackled using the concept of nonlinear control:

1. Discuss the issues associated with this specific problem in the context of practical nonlinear control.
2. Using the HYSYS model provided, implement a scheduling PI controller defining low, middle, and high ranges, which cover the operating regime between 96° and 108° for the controlled variable.
3. Compare the response with a constant PI controller such as the ones obtained in Exercise IV.15.

# Section VII

## Control in Modern Manufacturing

# 20 Plantwide Process Control

The scope of plantwide automation and control varies widely. To some, it is the control scheme or strategy of a particular plant unit. To others, it is the technology that provides a complete picture of plant operations within the plant site. These different viewpoints can be gleaned from the following definitions.

---

**Definition 1:**[1] Plantwide process control involves the systems and strategies required to control an entire chemical plant consisting of many interconnected unit operations. A control engineer is typically presented a process flowsheet containing several recycles streams, energy integration, and many different unit operations (distillation columns, heat exchangers, reactors, etc.). Given such a complex, integrated process, one must devise the necessary logic, instrumentation, and strategies to operate the plant safely and achieve its design objectives.

**Definition 2:**[2] Plantwide process control is defined as plant systems that enforce control strategies, provide a process information base, and enable operating personnel to monitor fully and command the plant process equipment to fulfill good manufacturing policies, support product marketing and customer sales activities, and protect the greater community and the environment.

---

These definitions suggest, then, two possible scenarios to introduce the problem of plantwide control:

1. In the context of the first scenario, one focuses on and analyzes the problems associated with defining the control strategy and the control loops of a complete plant, and also why one must design a control system from the viewpoint of the entire plant and not just combine the control scheme of each individual unit.
2. In the second scenario, one views the control strategy and design of the control loops as part of a broader problem in which all factors influencing plant operation, such as monitoring, optimization, fault diagnosis, etc., are taken into account.

Our approach to deal with the challenges posed by the disparity between Definitions 1 and 2 is to address the issues arising from the first scenario as a natural extension or continuation of the topics in the previous chapters. Then, in later chapters, the role of process control in modern manufacturing processes will be addressed, in response to the issues raised by the second scenario.

## 20.1 FUNDAMENTALS OF PLANTWIDE CONTROL

The background material for this chapter relies heavily on the book by Luyben *et al.*[1] We will introduce the methodology discussed therein and illustrate it with examples.

---

***Goal*:** To explain that the control system must be designed with the entire plant in mind, and that just designing the control schemes for each individual unit independently and then combining them later would not be an effective strategy.

---

A typical flowsheet for a chemical plant reveals the complex nature of process operations as it displays a number of units operating sequentially and sometimes in parallel to produce the desired product. To simplify, one can classify the main sections in a flowsheet as belonging to a *reaction* section and a *separation* section. The *reaction* section may include feed processing steps such as heating and pressurizing to ensure that the feed streams, taken from their storage units, are brought to the conditions required by the reaction step. This section may also involve some separation units if one elects to remove impurities from a feed stream.[3] The reaction step itself can take place in one of many reactor configurations (continuous-stirred tank [CSTR], tubular, packed-bed, fluidized bed, and so on, or their combinations) to produce the vast array of reaction products, including the desired product(s) and by-product(s). This mixture is processed in the *separation* section to isolate the product(s) with the desired purity and also recover the unreacted reactants. This section can take advantage of a number of separation technologies ranging from distillation and evaporation to absorption, filtration, and crystallization.

If the unreacted reactants are considered to be valuable, they need to be recycled back to the reactor for reprocessing, so that effectively no reactant leaves the flowsheet boundaries (complete overall conversion). Naturally, recycle streams cause complications as they integrate the reaction and separation sections of the plant. As an additional complication, energy integration is also frequently performed to minimize utility consumption. This generates another layer of integration among units as they now may share streams exchanging heat.

The following features, then arise as a consequence of the integrated operation of a chemical plant, and motivate consideration of the entire plant when designing a control system:

- Material recycle
- Energy integration
- Chemical component inventories

### 20.1.1 Material Recycle

If process units are arranged in series, where the product of each unit feeds a downstream unit and there is no material or energy recycle, the control problem for the whole plant is greatly simplified. To handle load disturbances, a control scheme can be configured for each individual unit operation, and this may be sufficient in most cases. Most industrial processes, however, contain recycle streams. In this case, the overall plant control problem becomes much more complex and its solution is not intuitively obvious. There are six basic reasons for material recycle:

- To increase conversion when dealing with reversible reactions
- To improve economics by designing a smaller reactor with incomplete conversion
- To improve yields when there are competing reactions
- To provide thermal sink with excess reactants that will then need to be recycled
- To prevent side reactions by using one reactant in excess
- To control product properties by managing the extent of the reaction

The presence of recycle streams seriously alters the dynamic and steady-state behavior of a plant. There are two key consequences of a process recycle:

1. *Influence on process dynamics*: One cannot expect that the time constant of a process with the recycle remains the same as the process without the recycle.
2. *Cascade effect*: If the feed composition changes even so slightly, this can result in a large change in steady-state recycle stream flow rates. Such a disturbance can result in even larger changes in flow rates that propagate along the recycle loop.

### 20.1.2 Energy Integration

Increased energy costs, growing concerns over environmental impact, and international competition during the last decade have all contributed to calls for improving plant capacity and efficiency. By improving the thermodynamic efficiency of the process, energy integration helps in reducing utility consumption. For energy-intensive processes, such cost savings can be significant.

Process integration is clearly motivated by economics, but it also changes the operational characteristics of the plant. Instead of addressing each process unit independently, the plant has to be investigated as a whole (a systems approach). This issue should be particularly emphasized when considering operation and control of the plant, mainly due to the following points:

- Process integration increases interactions in the plant. In other words, a change in a disturbance or a manipulated input may not only yield local effects, but can spread to other parts of the plant.

- Process integration may dramatically change the plant dynamics. As a critical consequence, integration of units that are all stable may result in an unstable plant.

For the reasons cited above, it is clear that operation of integrated processes can be difficult yet important for the total economy of the plant. This underscores the fact that simply combining the control systems of individual unit operations to design a plantwide control system may not be possible.

### 20.1.3 Chemical Component Inventories

In a chemical processing plant, one has to keep track of all chemical species and be capable of accounting for them as they enter and leave the flowsheet. One can categorize each species as either a reactant, a product, or as an inert. Then, a material balance for each species needs to be conducted to show how they are managed within the process. This is often straightforward for products and inerts, yet problems may arise when reactants need to be recycled and their inventories in the plant need to be accounted for. Every molecule of reactants fed into the plant must either be consumed via reaction or leave as an impurity or through a purge.[3] As mentioned before, our goal should be to consume all the reactants within the process boundaries (overall complete conversion), so that we do not incur any yield penalties for reactants lost as part of product or other terminal streams.

When considering the control of individual unit operations, balancing of chemical components is not an issue since exit streams automatically adjust their flow rates and compositions to satisfy performance targets. This will not be true if the unit is part of a recycle loop. For example, if more reactant enters the plant, changes have to be made in the reactor conditions so that the recycled component (reactant) will not accumulate in the loop.

#### Example 20.1

To demonstrate the ideas discussed above, a case study is presented for the production of vinyl chloride monomer (VCM).[4] The production of VCM has recently witnessed a rapid growth worldwide as it has been closely related with the polyvinyl chloride (PVC) industries, which are the primary customers of VCM.

Figure 20.1 is a simplified flowsheet of a typical VCM plant.[5] This integrated process produces VCM from ethylene, chlorine, oxygen, and the recycled by-product hydrogen chloride (HCl). Modern large-scale production of VCM can be broken down into five key stages: (1) direct chlorination of ethylene to produce ethylene dichloride (EDC), (2) oxychlorination to produce EDC by reacting ethylene with oxygen and recycled HCl, (3) two crude EDC streams mixed with the recycled unconverted EDC to be purified in a pair of distillation columns (essentially to remove water and unwanted reaction by-products), (4) pure EDC undergoing (partial) thermal cracking in a pyrolysis furnace to yield VCM and HCl, and (5) VCM separation from the HCl and EDC in another pair of distillation columns, where HCl is recycled to the oxychlorination section and EDC is recycled to the purification section.

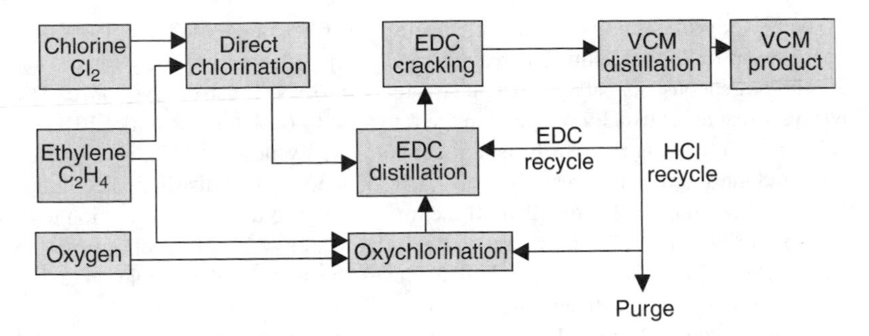


**FIGURE 20.1** A simplified conceptual block diagram of a typical VCM plant. *Source*: Alhammadi, H. and Romagnoli, J.A., in *The Integration of Process Design and Control Series: Computer-Aided Chemical Engineering, No. 17*, Seferlis, P. and Georgiadis, M.C. (Eds.), Elsevier, London, 2004.

The main reactions in a typical VCM plant are the following:
Direct-chlorination reaction:

$$C_4H_4 + Cl_2 \rightarrow C_2H_4Cl_2\ (EDC) + \text{impurities}$$

Oxychlorination reaction:

$$C_2H_4 + 2HCl + \tfrac{1}{2}O_2 \rightarrow C_2H_4Cl_2\ (EDC) + H_2O + \text{impurities}$$

Pyrolysis cracking:

$$C_2H_4Cl_2\ (EDC) \rightarrow HCl + C_2H_3Cl\ (VCM) + \text{impurities}$$

The direct chlorination of ethylene to EDC is always carried out in a liquid-phase reactor by mixing ethylene and chlorine in EDC. Cooling water is used to remove the heat produced by this exothermic reaction. Direct chlorination reactions may be run rich in either reactant: ethylene or chlorine, and usually the conversion of the lean component is 100% with selectivity for EDC exceeding 99%.

The oxychlorination unit aims to make use of the available process materials, HCl and ethylene, to optimize the VCM production process and operates at 220 to 330°C and 1 to 15 atm. The oxychlorination reaction is highly exothermic and requires good temperature control for a successful production of EDC. Typical results for the oxychlorination unit are 94 to 97% ethylene conversion, 95 to 97% HCl conversion and 94 to 96% EDC selectivity.

The EDC produced from the direct chlorination and oxychlorination and the EDC recovered from the cracking step are required to be treated to reach more than 99.5% purity before entering the pyrolysis unit. The by-products are removed in a sequence of two distillation columns. The first column removes the light wastes while the second column removes the heavy wastes, which mainly contain $C_2H_3Cl_3$ (1,1,1-trichloroethane).

The EDC pyrolysis unit operates at temperatures in the range of 500 to 550°C and pressures up to 25 to 30 atm. The reaction is endothermic and normally carried out as a homogeneous noncatalytic gas-phase reaction in a direct-fired furnace. The pyrolysis reactor is usually operated in the range 50 to 60% conversion of EDC.

The stream leaving the pyrolysis unit contains the by-product HCl, unconverted EDC, and the VCM. This stream is treated in a sequence of two distillation columns. In the first column, HCl is distilled off the top and sent to the oxychlorination unit. The bottom product is fed to the second column to purify VCM product from the unconverted EDC. The unconverted EDC leaves the bottom of the column and is recycled to the EDC purification section.

The integrated VCM plant has some undesirable reactions that must be accounted for in the process design. The main by-products (wastes) of the VCM plant are HCl, $CO_2$, $C_2H_4$, EDC, and 1,1,2 trichloroethane.

We are now ready to tackle the issues discussed above in connection with the plantwide control of the plant.

### *Material Recycle*

There are two recycle streams in the VCM plant. The first recycle stream contains mainly HCl with a purity of 99% and is produced as a by-product in the pyrolysis cracking reaction. HCl is recovered from the produced vinyl chloride and the unconverted EDC. This stream is recycled back to the oxychlorination section to be used as one of the feed streams to produce more EDC, and subsequently increase the production rate of vinyl chloride. The second recycled stream returns the unconverted EDC back to the upstream processes, where it is treated for the production of vinyl chloride. This unconverted EDC stream is recovered as a bottom product of the vinyl chloride separation column.

### *Energy Integration*

The effectiveness of designing energy-intensive processes will be illustrated through the VCM production plant. Energy is required to heat up the pyrolysis reactants and utilized in the distillation column reboilers (*sinks*). On the other hand, heat content of the exothermic reactors and hot streams to be cooled, and column condensers constitute the energy *sources*. A thermally integrated VCM plant has a great opportunity in the reduction of utility consumption as it matches the appropriate sinks and sources.

### *Chemical Component Inventories*

For the VCM plant, the side-reaction products and by-products are removed as light and heavy impurities in the purification sections. For the reactant, the design goal is to consume as much reactant as possible via the reactions, or the second option is to remove them from the plant as impurities or purge. For the direct chlorination reactants, unreacted chloride and most of ethylene components are removed as impurities as they leave the reactor in small amounts due to the high conversion that exceeds 99%. The conversion of the oxychlorination reaction exceeds 96% and the unreacted oxygen is removed with the light impurities due to its small amount and low cost. The unreacted HCl is recycled back to the reaction with some of the recovered ethylene.

## 20.2 PLANTWIDE CONTROL DESIGN PROCEDURE

In this section, the basic steps of a general heuristic plantwide control design procedure[1] are briefly outlined and applied to the VCM plant described in Example 20.1. The procedure essentially decomposes the plantwide control problem into various levels and tries to satisfy the two fundamental chemical engineering principles, i.e., the overall energy and mass balances. The proposed heuristic design procedure consists of the following steps:

***Step 1*:** Establish control objectives.

As we have seen in Chapter 2, the control objectives dictate the structure of the control problem. Therefore, it is crucial to articulate the control objectives commensurate with the design and control criteria established.

***Step 2*:** Determine control degrees of freedom.

As discussed in Chapter 3, this is related to the number of variables that need to be specified to yield a well-defined problem. The control degrees of freedom are often associated with the number of control valves available in the plant and their location. Mostly, these valves are already placed by virtue of the plant design and the control engineer may not have the flexibility to determine their number and location. In cases where this is feasible, one can allocate valves to specific units and streams to improve dynamic performance. Most control valves are used to perform basic regulatory control tasks, and any valve that remains can be assigned to enhance overall plant operation (e.g., optimization).

***Step 3*:** Establish energy management system.

The key issue in this step is that the control system should be designed in such a way that the energy disturbances are not allowed to propagate to the utility plants. In other words, the control system should be able to manage energy utilization and minimize its variability within the plant. For example, the reactor control system should be able to remove and provide exothermic or endothermic heats of reaction through the utilities. If the unit being controlled is part of an energy integration network, then, the goal of the control system would be to minimize the propagation of disturbances to other units by managing properly the stream and utility flows.

Energy (heat) integration is widely used in chemical plants to reduce utility consumption. However, heat transfer between process streams can create significant interactions, and while these may generate favorable steady-state economics, the dynamic performance may be adversely affected.

***Step 4*:** Set production rate.

To control the production rate, we first have to identify the variables that have a large influence on productivity. Then, we need to select the manipulated variables (control valves), which can be used to affect these variables. In other words, we identify a variable that dominates productivity, and maintain it at a target (set-point) using an appropriate manipulated variable. The set-point can be adjusted to satisfy production rate objectives as well as other objectives with economic consequences.

***Step 5*:** Control product quality and handle safety, operational, and environmental constraints.

The control system is expected to ensure consistent delivery of product quality while keeping the operation within constraints imposed by safety, operational, and environmental requirements. These are critical objectives, thus, one needs to select the manipulated variables so that the controlled variables respond rapidly (small time constants and delays and large steady-state gains).

***Step 6*:** Fix a flow in every recycle loop and control inventories (pressures and levels).

A flow controller is placed in all liquid recycle loops to prevent flow rate disturbances from propagating around the recycle loop. Once the flow in each recycle loop is fixed, we must, then, determine what valve should be used to control each inventory variable. Inventories include all liquid levels and gas pressures.

***Step 7*:** Check component balances.

Component balances are particularly important in processes with recycle streams because chemical components can accumulate in the loop, resulting in an integrating behavior (Chapter 8). This behavior can be prevented by monitoring chemical component inventories (reactants, products, and inerts) within the plant.

***Step 8*:** Control individual unit operations.

We need to determine appropriate control loops to operate individual units reliably. For many traditional unit operations such as distillation columns and CSTRs, there are already well-established procedures to design control systems.

***Step 9*:** Optimize economics or improve controllability.

As mentioned in Step 2, additional degrees of freedom remain after all regulatory tasks have been assigned, leaving a number of unallocated control valves. These degrees of freedom can be exploited either to improve steady-state economic performance or dynamic process response.

### Example 20.2

The plantwide process control procedure will now be demonstrated through the application to the VCM plant that contains a large number of control loops.[4]

***Step 1*:** Establish control objectives:

- To meet vinyl chloride production rate and quality
- To maximize recovery of unconverted EDC
- To maximize recovery of the by-product HCl
- For safe operating conditions, to maintain the oxygen concentration in the gas loop below the explosive region for ethylene (below 8 mol% anywhere in the gas loop)
- To maintain good temperature control for the sensitive oxychlorination reactions
- To reject process disturbances and to manage set-point changes

***Step 2*:** Determine control degrees of freedom

There are 38 control degrees of freedom in this process:

- Four feed valves [4]
- Direct reaction and oxy-reaction coolers [2]

- Direct reaction and oxy-reaction product valves [2]
- Oxy-quench [1]
- Decanter product valves [3]
- Pyrolysis preheater and heater [2]
- Pyrolysis product valve [1]
- Pyrolysis quench [1]
- Steam and cooling systems of four distillation columns [8]
- Base, top, and reflux valves of four distillation columns [13]
- HCl heater [1]

***Step 3*:** Establish energy management system.

The direct chlorination and oxychlorination reactions are exothermic reactions, and good temperature controllers are required to keep the reactions at the optimum conditions. For the case of no heat integration within the plant processes, the temperatures of the reactors are controlled by the flow rate of the cooling water streams. The temperature of the endothermic cracking reaction is controlled by the fuel gas (heating utilities) at the optimum temperature of 500°C. For the quench processes, cooling water is used first to cool the hot reactor streams before they proceed to the refrigeration sections so that the refrigeration load is reduced.

***Step 4*:** Set production rate

Ethylene, chlorine, and oxygen feeds are supplied from headers and supply tanks. Therefore, no design constraint is required to be set for the production rate control. In terms of the relationships between the reactor conditions and the production rate, the pyrolysis has the most influence on the production rate through the reaction conversion by manipulating the reaction temperature. However, this manipulation needs great attention due to the trade-off between the reaction conversion, coke formation, and by-product production.

***Step 5*:** Control product quality and handle safety, operational, and environmental constraints

The principal role of the two distillation columns that follow the pyrolysis section is to recover all the produced vinyl chloride, recycle the by-product HCl, and recover the unconverted EDC. HCl and EDC are recovered in the first and second columns, respectively. Therefore, the presence of HCl and EDC in the vinyl chloride product stream is required to be reduced as much as possible to prevent the yield loss as well as the presence of vinyl chloride in the recycled streams of HCl and EDC. Therefore, three control objectives are considered:

1. Vinyl chloride composition in HCl recycle stream
2. Vinyl chloride composition in EDC recycle stream
3. Impurities (HCl and EDC) compositions in vinyl chloride product stream

The recommended manipulated variables for these control objectives are the reflux flow or the distillate flow to control the top stream composition and the reboiler duty, or the bottom flow for controlling the bottom stream composition.

For operational and safety considerations of this process, the oxygen concentration in the gas loop should be kept below the ethylene explosivity region. The oxygen concentration can be controlled through the oxygen feed flow or the conversion of the oxychlorination reaction by manipulating the reactor temperature.

***Step 6*:** Fix a flow in every recycle loop and control inventories (pressures and levels)
In the VCM plant, there are eight pressures to be controlled:

- Four distillation column pressures: The direct way to control a column pressure is by manipulating the vent stream from the condensation section. However, for the HCl separation column, a flow controller already controls the vent stream that is recycled to the oxychlorination section. Therefore, the pressure of this column can be controlled by manipulating the condenser duty or the reflux flow rate.
- The pressure of the gas loop can be controlled by the ethylene flow, the oxygen flow or the pressure of the recycled HCl stream. Since the oxygen flow has been selected earlier, and the composition of both oxygen and ethylene is very small in the recycled HCl stream, the pressure of the gas loop is controlled consequently by controlling the top pressure of the HCl recovery column.
- The pressure of the decanter can be controlled in a straightforward manner by manipulating the vent flow rate.
- The pressures of the oxychlorination and pyrolysis can be controlled by manipulating the flow rate of the gaseous products streams.

For level control, there are 11 liquid levels to be controlled:

- There are four distillation columns and in each one, two liquid levels are to be controlled at the column base and the condenser. The most direct way to control these levels would be through manipulating the valves of the distillates and the bottom streams, respectively. However, the problem of the fixed flow recycled stream rises again in the last distillation column where the bottom stream is recycled back to the upstream section of the plant and it is the only effective manipulated variable to control the liquid level of the column base. Therefore, a cascade control system (Chapter 13) is designed over the flow of the recycled bottom stream to control both the recycled stream flow and the column liquid level.
- Controlling the liquid level of the direct chlorination reactor is performed by manipulating the bottom stream flow rate.
- The control of the decanter levels is done as a standard control structure. The EDC product flow controls the organic-phase level, while the aqueous flow controls the aqueous-phase level.

***Step 7*:** Check component balances.
Owing to the high conversion of the direct and oxychlorination reactions, a small amount of the reactants pass over the purification section where they are removed as impurities with the unwanted by-products. Carbon dioxide is an unwanted by-product and removed from the process within the top of the light removal column. Similarly, 1,1,2-trichloroethane is removed as a heavy by-product at the bottom of the heavy removal column. The first column in the purification section removes 99% of the water as the top product. The temperature control in this column achieves EDC–water separation control. The bottom product stream of this column is fed into the heavy column where the top product stream of this column is purified to 99% EDC through the column temperature control. The VCM produced in the pyrolysis section is separated

in the VCM purification section. In the first column, called HCl column, temperature control is used to distil HCl off the top of the mixed feed containing mainly EDC, VCM, and HCl. The bottom product is fed to the VCM column, where the temperature is controlled to purify VCM as the overhead product, and the recovered EDC is recycled back to the EDC purification section.

***Step 8*:** Control individual unit operations.

After accomplishing the above steps, a number of control valves remain unallocated. The cooling water streams associated with the direct-chlorination and oxychlorination reactors are used to control the reaction temperatures, so that optimum conversion is maintained. The flow of the heating fuel to the furnace is used to control the pyrolysis conversion through the temperature controller. For the quench systems, after the oxychlorination and pyrolysis reactors, the cooling water flow rate and the refrigeration duties are used to control the temperatures of the product streams.

***Step 9*:** Optimize economics or improve controllability

Up to this point, the basic regulatory plantwide control approach has been established on the VCM production processes. The recycle-HCl flow is to be maximized to increase the product production rate. However, this target may conflict with other environmental and operational objectives. Economical, environmental, and operational objectives may play an important role in the optimization of several controller set-points. Moreover, the production rate can be maximized through the pyrolysis temperature controller, where a trade-off rises due to the coke formation and the increase in the by-product flow rates. Therefore, a number of specified objectives, constraints, and external factors, such as raw materials, energy costs, and the products prices are required to be considered during the optimization process of the entire plant.

Table 20.1 summarizes the developed regulatory plantwide approach for the VCM plant through the pairing process between the available manipulated variables and the required variables to be controlled. Figure 20.2 displays this pairing process between the controlled and manipulated variables in the process flow diagram of the VCM plant.

## 20.3 IMPORTANCE OF SIMULATIONS FOR PLANTWIDE CONTROL

The role of simulations in evaluating process behavior has been discussed in detail in Chapter 4. In the context of plantwide control, simulation of highly integrated processes provides critical insight toward the understanding of possible interactions among process units, and how one can influence the dynamic performance of the plantwide control system.

Using the steady-state simulation capabilities, the design engineer can investigate different design alternatives to choose the best one. Therefore, the steady-state simulation can facilitate the optimization of the designed process in terms of different objectives such as economical, environmental, operational, social, etc. On the other hand, the dynamic simulations enable operators and control engineers to improve the control system design, and investigate the operability and controllability characteristics of the plant. With a dynamic model, individual and plantwide

**TABLE 20.1**
**Pairing the Manipulated and Controlled Variables of the VCM Plant**

| No. | Manipulated Variable | Controlled Variable | Type |
|---|---|---|---|
| 1 | Direct chlorination cooler duty | Direct-chlorination temperature | TC |
| 2 | Direct chlorination liquid product flow | Direct-chlorination liquid level | LC |
| 3 | Oxy-chlorination cooler duty | Oxy-chlorination temperature | TC |
| 4 | Oxy-chlorination quench duty | Oxy-chlorination quench temperature | TC |
| 5 | Oxy-chlorination gas product flow | Oxy-chlorination pressure | PC |
| 6 | Fuel gas flow | Pyrolysis temperature | TC |
| 7 | Pyrolysis quench duty | Pyrolysis quench temperature | TC |
| 8 | Pyrolysis gas product flow | Pyrolysis pressure | PC |
| 9 | Oxygen feed flow | Oxygen concentration in gas loop | CC |
| 10 | Recycled HCl flow | Recycled HCl flow | FC |
| 11 | Recycled HCl heater duty | Recycled HCl temperature | TC |
| 12 | Recycled EDC flow | Recycled EDC flow | FC |
| 13 | Decanter vent flow | Decanter pressure | PC |
| 14 | Decanter organic flow | Decanter organic-phase level | LC |
| 15 | Decanter aqueous flow | Decanter aqueous-phase level | LC |
| 16 | Light-column vent flow | Light-column top pressure | PC |
| 17 | Light-column distillate flow | Light-column condenser liquid level | LC |
| 18 | Light-column bottom flow | Light-column base liquid level | LC |
| 19 | Light-column reboiler duty | Light-column temperature | TC |
| 20 | Heavy-column condenser duty | Heavy-column top pressure | PC |
| 21 | Heavy-column distillate flow | Heavy-column condenser liquid level | LC |
| 22 | Heavy-column bottom flow | Heavy-column base liquid level | LC |
| 23 | Heavy-column reboiler duty | Heavy-column temperature | TC |
| 24 | HCl-column condenser duty | HCl-column top pressure | PC |
| 25 | HCl-column reflux flow | HCl-column condenser liquid level | LC |
| 26 | HCl-column bottom flow | HCl-column base liquid level | LC |
| 27 | HCl-column reboiler duty | HCl-column temperature | TC |
| 28 | VCM-column condenser duty | VCM-column top pressure | PC |
| 29 | VCM-column distillate flow | VCM-column condenser liquid level | LC |
| 30 | VCM-column Recycled EDC flow | VCM-column base liquid level | LC |
| 31 | VCM-column reboiler duty | VCM-column temperature | TC |

*Source*: Alhammadi, H. and Romagnoli, J.A., in *The Integration of Process Design and Control Series: Computer-Aided Chemical Engineering, No. 17*, Seferlis, P. and Georgiadis, M.C. (Eds.), Elsevier, London, 2004.

control strategies can be designed and tested, and even the control loops can be tuned beforehand.

### Example 20.3

The simulation of the VCM plant is developed in HYSYS.PLANT in both steady-state and dynamic modes, and can be used for economical, environmental, and operational

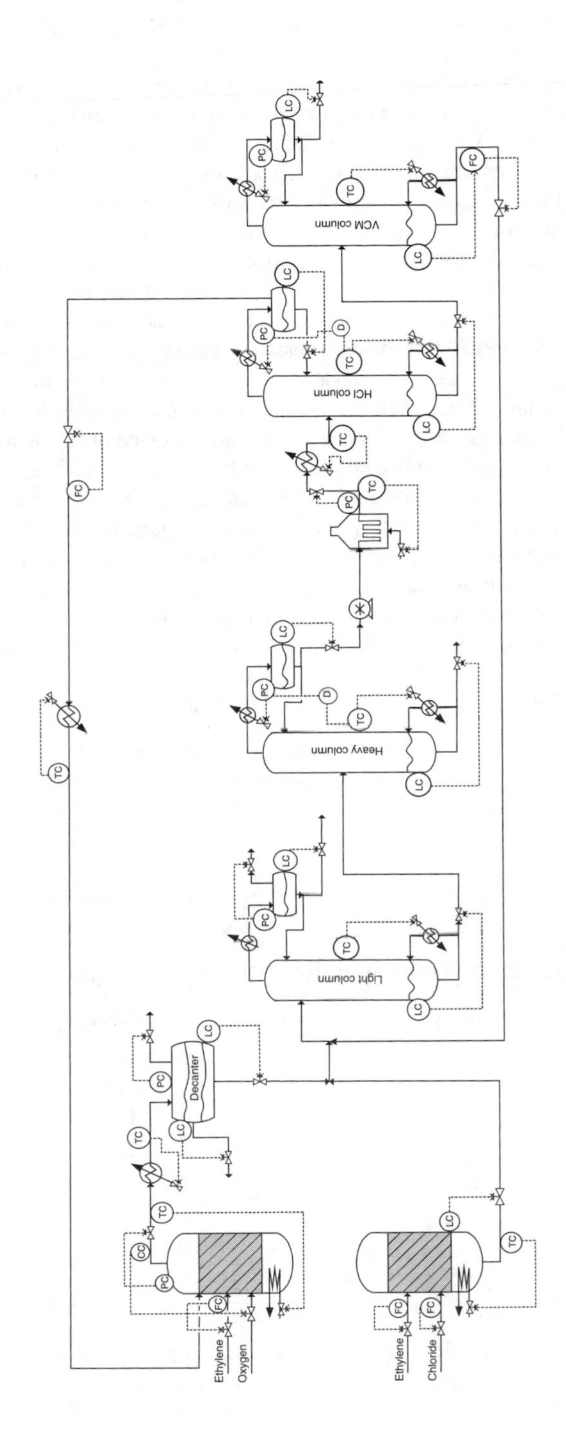


**FIGURE 20.2** Final control strategy for the VCM plant.
*Source*: Alhammadi, H. and Romagnoli, J.A., in *The Integration of Process Design and Control Series: Computer-Aided Chemical Engineering, No. 17*, Seferlis, P. and Georgiadis, M.C. (Eds.), Elsevier, London, 2004.

evaluations.[4] Table 20.2 depicts the characteristics of the VCM plant model and the specifications of the main process variables.

First, a steady-state model is built. The three reactors are modeled as CSTRs and plug flow reactors (PFRs), while the reaction kinetics is modeled with the Arrhenius kinetic expressions in HYSYS.PLANT with the kinetic data available in the literature. The four distillation columns are modeled, at steady-state, based on the specifications of the inlet streams, columns pressure profiles, required number of trays, and feed tray location. Moreover, two more specifications are required for each column with both reboiler and condenser. These specifications could be the duties, reflux flow rate, draw streams rates, composition fractions, column recovery, etc.

The integrated steady-state model is switched to the dynamic simulation environment provided by HYSYS.PLANT, where it shares the same physical properties and flowsheet topology as the steady-state model. Unlike the steady-state model, the dynamic model uses a different set of conservation equations that account for changes occurring over time. Within the dynamic mode, an advanced method of calculating the pressure and flow profile of the simulated model is used as the user states the required number of pressure–flow (P–F) specifications. These equations are solved simultaneously to find the unknown pressure or flow rates. As the last stage in the steady-state model before its transition to the dynamic mode, the developed plantwide control strategy is implemented within the designed VCM production processes in HYSYS.PLANT. The stability and controllability of the entire integrated and controlled plant can now be tested within the developed closed-loop nonlinear dynamic model.

### *Validation of the Control Strategy for the VCM Plant*

The final control structure in Figure 20.2 is implemented and the model in HYSYS is adjusted and modified accordingly so that it can be switched from the steady-state mode to the dynamic mode.

**TABLE 20.2**
**VCM-Plant Model Conditions and Specifications**

| **Reactors** | **Direct** | **Oxy** | **Pyrolysis** | |
|---|---|---|---|---|
| Temperature (°C) | 65 | 90 | 500 | |
| Pressure (kPa) | 200 | 717 | 2645 | |
| Structure | CSTR | CSTR | PFR | |
| **Columns** | **Light** | **Heavy** | **HCl** | **VCM** |
| No. of trays | 7 | 26 | 11 | 9 |
| Feed tray | 5 | 3 | 6 | 4 |
| Reflux Ratio | 10.3 | 0.9 | 1.26 | 0.5 |
| Pressure top (kPa) | 165 | 145 | 1200 | 471 |
| Pressure bottom (kPa) | 173 | 170 | 1210 | 477 |

*Source*: Alhammadi, H. and Romagnoli, J.A., in *The Integration of Process Design and Control Series: Computer-Aided Chemical Engineering, No. 17*, Seferlis, P. and Georgiadis, M.C. (Eds.), Elsevier, London, 2004.

A number of case studies have been performed for the closed-loop system; however, only two of these studies are summarized here:

- A set-point change in the temperature controller of the pyrolysis reactor
- A perturbation on the feed flow rates to the direct chlorination reactor

Owing to the size of the plant, there are a large number of possible variables to be plotted; however, only the response of a few key variables will be shown for demonstration purposes.

Figure 20.3 shows the dynamic responses of some process variables for a step change in the set-point of the temperature controller in the pyrolysis reactor from 500 to 502°C. As the pyrolysis temperature increases, more EDC is cracked to VCM and HCl as the conversion of reaction increases. However, the production rate of VCM increases at the expense of more coke formation and by-product generation. This temperature change causes disturbances in the HCl and VCM columns and, accordingly, the temperature and pressure controllers adjust the manipulated variables to bring the controlled variables back to the set-point values.

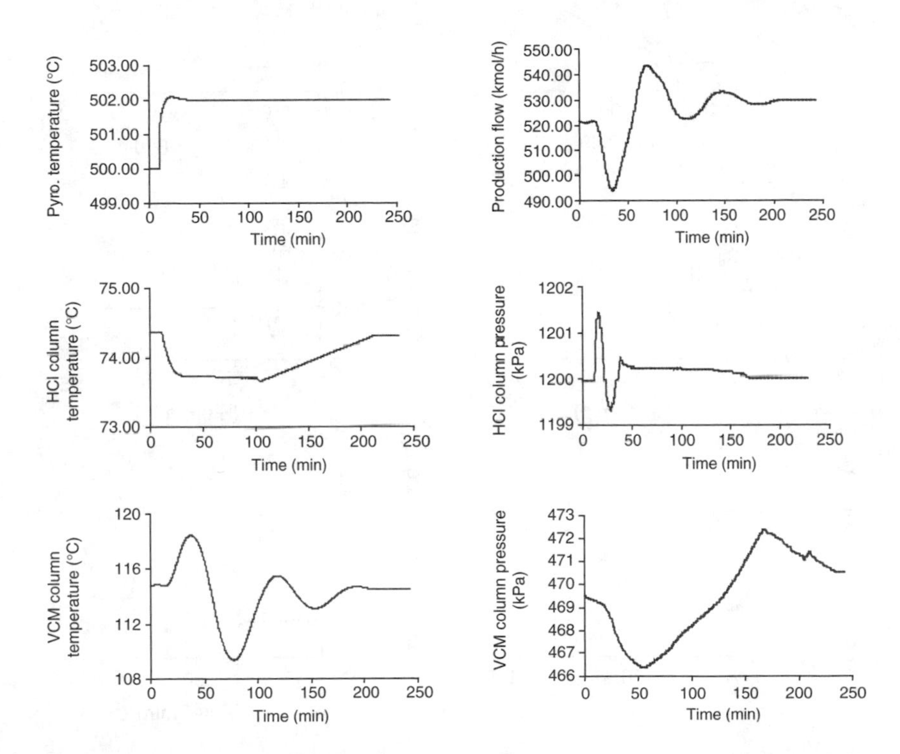


**FIGURE 20.3** Dynamic responses of process variables to a step change in pyrolysis temperature by 2°C.
*Source*: Alhammadi, H. and Romagnoli, J.A., in *The Integration of Process Design and Control Series: Computer-Aided Chemical Engineering, No. 17*, Seferlis, P. and Georgiadis, M.C. (Eds.), Elsevier, London, 2004.

Figure 20.4 displays the responses for a 5% step change in the feed flow rate to the direct chlorination reactor. In response to this disturbance, the temperature and pressure controllers of the direct chlorination reactor help bring the reactor operation to the desired level. This increase in the feed flow rate results in an increase in the EDC formation and consequently increases the feed flow rate to the purification columns. Figure 20.4 indicates that the response of the light column temperature is faster compared with that of the heavy column as the disturbance moves sequentially through the units.

Through these nonlinear dynamic simulations, we can show that the process under the proposed plantwide control structure is operable and controllable as it holds the system at the desired optimal operating conditions (set-points) and displays satisfactory disturbance rejection capabilities. Note that the dynamic responses are more complicated than the simple second-order systems studied previously.

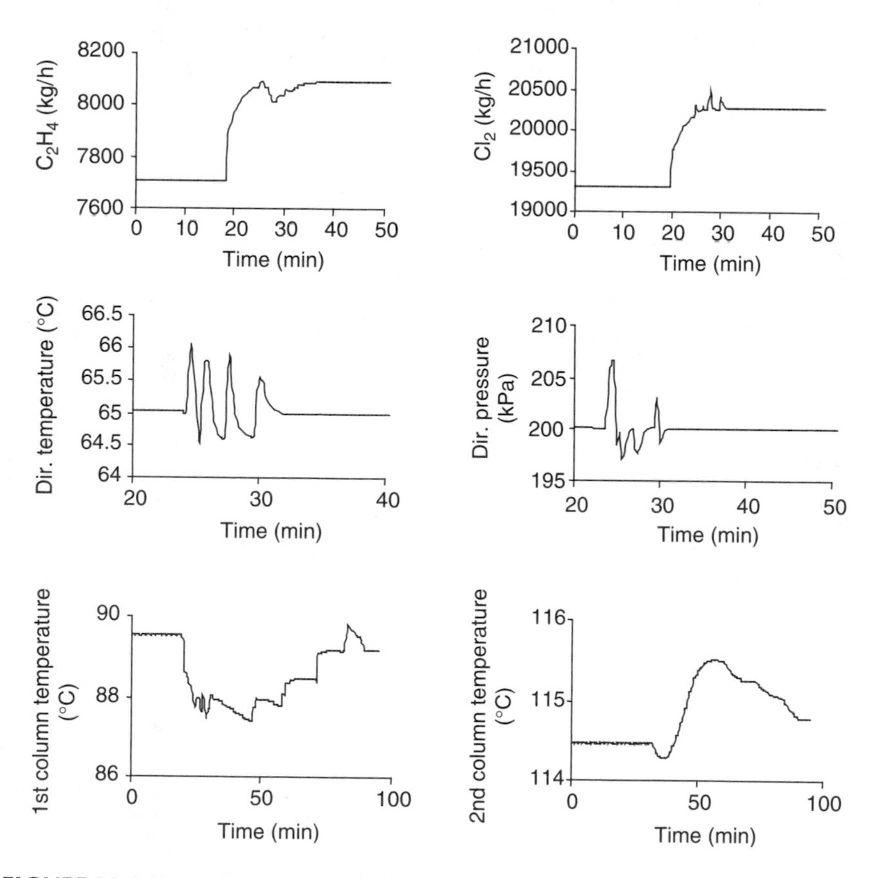


**FIGURE 20.4** Dynamic responses of the process variables to a step change in direct chlorination feed flow rate by 5%.
*Source*: Alhammadi, H. and Romagnoli, J.A., in *The Integration of Process Design and Control Series: Computer-Aided Chemical Engineering, No. 17*, Seferlis, P. and Georgiadis, M.C. (Eds.), Elsevier, London, 2004.

## 20.4 TARGETING ECONOMICS AND ENVIRONMENT

Improving the environmental performance of chemical processes has been a growing awareness in the process industries. This has led to the development of a variety of integration technologies and design tools. However, this trend toward process integration as a means of improving economic and environmental performance results in complex plant designs, and may have notable effects on the operational performance and on the control structure of the plant. Therefore, modern design of chemical processes is expected to reconcile multiple objectives, including economical and environmental objectives, as well as operational and control objectives.

This section discusses the issues that revolve around this multiobjective problem, and studies the incorporation of integration strategies for simultaneous improvement as a prelude to analyze their impact on the plant control strategy and performance.

### 20.4.1 Economically Conscious Processes

In almost every problem encountered by a chemical engineer, there are several alternative strategies which can be followed for a given product, process, or operation. It is the responsibility of the chemical engineer to choose the best process and to incorporate it into the design of the equipment and the methods, which will ultimately yield the best results. If there were two or more methods for obtaining exactly equivalent results, the preferred method would be the one involving the least total cost (or the maximum profit). This is the basis for the *optimum economic design*. In many cases, however, alternative designs do not give final products or methods that are equivalent. It is, then, necessary to consider other aspects, such as quality of the product, the operational performance, as well as the total cost.

Furthermore, for a given process or technology, we may require specific operational conditions (temperatures, pressures, etc.) for the best results to be achieved. In these cases, the problem is usually referred to as *optimum operational design* or *optimal operating condition design*. In any case, economic considerations will ultimately be the deciding factor. Thus, optimum operation design is usually merely a tool or a step in the development of an optimum economic design.

### 20.4.2 Environmentally Conscious Processes

Previously, chemical plants were designed chiefly to maximize reliability, product quality, and profitability objectives. Issues such as toxic emissions, waste disposal, and process safety have often been treated as secondary factors. Recently, as a result of growing environmental awareness and stringent regulations, process system engineers are expected to improve the environmental performance of chemical processes as one of their primary objectives. One of the key obstacles, however, has been the difficulty of formulating the environmental performances as part of traditional design objectives and then presenting to the environmental stakeholders and decision-makers in a transparent way.

Process design engineers should be concerned not only about the environmental impact that is a direct consequence of the designed process, but also consider the environmental impact associated with the provision of raw materials and services they specify. To quantify such an impact, life cycle assessment (LCA) has been offered as an indicator of environmental benefit to chemical processes.[7–9] The LCA is a comprehensive technique that covers both "upstream" and "downstream" effects of the activity or product under examination, and often referred to as a "cradle-to-grave" analysis.

---

It is interesting to note that economic and environmental objectives may appear to conflict with each other, i.e., if we attempt to optimize the economics of the process, we may be deteriorating the environmental performance and vice versa. This is mostly due to our inability to rigorously account for the potential cost of the environmental impact (short and long term) of our design decisions as part of, for instance, a discounted cash flow analysis. Thus, the design engineer needs to be aware of this problem and carefully evaluate all the trade-offs involved. This calls for multiobjective optimization methods to seek for the optimal solution. The concept of *noninferiority* (also called Pareto optimality)[10] characterizes the multiple objectives. By definition, a noninferior solution is one in which an improvement in one objective requires a degradation of another. It is important to identify methodologies that can be used to improve all objectives simultaneously. *Process integration* is such a methodology and will be discussed next.

---

### 20.4.3 Process Integration for Economic and Environmental Improvement

Energy integration techniques provide the means for improving the process economics, while at the same time reducing the environmental impact of the process. Among these, *pinch analysis* (PA) is a simple, thermodynamically based technique for the design of heat and power systems. It identifies the heat sources and sinks in a process and assigns the streams into temperature intervals to maximize the heat recovery by using process-to-process heat exchangers, thus minimizing heating and cooling utility requirements.[11]

The PA provides the targets for a heat-exchanger network (HEN) design associated with the process under study. The HEN design problem is complex and involves a combinatorial problem in pairing the hot and cold streams to enhance heat recovery. PA is performed to determine the minimum achievable utility usage for any given operating point as a heating recovery target for the HEN designs. These values are used in economic and environmental models as *best estimates* for the energy-integrated version of the process.

Process integration is motivated by both economic and environmental benefits, and can produce significant improvements. Yet, the main drawback is that it also alters the characteristics of the plant and thus has strong implications on the control structure and performance. There are techniques, however, which provide the means for selecting an integrated structure (HEN scheme), which is optimal from an operability point of view and will also minimize the impact on the global plant control system performance.[4]

### 20.4.4 Plantwide Control of Integrated Processes

Once the optimal HEN configuration has been decided, the next step is to implement the selected HEN configuration by modifying the plant structure to allow for energy integration, and further to incorporate or blend the HEN control system into the existing plant control system.[4]

The plantwide control for an integrated process can be performed in two stages.

***Stage 1*:** The plantwide control system is developed for the base-case design that has no heat integration. This is done first to make sure that the designed process is controllable and satisfies dynamic performance requirements.

***Stage 2*:** The design with the selected HEN configuration, based on the preferences of the decision-makers, is integrated or blended within the entire plant, and the plantwide control structure is modified accordingly. Most of the bypass control schemes within the HEN control system can be incorporated in a cascade configuration with the base-case control scheme for the plant with no heat integration.

Once the base-case model is modified to incorporate the HEN structure, the stability and controllability, as well as the performance of the entire integrated plant could be tested using dynamic simulations in open-loop as well as in closed-loop.

**Example 20.4**

We will continue to study the VCM plant introduced in the preceding examples.[4] The economical model for the plant is based on the operating profit, i.e., the difference between the total value of the products and the total cost of the raw materials and the utilities:

$$\text{Profit} = \sum \text{product values} - \sum \text{raw material costs} - \sum \text{utility costs}$$

Within the environmental model, the LCA analysis not only assesses the environmental performance of the modeled process, but also all the upstream and external activities. In this application, the LCA study is carried out as a "cradle-to-gate" analysis, and the final usage and disposal phases of the various products are not considered. The inventory analysis evaluates not only the environmental burden of the VCM process, but also assesses the burdens incurred by the production of raw materials ($C_2H_4$, $Cl_2$, and $O_2$) and generation of utilities (heating, cooling, and refrigeration), as well as waste treatment. The analysis includes six environmental impact factors: global warming potential (GWP), ozone depletion potential (ODP), eutrophication potential (EP), acidification potential (AP), summer smog potential (SP), and human toxicity potential (HTP).[9]

Figure 20.5 presents the optimization results through the Pareto curves for all of the designs. In this case, we are displaying two objectives: the economic objective and the environmental impact factor associated with GWP. We note that the range of each normalized objective function is [0,1], where 0 represents the best achievable

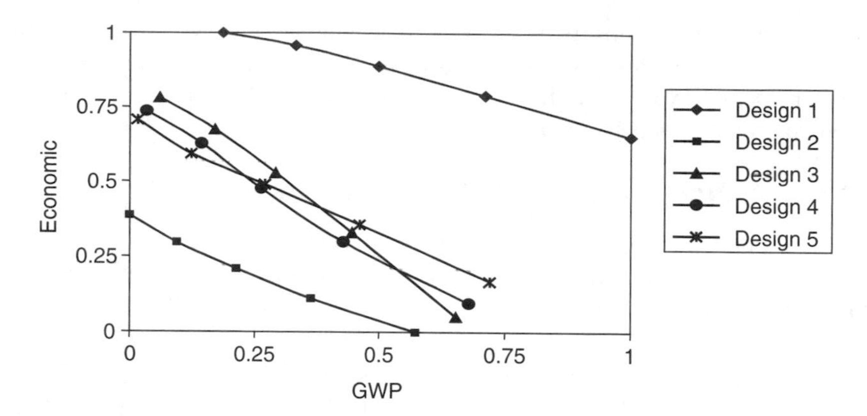


**FIGURE 20.5** Pareto curves for different designs.
*Source*: Alhammadi, H. and Romagnoli, J.A., in *The Integration of Process Design and Control Series: Computer-Aided Chemical Engineering, No. 17*, Seferlis, P. and Georgiadis, M.C. (Eds.), Elsevier, London, 2004.

value and 1 represents the worst achievable value, respectively. The curves for Design 1 (*no* heat integration) and Design 2 (*optimal* heat integration) provide the lower and upper bounds for all possible levels of heat integration at all operating points. The practical HEN designs shift the Pareto curve of no heat integration toward the one corresponding to the optimal heat integration. Design 3 shows an improvement of the objectives for the end point while Design 4 seems to be the best one for the whole range of the process variable. Design 1 shows the maximum possible reduction achievable for the objectives. However, as mentioned previously, this level of heat integration may be more difficult to operate and control as the HEN requires more heat exchangers and results in more process interaction.

The example shows that the heat integration improves the process economics and minimizes the environmental impact by reducing the utility loads as a result of heat recovery through process integration. However, such a strategy also creates interactions within the process that would not have existed, if all loads were satisfied by the utilities. It is clear that improved energy efficiency, generally, increases plant complexity and may well have a significant impact on the process operability and controllability. Thus, it will be necessary to incorporate explicitly the effects of the process dynamics and control as we move between the two *bounds* of the Pareto curves. For this specific example, Design 5 is chosen as the final HEN configuration, as it presents an acceptable trade-off among economics, environment, and controllability objectives.[4]

The last stage involves defining the overall plantwide control strategy and its validation based on the dynamic behavior of the entire plant as discussed before. Figure 20.6 shows the final regulatory plantwide control strategy for the integrated VCM plant through the pairing process between the available manipulated variables and the required variables to be controlled.

### *Dynamic Validation using Plant Simulation*

For the dynamic model with the final plantwide control strategy, a step change is introduced in the set-point of the temperature controller of the pyrolysis reactor

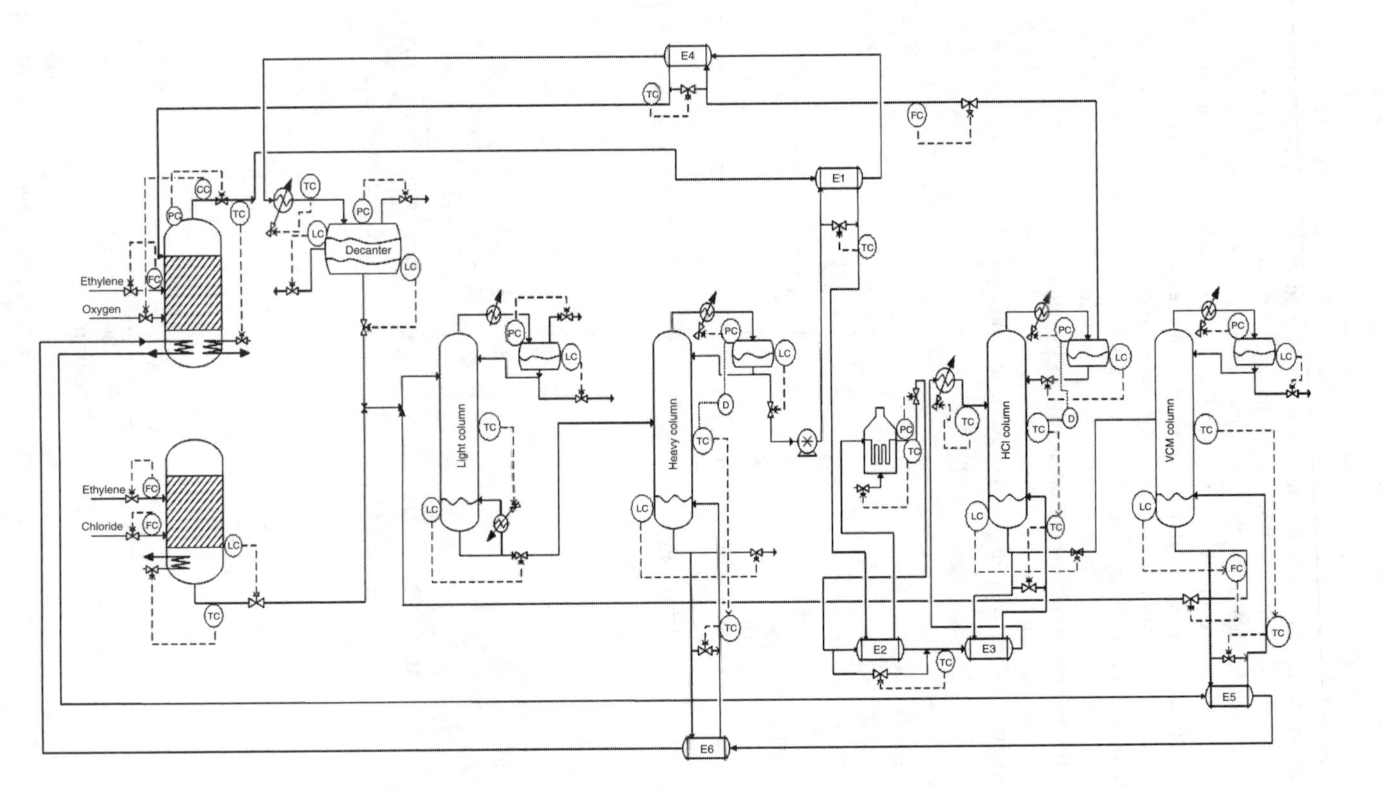


**FIGURE 20.6** Plantwide control strategy of the integrated VCM plant.

from 500 to 501°C. Figure 20.7 and Figure 20.8 show the dynamic behavior of the key process variables as they respond to this disturbance. As shown in the figures, the conversion rate of the pyrolysis reaction is increased as its temperature increases and, consequently, the production rate of the VCM product is increased. However, this increase is at the expense of producing more waste and by-products. Furthermore, it is shown that the temperature of the feed stream to the pyrolysis reactor is increased, cold stream to E2 (see Figure 20.6), as the temperature of the stream exiting the pyrolysis reactor is increased, hot stream to E2. The disturbances in the feed streams to the HCl and VCM columns affect their pressures and temperatures, and the corresponding controllers adjust the manipulated variables to bring the controlled variables back to their set-point values. The duty of the reboiler in the HCl column is increased and this reduces the temperature of the hot stream out of E3. Similarly, the reboiler in the VCM column reduces the temperature of the hot stream out of E5 from its steady-state value. Consequently, the amount of the

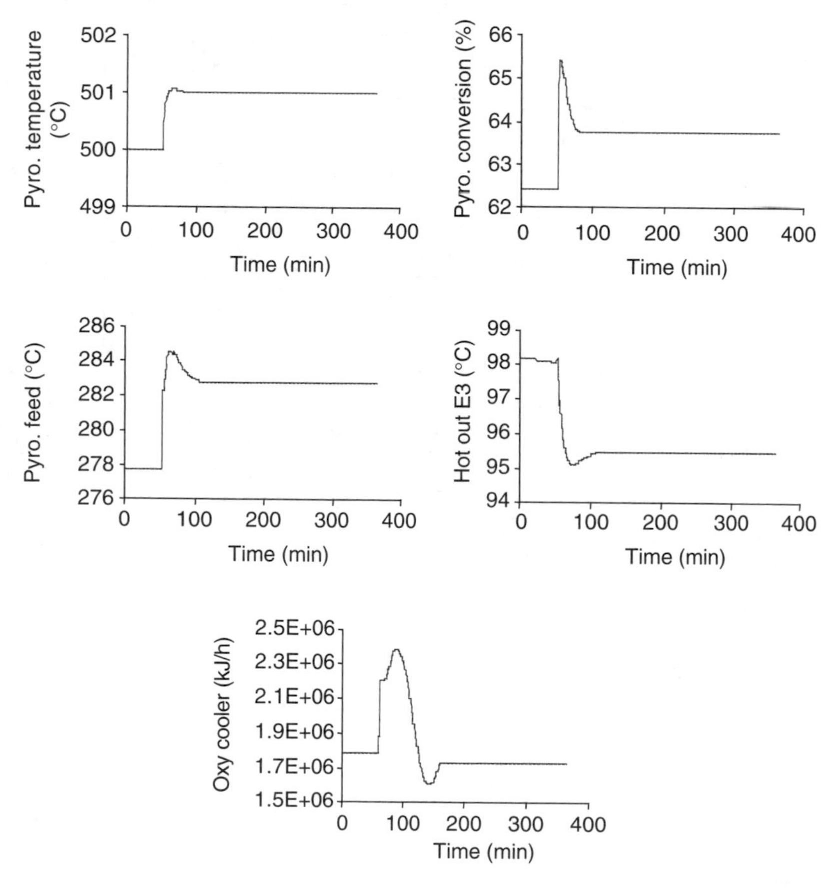


**FIGURE 20.7** Dynamic responses of the process variables in the integrated plant to a step change in the pyrolysis temperature by 1°C.

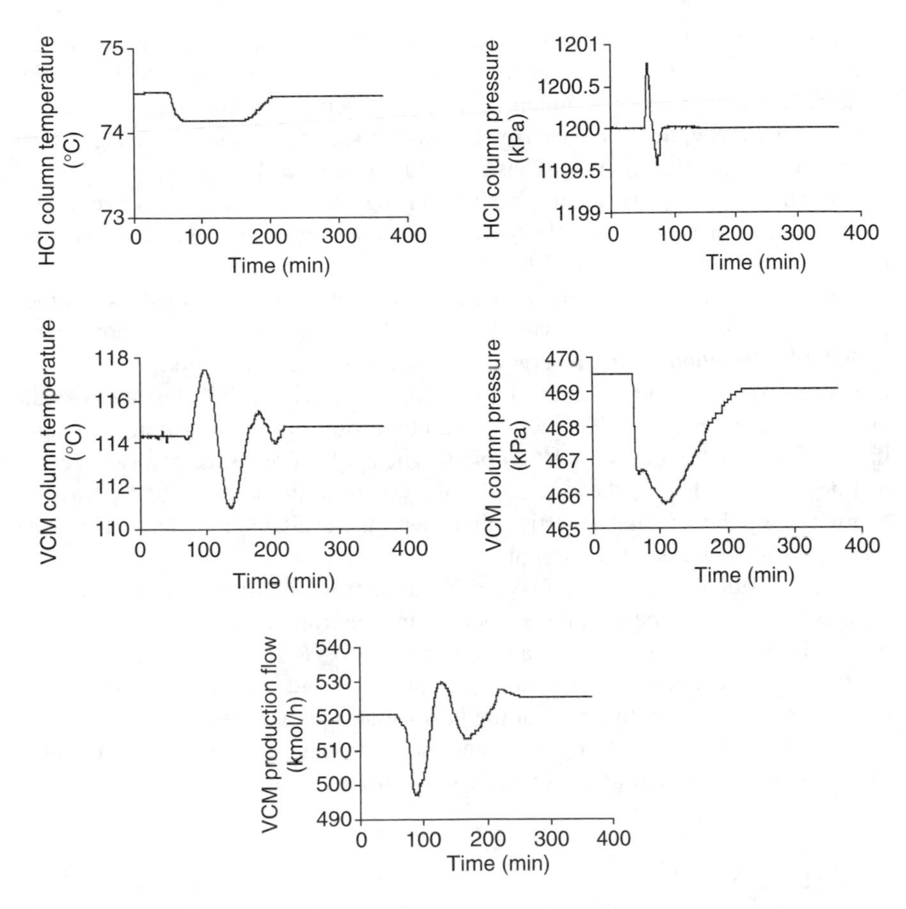


**FIGURE 20.8** Dynamic responses of the process variables in the integrated plant to a step change in the pyrolysis temperature by 1°C.

cooling utility required in controlling the temperature of the oxychlorination reactor is slightly reduced.

From the above discussion, one can see how the disturbances affect the integrated process, and how they are amplified and propagated through the entire plant. Therefore, we can appreciate the importance of a satisfactory and integrated plantwide control structure that can maintain the designed process within the required operability region.

## 20.5 SUMMARY

Plantwide process control aims to achieve the design goals through automation of decision support and direct control. Process engineers are required to develop an effective plantwide process control strategy so that a chemical plant can attain its production and quality goals set at the design stage.

The traditional process control techniques focused on the optimal control strategies of the individual unit operations to develop the overall plantwide control structure, assuming no or minimum level of interactions among the processes in the plant. However, this approach is not feasible today due to the increasing trends toward integrating the unit operations for both economical and environmental purposes. The resulting interactions, mainly through heat integration and recycled streams, make designing the plantwide control system a more challenging and demanding task.

In this chapter, we briefly introduced and summarized the ideas behind plantwide process control following Luyben's heuristic approach. The procedure essentially decomposes the plantwide control problem into various levels, and tries to satisfy the two fundamental chemical engineering principles: the overall energy and mass balances. The step-by-step procedure was demonstrated through the application to an industrial VCM plant. The application of the plantwide control design procedure to the case study illustrated the importance of employing engineering judgment and expertise along with the available systematic analysis tools for ensuring a practical design.

We also studied the trade-off issues that arise from multiple design objectives such as those associated with the economic and environmental concerns. A number of alternative designs arise as a result of the effort to reconcile such (possibly conflicting) objectives, and they can be evaluated based on their plantwide control performance to finally arrive at the best (or acceptable) design.

The case study also clearly indicates the usefulness of dynamic simulations in designing and evaluating plantwide control strategies.

## REFERENCES


1. Luyben, W.L., Tyreus, B.D., and Luyben, M.L., *Plantwide Process Control*, McGraw-Hill, New York, 1999.
2. Erickson, K.T. and Hedrick, J.L., *Plantwide Process Control*, Wiley, New York, 1999.
3. Douglas, J.M., *Conceptual Design of Chemical Processes*, McGraw-Hill, New York, 1988.
4. Alhammadi, H. and Romagnoli, J.A., Process design and operation: Incorporating environmental, profitability, heat integration and controllability considerations, in *The Integration of Process Design and Control Series: Computer-Aided Chemical Engineering, No. 17*, Seferlis, P. and Georgiadis, M.C. (Eds.), Elsevier, London, 2004.
5. McPherson, R., Starks, C., and Fryar, G., Vinyl chloride monomer — What you should know, *Hydrocarbon Process*, 3, 75–88, 1979.
6. Cowfer, J. and Gorensek, M., *Vinyl Chloride: Encyclopedia of Chem. Technol.*, 24, 851–882, 1997.
7. Azapagic, A. and Clift, R., The application of life cycle assessment to process optimization, *Comput. & Chem. Eng.*, 23, 1509–1526, 1999.
8. Burgess, A. and Brennan, D., Application of life cycle assessment to chemical processes, *Chem. Eng. Sci.*, 56, 2589–2604, 2001.

9. Heijuhes, R., Guinée, J., Huppes, G., Lankreijer, R., Udo de Haes, H., and Sleeswijk, A., Environmental life cycle assessment of products — Background, Center of Environmental Science, Leiden, 1994.
10. Edgar, T.F. and Himmelblau, D.M., *Optimization of Chemical Processes*, McGraw-Hill, New York, 1988.
11. Linnhoff, B., Use pinch analysis to knock down capital costs and emissions, *Chem. Eng. Prog.*, 90, 32–57, 1994.

# 21 Industrial Control Technology

There are many factors that influence and shape the nature of modern process control today. First, developments in computer technology resulted in tremendous reduction in hardware costs, while computational speeds and storage capabilities advanced dramatically. Furthermore, user-oriented programming languages helped relieve the man–machine interaction problem, so that the operators can navigate through the control system software with ease. Finally, reliability improved substantially as a result of more dependable components and the increased feasibility of fault-tolerant designs, redundancy, and diagnostic routines in communication networks.

Advances in real-time applications of microprocessors and personal computers (PCs) had a profound effect on the directions of current efforts in industrial systems control. Specifically, these created new opportunities for system configurations based on distributed data acquisition and control, as well as hierarchical computer control where each computer performs selected tasks.[1]

Furthermore, new open control system architectures enable a particular vendor system to mesh seamlessly with other systems, eliminating the need for customized, inflexible solutions for a given application.

This chapter describes the current technology and the main components that supply the control functions typically found in an industrial application. This will provide the basis to analyze the expanded role of control in modern manufacturing in the next chapter.

## 21.1 EVOLUTION OF INDUSTRIAL CONTROL TECHNOLOGY

The advent of the digital control computer in the 1950s initiated a revolution in the control of industrial plants. This trend to employ computers in an increasing variety of functions in industrial control systems continued in the 1960s, and accelerated in the 1970s with the emergence of the programmable logic controllers (PLCs) and the introduction of the microprocessors in direct control applications.

The 1980s were marked by the introduction of the PCs as it became more and more common to employ them as user interfaces in the plant control rooms. The key element of the technology puzzle was put in place in the 1990s with the development of the plantwide communication networks as they connected the computers with the control systems.

The operator interface is an essential component of a control system for a large industrial plant. The need for displaying process information in a timely and appropriate manner is critical to achieving plant efficiency, security, and integrated control. Following the trends described above, the operator (human) interface has also advanced to the point where it became a natural extension of the computer technology allowing the use of touch-screens, mouse, and keyboards.

---

As a consequence of these developments, the computer and the computer knowledge have developed into core components of modern industrial control systems.

---

## 21.2 GENERIC INDUSTRIAL CONTROL SYSTEMS ARCHITECTURE

The modern architecture of an industrial process control system is basically a collection of networked computers, including components and technology that make the overall system cost-effective, open, and easily scaleable. A typical system topology for an all-digital control system is illustrated in Figure 21.1 (from Figure 1.4).

The overall system can be segmented into a basic set of hardware component platforms:

- *Supervisory environment* includes, in most cases, nonproprietary computing machines running, typically, Windows operating systems, and serving as server and client stations.
- *Controller environment* consists of the control modules and the I/O modules interconnected through the control network.
- *Communication environment* allows communication among different environments utilizing, in general, open network standards.

Associated with the hardware, as in classical computer technology, there is the *software component* consisting of embedded software and application software. In an industrial control application, all these components must work together in a reliable manner. Embedded software refers to the control system functions provided by the supplier. The key components of the embedded software are the operating system and the programs that execute the application software.

### 21.2.1 SUPERVISORY ENVIRONMENT

The supervisory environment consists of one or more servers and a number of client PCs running user-interface applications providing the infrastructure for engineering and operation software.

The *server* acts as the central repository for all system data and runs all the core system functionalities. Some of the most important ones are:

- Alarm and event management
- History collection

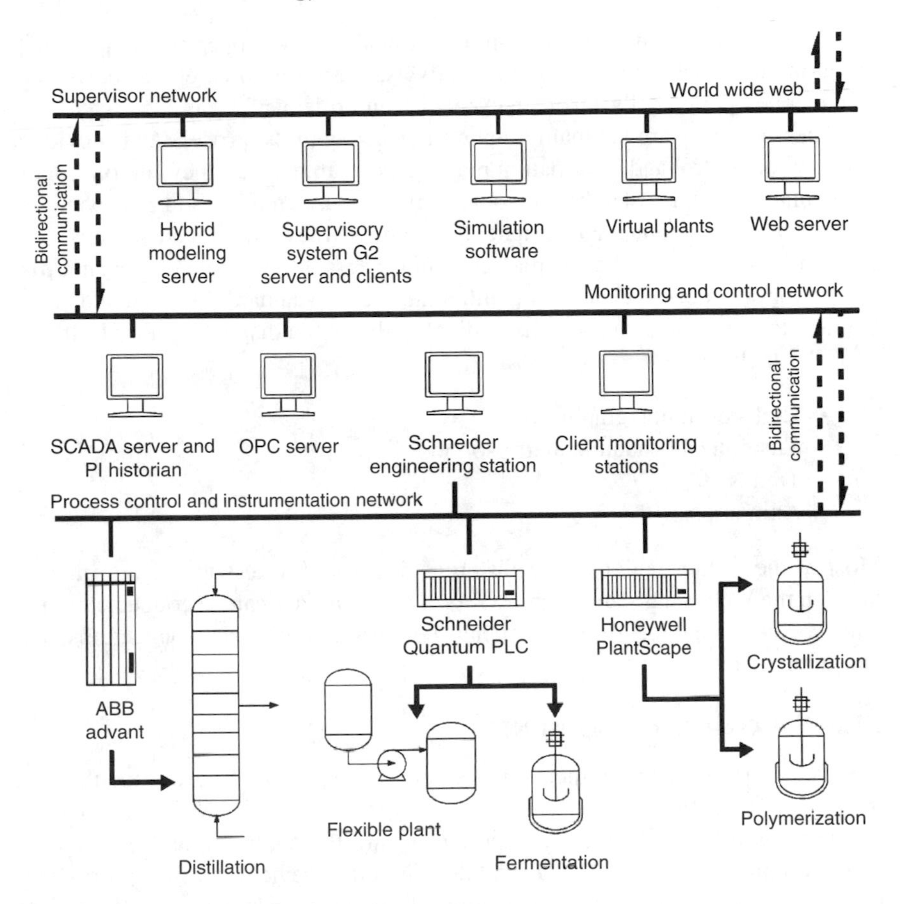


**FIGURE 21.1** A typical control system architecture.

- Trending
- Reporting
- Specialist and user applications

*Alarm and event management* provides comprehensive alarm and event detection, management, and reporting facilities to target the source of the problem, allowing the operator to focus on the data of interest in times of urgency.

*History collection* contains historical data over a range of frequencies in both average and snapshot or production formats. A large amount of historical data is retained online, via automatic archiving, allowing retention of and access to unlimited historical information.

*Trending* provides advanced trending facilities in a number of formats through simple configuration and can be configured online through standard trend displays. Some vendors support copying and pasting functions, and the user can copy trend data to other applications such as Microsoft Excel.

*Reporting* has built-in functionalities for analyzing key operational data, such as alarm and event report, downtime analysis, alarm duration, etc. Reports may be generated periodically, or on an event-driven or demand basis.

The *client* PCs are the main human interface with the process and provide a window for control and information management. In general, they run on a standard computer, with a standard keyboard, monitor, and mouse, and most vendors support Windows-compliant peripherals such as trackballs, touch-screens, and specialized keyboards. They display current process information in a manner that is easy for operators to understand with a number of standard displays, and allow the user to create its own specialized display through a display builder. Displays typically include:

- High-resolution graphics
- Customizable audible alarm sounds
- Trends
- Animations

Most of the vendor options offer different degrees of access to the information through the Web using standard browsers. The client PCs can be connected to the main server in a number of ways, such as networks, direct serial connections, and dial-up modems.

### 21.2.2 Control Environment

Typical components within the control environment are the control modules and the input/output (I/O) Modules.

The *control module* is the component where the control strategies are executed. It communicates with I/O modules and other devices via the control protocol. In general, depending on the vendor, the control module may handle a wide variety of requirements, including continuous processes, batch processes, and discrete operations.

Control functions are often supplied through a library of templates called function blocks, and control strategies are built using graphical engineering tools called control builders.

*I/O modules* provide the terminal and processing power to accept input signals from transmitters, thermocouples, etc., and send output signals to final control elements such as control valves. Different I/O modules are available for analog and digital I/O.

### 21.2.3 Communication Environment

The communication environment provides the link for data transfer between the different components of the system serving as the network technology for:

- Controller-to-server communications
- Controller-to-controller communication
- Controller-to-I/O communications

The communication environment must provide a high degree of throughput performance and fault tolerance, and utilize open network standards, including:

- Ethernet-based plant information network, linking servers and clients for the purpose of supervisory level communications.
- Control network, providing the communications link between the controllers and the supervisory level, peer-to-peer communication between controllers and between the controllers and remote I/O.

### 21.2.4 Data Exchange

Often built on the Windows operating systems and object linking and embedding (OLE), OPC (OLE for Process Control) is a new standard that provides interoperability among control applications, devices, and business applications. OPC allows the integration of the control system with other systems. There are two components:

- *OPC server* provides a standardized interface for OPC clients to query data
- *OPC client* provides an interface to request and write data to an OPC server

**Example 21.1**

A crystallization pilot plant was first introduced in Example 4.4, and a schematic representation of the cooling circuit was given in Figure 13.8. This circuit permits the mixing of the hot and cold splits under the action of a ratio controller. The boiler and chiller must provide enough heating and cooling to maintain the split temperatures at desired values. Ideally, the temperature of the hot side will be set to a high value around 110°C and the cold side will be set to a low value, ~15°C. If this is achievable, it would then be possible to control the temperature within the range 15 to 110°C by varying the ratio of the hot and cold streams. It is possible to overcome any temperature irregularities during operation as heating and cooling can be invoked at any time using the ratio controller.

The pilot facility is interfaced to a Honeywell DCS, and the Honeywell's PlantScape R300 software residing on a server allows the configuration and implementation of control schemes; a client station is used for operator manipulations.

Figure 21.2 is a schematic representation depicting the data acquisition, monitoring, and control configuration activities and illustrates the control layout for the pilot plant. It involves a controller (Figure 21.3), which is the *brain* of the control system that receives as input the process variable (PV) signals from the instruments such as thermocouples, impeller motors, etc. These instruments relay an analog signal in the form of electric current or voltage into the digital control panels before eventually ending up in the controller. The controller, in turn, sends the variables to the server, which is a Windows-based PC. The data are then displayed as either numeric characters or transformed into an action for activating an animation or alarm or display in charts for trend monitoring.

Control strategies are built using the control builder, a graphical, object-oriented software tool that supports the hybrid controller system. Control builder provides a

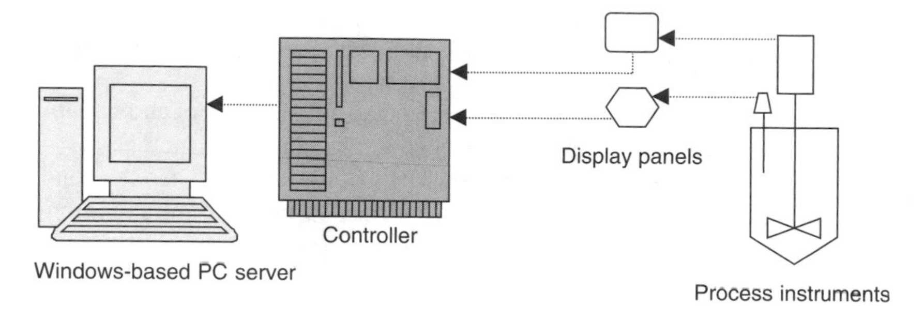


**FIGURE 21.2** DCS configuration of crystallization pilot facility.

**FIGURE 21.3** The Honeywell PlantScape hybrid controller. (Courtesy of Honeywell, Inc.)

comprehensive handling of I/O and covers continuous, logic, motor, sequential, batch, and advanced control functions through a library of function blocks (FBs). Function blocks are built-in objects provided by Honeywell to execute different control functions. Each block supports parameters that provide an external view of what the block is doing. The FBs are interconnected via "soft wires" to construct control applications or strategies. A library consisting of different unique FBs for various applications can be used by *dragging-and-dropping* into the main project-editing screen. An example of this screen is depicted in Figure 21.4. The definitions of the FBs used in Figure 21.4 are as follows:

1. **AICHANNEL1**: Analog input channel for receiving on-line PV values.
2. **DACA**: Data acquisition block with the primary functions of filtering, fixing PV values, and limiting maximum and minimum alarm values.
3. **PIDA**: Regulatory control function block with the primary feature of setting PID loops for this particular control scheme.
4. **AOCHANNEL4**: Analog output channel for receiving controller output signals (OP) and then performing the necessary control actions on the physical plant devices.

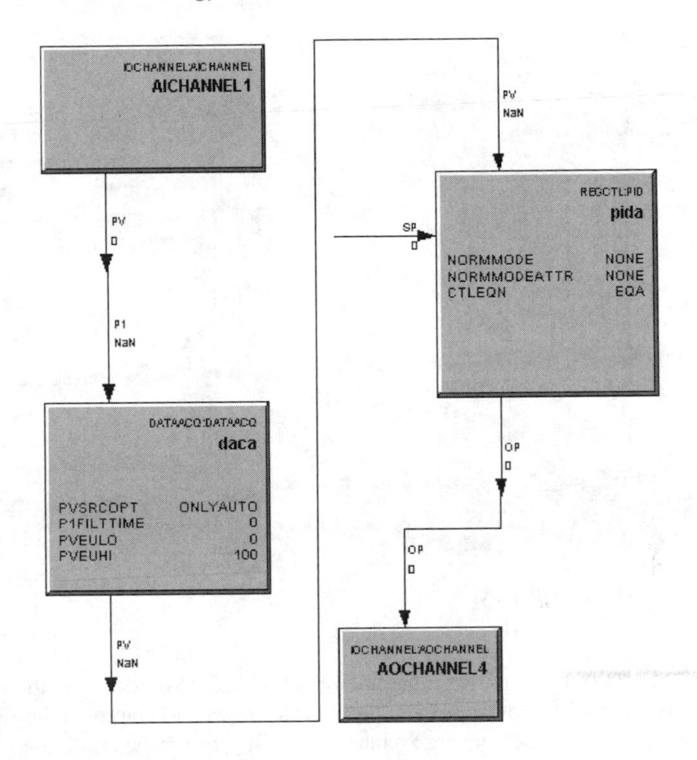


**FIGURE 21.4** FBs used to develop a PID control loop.

PlantScape has an object-based, custom display builder for development of application-specific graphics unique to any particular process. Animation of displays can be done with the included programming tool or simply by pointing and clicking. A library (shape gallery) is also available, providing an array of the most commonly used plant equipments such as vessels, piping, valves, tanks, etc. This facilitates graphic display building and development as well as providing uniformity throughout the GUI.

The user interface for PlantScape is a platform referred to as the PlantScape Station (Figure 21.5). It allows the operator(s) to view, manipulate, and analyze all data on any node in a control network. In addition, Station indicates potential plant failures including minor equipment faults. Security access is configured such that different operator areas may have access restrictions. Trends can be configured online by entering the point and selecting a parameter from the database that was named and created in control builder. Real-time and archived historical data are presented together on the same trend.

### *Regulatory Control of Crystallizer Temperature via Ratio Control*

The temperature of the crystallizer is the main variable that dictates the course of any cooling crystallization operation. The control scheme consists of two feedback loops and is implemented in the Honeywell DCS environment and controls the crystallizer

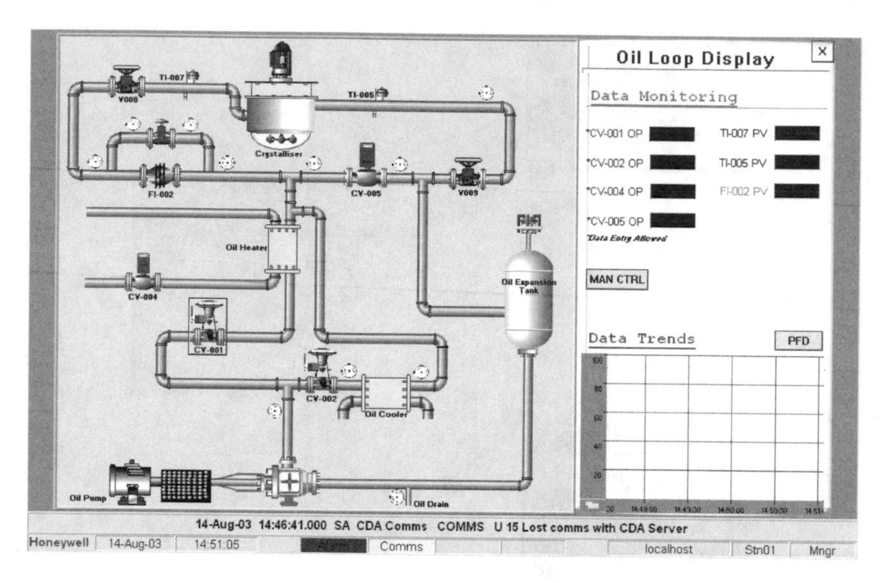


**FIGURE 21.5** Oil loop display.

temperature, TI-001, by adjusting the ratio of hot-oil to cold-oil flow into the jacket. The temperature of the hot stream, TI-012, is regulated by a PI controller that adjusts the steam flow rate to the hot heat exchanger. This suppresses the oscillations in the jacket inlet by providing a constant-temperature heating source, thus providing better control.

The oil circulating through the crystallizer jacket is split and each stream passes through a hot and a cold heat exchanger, respectively. The position of the hot stream control valve, CV-001, is determined by a PI controller block measuring TI-001 and minimizing the error between this measurement and a given set-point. The position of the cold stream control valve, CV-002, is fixed by the equation

$$\text{CV-002 (OP)} = 100 - \text{CV-001 (OP)} \tag{21.1}$$

The implementation of this ratio control is achieved by a REGCALC block that is responsible for calculating the position of CV-002 from Eq. (21.1). A sample response showing the actions of CV-001 and CV-002 was extracted from the monitoring trend display and is displayed in Figure 21.6. The trend indicates how the temperature of the crystallizer is oscillating around the temperature set-point (65°C) under the regulatory actions of the control valves.

The PI controller was tuned by traditional techniques. A sensitive response was sought, meaning that the control valves would respond in a slightly aggressive manner to set-point changes. The response of the crystallizer temperature during the whole batch cycle is displayed in Figure 21.7. In this figure, the quality of the control system to follow the required set-point trajectory along the batch cycle as well as the movement of the control valves both for heating and cooling to provide the appropriate mixing of the oil streams can be observed, thus controlling the temperature of the jacket and, consequently, the crystallizer temperature.

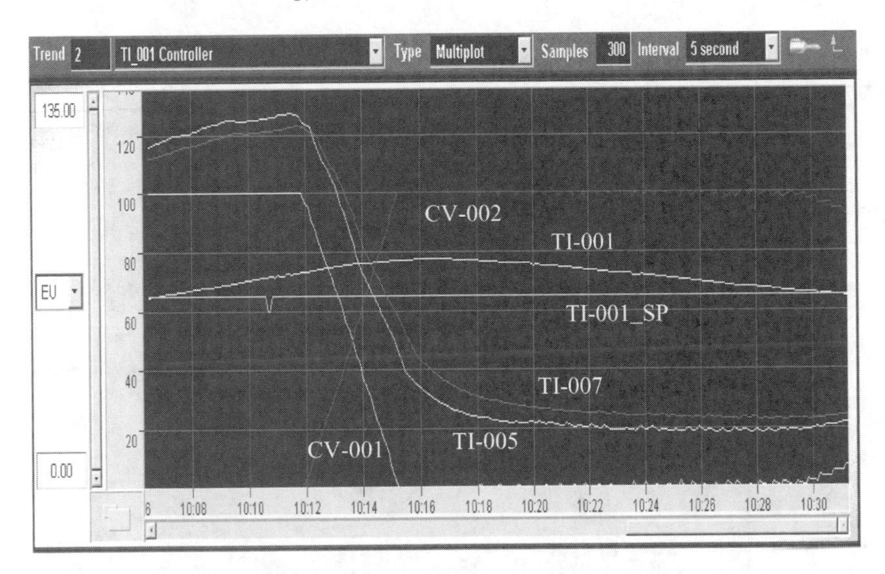


**FIGURE 21.6** The response of the ratio control scheme (from DCS).

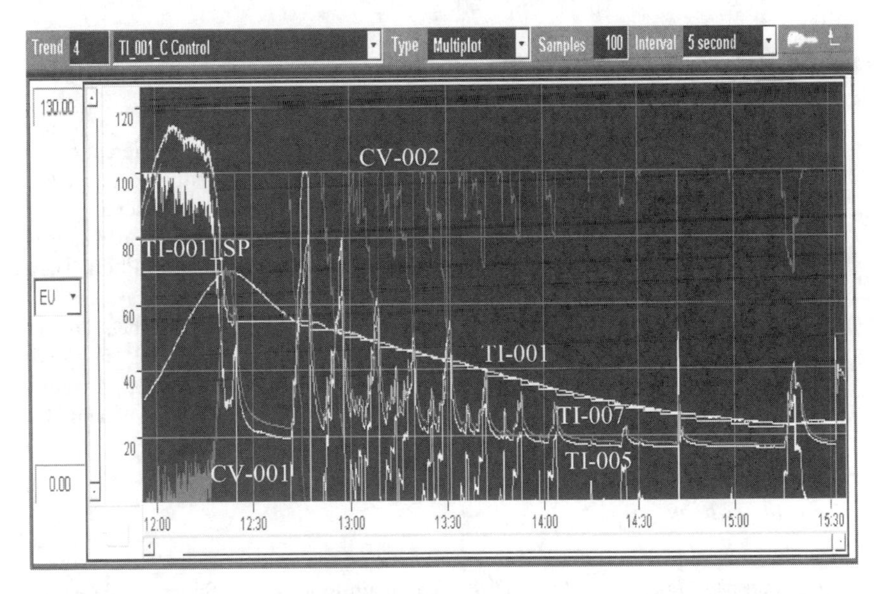


**FIGURE 21.7** The response of the crystallizer temperature during the batch cycle (from DCS).

### Example 21.2

A flexible pilot plant (see Figure 21.8) at the Department of Chemical Engineering, The University of Sydney, consists of two CSTRs, a mixer, and a number of heat exchangers. The total number of measured variables is 25, including 14 temperatures,

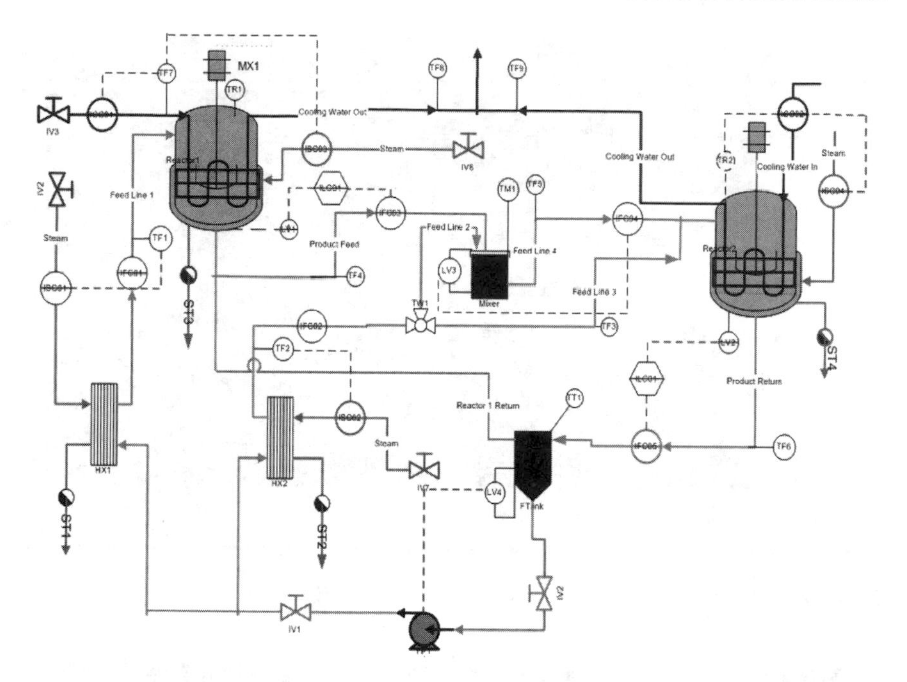


**FIGURE 21.8** The flexible pilot plant.

7 flow rates, and 4 levels. The facility is flexible allowing a number of different configurations in terms of running the reactors in series, parallel, or totally independent. The facility is interfaced to Schneider Quantum PLC industrial control system.

The control strategies implemented on the quantum system were developed using *Concept* (commercially supplied developing software for the Quantum PLC). The software provides a graphical programming environment, which allows users to combine structure text, FBs, user-defined FBs, ladder logic, and C++ when developing system control architecture. This object-oriented software tool supports hybrid controller system design, documentation, and monitoring. *Concept* provides a comprehensive handling of I/O and covers continuous, logic, motor, sequential, batch, and advanced control functions through a library of FBs.

## *Control Strategy*

A number of loops are configured and implemented within the control system for a fully automated operation of the facility. The main loops as configured in the system are displayed in Figure 21.8. They involve level controllers for the different vessels and temperature controllers for vessels and preheating and cooling of the feed streams. Among them, of special interest, are the temperature control of both reactors and the level control of "reactor one" (R-1), since they involve advanced configurations described in Chapter 13. In the following, we focus on the override control implementation of unit R-1.

**R-1 Level Control:** An override control scheme is implemented to control the level of R-1 allowing switching between manipulation of the outlet and inlet reactor flow rate to achieve full controllability and safety for a range of operational scenarios. This scheme is previously discussed in Chapter 13.

The level override control scheme consists of both FB and structure text to monitor both the reactor level and the associated set-point. Figure 21.9 shows the logic required to monitor the level and implement an action based on the plant state.

Once the logic has been executed, several FBs implement the action. Figure 21.10 shows the controller implementation for maintaining the R-1 level.

```
Safety_Interlocks
»»»»»»»»»»»»»»» Structured Text Start «««««««««««««««
(*Interlock 1 : Over ride control for level of reactor 1 *)

(*Checks the tank level, if it is above setpoint + constant, then use
the feed flow control valve to control level. If the level is less th
the setpoint - constant then use the product valve.*)

IF ILV001 > ILC01_SP + ILC01_OS  THEN IWLC01B := 0; IWLC01A := 1;

    ELSE IWLC01B :=1 ; IWLC01A := 0;

END_IF;

IF ILV001 < ILC01_SP - ILC01_OS  THEN IWLC01B := 1; IWLC01A := 0;

    ELSE IWLC01B := 0; IWLC01A := 1;

END_IF;
```

**FIGURE 21.9** Logic required for override control.

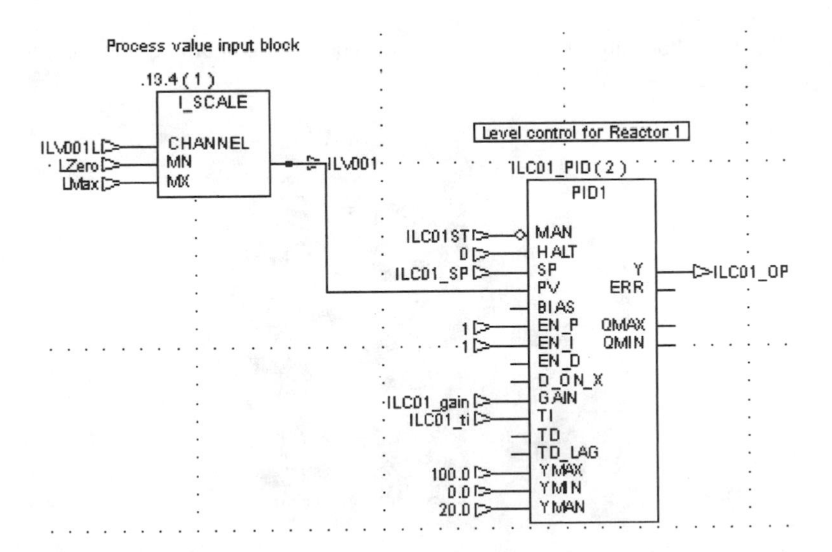


**FIGURE 21.10** Illustration of the implementation of R-1 level controller in Quantum.

Figure 21.11 shows the logic required to switch the controller output between the feed valve and the reactor product valve.

Figure 21.12 shows the block corresponding to the level control for R-1. By clicking the button on the left upper corner, one can choose between alternative configurations for the level: control by feed or the override control. The dot next to IFC01 indicates override is on, in this case, and since level reaches above the set-point (60%) and the outlet flow valve is already fully open, the inlet control valve has already taken over the control of the level. The outlet flow valve control system is placed in manual and the valve is set to be fully open.

A sample response showing reactor level (ILV01-PV), feed flow (IFC01_PV) and reactor outlet flow (IFC03_PV) is extracted from the monitoring trend display and shown in Figure 21.13. The trend shows that when the level exceeds 60% and the outlet flow is at the maximum admissible value (control valve saturated at 100%), the feed flow-level control scheme overrides the outlet flow control by reducing the inlet flow rate to bring the reactor level to the set-point value of 60%.

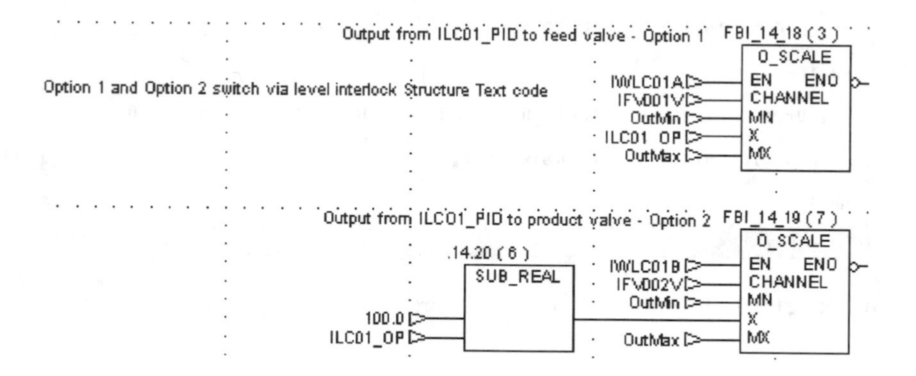


**FIGURE 21.11** Illustration of logic to switch the controller output in Quantum.

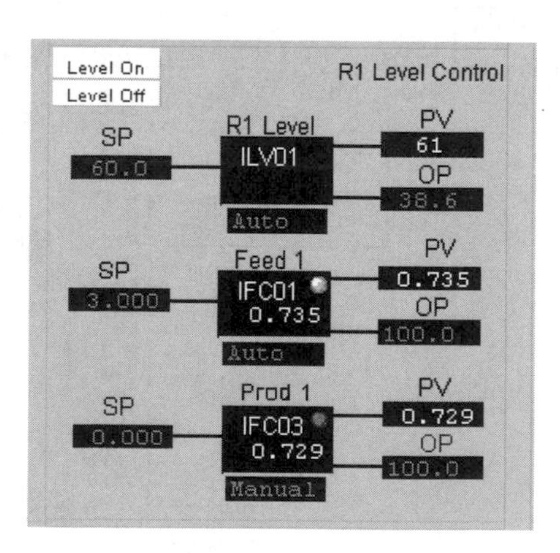


**FIGURE 21.12** R-1 level control display.

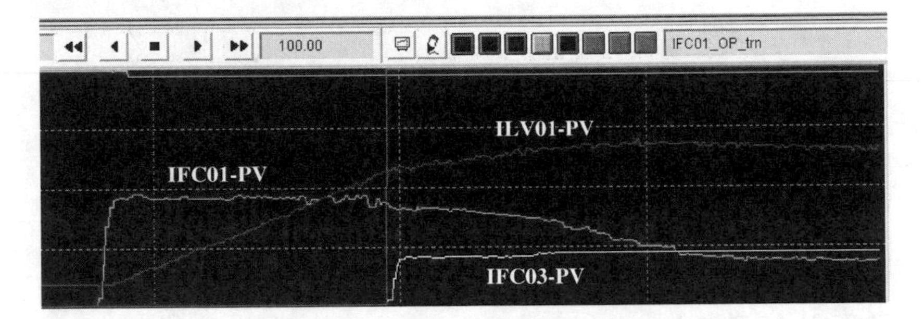


**FIGURE 21.13** Trends for override level control system.

## 21.3 SUMMARY

In this chapter, we have briefly introduced the current technology as well as the main components that support the control functions typically needed in any industrial application. Modern control architectures allow cost-effective open control systems for batch, semi-continuous, and continuous applications. The resulting system can be segmented into a series of hardware-interconnected component platforms namely the supervisory, controller, and communication environments connected through open network standards. The OPC, as the new standard that provides integration of control systems with other systems, is also briefly introduced as a tool for total open system manufacturing. Finally, as practical examples, the implementation of such computer control system architecture on two pilot-scale also facilities is discussed.

## CONTINUING PROBLEM

To complete the discussion on the continuing problem, we show how to implement the feedback and the feedback–feedforward control configurations for the blending process, studied earlier, within an industrial DCS environment.

A virtual environment is developed within the DCS system to evaluate and compare different control strategies and explore the potential difficulties that may be encountered when such strategies are implemented within an industrial control setup. The virtual environment consists mainly of a simulation of the blending process running on gPROMS, the control system and MS Excel. Between gPROMS and Excel, the gFPI (gPROMS foreign process interface) is used. This is a product of gPROMS, which allows gPROMS to communicate with foreign processes such as MS Excel via a dynamic library link (DLL) file. The other link between PlantScape and MS Excel is constructed by a Network application programming interface (NAPI) protocol. NAPI is a communication protocol for establishing links between applications with VBA, VB, C and C++ coding capabilities. Figure 22.14 illustrates the Control area network (CAN) structure.

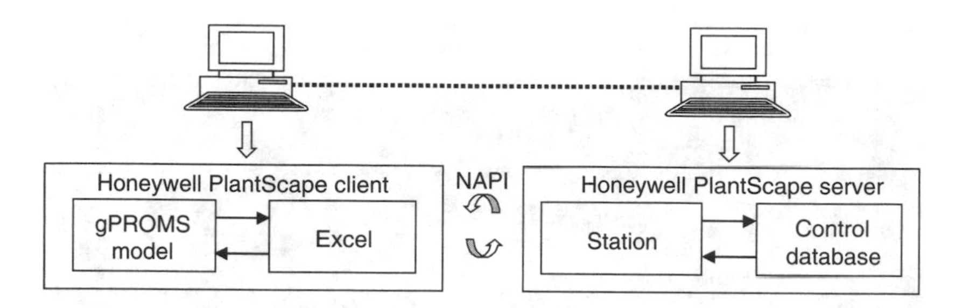


**FIGURE 21.14** The CAN structure.

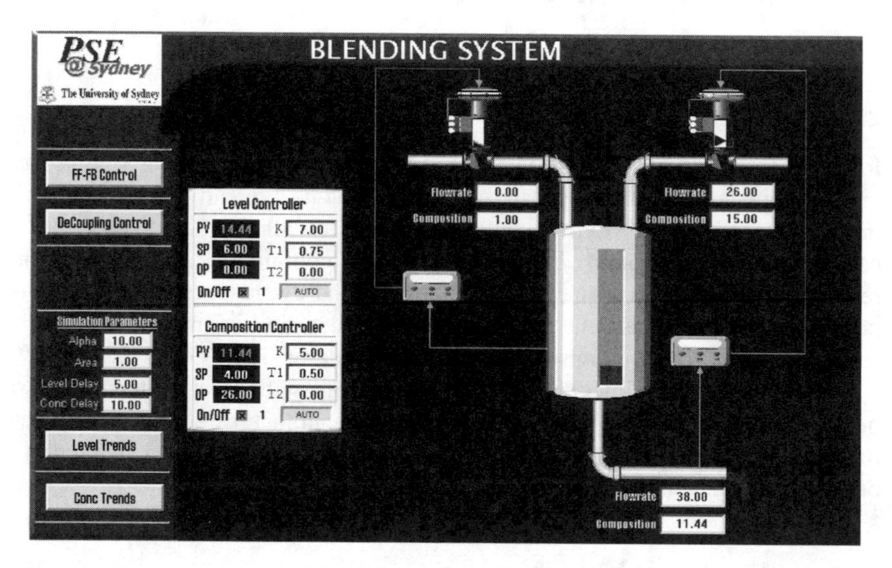


**FIGURE 21.15** Blending system with conventional feedback control configuration.

## Conventional Feedback Controller

A flexible user interface for process operator control (Figure 21.15) for the blending process is developed in display builder. It allows the operator(s) to view, manipulate, and analyze all data on any node in a control network, but also, in this case of a virtual plant, to change plant parameters, such as the area of the tank, resistance to flow, as well as the delay associated with the key process variables to be monitored and control. Trends are configured for the level, outlet concentration, and all main input variables so they can be monitored online. Real-time and historical data are presented together on the same trend so that the users can easily download historical data from MS Excel for model identification and further dynamic analysis tasks.

A multiloop feedback control strategy for controlling both the level and the outlet concentration is configured using the control builder. The basic structure as shown in the main project-editing screen is displayed in Figure 21.16. The data

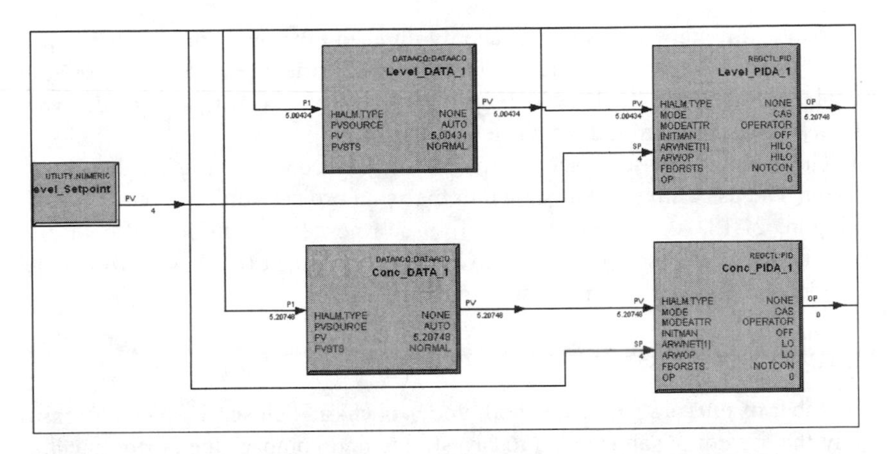


**FIGURE 21.16** Blending system with conventional feedback control configuration implemented in the control builder.

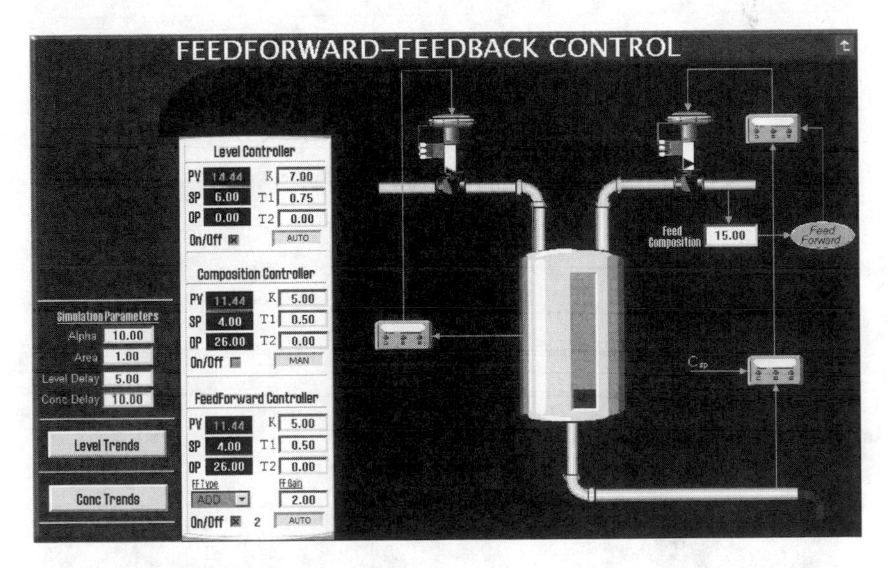


**FIGURE 21.17** Blending system with feedforward–feedback control configuration.

acquisition block and the regulatory control function block with the primary feature of setting PID loops for this particular control scheme for both level and outlet concentration are shown.

## Feedforward–Feedback Control Configuration

A feedforward loop is incorporated into the previous feedback configuration to compensate for the changes in the feed composition of stream 2 (Figure 21.17). It is worth noting that any of the implemented configurations can be activated or deactivated through the main interface by choosing the on or off button, respectively. The new interface for the alternative control structure also allows for additional functionalities such as the operator having access to the feedforward (model) parameters from the main screen.

The feedforward–feedback control strategy is configured using the control builder. The basic structure as shown in the main project-editing screen is shown in Figure 21.18. As illustrated in the figure, a new PID feedforward controller block (PIDFFA) is used in addition to a lead-lag element (LEADLAGA), which provides the necessary elements for the configuration.

## Virtual Experiments

A number of runs are performed both under open- and closed-loop conditions to study the system dynamics, and to investigate and compare the performance of the process under different control schemes.

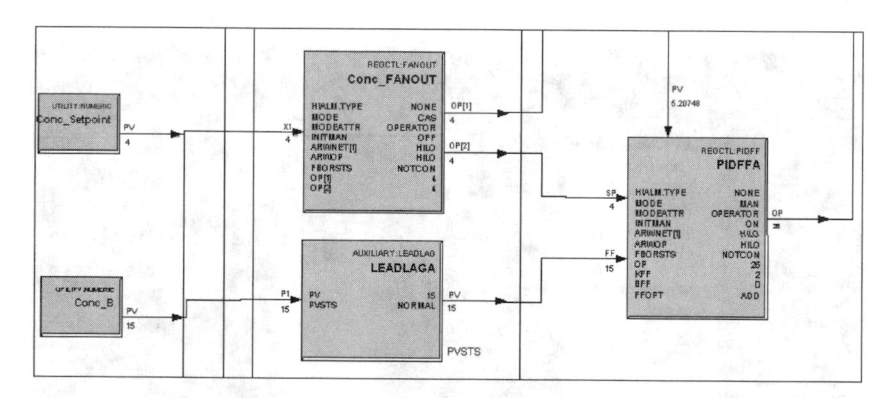


**FIGURE 21.18** Blending system with feedforward–feedback control configuration implemented in the control builder.

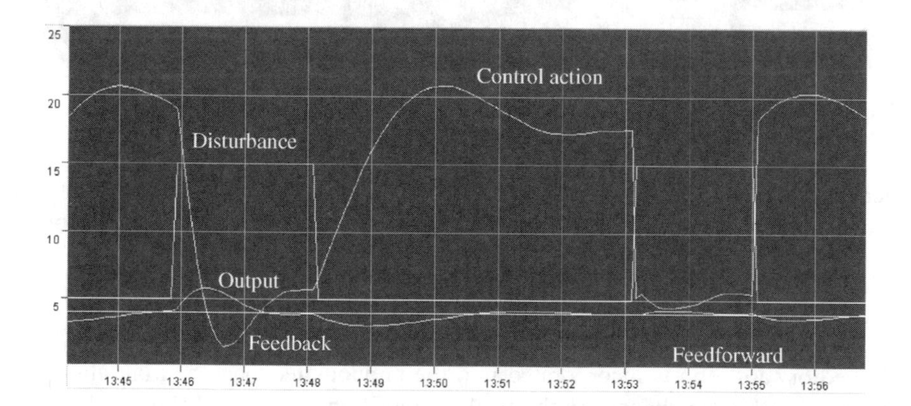


**FIGURE 21.19** Comparison between feedback and feedforward–feedback strategies.

It is worth noting that all historical information in the trends can be dragged-and-dropped from the trends screen. The trends can then be downloaded to a spreadsheet; a macro is developed to convert the information into a data file format, which can be accessed by identification software so that the process model can readily be obtained. Figure 21.19 illustrates a comparison between feedback and feedforward–feedback control strategies under similar disturbance changes.

## REFERENCES

1. Erickson, K.T. and Hedrick, J.L., *Plantwide Process Control*, Wiley, New York, 1999.

# 22 Role of Process Control in Modern Manufacturing

Supported by the availability of system architectures as described in the previous chapter, we witness a vast broadening of the domain of what is technologically and economically achievable in the application of computers to control industrial processes. This domain now includes process information and data gathering, control, online optimization, even production scheduling, and maintenance planning functions. Modern process control should then be viewed as an efficient integration of real-time information management with the traditional concept of control. This greatly facilitates the realization of the concept of integrated control systems (ICS) in which all factors influencing plant performance are articulated.

This chapter is devoted to an overview of the expanded role of control in modern manufacturing that exploits the current capabilities of industrial control systems. In recent years, we have also witnessed the development of new support strategies that help increase the available functionalities of computer control systems. These strategies allow for processing large amounts of data, transforming raw plant data into consistent information, predicting trends to detect possible abnormal situations, monitoring unmeasured quality variables from measured quantities, detecting or identifying malfunctioning instruments and process equipment, etc., among others. Such strategies provide the basis for implementing the ICS vision for a complete automated operation of the plant and they will be introduced in this chapter.

## 22.1 EXPANDED ROLE OF CONTROL IN MODERN MANUFACTURING

The conventional role of process control in industrial plants has been the implementation of control strategies through closed-loop automation. Today, this still remains to be the primary function of a control system. However, as discussed before, the advances in computer technology allowed the expansion of functionalities that can be simply referred to as *information management* at the plantwide scale.

The processing and reporting of plant information can be crucial for plant operations as well as planning activities. Erickson and Hedrick establish the objectives of a control system in modern manufacturing to:[1]

- Enforce plant control strategy
- Report plant performance
- Provide a proper window to the process

These activities are carried out using the control system technology that consists of a number of functionalities performed in a coordinated manner. It is noted that the functionalities included in the control system strongly depend on the complexity of the control actions as well as the analysis and reporting demands of the plant operators, the engineers and the managers. The control strategy then can be described through a hierarchical decomposition, referred to as the *control layers*. The goal of these control layers is to manage the inherent complexity in the plantwide industrial control architecture. They are conceived not only to address the primary role of the control system but also to be able to accomplish the expanded role of modern control for advanced manufacturing. The appropriate combination and the complexity of each layer are determined by the control engineer.

Using the definitions suggested by Erickson and Hedrick,[1] a natural decomposition for a typical plantwide control application could be:

1. Layer_1 control:
   - Basic regulatory control
   - Discrete control
   - Procedural control
2. Layer_2 control:
   - Coordination control
   - Supervisory control
3. Layer_3 control:
   - Exception handling
   - Alarm management
   - Trend and event recording
   - Production run and batch recording

Each layer should also be equipped with proper user-interfaces commensurate with the provided functionalities. It should be noted that this decomposition is somewhat arbitrary and different activities among different layers may be superimposed for a particular application. Nevertheless, such a hierarchical decomposition allows for a step-by-step procedure in building the plantwide control strategy and, in our view, provides a clearer understanding of various activities and functionalities.

***Layer_1 control***: This is the basic control layer utilized during the start-up of the plant and allows the plant to be operated around the design conditions. It is the foundation of the plantwide control system and the controllability of the process depends on it.

- *Basic regulatory control* comprises the conventional feedback loops (e.g., PID control) for key process variables.
- *Discrete control* is often the conventional safety system that prevents undesirable events (referred to as interlock control).
- *Procedural control* represents the sequential action of the control system that follows a procedure or a recipe for accomplishing a task. This is often used in batch process control.

***Layer_2 control***: It is implemented sometime after the plant is in operation and a reasonable level of consistency in operation is reached. This layer is aimed at the integration of the production process and to improve process efficiency and profitability. A typical application is in handling production rate changes in an optimal and coordinated fashion. This layer is particularly important in integrated processes (Chapter 20), where coordination of different sections of the plant is essential.

- *Coordination control*, as the name suggests, manages and coordinates the access to shared resources among the control system functionalities.
- *Supervisory control* is often referred to as set-point control and involves the use of a process or plant model to determine the correct (optimal) set-points that the lower layers would follow.

***Layer_3 control***: This layer is associated with the handling of abnormal operational conditions. Some of the basic functionalities are implemented from the beginning of the plant operation, since they may be needed during normal operational procedures. A typical example is a *basic* alarm system for the plant. However, more advanced functionalities would be implemented after the plant is fully operational. An example of this could be the implementation of an *advanced* alarm management system.

- *Exception handling* refers to the management of failure and recovery actions when an abnormal process event occurs.
- *Alarm management* helps the operators to identify abnormal events. There needs to be a procedure to deal with an avalanche of alarms by effectively suppressing secondary (nuisance) alarms and focusing on the primary (root) alarms.
- *Trend and event recording* helps save information from process sensors and presents them in a form that may be used for *post mortem* analysis of abnormal events. This is a critical task for statistical process (quality) control functions (see Chapter 24).
- *Production run and batch recording* function keeps time-based records of the production process that can be used for validation and long-term histories.

### Example 22.1

A typical paper mill is a complex and highly integrated operation.[2] It may be divided into several process areas, such as wood handling, the fiberline, the paper machine, the chemical recovery, and auxiliary systems. The evaporation plant is the first step in the chemical recovery process, located between the digester and the recovery boiler.

The main purpose of the evaporation plant is to increase the solids concentration of the weak black liquor (BL) exiting the digester, transforming it into heavy BL that, in turn, becomes a fuel source for the recovery boiler. To achieve this, the weak

BL is passed through a series of evaporators, removing excess water and increasing the weak BL solids content from 13 to 75 wt%.

The main unit operations within the plant are six evaporators, seven flash tanks, twelve tanks, nine heat exchangers and one stripping column (see Figure 22.1). The control objective is to maintain a constant final solids concentration, regardless of the initial solids concentration of the weak BL entering the process.

The control activities in the evaporation plant are decomposed into the following levels:

- *Layer_1 control* (*base-layer control*). Base-layer control deals with maintaining and manipulating tank heights, flow rates out of condensate tanks, and the temperature of the cooling water recycled to the cooling towers.
- *Layer_2 control* (*supervisory control*). Coordinating or supervisory higher layer control for the evaporation plant would assist the operators in setting several key variables.

Base-layer control is always active within the plant and allows the operator to adjust the process conditions. Supervisory control, on the other hand, should be able to be reset to manual mode when the plant is undergoing large disturbances, or during start-ups, shut-downs and inactivity periods.

*Base-layer control* (*L1 control*) It is implemented within the evaporation plant and includes the following loops:

- Level of different effects
- Level of the foul condensate and intermediate flash tank
- Level of all three sections of the clean condensate and intermediate flash tank
- Level of the clean condensate tank
- Level of the clean stripping column clean condensate tank

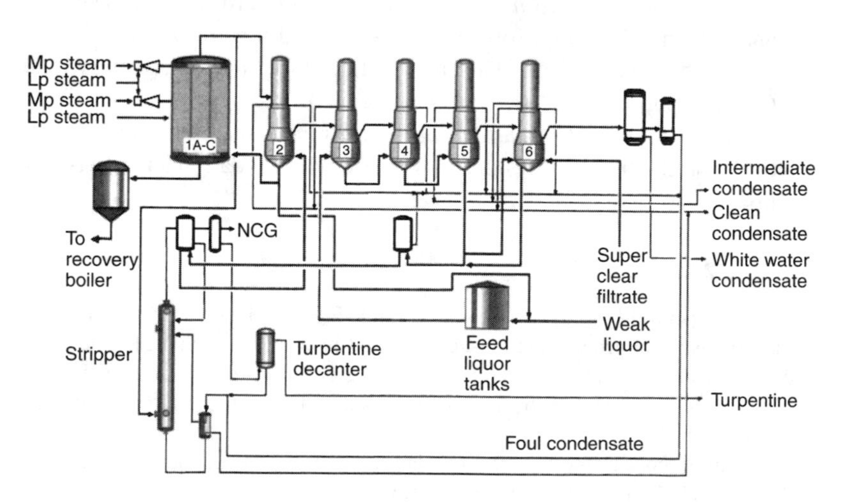


**FIGURE 22.1** Simplified schematic diagram of an evaporation plant.

*Advanced supervisory control* (*L2 control*) A multilayer control hierarchy is developed to manage variations in the feed conditions and maintaining the target heavy liquor dry solids. A change in the required flow may be a result of increased spills, variations in weak BL solids contents from the cooking plant, and increased weak BL flow from the cooking plant. This control layer can be associated with the coordination control described previously (coordinating operation of the cooking plant with the evaporation plant) in conjunction with the supervisory quality control that maintains the solids target.

The L2 Control architecture consists of the following components:

- *Mass balance control* (*inventory control*) for both the intermediate BL storage (BLST) and the medium (MBLST) tanks.
- *Energy balance control* where the steam flow target is automatically adjusted based on a feedforward compensation scheme. The calculated value is a function of the weak liquor flow, the spill liquor flow, the white water flow, the deviation in the feed liquor tank level, and the amount of dry solids. Whenever there is a change in the liquor flow rates, the required amount of steam is adjusted instantaneously. The medium strong liquor flow to concentrator 1 is also adjusted (feedforward compensation) based on the basic weak liquor flow to the evaporation plant and the amount of change.
- *Quality control* to maintain a constant solids concentration. The heavy liquor dry solids target is a key variable in the evaporation plant. It is introduced as a target set-point for most of the calculations described before. The final (fine-tuned) heavy liquor dry solids content can be controlled with the proposed strategy using both the medium strong liquor flow into concentrator 1 and by adjusting the final amount of steam into the system, through a multi-input–single-output (MISO) MPC (see Chapter 18) configuration.

A dynamic model for the evaporation plant is developed within the MATLAB/Simulink environment and used for validation of the proposed multilayer control configuration. Figure 22.2 and Figure 22.3 illustrate some of the results for disturbance rejection and set-point tracking.

Similar configurations can be developed for different sections of the plant in such a way that the whole plant acts in a coordinated fashion when subject to changes such as in production rate. Furthermore, a third control layer (Layer_3) can also be developed (for each section) aiming at supporting safe and consistent operation, as well as detecting abnormal operating conditions or faulty sensors. This additional layer acts as a decision support system with functionalities such as data processing and reconciliation, process monitoring, fault-diagnosis transition planning, etc. The definition of such tools will be the subject of next section.

## Example 22.2

In this example,[3,4] an advanced (multilayer) control architecture for the crystallization plant (see Example 21.1) is discussed. We consider the integrated utilization of various technologies including advanced modeling software (both off-line and real-time), MPC and distributed computer control technology in a collective environment.

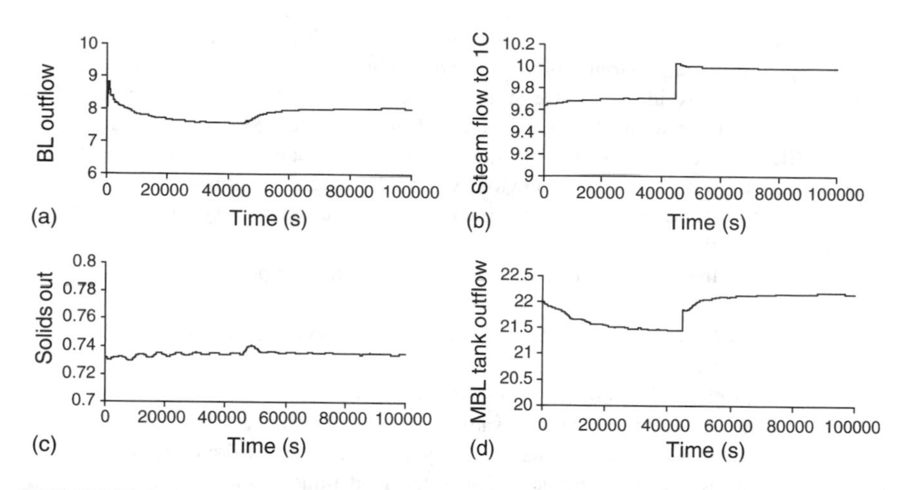


**FIGURE 22.2** System response against disturbance rejection: (a) final BL flow out of the system, (b) steam flow, (c) final dry solids content, and (d) flow out of MBLST.

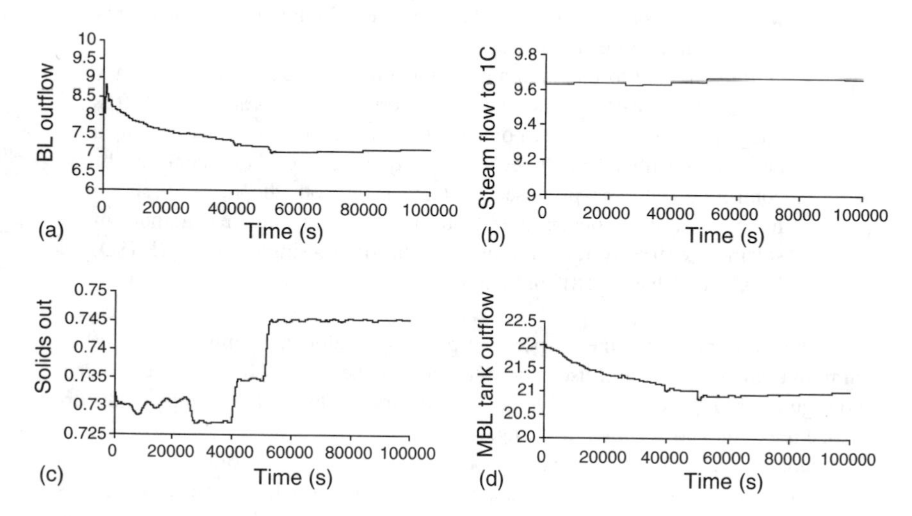


**FIGURE 22.3** System response for set-point changes in the solids content. (a) final BL flow out of the system, (b) steam flow, (c) final dry solids content, and (d) flow out of MBLST.

The outcome will be a structured framework for the control of complex units, specifically, of particulate processes in general. The following multilayer control hierarchy (Figure 22.4) is formulated:

- ***Lower layer_1*:** A regulatory control layer, where the process variables, namely the crystallizer temperatures, are kept at their set-points. In essence, the role of the regulatory layer is to follow the directives from the upper layer.

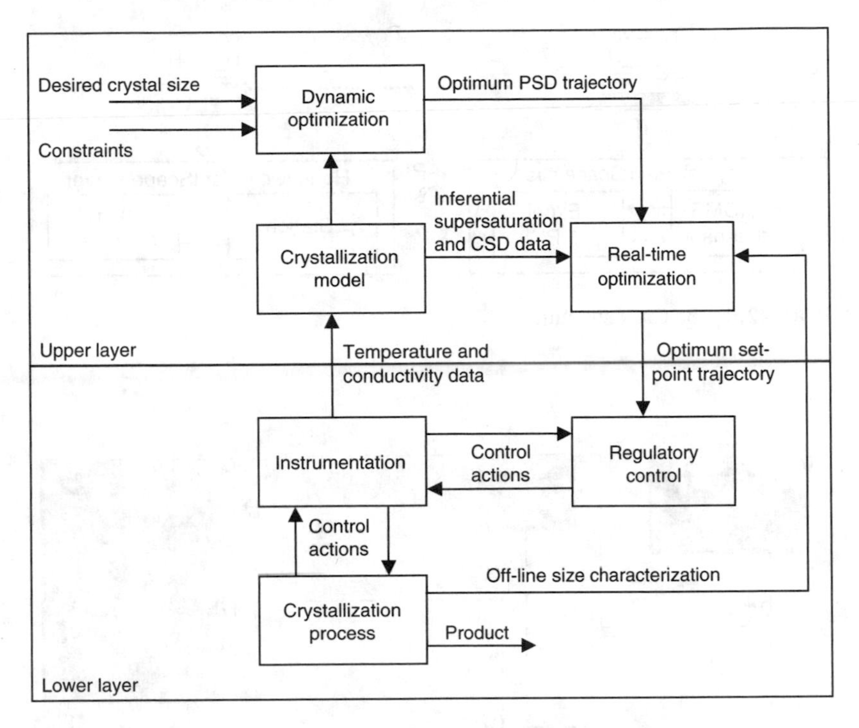


**FIGURE 22.4** Online optimal control framework for crystallization plant.

- ***Upper layer_1*:** The MPC layer determines the optimum temperature set-point trajectory of the plant given the production requirements (PSD distribution) and operational constraints, and keeps the process operating near optimum efficiency by constantly adjusting the set-points and responding to plant disturbances. This layer utilizes the process model to guide its actions.
- ***Upper layer_2*:** The off-line optimization uses the crystallizer model and process optimization to calculate the optimal PSD, which becomes the set-point for the lower layer.

Figure 22.5 illustrates the CAN structure. This configuration allows the communication between the model in gPROMS and the control facilities in PlantScape, and it is worth noting that it is possible to incorporate software packages (e.g., HYSYS) to be used as a virtual plant alongside the real plant. This has the added benefit of having a test facility where the simulation experiments can be performed to study plant behavior prior to execution in real time. This platform could also serve as a tool for operator training. Furthermore, such a facility has the potential to be applied as a teaching tool using the internet capabilities of PlantScape.

The configuration described in Figure 22.5 allows the model to be used as an online sensor for inferring the mean crystal size. It is useful to have the model-generated crystal size data available online as this would circumvent the problems

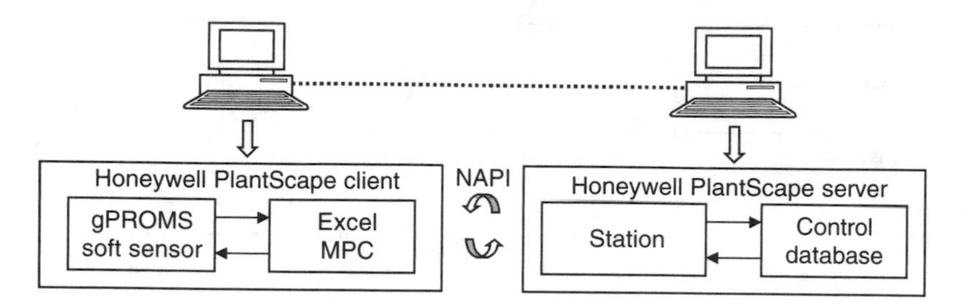


**FIGURE 22.5** The CAN structure.

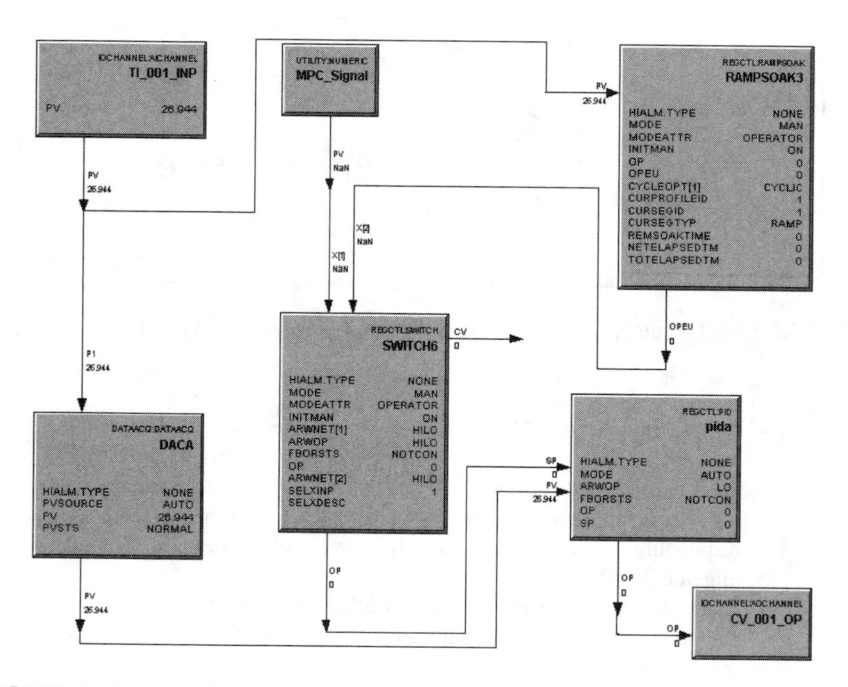


**FIGURE 22.6** Control builder environment showing the FBs used in the construction.

associated with unreliable or expensive on-line particle size instrumentation. Given the responsibility of solving the dynamic model, gPROMS communicates with the Honeywell PlantScape DCS by way of the NAPI protocol. The MPC code written in MS Excel and VBA resides on the same machine as the gPROMS model. The effect of disturbances is usually accounted for by taking off-line measurements of the output variable. This effectively takes care of the modeling errors as well.

The control strategy, in its corresponding control builder format, is shown in Figure 22.6. The implementation of the overall control strategy within the DCS environment is simplified by the available functionalities of the control builder. The input to the PI controller is the set-point provided from either one of the following three sources:

1. *Manual set-point* is used when the temperature of the crystallizer is required to be set at a constant temperature.
2. *RAMPSOAK set-point* is a set-point varying with time and changes according to a built-in temperature profile. A RAMPSOAK FB is used for implementation.
3. *External set-point* is used when an external program provides the set-point. An MPC, for example, could be used as the external set-point source.

The control scheme is designed to allow for the choice between a predefined set-point profile delivered by a RAMPSOAK FB or an external set-point signal delivered by an MPC controller as will be explained later. A switch FB is used to divert the signal from one set-point source to another.

Once the relationship between the temperature and the crystal size is elucidated, the next logical step is to develop a strategy to manipulate the temperature to achieve a crystal product of desired size. The control scheme consists of two control loops and was briefly discussed in Example 22.1. It is implemented in the Honeywell DCS environment and controls the crystallizer temperature, TI-001, by adjusting the ratio of hot oil to cold oil flow into the jacket. The temperature of the hot stream, TI-012, is kept constant using a PI controller that manipulates the steam flow rate to the hot heat exchanger.

The control strategy utilizes off-line process optimization to calculate the optimal trajectories of the PSD to be followed by the regulatory control layer. An optimal solution is found by maximizing the mean size by off-line optimization of a crystallizer model using gPROMS software. The corresponding optimal cooling and particle size profiles are shown in Figure 22.7 and Figure 22.8:

- An interval of fast cooling resulted in the initial stages of the batch followed by slower cooling as the run proceeds.
- The slow cooling is representative of a scenario where the growth of the crystals is dominant over the nucleation favoring the formation of larger crystals and hence maximizing the mean size.

A full batch cycle is then implemented in the system (Figure 22.9) excluding the feeding time, but including heating, soaking, and cooling stages. The cooling stage of the cycle represents an optimal cooling profile as discussed before. The regulatory control scheme accomplished the task of maintaining the temperature at set-point quite satisfactorily and as expected. A problem, however, is exposed here where the batch cooling terminates near the cooling tower cooling limits. The control is constrained by the physical limitation in the form of the minimum temperature available from the cooling tower. As anticipated, the diminishing cooling temperature driving force at the low end of the cooling profile constrains the cooling.

## 22.2 INTEGRATED CONTROL SYSTEMS

It is clear from the discussion in the previous section that, today, control systems can do more than just performing closed-loop automation. The advances in data

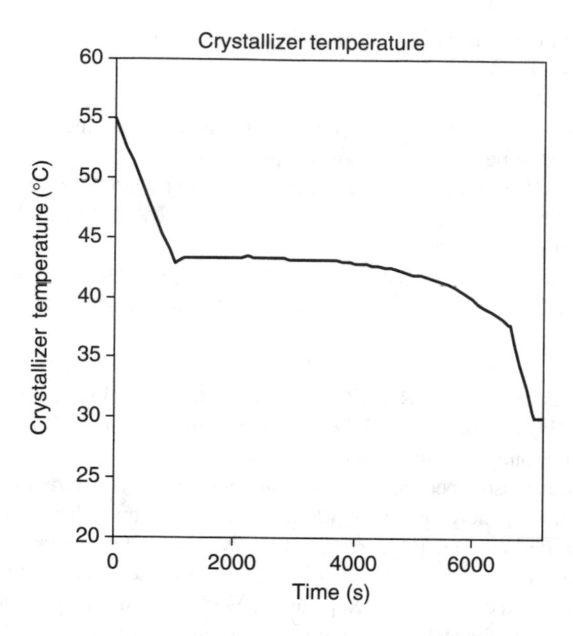


**FIGURE 22.7** The optimal solution for the cooling profile.

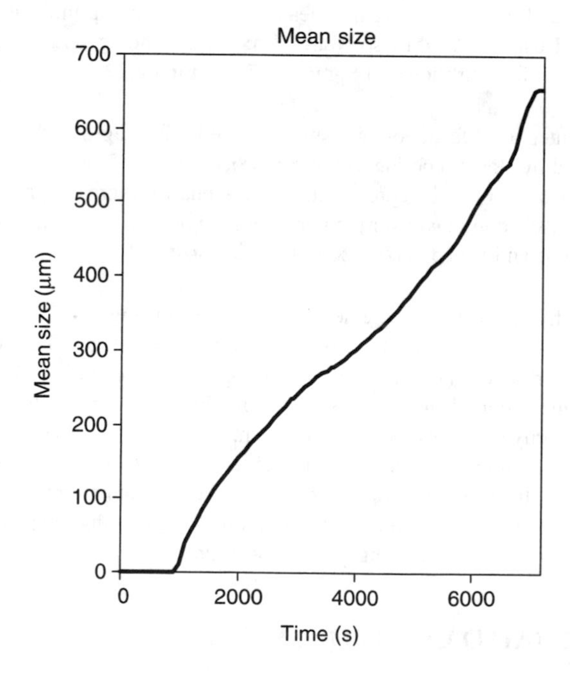


**FIGURE 22.8** The mean size trajectory simulated with the optimal cooling profile of Figure 22.7.

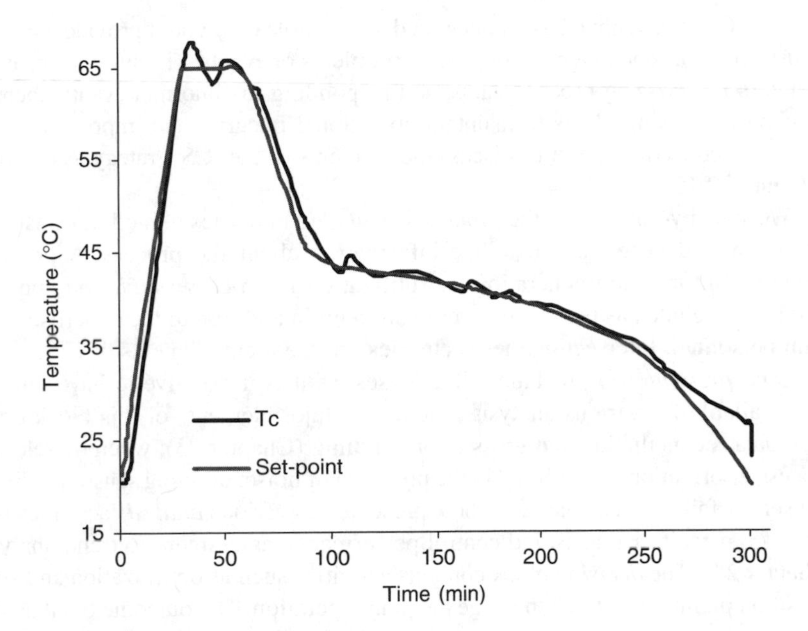


**FIGURE 22.9** Illustration of the regulatory control action for the full batch cycle with optimal cooling.

exchange technologies, as shown in Chapter 21, opening the control systems to third-party applications software allow new developments and tools from different vendors to be readily incorporated into the configuration of a particular control system. As such, this creates the opportunities for ICS, in which all aspects of plant operation can be considered as part of the control activities. An ICS environment can interpret and summarize raw data, present the refined information in forms that could be interpreted easily, highlight critical events, and suggest recovery procedures to facilitate the supervision tasks of plant personnel. There are several factors that motivate the ICS vision:

- Safer plant operation
- Proactive response to abnormal situations
- More consistent and reliable production of high-quality products
- Better management and utilization of resources through improved supervision
- Effective use of process data for monitoring, event detection and diagnosis
- Economic benefits of reduced downtimes and rational organization of plant maintenance
- Integration of data processing, process monitoring, and supervisory tasks
- Flexible software environment for development, implementation, and maintenance

The ICS is envisioned as an integrated set of tools that would provide a timely, robust detection, and diagnosis of process problems or events. The system can assist operators in assessing process status, and responding to abnormal events, thereby enabling processing plants to maintain operational integrity and improve product quality at reduced cost. Let us discuss the elements of the ICS strategy as depicted in Figure 22.10.

We start by noting that the *process knowledge* underlies all activities as each activity would generate and utilize information about the process. A possible venue for information generation and utilization is *process simulation* and this activity is included as part of the overall strategy in addition to the expertise from plant personnel. Intersecting these activities are the main elements of ICS.

*Data pretreatment* precludes all analyses, as it is imperative to have reliable data available for various analysis functions. Major elements of this block comprise data reconciliation and gross error handling (Chapter 23), whereby relevant process information is preserved in the presence of noise, outlying sensor readings, and sensor failures. The data are, then, presented to various *analysis* activities that range from fault diagnosis and control performance assessment to trend analysis (Chapter 24). The *decision* block contains activities such as optimization and state transition planning that help manage the plant operation. The outcome of all analyses is the action to be taken by the operator or the plant management in response to the observed event.

### Example 22.3

The polymerization laboratory at the University of Sydney, Australia, consists of a 5-L jacketed stirred reactor; a Julabo heating circulator to provide heat to the reactor through the jacket; 4 solenoid dosing pumps for providing the monomers,

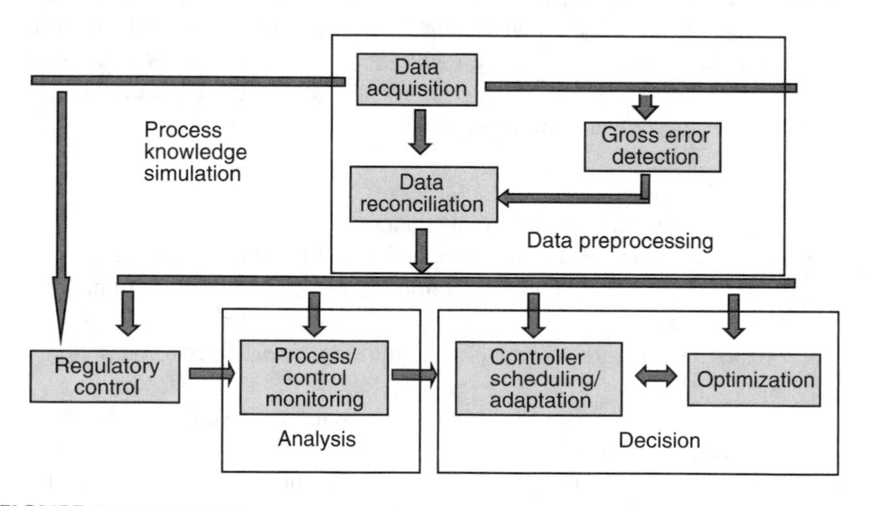


**FIGURE 22.10** ICS Vision.

surfactant and initiator to the reactor; 3 RTDs for monitoring temperature; and 4 precision balances to determine the quantities of the reactants used. The system is controlled by a Honeywell PlantScape C200 controller on a Honeywell PlantScape DCS. The uploading and configuration of the control schemes to the controller are done using the control builder (see Chapter 24). The 4-port serial to Ethernet converter is used in conjunction with software to send flow rates or differential weight readings from the balances to the controller. All the manipulated variable profiles are sent to the controller to be operated automatically.

The setup allows the configuration and implementation of a multilayer control scheme (see Figure 22.11) for advanced operation and control of the process similar to the one described before for the crystallization unit in Example 22.2 and a client station is used for operator manipulations.[5]

The lower level control consists of a series of conventional PID controllers to control the monomer feeds, the surfactant, and the initiator flow rates as well as the temperature. The inputs to the PID controllers are the set-points (upper-layer) provided from the three sources as discussed in Example 22.2.

Our goal in introducing this example here is to illustrate the implementation of another level within the control hierarchy as shown in Figure 22.12 that consists of a decision support system for process monitoring, diagnosis, and supervision[6] that conform to the ICS vision. This involves the development and implementation of an *expert* system, acting as a high-level supervisory and coordination agent in the operation of the process using Gensym G2 commercial software, which is an object-oriented environment for building and deploying real-time expert system applications to improve complex business and industrial operations.

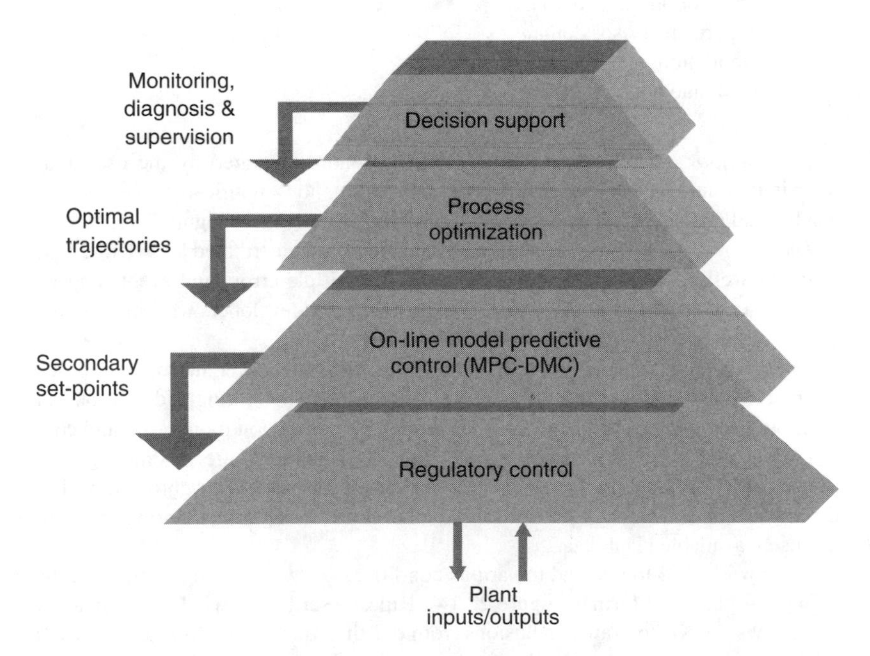


**FIGURE 22.11** Multi-layer control strategy for the polymerization laboratory.

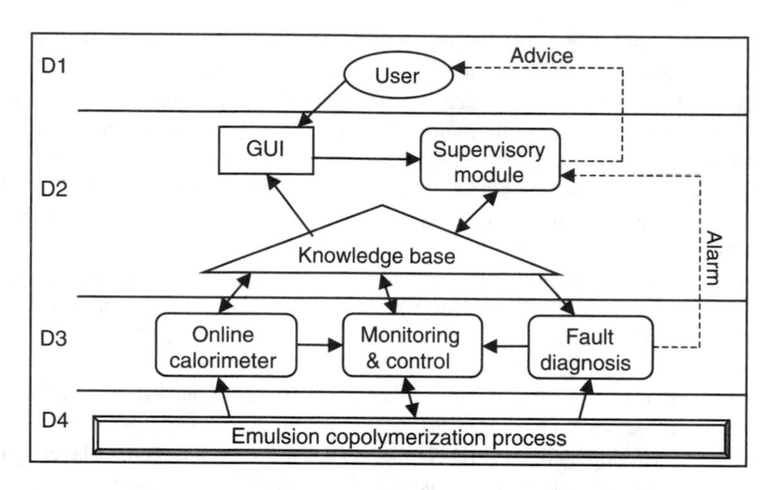


**FIGURE 22.12** Schematic representation of the supervisory architecture.

Figure 22.12 illustrates the schematic representation of the supervisory architecture for the polymer laboratory.

The proposed architecture, or the expert system shell, is composed of four different domains:

- operator and user domain,
- expert supervisor domain,
- simulation and processing domain,
- plant and action domain.

*The operator domain* comprises a DCS workstation operated by the user, thus allowing for human intervention of the process. Suggestions, notifications, and alarms can be read and acknowledged by the user through an interface (Figure 22.13).

*The expert supervisor domain* acts as a manager of the distributed knowledge base (KB), as well as managing prioritized alarms if multiple errors and events appear. The artificial intelligence (AI) comes solely from the developer, who imparts the extensive knowledge and rules into the KB.

*The simulation domain* provides interactions with the different software-based inferential components built in the system for advanced monitoring and control. For example, the online calorimeter provides inferential conversion monitoring and control through calorimetric measurements of the reaction temperature. Additionally, the online particle size and molecular weight *soft* sensor (that uses the rigorous model of the polymerization reactor) provides inferential information for monitoring and control using available plant data.

To allow the KB to respond to various conditions, *rules* are used and they can be defined in plain text format (Figure 22.14). Rules describe knowledge in a manner that allows the KB to draw conclusions from existing knowledge, to react to certain kinds of events, and to monitor the passage of time. A total of 23 rules were updated into the KB System.

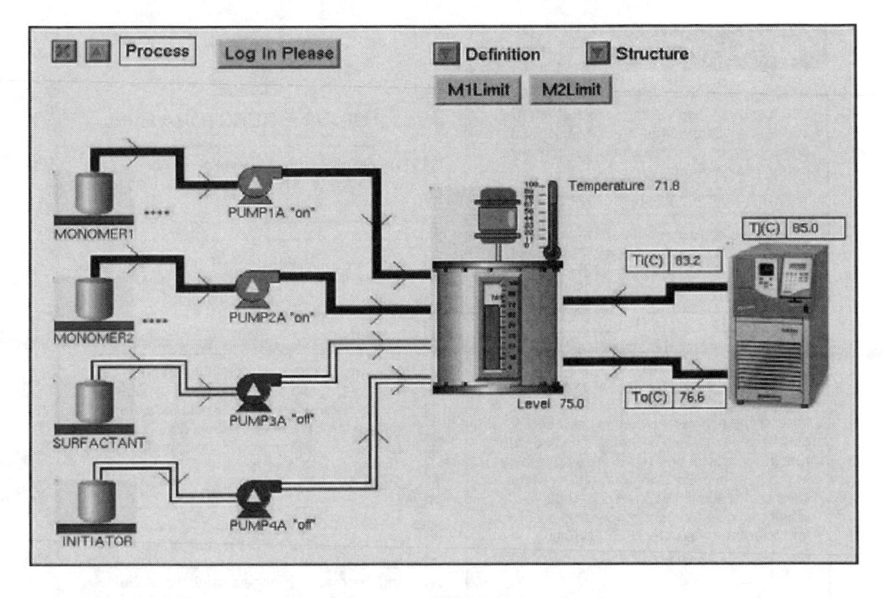


**FIGURE 22.13** The user-interface for the polymerization laboratory.

A number of key functionalities are also incorporated into the decision support system:

- An automated software launcher developed in G2, which eliminates the complicated startup an operator has to face before running the experiment.
- A recipe scheduler that automates the task of selecting the optimal feed profiles in control builder. The user or customer only needs to enter his or her preferred polymer properties, press *Enter* and the recipe formula would be selected accordingly to the matched logic structure developed in G2 diagnostic assistant (GDA).
- A diagnostic support tool for early fault detection and execution of corrective measures. Preanalysis problems were targeted to test-run this prototype, which includes
  - Detection in pump flow deviations
  - Pump or pipe blockages
  - Off-spec products due to rate of reaction and abnormal reaction temperature
  - Early detection of monomer feed run out

## 22.3 SUMMARY

The availability of modern industrial computer control system architectures has made possible the expansion of the functionalities of the plant control systems, broadening the domain of what is technologically and economically feasible to achieve in the application of computers to control industrial systems. We have discussed in this chapter, an expanded role of the plant control system as compared

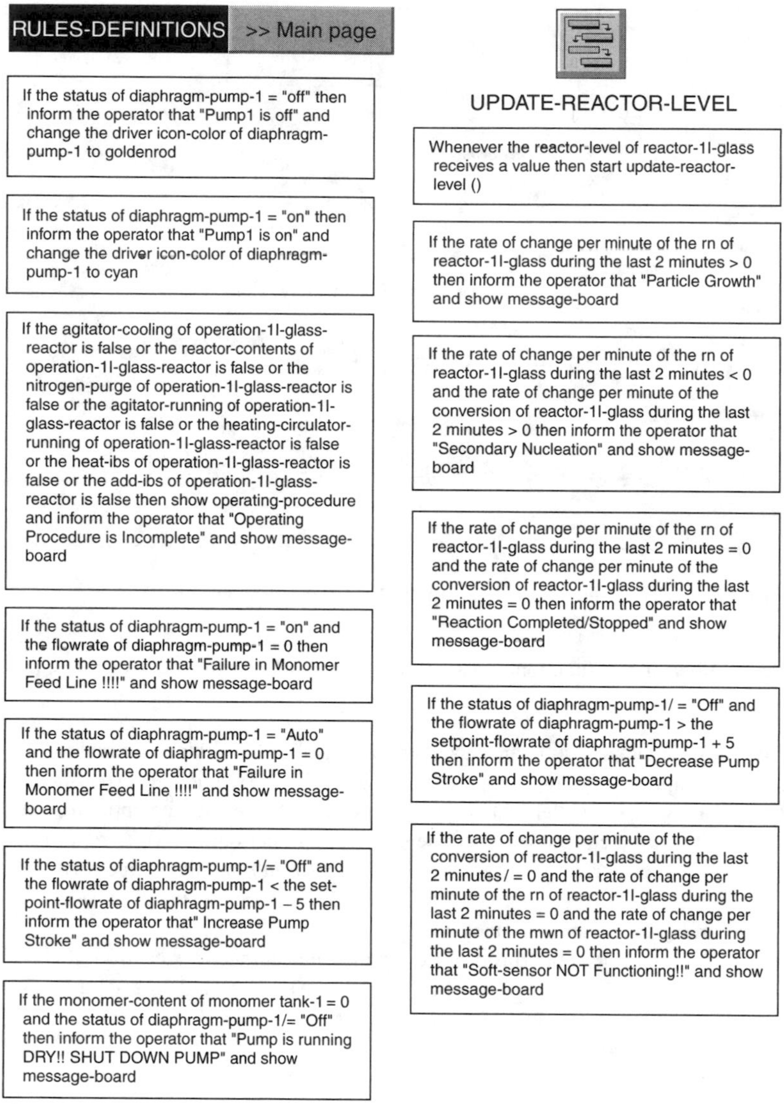

RULES-DEFINITIONS >> Main page

If the status of diaphragm-pump-1 = "off" then inform the operator that "Pump1 is off" and change the driver icon-color of diaphragm-pump-1 to goldenrod

If the status of diaphragm-pump-1 = "on" then inform the operator that "Pump1 is on" and change the driver icon-color of diaphragm-pump-1 to cyan

If the agitator-cooling of operation-1l-glass-reactor is false or the reactor-contents of operation-1l-glass-reactor is false or the nitrogen-purge of operation-1l-glass-reactor is false or the agitator-running of operation-1l-glass-reactor is false or the heating-circulator-running of operation-1l-glass-reactor is false or the heat-ibs of operation-1l-glass-reactor is false or the add-ibs of operation-1l-glass-reactor is false then show operating-procedure and inform the operator that "Operating Procedure is Incomplete" and show message-board

If the status of diaphragm-pump-1 = "on" and the flowrate of diaphragm-pump-1 = 0 then inform the operator that "Failure in Monomer Feed Line !!!!" and show message-board

If the status of diaphragm-pump-1 = "Auto" and the flowrate of diaphragm-pump-1 = 0 then inform the operator that "Failure in Monomer Feed Line !!!!" and show message-board

If the status of diaphragm-pump-1/= "Off" and the flowrate of diaphragm-pump-1 < the set-point-flowrate of diaphragm-pump-1 – 5 then inform the operator that" Increase Pump Stroke" and show message-board

If the monomer-content of monomer tank-1 = 0 and the status of diaphragm-pump-1/= "Off" then inform the operator that "Pump is running DRY!! SHUT DOWN PUMP" and show message-board

UPDATE-REACTOR-LEVEL

Whenever the reactor-level of reactor-1l-glass receives a value then start update-reactor-level ()

If the rate of change per minute of the rn of reactor-1l-glass during the last 2 minutes > 0 then inform the operator that "Particle Growth" and show message-board

If the rate of change per minute of the rn of reactor-1l-glass during the last 2 minutes < 0 and the rate of change per minute of the conversion of reactor-1l-glass during the last 2 minutes > 0 then inform the operator that "Secondary Nucleation" and show message-board

If the rate of change per minute of the rn of reactor-1l-glass during the last 2 minutes = 0 and the rate of change per minute of the conversion of reactor-1l-glass during the last 2 minutes = 0 then inform the operator that "Reaction Completed/Stopped" and show message-board

If the status of diaphragm-pump-1/ = "Off" and the flowrate of diaphragm-pump-1 > the setpoint-flowrate of diaphragm-pump-1 + 5 then inform the operator that "Decrease Pump Stroke" and show message-board

If the rate of change per minute of the conversion of reactor-1l-glass during the last 2 minutes/ = 0 and the rate of change per minute of the rn of reactor-1l-glass during the last 2 minutes = 0 and the rate of change per minute of the mwn of reactor-1l-glass during the last 2 minutes = 0 then inform the operator that "Soft-sensor NOT Functioning!!" and show message-board

**FIGURE 22.14** The page of rules for the polymerization process.

with the conventional view of the first part of this book. It was shown that now all aspects involved in the operation of a plant such as information processing, data gathering, process control, on-line optimization – advancing even to scheduling and production planning functions – may be included in the range of tasks to be carried out by the computer control system. Following these ideas, the concept of ICS was introduced together with the key components and tools involved in such an integrated strategy.

Among these key components, data processing and reconciliation, and intelligent monitoring are intimately related to the performance of the plant regulatory control. Consequently, an introduction to these topics is the subject of the next two chapters.

## REFERENCES


1. Erickson, K.T. and Hedrick, J.L., *Plantwide Process Control*, Wiley, New York, 1999.
2. Smook, G.A., *Handbook for Pulp and Paper Technologists*, 3rd Ed., Angus Wilde Publications, Vancouver, Canada, 2002.
3. Abbas, A. and Romagnoli, J.A., A modeling environment for the advanced operation of crystallization processes, *Proceedings of PSE 2003*, Kunming, China, 2003.
4. Abbas, A. and Romagnoli, J.A., Advanced optimal control of a batch crystallization operation, *Proceedings of BatchPro*, Poros, Greece, 2004.
5. Alhammadi, B., Willis, R., Romagnoli, J.A., and Gomes, V.G., Optimal control for emulsion copolymerization: application within a DCS environment, *Proceedings of DYCOPS-7*, Cambridge, MA, 2004.
6. Chew, R., Romagnoli, J.A., and Gomes, V.G., Knowledge-based supervisory and diagnostic system for a fedbatch pilot-scale emulsion polymerization facility, *Proceedings of BatchPro*, Poros, Greece, 2004.

# 23 Data Processing and Reconciliation

With the proliferation of computer control systems in process plants as discussed in Chapter 21, collection and storage of data from plant instrumentation has become a routine activity. However, such data are subject to variation in quality. This is inherent in all process data collection as the sampling and testing equipment, schedules and techniques are exposed to a wide range of influences. Such influences will always give rise to a raw data set containing missing points, gross errors and outliers, all of which need to be eliminated to obtain a useable data set.

A snapshot of a typical raw data set is shown in Figure 23.1. The selection depicts ten variables recorded as part of a daily monitoring activity in a sewage treatment plant, and illustrates some of the problems associated with raw plant data. From Figure 23.1, two types of errors can be easily identified. First are the *constant values*, which require some knowledge of the process to determine if they are erroneous or not. The other is the presence of the *nan* (Not a Number) term, which signifies a lack of numerical data. This could mean that the data point is below the detection limits, however, it may also be a *missing point*. As a distinction between the two may be difficult to make, such a point is always regarded as a missing point.

In this chapter, our goal is to study a set of methodologies that can help engineers obtain a meaningful data set from raw data, so that the conclusions based on the data set are sound and reliable.

## 23.1 DEALING WITH MISSING POINTS

Missing points may include an isolated point or a series of points. An isolated missing point may not have a noticeable impact on the subsequent analysis of the data. However, for a series of missing points, the impact is largely dependent on the length of the series or the frequency of recurrence. If one has enough information, often the missing points could be ignored and the data set would be truncated. In other cases, one attempts to *fill in* the missing points by using interpolation techniques. The interpolation techniques should be used with caution, however, as they may change the nature of the variations that the data exhibits if the points were not missing. The next example illustrates this point.

### Example 23.1

We shall consider the plot of data given in Figure 23.2 that corresponds to the $NO_x$ levels in a sewage treatment plant.

| Date | V1 | V2 | V3 | V4 | V5 | V6 | V7 | V8 | V9 | V10 |
|---|---|---|---|---|---|---|---|---|---|---|
| 950 101 | 573 | nan | 13 | 433 | 168 | 7.4 | 248 | 5 60 | 33.3 | 27.3 |
| 950 102 | 489 | nan | 4.3 | 372 | 245 | 2.1 | 182 | 5 60 | 86.5 | 27.3 |
| 950 103 | 686 | nan | nan | nan | nan | nan | nan | 5 60 | nan | 16.1 |
| 950 104 | 696 | nan | 10.1 | 544 | 140 | 7 | 379 | 5 60 | 53.9 | 16.1 |
| 950 105 | 1194 | nan | 7.1 | 554 | 124 | 8.5 | 661 | 5 60 | 78 | 16.1 |
| 950 106 | 968 | nan | nan | 709 | 153 | nan | nan | 5 60 | nan | 16.1 |
| 950 107 | 1504 | nan | 14.7 | 902 | nan | 22.1 | 1357 | 5 60 | 61.4 | 16.1 |
| 950 108 | 860 | nan | 8.7 | 427 | 148 | 7.5 | 367 | 5 60 | 49.1 | 16.1 |
| 950 109 | 864 | nan | 7.2 | 886 | 440 | 6.2 | 766 | 5 60 | 123.1 | 16.1 |
| 950 110 | 1236 | nan | 7.6 | 611 | 111 | 9.4 | 755 | 5 60 | 80.4 | 16.1 |
| 950 111 | 984 | nan | 8.7 | 624 | 178 | 8.6 | 614 | 5 60 | 71.7 | 16.1 |
| 950 112 | 778 | nan | 8.8 | 742 | 82 | 6.8 | 577 | 5 60 | 84.3 | 16.1 |
| 950 113 | 519 | nan | 8.7 | 654 | 77 | 4.5 | 339 | 5 60 | 75.2 | 16.1 |
| 950 114 | 663 | nan | 7.9 | 791 | 100 | 5.2 | 524 | 5 60 | 100.1 | 16.1 |
| 950 115 | 680 | nan | 7.7 | 513 | 40 | 5.2 | 349 | 5 60 | 66.6 | 16.1 |
| 950 116 | 719 | nan | 8.4 | 746 | 232 | 6 | 536 | 5 60 | 88.8 | 16.1 |
| 950 117 | 937 | nan | 8.1 | 579 | 196 | 7.6 | 543 | 5 60 | 71.5 | 16.1 |
| 950 118 | 1174 | nan | 9.7 | 1160 | 280 | 11.4 | 1362 | 5 60 | 119.6 | 16.1 |
| 950 119 | 1398 | nan | nan | nan | nan | nan | nan | 5 60 | nan | 16.1 |
| 950 120 | 968 | nan | nan | nan | nan | nan | nan | 5 60 | nan | 16.1 |
| 950 121 | 3073 | 21 | 5 | 339 | 221 | 15.4 | 1042 | 5 60 | 67.8 | 16.1 |
| 950 122 | 723 | 22 | 6.7 | 475 | 378 | 4.8 | 343 | 5 60 | 70.9 | 16.1 |
| 950 123 | 636 | 18 | 7.6 | 946 | 193 | 4.8 | 602 | 5 60 | 124.5 | 16.1 |
| 950 124 | 838 | 19 | 10.8 | 685 | 288 | 9.1 | 574 | 5 60 | 63.4 | 16.1 |
| 950 125 | 955 | nan | 10.1 | 1001 | 173 | 9.6 | 956 | 5 60 | 99.1 | 16.1 |
| 950 126 | 890 | nan | 12.3 | 548 | 333 | 10.9 | 488 | 5 60 | 44.6 | 16.1 |
| 950 127 | 836 | nan | 14.4 | 543 | 231 | 12 | 454 | 5 60 | 37.7 | 16.1 |

**FIGURE 23.1** Typical plant data structure.

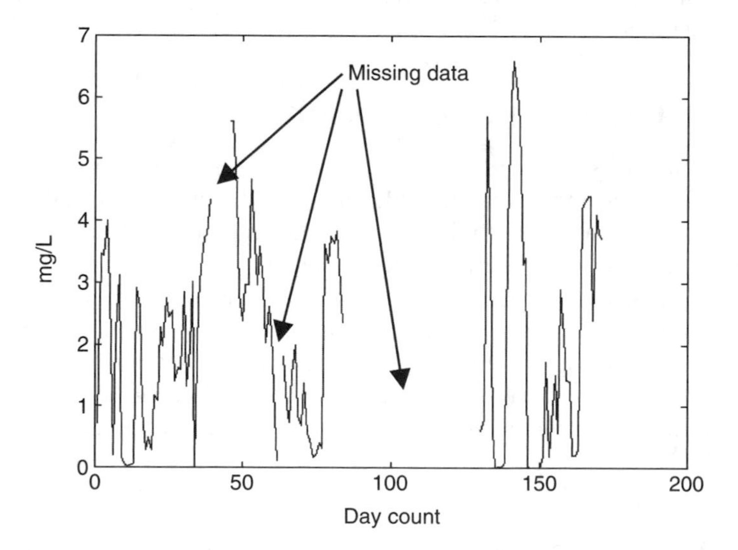


**FIGURE 23.2** Process data for $NO_x$ level in a treatment plant.

If the data points are interpolated using simple linear interpolation, the raw data in Figure 23.2 becomes a continuous set, as depicted in Figure 23.3, and enables a continuous data set to be used for subsequent analyses.

One has a choice in the application of interpolation techniques, and the outcomes can be seen in the two different lines plotted for the missing series from 85 to 130 in Figure 23.4. For a missing data section, one can apply interpolation by assuming either that the last data point is valid, or the next available data point is valid. In Figure 23.4,

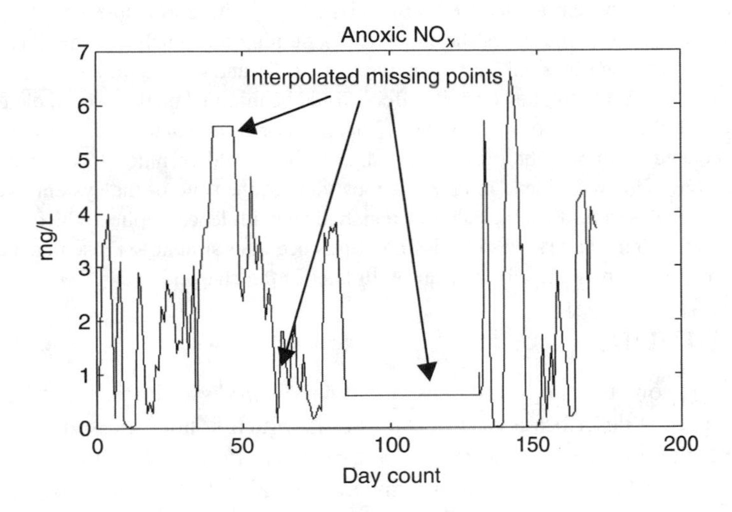


**FIGURE 23.3** New data set after interpolation.

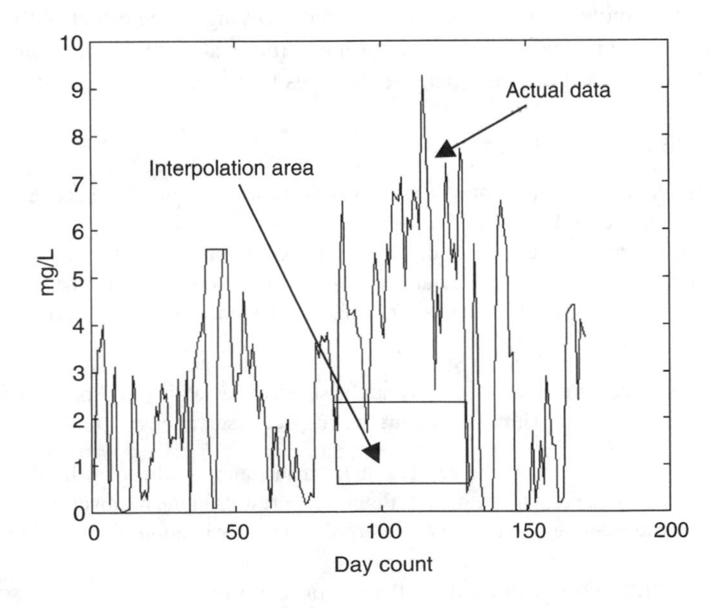


**FIGURE 23.4** Interpolation area and the comparison with real data.

we can see that the lower line shows the case where one uses the value of the next data point and the upper line represents the case where the value of the last real data point before the missing section is used. One can see that the interpolation result could lie between either of these lines. Nevertheless, any interpolation may still be wrong and if the interpolated section is sufficiently large compared to the length of the overall data set being used for analyses, it may be worthwhile checking the statistical relevance of any decisions made, which are based on such a modified data set. This is illustrated in Figure 23.4, where the modified continuous data are compared with the true data that would have been obtained if the missing points were available.

Figure 23.4 also illustrates the impact of assuming too much information. The first interpolation (the small section between days 40 and 45) would be a misrepresentation of the system; however, its effect would be minimal on the overall data statistics. If the large portion of missing data, from days 85 to 130, were to be interpolated as shown, the mean of the data series would be much lower than the actual one. This would lead to an erroneous view of the state of the system, which may be more serious than the calculated statistical confidence implies. The decision made with such a large series of missing points becomes similar to those made concerning gross errors and will be dealt with later in the chapter.

## 23.2 OUTLIERS

Outliers are sometimes referred to as *statistical anomalies*. The term outlier, as the name suggests, describes a type of data point, which is not representative of the sample being considered. The term *gross error* is often used in place of *outlier*. Outliers are influential observations; in other words, their deletion often causes major changes in estimates, confidence regions, and other such tests associated with the data. The results of subsequent data analysis can be seriously affected by the failure to remove outliers from process data. Owing to the effect outliers have on the *integrity* of a data set, their detection and removal is vital before any further analysis takes place. There are many techniques to detect and remove outliers.[1]

### Example 23.2

An example of a data set containing outliers is shown in Figure 23.5 for the same treatment plant as in Example 23.1.

A common outlier detection method is a standard deviation test. In this procedure, one examines the changes of a parameter estimating the dispersion of the data. This parameter is the standard deviation or the variance. The procedure for this test is simple:

1. Rank order the data
2. Calculate the cumulative mean and the standard deviation (see Appendix D for definitions) with and without the suspect observations

An increase in the variance can result from including outliers in the data set. Applying this procedure to the data in Figure 23.5 results in the removal of outliers, and their values are replaced by their interpolated approximations (Figure 23.6).

A conceptual representation of the preprocessing steps discussed so far is shown in Figure 23.7.

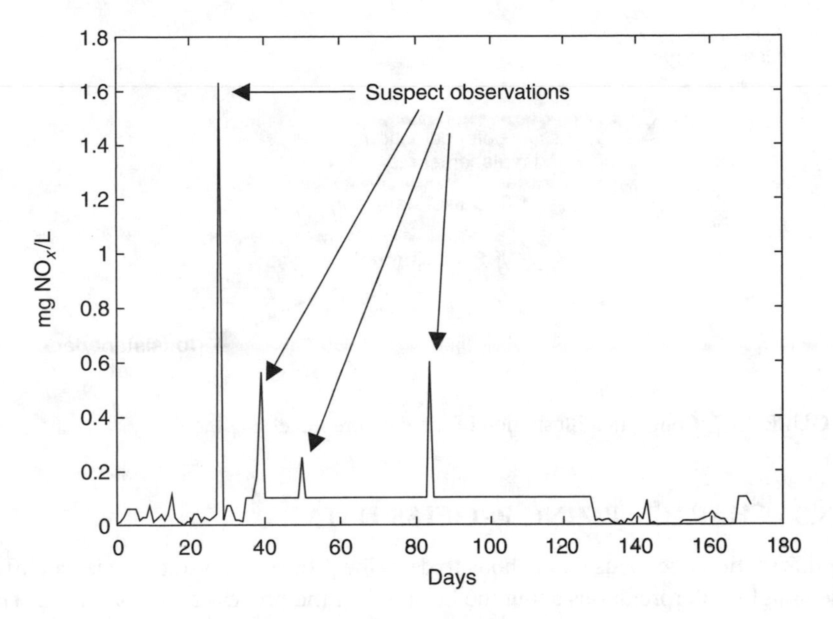


**FIGURE 23.5** The data set with suspect observations (possible outliers).

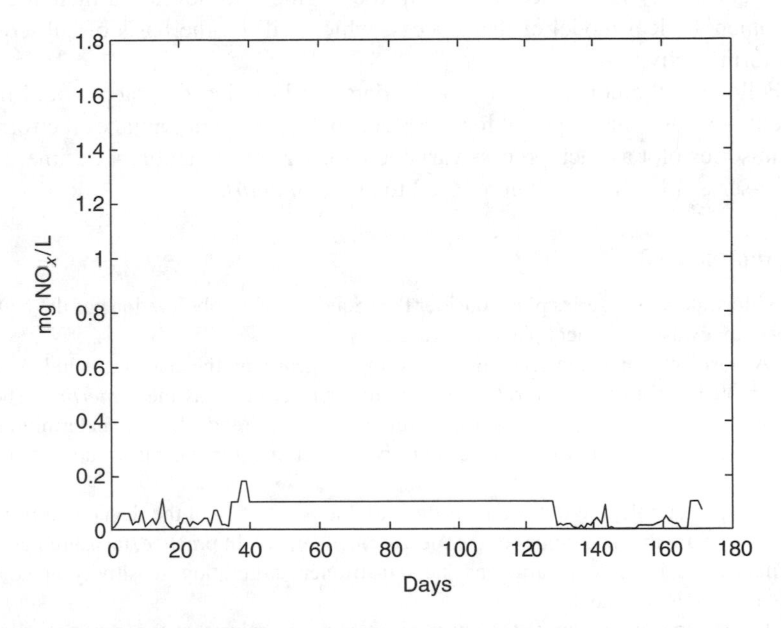


**FIGURE 23.6** The data set without outliers.

For some data analysis techniques, further processing is required. For instance, if a homogenous data set is required, mean centering and scaling to unit variance can also be conducted (see Chapter 24).

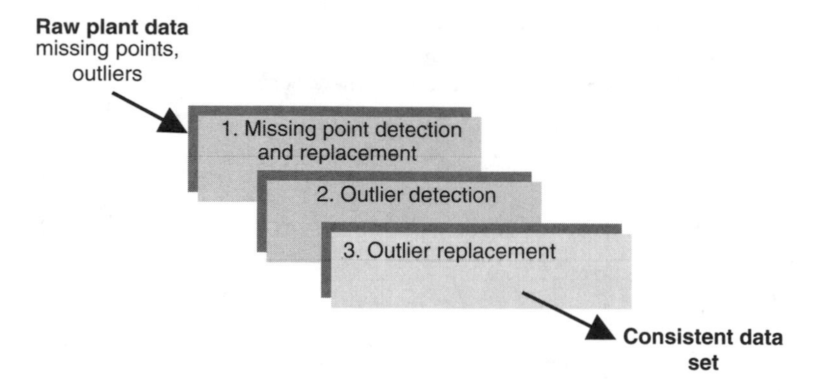


**FIGURE 23.7** Conceptual illustration of the data preprocessing tasks.

## 23.3 CHARACTERIZING PROCESS DATA

In this section, we focus on methods to describe process data so that one can offer meaningful interpretations about the behavior of the process being observed. The data presentation methods include visual displays (qualitative) as well as numerical (quantitative) summaries. They help the engineer to develop a mental and a phenomenological model of the process, which will be the basis of subsequent monitoring activities.[2]

Following the collection of process data, an intuitive first step would be to present this data in a graphical form. Often, such a presentation takes the form of a time-series plot as each process variable is observed sequentially in time. Such a time-series plot is also often referred to as a *run chart*.

### Example 23.3

To illustrate a time series plot, consider the sequence of 50 observations of the $NO_x$ level in a waste treatment plant in Figure 23.8.

A horizontal line representing the average (mean) of the data is included in the plot. In control chart vernacular, this line is referred to as the *centerline*. The centerline is very important since it serves as a reference line in determining whether patterns exist in the data and gives a feeling for the magnitude of the process data.

In this example, the true value of the variable is ~3.76, and the deviations from this value are the magnitudes of the measurement errors. In practice, measurements will unavoidably exhibit variations due to instrument calibration, sensitivity, and the data transmission medium.

In the time-series plot of Figure 23.8, we can recognize that the measurements exhibit clear variations around the centerline, referred to as the *measurement error*, which is defined as any deviation from the true value of the variable being measured.

When dealing with measurements, two characteristics are important and they are known as *accuracy* and *precision*:[2]

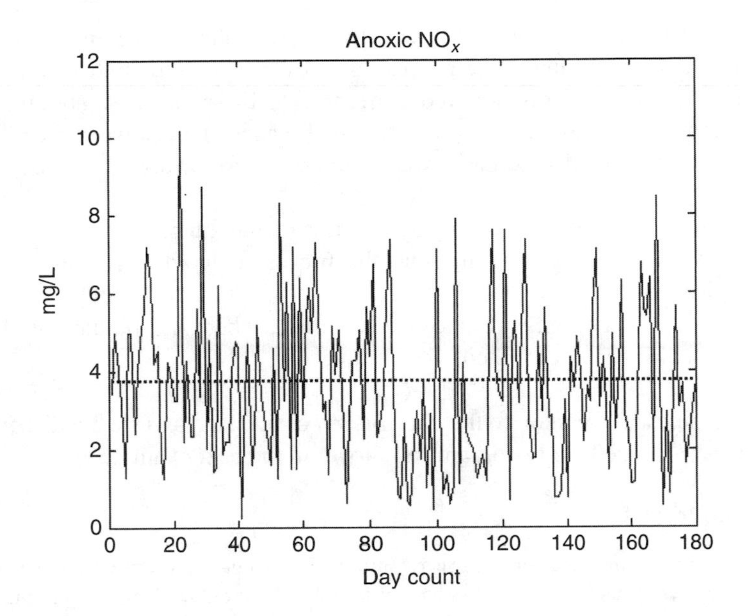


**FIGURE 23.8** A time-series plot of data for $NO_x$ level in a waste treatment plant.

- *Accuracy* refers to the extent to which the average of repeated measurements on the same instrument differs from the true value. The difference between the average and the true value is known as the *systematic error* (or *bias*).
- *Precision* refers to the ability of a measurement system to replicate its own measurements. *Precision* is inversely related to the variations or dispersion of the replicated measurements.

As we discussed before, the time-series plot enables us to recognize the presence of variability in the data set. This leads us to the definition of a random process.

---

The characteristics of a constant process level (no systematic pattern) of observations around the centerline, and constant process dispersion define a *random process*. In traditional statistical process control (SPC), a random process is considered as a process in a *statistical state of control*.

---

The term nonrandom process behavior is typically used when dealing with data and deserves some clarification. When using the term *nonrandom process*, it is not meant to imply that the process is not subject to any random variability. The term nonrandom process refers simply to process behavior affected by phenomena other than *pure* random variation.

Previously, we have studied how a time-series plot enables us to discover some key features (patterns) of process data over time. Although a time-series plot provides good information about the process behavior, there are still some other important process characteristics, which cannot be easily extracted from this type of analysis. For example, we could ask the following questions:

- Are the observations evenly spread in a certain range?
- Are the chances the same that the process outcome will fall in any given interval?
- If the observations are not evenly spread, where do the data tend to group or be concentrated?

These questions pertain to the *distribution* of the data, and such a distribution can be characterized via histograms as shown in the next example.

### Example 23.4

Consider the same data set used in Example 23.3. To gain a different perspective from the set of data, consider the display found in Figure 23.9. This graph, known as a *histogram*, makes clearer the dispersion of the measurements, as well as the relative frequency of the various values.

In the construction of a histogram, the value axis is divided into adjacent and equal-width class intervals or cells. The number of observations that fall within each class interval is then counted, and displayed.

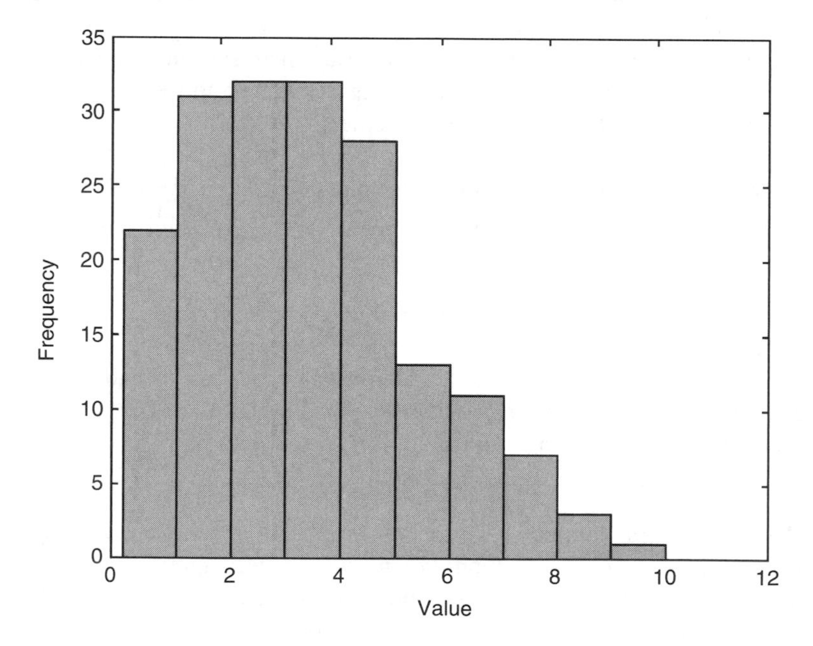


**FIGURE 23.9** Histogram of data in Figure 23.8.

Now, if the number of observations were very large and the groups were made much narrower, the previous irregular step form of the histogram could be represented as a smoothed curve, which is known as the frequency curve. Sometimes, it is preferable to express the frequencies in terms of proportional frequencies instead of actual frequencies (probability distribution) or probability density function (PDF) curve. The PDF curve for the data in Figure 23.8 is given in Figure 23.10.

This example illustrates how a simple distribution diagram (histogram or PDF curve), provides insight about the data concentration or distribution around the centerline without considering the time dimension.

## 23.4 MODELING PROCESS DATA

Motivated by the previous discussions, data generated by a random process can be represented by the following conceptual equation:[2]

$$\text{observed data value} = \text{actual (mean) level} + \text{random deviation} \quad (23.1)$$

In other words, any observation or measurement comprises the contribution of the actual state of the process (mean level) and a random deviation component. The latter describes the random phenomena that might push what is actually observed above or below the mean level and, therefore, can have a positive or a negative value.

It is often assumed that the average of a given set of random deviations is close to zero. Actually, one can prove that this average equals to zero exactly for a very large (infinite) number of observations. This is one evidence that will allow us to assume that the random deviations follow the normal (Gaussian) distribution (see Appendix D).

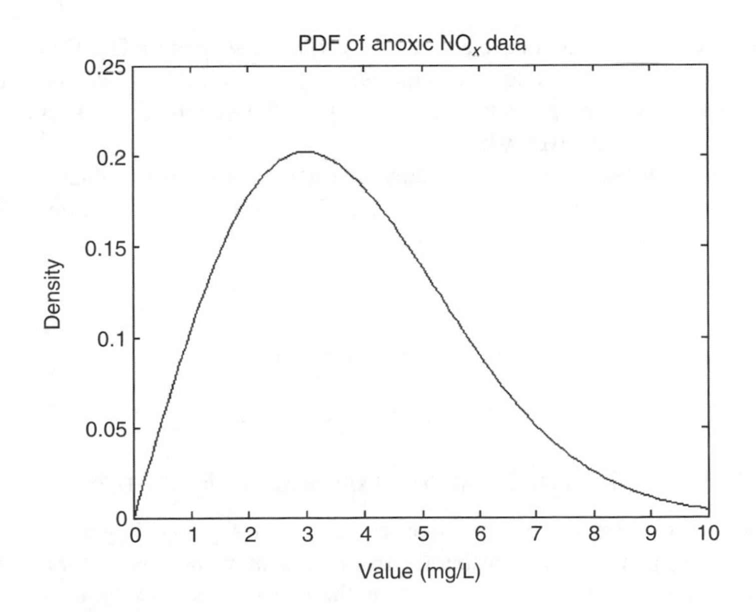


**FIGURE 23.10** PDF of data in Figure 23.9.

Often, the normal distribution is a good assumption to start with, but not something that should be taken for granted. In other words, if the data conforms to the normal distribution, it can be easily characterized by two parameters, namely, the mean and the variance. This results in a very simple model. However, for most process signals, such a model may be overly simplistic.

Let us express the model for a random variable (process) $y$ as follows:

$$y(i) = \bar{y} + \varepsilon(i), \quad i = 1, 2, \ldots \tag{23.2}$$

In this equation, $\varepsilon$ denotes the random deviation at the sampling instant $i$ and $\bar{y}$ is the underlying mean level. The random deviations $\varepsilon$ are also commonly referred to as *random errors* or *disturbances.*

To facilitate the development and interpretation of data analysis techniques that will follow, one has to make the following assumptions about random errors:[2]

- The random errors are independent. Keeping in mind that the random errors occur sequentially over time, this assumption basically implies that the errors are generated from a random process.
- The random errors have an expected value (or mean) of zero, i.e.,

$$E(\varepsilon) = 0 \tag{23.3}$$

where $E(\cdot)$ denotes the mean or expected value of a random variable.

- The random errors have a variance of $\sigma_\varepsilon^2$ (see Appendix D). This specification is often called the *assumption of constant dispersion*. The idea is that the vertical scatter of points about the systematic pattern tends to be the same everywhere.
- The random errors are normally distributed random variables. Using the notation introduced in Appendix D, this specification along with the other assumptions implies that

$$\varepsilon \rightarrow N(0, \sigma_\varepsilon^2) \tag{23.4}$$

Accordingly, we can rephrase the process model as

$$y(i) = \text{true value} + \varepsilon(i), \ i = 1, 2, \ldots \tag{23.5}$$

### 23.4.1 Model Fitting Based on Least-Squares Estimation

We learned that observations from a process could be expressed by a contribution of the true model, and a contribution from random variations. In practice, the measurements reflect the combination of these two effects, without any information as to the relative contributions. The challenge is, then, to understand and estimate the contribution of the true model. To address this problem, we define the *residual* as

$$\text{residual} = \text{observed data value} - \text{fitted model value} \tag{23.6}$$

Let $\hat{y}(i)$ be the fitted model value at sample time $i$. Then the residual can be expressed mathematically as

$$\varepsilon(i) = y(i) - \hat{y}(i) \tag{23.7}$$

One can circumvent the problem of positive and negative residuals by defining the best-fit model as the model for which the sum of the squares of residuals is minimized. This procedure is called the *method of least squares* (first mentioned in Chapter 6) and is by far the most widely used guiding principle for statistical model fitting. The sum of the squared errors (residuals) (SSE) is defined as

$$\text{SSE} = \sum_{i=1}^{n}(y(i) - \hat{y}(i))^2 = \sum_{i=1}^{n}\varepsilon^2(i) \tag{23.8}$$

The minimum of SSE can be found analytically (for linear models) or, in general, for nonlinear models, numerically, using nonlinear optimization codes.

## 23.5 DATA RECONCILIATION

In the presence of random errors, it would not be expected that process measurements would necessarily conform to conservation laws. Due to the presence of both random and systematic errors, the data collected from a process (e.g., flow rates, temperatures, etc.) would not provide consistent information. Thus, material, and energy balance equations at steady-state may not be satisfied. To reduce the discrepancy in these balance equations, one needs a method that corrects and adjusts measured values to be consistent with the balance equations. This is known as *data reconciliation.*

---

Data reconciliation is the process of adjusting or reconciling the process measurements to obtain more accurate estimates of flow rates, temperatures, compositions, etc., that are consistent with material and energy balances. It takes raw data from a process plant to match material and energy balances, and is based on the minimization of the sum of the weighted squared error of the deviation between the measured variables and the estimated variables.

---

In the absence of gross errors, using Eq. (23.5), let us express the measurement vector as

$$y(i) = x(i) + \varepsilon(i) \tag{23.9}$$

where $y$ is the $(n \times 1)$ measurement vector, $x(i)$ represents the $(n \times 1)$ vector of true values and $\varepsilon$ is the $(n \times 1)$ vector of random measurement errors, with $i$ representing the sampling time scale, $i = 1, 2, \ldots, T$. It is usually assumed that:

- The expected value of $\varepsilon$ is the null vector, i.e., $E(\varepsilon) = 0$.
- The successive vector of measurements are independent, i.e.,

$$E(y^{\mathrm{T}}(i)y(i+1)) = 0, \quad \text{for any } i$$

- The covariance matrix of the measurement errors is known and positive-definite, i.e.,

$$\Psi = \mathrm{cov}(\varepsilon) = E(\varepsilon\varepsilon^{\mathrm{T}}) \tag{23.10}$$

For most applications, we also need to introduce a set of *constraints* that represent the additional information that must be taken into account through the process model equations (constraint equations).

$$\varphi(x_{\mathrm{m}}, x_{\mathrm{u}}) = 0 \tag{23.11}$$

where $x_{\mathrm{m}}$ and $x_{\mathrm{u}}$ denote the vectors of *measured* and *unmeasured* process variables, respectively, $x = [x_{\mathrm{m}}^{\mathrm{T}} \quad x_{\mathrm{u}}^{\mathrm{T}}]^{\mathrm{T}}$. In practice, these constraints are formed when some or all of the process variables are restricted to some relationships arising from the physical phenomena described by the model.

The data reconciliation problem can be stated as a weighted least-squares estimation problem subject to constraints:

$$\begin{aligned} &\min_{x_{\mathrm{u}}, x_{\mathrm{m}}} (y - x_{\mathrm{m}})^{\mathrm{T}} \Psi^{-1} (y - x_{\mathrm{m}}) \\ &\qquad \varphi(x_{\mathrm{m}}, x_{\mathrm{u}}) = 0 \\ &\qquad x_{\mathrm{m}}^{\mathrm{L}} \le x_{\mathrm{m}} \le x_{\mathrm{m}}^{\mathrm{U}} \\ &\qquad x_{\mathrm{u}}^{\mathrm{L}} \le x_{\mathrm{u}} \le x_{\mathrm{u}}^{\mathrm{U}} \end{aligned} \tag{23.12}$$

If the measurement errors are normally distributed, the solution of the above problem yields the *maximum likelihood estimates of process variables*, which means that they are of minimum variance and unbiased estimators.

For a process with simple linear model equations, and assuming all variables being measured, the data reconciliation problem can be reformulated as

$$\begin{aligned} &\min_{x} (y - x)^{\mathrm{T}} \Psi^{-1} (y - x) \\ &\qquad Ax = 0 \end{aligned} \tag{23.13}$$

where $A$ is a $(m \times g)$ matrix of known (model) constants. The solution is obtained using the Lagrange multiplier method.[3] The Lagrangian, $L$, for this problem is given as

$$L = \varepsilon^{\mathrm{T}} \Psi^{-1} \varepsilon - 2\lambda^{\mathrm{T}}(Ay - A\varepsilon) \tag{23.14}$$

Since $\Psi$ is positive-definite and the constraint (model) equations are linear, the necessary and sufficient conditions for minimization are

$$\begin{aligned}\frac{\partial L}{\partial \varepsilon} &= 2\Psi^{-1}\varepsilon + 2A^{\mathrm{T}}\lambda = 0\\ \frac{\partial L}{\partial \lambda} &= A(y - \varepsilon) = 0\end{aligned} \tag{23.15}$$

The solution yields

$$\begin{aligned}\lambda &= -(A\Psi A^{\mathrm{T}})^{-1} Ay\\ \varepsilon &= -\Psi\Psi^{\mathrm{T}}\lambda\end{aligned} \tag{23.16}$$

Finally, the estimate of the process variable ($\hat{x}$) is obtained as

$$\hat{x} = y - \Psi A^{\mathrm{T}}(A\Psi A^{\mathrm{T}})^{-1} Ay \tag{23.17}$$

**Example 23.5**

To illustrate the application of data reconciliation to linear systems, let us consider the problem presented by Romagnoli and Sánchez.[3] Four mass flows around a chemical reactor are measured. Two stream flows ($f_1, f_2$) are entering and two ($f_3, f_4$) are leaving the reactor. Three elemental balances are considered:

$$\begin{aligned}0.1f_1 + 0.6f_2 - 0.2f_3 - 0.7f_4 &= 0\\ 0.8f_1 + 0.1f_2 - 0.2f_3 - 0.1f_4 &= 0\\ 0.1f_1 + 0.3f_2 - 0.6f_3 - 0.2f_4 &= 0\end{aligned} \tag{23.18}$$

The data for this problem are given in Table 23.1. From the balance equations and the random characteristics of the measuring devices, the matrices $A$ and $\Psi$ are given by

$$A = \begin{bmatrix} 0.1 & 0.6 & -0.2 & -0.7\\ 0.8 & 0.1 & -0.2 & -0.1\\ 0.1 & 0.3 & -0.6 & -0.2 \end{bmatrix},$$

$$\Psi = \begin{bmatrix} 0.000289 & 0 & 0 & 0\\ 0 & 0.0025 & 0 & 0\\ 0 & 0 & 0.000576 & 0\\ 0 & 0 & 0 & 0.04 \end{bmatrix} \tag{23.19}$$

By applying Eqs. (23.16) and (23.17), we have the following measurement errors and variable estimates

$$\varepsilon = \begin{bmatrix} 0.0182\\ -0.0659\\ 0.0565\\ 0.026 \end{bmatrix}, \quad \hat{x} = \begin{bmatrix} 0.1676\\ 4.8594\\ 1.173\\ 3.854 \end{bmatrix} \tag{23.20}$$

A comparison between the original measurements and the new estimates, and the relation between measurement adjustment and the standard deviation (relative error found by data adjustment) is given in Table 23.2.

**TABLE 23.1**
**Data for Example 23.5**

| Flow Rates | Measured Values | True Values | Variances |
|---|---|---|---|
| $f_1$ | 0.1858 | 0.1739 | 0.000289 |
| $f_2$ | 4.7935 | 5.0435 | 0.0025 |
| $f_3$ | 1.2295 | 1.2175 | 0.000576 |
| $f_4$ | 3.88 | 4.00 | 0.04 |

**TABLE 23.2**
**Measured and Reconciled Values for Example 23.5**

| Flow Rates | Measured Value | Estimated Value | Relative Error |
|---|---|---|---|
| $f_1$ | 0.1858 | 0.1676 | 1.07 |
| $f_2$ | 4.7936 | 4.8594 | −1.31 |
| $f_3$ | 1.2295 | 1.1730 | 2.35 |
| $f_4$ | 3.88 | 3.854 | 0.13 |

## 23.6 ISSUES IN DATA RECONCILIATION

### 23.6.1 Process Model

Let us consider a chemical process containing $K$ units denoted by $k = 1, 2, \ldots, K$ and $J$ streams by $j = 1, 2, \ldots, J$. Then, the multicomponent mass and energy balance equations can be represented by

$$\sum_{j=1}^{J} a_{jk} x_j = 0 \tag{23.21}$$

where $x_j$ represents a process variable of stream $j$ (i.e., component flow rate, total flow rate, enthalpy, so on), and $a_{jk}$ is the element of the incidence matrix which denotes the topology of the units and streams. $a_{jk} = 1$ if stream $j$ is an input to a unit $k$ and $a_{jk} = -1$, if stream $j$ is an output from a unit $k$. In matrix notation, we can express Eq. (23.21) as

$$Ax = 0 \tag{23.22}$$

In this equation, $x$ is defined as the $(n + m \times 1)$-dimensional vector of process variables, which can be partitioned into two vectors as before,

$$x = \begin{bmatrix} x_\mathrm{m} \\ x_\mathrm{u} \end{bmatrix}$$

with $x_\mathrm{m}$ being the $m$-dimensional vector of measured variables and $x_\mathrm{u}$ representing the $n$-dimensional vector of unmeasured variables.

### Example 23.6

Consider the process with three units and eight streams shown in Figure 23.11. Performing a material balance around each unit, we have

$$\begin{aligned} f_1 + f_2 + f_3 - f_4 &= 0 \\ f_4 + f_5 - f_3 - f_6 &= 0 \\ f_6 + f_7 - f_8 &= 0 \end{aligned} \tag{23.23}$$

In compact notation, Eq. (23.23) can be expressed as follows:

$$\begin{bmatrix} 1 & 1 & 1 & -1 & 0 & 0 & 0 & 0 \\ 0 & 0 & -1 & 1 & 1 & -1 & 0 & 0 \\ 0 & 0 & 0 & 0 & 0 & 1 & 1 & -1 \end{bmatrix} \begin{bmatrix} f_1 \\ f_2 \\ f_3 \\ f_4 \\ f_5 \\ f_6 \\ f_7 \\ f_8 \end{bmatrix} = \begin{bmatrix} 0 \\ 0 \\ 0 \end{bmatrix} \tag{23.24}$$

Previously, we have considered a system of simple linear balance equations, but practical situations may be different. For example, one device may provide the measurement of mass flow rate and separate analyzers may be used to determine the composition, etc. Thus, in general, the component balance equations around unit $k$, may take the following form:

$$\sum_{j=1}^{J} a_{jk} f_j x_{ji} = 0, \quad i = 1, 2, \ldots, I \tag{23.25}$$

where $f_j$ denotes the flow rate of stream $j$ and $x_{ji}$ the molar (or mass) fraction of the $i$th component in the $j$th stream. Often, the balance equations are expressed as

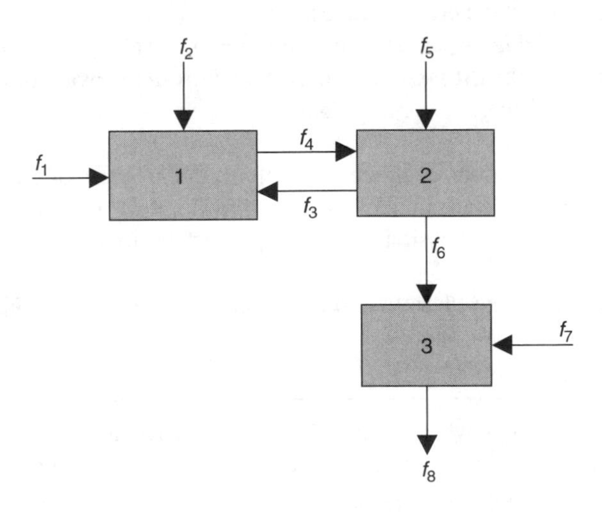


**FIGURE 23.11** Process flow diagram for Example 23.6.

a set of nonlinear relationships, i.e.,

$$\Gamma(v) = 0 \tag{23.26}$$

where $v$ is defined as

$$v = \begin{bmatrix} x_{\mathrm{m}} \\ f \end{bmatrix}$$

### 23.6.2 Classification of Process Variables

The number of measured variables in the process is determined as a function of cost, convenience, and technical feasibility by those responsible for instrumentation. Thus, variables throughout the plant may be classified as either measured or unmeasured variables (see Figure 23.12).[3]

Measured variables are classified as redundant and nonredundant variables:

- *Redundant measurements*: A measured variable is redundant if it is determinable from mass and energy balance in addition to being directly measured.
- *Nonredundant (just-measured) measurements*: A nonredundant variable is simply measured, and cannot be computed from balance equations using the other measured variables.

Unmeasured variables are classified as determinable and indeterminable variables:

- *Determinable variables*: An unmeasured variable is determinable or observable, if it can be evaluated from the available measurements using mass and energy balances.
- *Indeterminable variables*: An unmeasured variable is indeterminable or unobservable, if it cannot be evaluated from the available measurements using mass and energy balances.

Given these definitions of process variables, the problem of classifying variables and estimating determinable variables form part of the data reconciliation problem. A complete discussion on this topic can be found in Romagnoli and Sanches.[3]

Associated with the classification problem, we can define the concept of redundancy.

---

A system is redundant when the data (information) available exceed the minimum amount necessary for a unique determination of the independent (unmeasured) variables.

---

It is clear from the previous discussion that only the overmeasured variables provide a special redundancy that can be exploited for correcting the data. Redundancy is useful when there are biases in the measurements or imperfections in the model of the physical situation under consideration (e.g., unaccounted leaks in the units, etc.).

---

Any system, composed of measurements and balance equations, which is redundant admits decomposition in its redundant and nonredundant parts.

---

**Example 23.7**

In Example 23.6, consider that the flow rates of streams 1, 2, 5, 7, and 8 are measured. We can use the third balance equation in Eq. (23.23) to calculate $f_6$, but we cannot use the first and the second equations to calculate $f_3$ and $f_4$. However, there is one equation, which is a combination of the previous three equations (global material balance around the three units) that will contain only the measured variables. Thus, according to our classification, we have (see Figure 23.12):

- Overmeasured variables: $f_1, f_2, f_5, f_7$, and $f_8$
- Just-measured variables: none
- Determinable variables: $f_6$
- Indeterminable variables: $f_3$ and $f_4$

### 23.6.3 Linear Data Reconciliation with Unmeasured Variables

In most cases, some process variables will be unmeasured and would need to be estimated. The solution to the data reconciliation problem involving unmeasured variables is based on decomposing the problem into two subproblems,[3] considering measured and unmeasured variables. The problem is divided into estimation of redundant measurements and the calculation of unmeasured observable variables.

In terms of measured variables ($x_m$) and unmeasured variables ($x_u$), the constraint equations are written as follows:

$$A_1 x_m + A_2 x_u = 0 \tag{23.27}$$

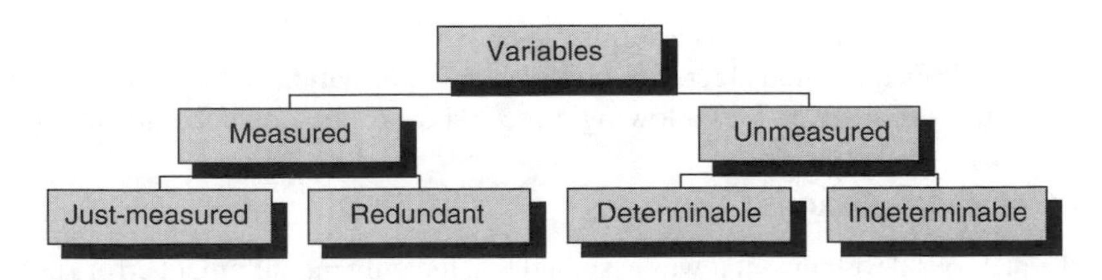


**FIGURE 23.12** Classification of process variables.

Decomposition of the data reconciliation problem with unmeasured variables is achieved by the Q-R decomposition.[3] Q-R decomposition is carried out on the matrix $A_2$ to decouple the unmeasured variables from the measured ones. The Q-R decomposition allows three new matrices to be defined: $Q_u$, $R_u$, and the permutation matrix $\Pi_u$ such that

$$A_2\Pi_u = Q_u R_u A_2 \tag{23.28}$$

Further, $Q_u$ (an orthogonal matrix) and $R_u$ (a non-singular upper-triangular matrix) may be decomposed as

$$Q_u = [Q_{u1} \quad Q_{u2}], \qquad R_u = \begin{bmatrix} R_{u1} & R_{u2} \\ 0 & 0 \end{bmatrix} \tag{23.29}$$

$R_u$ contains the topological information about the system in terms of the available measurements. Similarly, the vector of unmeasured variables is partitioned into two subsets, thereby classifying unmeasured variables into redundant and nonredundant variables.

$$\Pi_u^T x_u = \begin{bmatrix} x_{u,\,r_u} \\ x_{u,\,n-r_u} \end{bmatrix} \tag{23.30}$$

The first subproblem arising from the decomposition is to estimate the reconciled values for the redundant measured variables. Performing reconciliation on the measured variables and the constraints,

$$Q_{u2}^T A_1 x_m = G_{x_m} x_m = 0 \tag{23.31}$$

yields the solution

$$\hat{x}_m = y - \Psi G_{x_m}^T (G_{x_m} \Psi G_{x_m}^T)^{-1} G_{x_m} y \tag{23.32}$$

The second subproblem arising from decomposition is to estimate the unmeasured variables. Unmeasured variable estimation is dependent on instrument configurations around the plant and the structure of the process. For the unmeasured variables we have, in general,

$$x_{u,\,r_u} = -R_{u1}^{-1} Q_{u1}^T A_1 \hat{x}_m - R_{u1}^{-1} R_{u1} x_{u,\,n-r_u} \tag{23.33}$$

where the component $x_{u,n-r_u}$ is arbitrarily set.

The Q-R factorization algorithm provides the information about the estimability conditions of the variables allowing direct classification and decomposition.

### 23.6.4 Gross Errors

In the previous development it was assumed that the only measurement errors present in the data were normally distributed with zero mean and known covariance.

In practice, instrument bias may not be adequately compensated; measuring devices may malfunction, etc. We shall refer to these biases as *gross errors*.

---

The presence of gross errors *invalidates the statistical basis of the data reconciliation procedures* and it is impossible to prepare an adequate process model on the basis of erroneous measurements.

---

Consequently, we need to check for the presence of gross errors in the measurement data. One way to do this is by checking the closure of the balance equations. We can formulate the following test function

$$h = \delta^{\mathrm{T}}\phi^{-1}\delta \tag{23.34}$$

In this equation, $\delta$ is the residual in the unsatisfied balances, $\phi$ the covariance matrix of the residual ($\phi = \mathrm{A}\Psi\mathrm{A}^{\mathrm{T}}$) and A is the system Jacobian matrix. Since $\delta$ has $n$ elements ($n$ equations), $h$ will have a $\chi^2$ distribution with $n$ degrees of freedom (see Appendix D). Thus, at a specified level of significance, we have

$$\mathrm{Prob}(h > \chi^2_{1-\alpha}(n)) = \alpha \tag{23.35}$$

This means that, for a system containing $n$ degrees of freedom, the probability that a particular $h$ exceeds the critical $\chi^2$ value at a $(1 - \alpha)$ level of significance is $\alpha$. Hence, if $h$ exceeds the critical $\chi^2$ value of $\chi^2_{1-\alpha}(n)$, it is unlikely that the residual in the balances is due to random errors.

A number of techniques for detecting and identifying gross measurement errors have been developed during the last two decades and are described elsewhere.[2]

---

1. The Global Test provides us with a test for the consistency of a set of measurements in the presence of gross errors.
2. We need only to preassign an allowable error probability, which gives us a critical value of $h$.

---

**Example 23.8**

To illustrate the use of test functions for the detection of gross errors, we will consider the problem presented in Example 23.5. The residual in the balances and their corresponding variances are given by

$$\delta = \begin{bmatrix} -0.0672 \\ -0.0059 \\ -0.0571 \end{bmatrix}, \quad \varphi^{-1} = \begin{bmatrix} 483.7 & -242.0 & -1339.8 \\ -242.0 & 5890.6 & -2071.6 \\ -1339.8 & -2071.6 & 5506.0 \end{bmatrix}$$

By direct application of the global test defined in Eq. (23.34) we have

$$h = \delta^{T}\phi^{-1}\delta = 8.5347$$

If we consider an error probability of 0.10, the critical value for $h$ with three degrees of freedom is $h_c = 6.251$. Since, in this case $h > h_c$, we can say that the inconsistency is significant at an error probability level of 0.10, and gross errors are present in the data set.

## 23.7 SUMMARY

During operation of a chemical plant it is a common practice to obtain data from the process, such as flow rates, compositions, pressures, and temperatures. The numerical values resulting from the observations often *do not provide consistent information* (consistent with the conservation equations, i.e., mass and energy balances are not satisfied exactly). The treatment of plant data involves a set of tasks that allows the processing of data arising from different sources, such as the on-line data acquisition system and the laboratory, as well as direct reading from the operators and transforming them into reliable process information. This information can then be used by the company for different purposes.

In this chapter, we have discussed the issues associated with the processing of the data. The discussion was placed in the context of modeling process data (statistically) and based on these models we developed techniques to transform the data into reliable and consistent information. The goal was to introduce some of the typical statistical concepts, to distinguish typical anomalies on the process data, with emphasis on the problem of data reconciliation and rectification. It was shown that the ideas discussed in the chapter allow us not only to correct the data but also to detect and identify faulty sensors, which may ultimately lead to degradation, or violation of the product quality specifications. A number of concepts were introduced, starting from the well-established ideas of missing points and outliers, moving up to the least-squares techniques for process data reconciliation, and statistical tests for sensor validation and gross error detection. These will be further supplemented with the introduction, in the next chapter, of process and control monitoring ideas, which, together, as discussed in the previous chapter, constitute the backbone of modern integrated control systems.

## REFERENCES

1. Rousseeuw, P.J. and Leroy, A.M., *Robust Regression and Outlier Detection*, Wiley, New York, 1987.
2. Alwan, L.C., *Statistical Process Analysis*, McGraw-Hill, New York, 2002.
3. Romagnoli, J.A. and Sánchez, M., *Data Processing and Reconciliation for Chemical Process Operation*, Academic Press International, New York, 2000.

# 24 Process Monitoring

The recent strides in the understanding of advanced statistical techniques resulted in the revival of efforts to apply such knowledge across a wide range of fields. The chemical process industry has been on the receiving end of such advances under the rubric of *technology expansion*. The goal is, for the most part, to monitor the performance of a process over time, with emphasis on the detection of unnatural events, which ultimately lead to degradation or violation of the product quality specifications. The isolation of these events leads to the allocation of a cause, which, if eliminated, results in an improved reliability and control of the final product.

This chapter presents a study of the techniques used to assess the status of process operation. These techniques depend largely on statistical concepts and help quantify the probability of observing process behavior that does not conform to the expected design behavior. In this effort, it is crucial to make best use of the process data to detect process faults and identify their causes. While some abnormal process behavior may be the result of unexpected changes in the operating conditions, others may arise from the poor performance of the control system that is designed to counteract external influences. In other words, the control system may have been poorly designed or become invalid when operating conditions vary. This, then, may require the modification of the control system or some of its components (e.g., parameter tuning). In this chapter, both process monitoring and control system monitoring techniques are reviewed, as they are complementary functions to guide the process engineers toward safe and reliable operation.

## 24.1 STATISTICAL PROCESS CONTROL

The discovery of SPC concepts is credited to Dr. Walter Shewhart.[1] Using sampled data and a simple statistical tool, he developed the *control chart*, and demonstrated that it was easy to determine whether a process was being affected by extraneous influences. If the source of these influences could be identified and subsequently eliminated, the process would be restored to its normal state.

Shewhart noted that every process has some average condition and also some random variations around that average. Such a condition can be represented on a histogram showing its related frequency (probability) distribution. If the collected data conform to a *normal* (Gaussian) distribution curve (see Appendix D), then, the majority of the observations are made around the mean, and observations that are far from the mean appear less frequently.

While not all sampled data from a process may produce a normal distribution, Shewhart noted that the *averages* of the data do indeed conform to a normal distribution. Such distribution curves can deviate from normal as to their mean value, spread, and shape. To quantify these terms (spread and shape) the statistical term *standard deviation* is used (see Appendix D). The American Society of Quality Control suggests using the sample standard deviation (denoted by $s$) in place of the root-mean-square standard deviation (i.e., $\sigma$) for SPC applications.

When referring to the traditional Gaussian distribution, $\sigma$ has a well-defined meaning; in fact, $\sigma$ and the sample mean define the distribution curve explicitly. In addition to this curve definition, some sample population proportions have also been established. It is known, for instance, that 68.26% of all the samples of a population will fall within $\pm 1\sigma$ (1 sigma) of the mean, 95.46% with $\pm 2\sigma$ (2 sigma) and 99.73% within $\pm 3\sigma$ (3 sigma) of the mean (Appendix D). Shewhart demonstrated that plotting sample averages would produce a normal distribution, and one can then establish a strategy for quality monitoring:

---

Once statistical control is established, any data exceeding 3 sigma (or any other predefined threshold) represent a situation where the process is out of target (statistical control).

---

The observations made by Shewhart led him to the development of what is known today as *control charts*.

### 24.1.1 Control Charts

A control chart is a plot of sampled data over time in a format that renders an easy identification of *in-control* and *out-of-control* states. Principally, there are two types of control charts: those that deal with continuous measured variables and those that are based on attribute (discrete) data. Only control charts for continuous measured variables will be discussed here.

From a historical point of view, the most common control chart for continuous measured variables is the *XBAR-R* or the Shewhart chart. The terminology *XBAR-R* is derived from *XBAR* being the mean of the data, $\overline{X}$, and $R$ the range, or difference between the highest and lowest values in a data set. A graph of each (*XBAR* and $R$) is plotted with respect to the same time basis on the same chart. Control limit lines, displaying 3 sigma deviations from the mean are drawn on the plots, and, strictly speaking, for a process to be in statistical control, the data points must lie within the control limits.

**Example 24.1**

The control chart technique is applied to the plant data available from a biological nutrient water treatment pilot plant.[2] Figure 24.1 is a Shewhart chart for the measurements of anoxic $NO_x$ within the pilot plant. The upper and lower control limits (UCL and LCL) represent 3 sigma deviations from the sample mean.

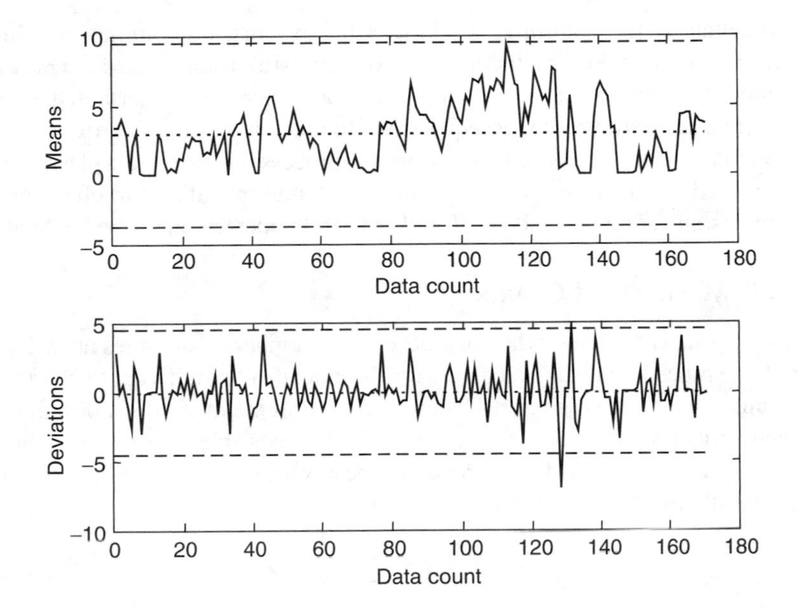


**FIGURE 24.1** Shewhart chart (*XBAR* and *R*) for anoxic $NO_x$.

Figure 24.1 demonstrates the ease with which trends, changing conditions, and out-of-control operation can be observed by simple visual inspection. Given the 3 sigma limit, one can conclude that this process is in-control.

The 3 sigma limits are sometimes considered to be too wide, because of the risk of not raising an alarm when the process is actually off-target. Some heuristic rules have been developed at Western Electric Company[3] to help interpret these charts more conservatively.

### 24.1.2 Control Chart Interpretation

Individual inspection of the *XBAR* and *R* charts highlights different aspects of the process being considered. The *XBAR* chart indicates the centering of the output data, while the *R* chart is indicative of data variability. As previously mentioned, if a point is located outside the control limits, then the process is considered as being in a state of statistically abnormal operation. The *R* chart usually flags first when a new event enters the sample space, and as a general rule-of-thumb, the *R* chart must be in control before the *XBAR* chart is considered.

Some of the early indications, which *XBAR*-*R* charts can give, can be summarized through the following heuristic rules:[3]

- A general upward or downward trend with eight or more points in succession. This indicates the effect of a drift in the operation.
- Having seven or more successive points on either side of the mean. This implies a shift in the direction of the process.
- Having a recurring cycling pattern. This is an indication of some repetitive abnormal influence that is affecting the process.

Although useful, control charts have a number of shortcomings as alluded earlier. For example, SPC methods are most successful when applied to processes that produce data, which are independent and normally distributed. These assumptions are hardly valid, especially due to the dynamic and highly nonlinear nature of chemical processes. In addition, for processes where multiple variables are measured and monitored (multivariate), using multiple Shewhart charts would be cumbersome and may lead to false alarms and undetected events.

### 24.1.3 Multivariate Charts

For multivariate processes, when one observes a number of variables at each time period, the procedure discussed above can be extended to multivariate charts. Let us assume that we have an $n \times 1$ vector of measurements $X$ of $n$ normally distributed variables with a covariance matrix $\Sigma$. It is possible to perform a statistical test to check whether the mean ($\mu$) of these variables is at the desired target $X_T$ by calculating the $\chi^2$ statistic (Appendix D):

$$\chi^2 = (X - X_T)^T \Sigma^{-1}(X - X_T) \tag{24.1}$$

This statistic will be distributed as a central $\chi^2$ distribution with $n$ degrees of freedom if $X_T = \mu$. This allows us to construct a multivariate $\chi^2$ control chart by plotting $\chi^2$ vs. time. UCL will be defined by $\chi_a^2$ where $\alpha$ is the level of significance to perform the hypothesis test (e.g., $\alpha = 0.05$ or 95% confidence limit).

There will be situations when the in-control covariance matrix is not known and has to be estimated.[4] The estimation uses the sample of $n$ past measurements as

$$S_{kj} = \frac{1}{n-1}\sum_{i=1}^{n}(X_{ik} - \bar{x}_k)(X_{ij} - \bar{x}_j)^T \tag{24.2}$$

where $\bar{x}$ represents the estimation of the mean through a finite number of previous samples (column-wise), and $x_i$ is the $i$th column of the data matrix $X$. The Hotteling statistic associated with the new sampled data can be expressed as

$$T^2 = (X - X_T)^T S^{-1}(X - X_T) \tag{24.3}$$

This statistic is plotted vs. time (sample count) with the UCL defined as follows (lower limit, of course, is zero):

$$T_{UCL}^2 = \frac{(n-1)(n+1)k}{n(n-k)} F_\alpha(k, n-k) \tag{24.4}$$

$F_\alpha(k, n-k)$ is the $100\alpha\%$ upper critical point of the $F$ distribution with $k$ and $n-k$ degrees of freedom (Appendix D).

For most practical applications, however, the number of variables involved is quite large. More importantly, the majority of these variables would be highly correlated with one another. This leads to an almost singular covariance matrix, $\Sigma$.

A possible solution to this problem is to use principal component analysis (PCA) techniques that also help reduce the dimensionality of the problem.

## 24.2 PRINCIPAL COMPONENT ANALYSIS (PCA)

PCA[5] belongs to a group of multivariate statistical techniques, which are broadly classified as *dimensionality reduction* methods. Recently, such methods generated a high level of interest from researchers and practitioners alike. One can visualize PCA as casting a shadow from a high-dimensional data space into a low-dimensional data space:[6]

---

> PCA can then be considered as a way of finding out the most useful and informative viewpoint from which the data points can be visualized. When found, it then casts a shadow giving an undistorted image from this viewpoint. Further shadows can also be cast, each of them using viewpoints that are 90° to all previous viewpoints.

---

Let us illustrate this concept using a two-dimensional data space, each axis representing one variable (Figure 24.2a). As shown in Figure 24.2b and Figure 24.2c, a new coordinate space will be created based on the overall mean of the data, and by rotating the axes so that the principal axis (first axis) runs along the cluster of observations. This rotation is performed to minimize the distance of observations to the axis.

The next step is to construct the second axis, which should be placed 90° to the first (principal) axis. This creates a set of *orthogonal* coordinate axes. It is noted that by placing the first axis along the direction of the maximum variance, we are able to account for the greatest percentage of the overall variance, thus, staying consistent with the notion of dimensionality reduction, while maintaining the maximum information content.

### 24.2.1 CALCULATION OF PCA

As discussed above, the extent of rotation for each variable needs to be determined. The extent of rotation is referred to as the *loadings* (or weightings) for each variable.

---

QUESTION: How do we find the best possible values for the loadings given to each variable?

ANSWER: They are given based on the eigenvectors and eigenvalues of the data covariance matrix.

---

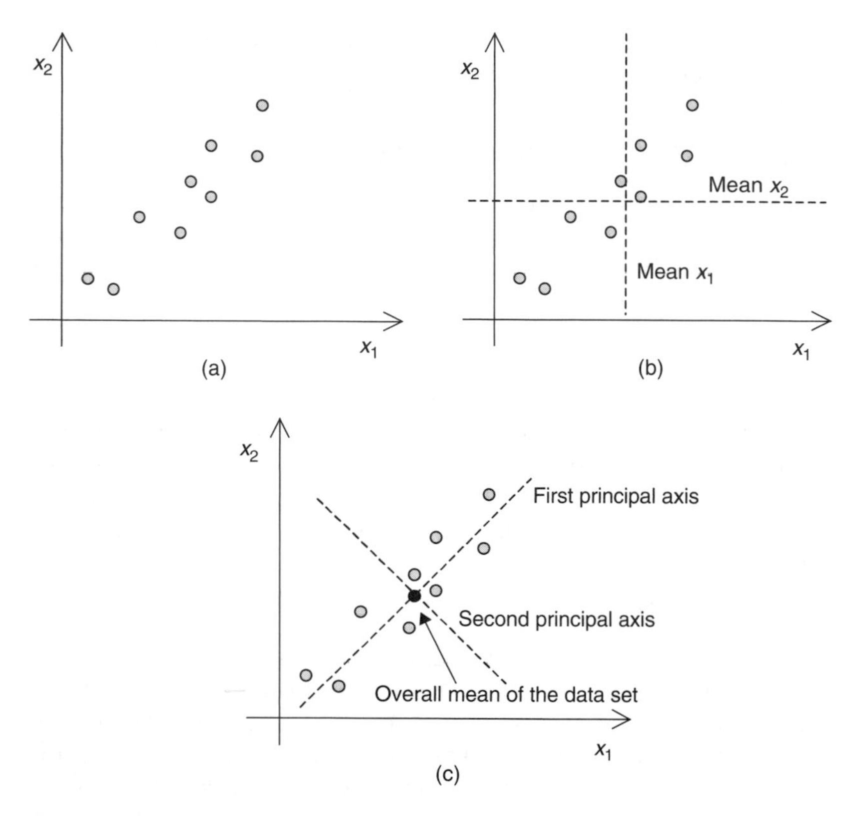


**FIGURE 24.2** Rotation of major axes in PCA.

Let us start with the sample covariance matrix $S$ for an $n$-dimensional problem. If the covariance matrix elements are not zero, a linear relationship exists between the two variables. Given an $n \times 1$ vector of measurements $X$ and their covariance matrix ($S$) the eigenvalue decomposition of the covariance matrix can be stated as

$$S = P\Lambda P^{\mathrm{T}} \tag{24.5}$$

where $P$ represents the matrix of eigenvectors and $\Lambda$ the diagonal matrix of eigenvalues, in descending order of magnitude. The $i$th column of $P$ is a unit vector ($|p_i| = 1$) defining the PCA coordinate system in which the data matrix will now be expressed. By projecting the data onto the new coordinates, we obtain the *score vector*, $t_i$, that measures the distance along the direction $p_i$:

$$t_i = p_i^{\mathrm{T}} X \tag{24.6}$$

The amount of variance captured by each PC direction is proportional to its corresponding eigenvalue. Accordingly, the first eigenvector direction (*loading*) $p_1$, associated with the eigenvalue having the largest magnitude, captures the most variance. The calculation of an element of the score vector is performed using a linear combination of the measured variables,

$$t_1 = p_{11}x_1 + p_{21}x_2 + \cdots + p_{n1}x_n \tag{24.7}$$

As can be seen, the weights used in the linear combination are taken from the corresponding column of the loading matrix $P$. The second PC direction, associated with the next largest eigenvalue, is then defined as

$$t_2 = p_2^{\mathrm{T}}X$$

and has the next greatest variance, subject to $|p_2| = 1$, with the second score vector, $t_2$. This direction should also satisfy the condition that it is uncorrelated with (orthogonal to) the first PC direction ($p_1 \perp p_2$). One can define subsequent PC directions (up to $n$) in a similar manner.

In summary, PCA decomposes the data (observation) matrix $X$ as

$$X = TP^{\mathrm{T}} = \sum_{i=1}^{n} t_i p_i^{\mathrm{T}} \tag{24.8}$$

If only the first $k$ PC directions are kept, this yields an approximation of the data matrix $X$, or the so-called PCA model:

$$X = \sum_{i=1}^{k} t_i p_i^{\mathrm{T}} + \sum_{i=k+1}^{n} t_i p_i^{\mathrm{T}} = \tilde{X} + E \tag{24.9}$$

The matrix $E$ contains the error (or the residual) information. In most practical applications, only a few PC directions are required ($k \ll n$) to capture the majority of the variation in the data set.

The number of PC directions $k$ should be chosen so that the PCA model captures an acceptable degree of variance. However, in an effort to improve the degree of variance captured, one needs to be cautious about including PC directions that only capture the noise variance. In other words, as $k \rightarrow n$, the model starts to include also the variance information that explains the measurement noise. The residual should only contain unstructured (uncorrelated) information associated with random variations such as noise, and the structured (correlated) information that pertains to the relationship between the measured variables and should be retained in the approximate PCA model. There are many methods that can guide the user in selecting the number of PC directions. Among those, one can cite the SCREE plots that suggest finding the place where the smooth decrease of eigenvalues of the covariance matrix appears to level off to the right of the plot. Another method is to calculate the PRESS (Prediction residual error sum of squares) for every possible PC. This is calculated by building a calibration model with a number of PCs, and then predicting some samples of known value (usually the training set data itself) against the model. The sum of the squared difference between the predicted and known values gives the PRESS value for that model.

### Example 24.2

Measurements of chloride, arsenic, and nitrates in groundwater over time are given in Table 24.1.[7] We shall use this data set to illustrate the calculation of principal components.

For this data matrix (with three variables and twelve observations), the vector of sample means is

$$\bar{x} = [\bar{x}_1 \quad \bar{x}_2 \quad \bar{x}_2] = [91.4167 \quad 0.0900 \quad 15.0833]$$

And the variable standard deviations are given as

$$\sigma = [\sigma_1 \quad \sigma_2 \quad \sigma_2] = [22.0102 \quad 0.0367 \quad 3.5280]$$

---

The PCA technique depends critically on the scales used to measure the variables. If the constructed data matrix contains variables of no visible relation, then, the structure of the latent variables derived from this data set would depend essentially on the arbitrary set of units of measurements. A common approach is to scale the variables to zero mean and unit variance.

---

In this example, the data matrix can be scaled using the variable means and the standard deviations found above. The unscaled and scaled variables are displayed in Figure 24.3. One can see that, without scaling, the variation in one of the variables would have dominated the modeling.

Using the scaled data matrix, we can now find the sample covariance matrix:

$$S = \begin{bmatrix} 1.0 & 0.6114 & 0.7207 \\ 0.6114 & 1.0 & 0.9413 \\ 0.7207 & 0.9413 & 1.0 \end{bmatrix}$$

**TABLE 24.1**
**Data for Testing**

| Sample No. | Chloride (mg/L) | Arsenic (mg/L) | Nitrate (mg/L) |
|---|---|---|---|
| 1 | 112 | 0.07 | 15 |
| 2 | 87 | 0.06 | 13 |
| 3 | 93 | 0.06 | 14 |
| 4 | 115 | 0.09 | 16 |
| 5 | 126 | 0.17 | 23 |
| 6 | 105 | 0.15 | 20 |
| 7 | 72 | 0.10 | 17 |
| 8 | 61 | 0.05 | 12 |
| 9 | 57 | 0.07 | 11 |
| 10 | 75 | 0.07 | 12 |
| 11 | 89 | 0.09 | 14 |
| 12 | 105 | 0.10 | 15 |

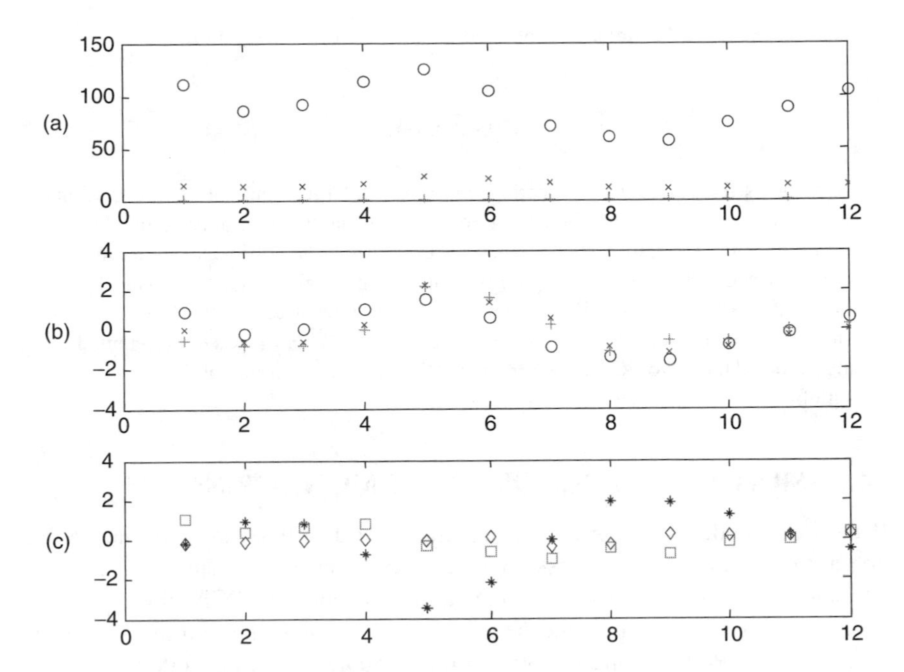


**FIGURE 24.3** Unscaled (a) and scaled (b) variables. (o) chloride, (+) arsenic, and (x) nitrate. The scores (c) with (∗) first score, (□) second score, and (◇) third score.

The next step involves the calculation of the eigenvalues of $S$. The eigenvalue decomposition of $S$ yields the following eigenvalues and the eigenvectors (loadings):

$$\Lambda = \begin{bmatrix} 2.5243 & 0 & 0 \\ 0 & 0.4293 & 0 \\ 0 & 0 & 0.0464 \end{bmatrix}$$

$$P = [p_1 \quad p_2 \quad p_3] = \begin{bmatrix} -0.5262 & 0.8358 & 0.1567 \\ -0.5895 & -0.4913 & 0.6411 \\ -0.6218 & -0.2450 & -0.7513 \end{bmatrix}$$

Generally speaking, the procedure just implemented is nothing more than a principal axis rotation. The results (load variables) are used in the transformation from measured data to latent variables or principal components. The score vectors can be calculated (see Eqs. [24.6] and [24.7]), and plotted in Figure 24.3 as well. These are the new variables in the transformed space. The amount of variance explained by each PC can be calculated using the eigenvalues:

$$\%\text{ variance explained} = \frac{\lambda_i}{\sum \lambda_j}$$

For this example, the variance explained by the first PC is given by:

$$\frac{2.5243}{2.5243 + 0.4293 + 0.0464} = 84.14\%$$

This shows that the first PC direction contains the largest amount of information. This is not unexpected since Figure 24.3c indicates that the general trend in all three variables is pretty similar; thus, a single variable may be sufficient to capture the trend in each. This is the main premise for dimensionality reduction using PCA. Figure 24.4 shows the SCREE plot along with the plot of cumulative percent variance captured that, together, would help decide the number of PCs to be retained in the model. Thus, two PCs appear to be sufficient to build a model that captures a significant amount of the data variance.

## 24.3 MULTIVARIATE PERFORMANCE MONITORING

If the PCA model is constructed using historical data that represent the normal operational status of the process, or in-control behavior, the current and future operational status can be referenced against the obtained PCA model. This is accomplished by projecting the new observations onto the coordinate system defined by the PCA loading vectors to obtain the new scores as follows:

$$t_{\text{new},i} = p_i^{\text{T}} X_{\text{new}} \tag{24.10}$$

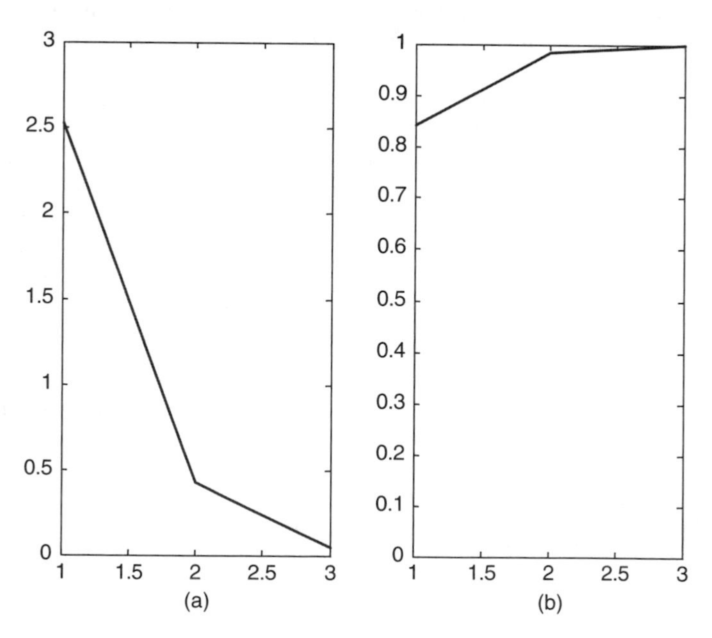


**FIGURE 24.4** SCREE plot (a) and percent variance captured (b).

The difference between the model prediction and the actual observation is defined as the error (residual),

$$E_{\text{new}} = X_{\text{new}} - \hat{X}_{\text{new}} \tag{24.11}$$

Then, by rearranging Eqs. (24.6) and (24.9), the PCA model prediction can be expressed as

$$\hat{X}_{\text{new}} = X_{\text{new}} PP^{\mathrm{T}} \tag{24.12}$$

Accordingly, the residual takes the form

$$E_{\text{new}} = X_{\text{new}} (I - PP^{\mathrm{T}})$$

As one can see in Eq. (24.12), if all the eigenvectors (loadings) are retained for the model, the error approaches a matrix of zeros and the data matrix is reconstructed perfectly.

The approach for a multivariate performance monitoring strategy is somewhat similar to univariate methods. Instead of bounding by UCL and LCL, a reference or a normal operating region (NOR) is identified through the use of PCA. In fact, the Hotteling's statistic (Eq. [24.3]) defines ellipses of constant probability for a multivariate Gaussian distribution. This region is based upon historical data recorded during known normal operating periods. This produces a fault-free region or a space that exhibits only common causal variations. The process of defining and constructing an NOR is initiated by splitting the available historical data into blocks of *good* and *bad* operation (a *start-up* block may also be included).

### Example 24.3

For the biological nutrient removal process, the measurements can be broadly split according to the nutrient removal type, i.e., nitrogen removal and phosphorus removal.[2] Considering the case of nitrogen removal first, the validation criteria for some key nitrogen removal parameters are:

- Clarifier effluent, $NH_3 - N$ $< 0.5$ mg/L
- Clarifier effluent, total $< 10$ mg/L
- Anoxic zone effluent, $NO_x$ $< 1$ mg/L
- Return activated sludge (RAS), $NO_x$ $< 0.5$ mg/L

The task of defining an NOR for nitrogen removal involves locating the days in which the above criteria were all met. This is done to ensure that the data used to construct the NOR does not contain any abnormal periods of time where the validation criteria are violated.

Following the scaling of the data matrix, PCA is used to project the measurement variables onto the new coordinate space that is spanned by the *latent* variables. Now that the PCs have been computed, the basis for multivariate monitoring charts can

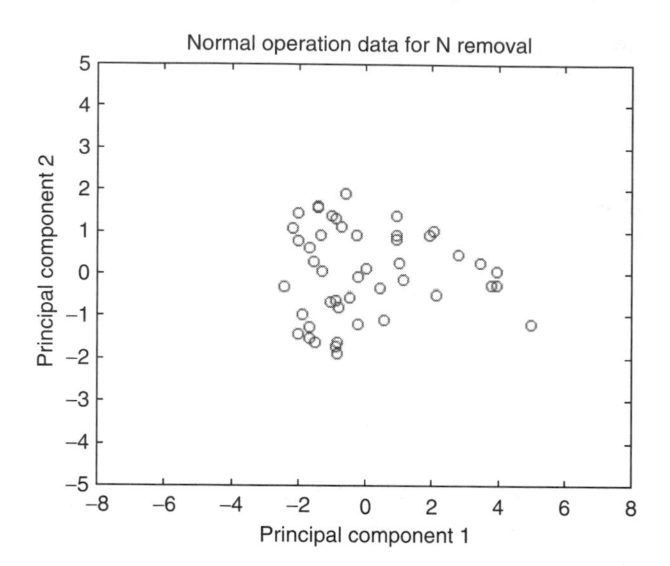


**FIGURE 24.5** Data projected onto first two PC directions.

be established. These multivariate charts are constructed using the first two PCs. PC 1 is generally plotted on the $x$-axis, with PC 2 along the $y$-axis. Figure 24.5 demonstrates the resulting plot.

Figure 24.5 reveals that the data points representing normal operational conditions are concentrated in a relatively small region, with no apparent outliers.

Similar to the UCL and LCL used in the univariate *XBAR-R* charts, an algorithm for defining and constructing a bounded region around the NOR is necessary. Next, one such approach is discussed.

### 24.3.1 Elliptical Normal Operation Region

A simple approach to construct a bounded region around the normal operating data is to draw an ellipse whose boundary encloses the *good* data. The construction and use of an ellipse is based around a few fundamental properties of elliptical functions, the first of these being the defining equation:

$$\frac{x^2}{a^2} + \frac{y^2}{b^2} = 1 \tag{24.13}$$

This is for an ellipse centered at the origin (0,0) of a Cartesian geometry, where $a$ is the semimajor axis length, and $b$ is the semiminor axis length ($a > b$) (Figure 24.6).

The problem of defining an ellipse suitable for any normal data (regardless of measurement type), thus, reduces to finding an expression that links $a$ and $b$ to the predetermined latent variables or principal components. In linking these two objects, it is expected that a statistical parameter (possibly the standard deviation)

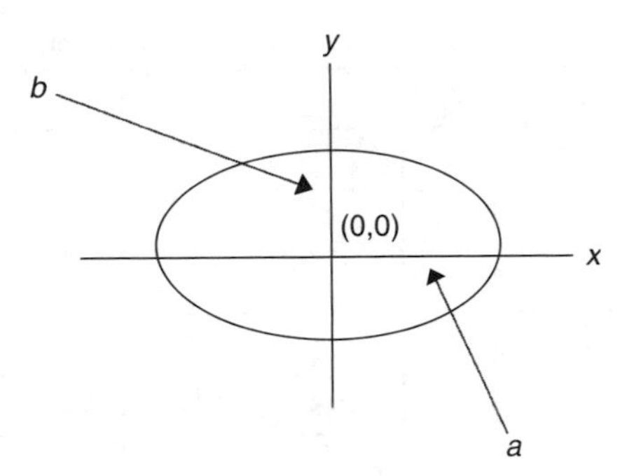


**FIGURE 24.6** An ellipse centered at the origin.

can be used to define the bounded region. The solution to this problem is the covariance matrix. The covariance matrix for a two-dimensional problem is stated as

$$S = \begin{bmatrix} s_1^2 & s_{21} \\ s_{12} & s_2^2 \end{bmatrix} \tag{24.14}$$

Note that we have $s_{12} = s_{21}$. Consider the matrix of latent variables, $Z = [z_1, z_2]$, then the covariance of $Z$ will produce a $2 \times 2$ matrix with the diagonal elements being the variance of the latent variables. Thus $a$ and $b$ can be defined in terms of these diagonal elements, such that

$$a = k\sqrt{s_1^2}, \quad b = k\sqrt{s_2^2} \tag{24.15}$$

where $k$ is a parameter associated with the *confidence* of the bounded region. As in the univariate example where 3 sigma was used to define the UCL and LCL, $k$ is also generally specified as 3. Given $a$ and $b$, the problem now becomes a simple rearrangement of the equation of an ellipse to get $y = f(x)$.

### Example 24.4

The application of the NOR definition to the previous example of normal data yields the plot in Figure 24.7. The plots show the results where three standard deviations are used as the bounding factor. The simplistic nature of this approach makes it excellent as a first estimate for determining the NOR.

The score plots demonstrate that while the method is effective in displaying the NOR, the region does not include all the data. This construction is described as parametric, i.e., it is assumed that the data belong to a known family of distributions; in this case, by setting 3 sigma as the bounded condition, a normal distribution is assumed.

Another method for defining the NOR is based upon estimating the PDF of the variables. A number of algorithms can be employed to achieve this result; for

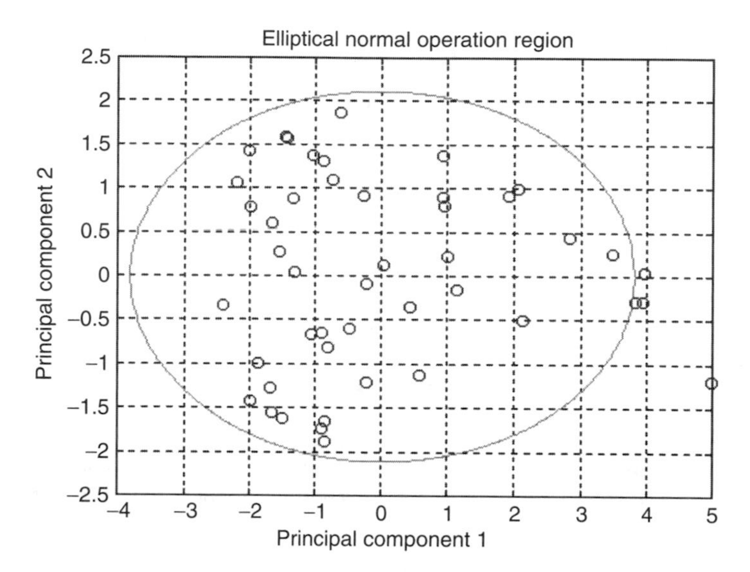


**FIGURE 24.7** Elliptical NOR for N-removal data.

example, a kernel estimator allows the definition of this region based on the available data.[8]

### 24.3.2 $T^2$ AND SPE CHARTS

Multivariate control charts based on the Hotelling's $T^2$ can be used to monitor the process status by plotting the $T^2$ based on the first $k$ PCs, where

$$T^2 = \sum_{i=1}^{k} \frac{t_i^2}{s_{t_i}^2} \tag{24.16}$$

and $s_{t_i}^2$ is the estimated variance of $t_i$. By scaling each $t_i^2$ by the reciprocal of its variance, each PC term plays an equal role in the computation of $T^2$, irrespective of the amount of variance it explains in the $X$ matrix.

Monitoring the process via $T_k^2$ based on the first $k$ PCs is, however, not sufficient. This will only detect whether or not the variation in the variables in the plane of the first $k$ PCs is greater than that explained by the normal operating condition. If a new type of event occurs, which was not present in the reference data used to develop the PCA model, then new PCs will appear and the new observations $X_{\text{new}}$ will move off the plane. Such new events can be detected by computing the squared prediction error (SPE) of the residuals of a new observation.

$$\text{SPE}_{\text{new}} = \sum_{j=1}^{n} (x_{\text{new},j} - \hat{x}_{\text{new},j})^2 \tag{24.17}$$

This is also often referred to as the $Q$ statistic or the distance to the model.

Sometimes, fault detection based on the information captured by the latent space is insensitive to changes in the sensor arrays or to small process upsets, thus, leading to false-negative situations. This can be seen by examining the expression to represent the $k$th latent variable

$$t_k = p_{1k}x_1 + p_{2k}x_2 + \cdots + p_{nk}x_n \tag{24.18}$$

Because each latent variable is a linear combination of all variables, a fault in one of the sensors or small process upsets may not be amplified sufficiently to trigger the alarm and give an indication of an out-of-control signal. On the other hand, the SPE measure is more sensitive to such changes compared to the $T^2$ or score plot. This is due to the fact that the error of any type will be propagated to all latent space, and thus, will be reflected in the demapping part of the PCA,

$$\hat{X} = t_1 p_1^T + t_2 p_2^T + \cdots + t_k p_k{}^T \tag{24.19}$$

All estimated variables would be influenced by any type of disturbance in the input sequence. Hence, at the time instant the disturbance occurs, it is more likely to manifest itself in the SPE by violating the UCL than that of the $T^2$ (Figure 24.8, region I). However, significant changes in the process or in the sensor characteristics can trigger the alarm in both these measures (Figure 24.8, region II). Nevertheless, there might be process upsets undetected by the SPE due to the extrapolating feature of the calibration model. In such cases, the latent space will capture these changes but no violation in the SPE will be observed (Figure 24.8, region III). This feature of the PCA model is the first one to be noticed and it plays an important role to uncover more of the underlying process upsets that may occur.

### 24.3.3 Contribution Plots

Once an event is detected, it is then left up to the process operators and engineers to try to diagnose an assignable cause using their process knowledge. However, multivariate analysis based on PCA provides additional capabilities for this task. By interrogating the underlying PCA model at the point where the event has occurred, one can extract *contribution plots* that reveal the process variable or group of process variables with the greatest contribution to the deviation in the scores and in the SPE. Although these plots will not unequivocally diagnose the cause, they will provide much greater insight into possible causes and consequently greatly narrow the search.

## 24.4 FAULT DIAGNOSIS AND CLASSIFICATION

Fault diagnosis or classification is a comparative process. Previously, we have seen how to build the NOR from historical data and use it to flag the occurrence of faults. The next step is the classification of faulty data from up-to-date measurements. Fault diagnosis makes use of the available plant historical data.

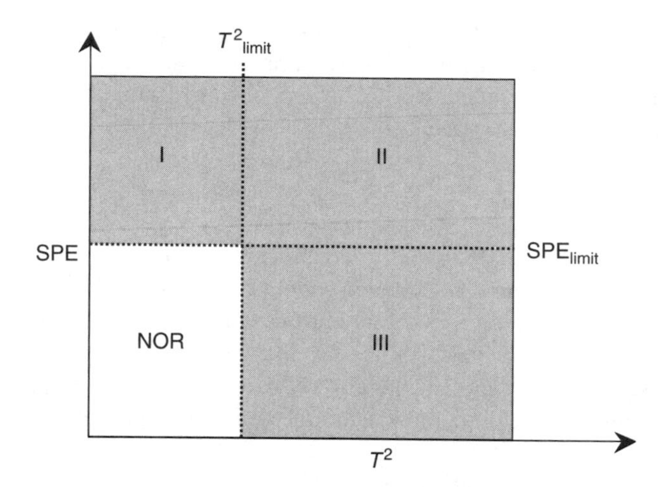


**FIGURE 24.8** Illustrative plot of SPE vs. $T^2$ for detecting events.

Conditions of poor plant operational status can be extracted from the historical data and used to construct *training clusters* for a particular known fault. When considering future on-line measurements, if a data point is located within a fault training cluster, then, an appropriate diagnosis can be made and an action can be promptly taken.

The first step in process fault diagnosis is the development of the training clusters associated with known faults. This step highlights whether or not it is possible to differentiate between normal and abnormal conditions, and furthermore whether it is possible to discriminate between process fault types. The terms *differentiate* and *discriminate* are used in a visual sense, i.e., is it possible to see (on a performance monitoring chart) the above-mentioned conditions?

## Example 24.5

Let us consider the data given in Table 24.2.[2]

Once the faulty points are located and assigned an index, a data matrix of faulty points can be constructed. This procedure is again congruent to the method used for constructing the NOR. The underlying factor is that the data matrix should only contain points of abnormal operation; if normal plant status points are present, then, the results will be adversely affected.

This step is principally the *feature extraction* stage and is of major importance. To transform the data matrix into its latent variables, one needs to find the eigenvector directions. The normal method would involve decomposing the *faulty* data matrix into its eigenvectors and eigenvalues, and using these values to transform the data matrix into its latent variables. However, as the objective is to identify this data matrix as abnormal operation when compared with the NOR, the eigenvalues used must be those obtained from decomposing the original *good* data matrix, i.e.,

$$Z_i = F \cdot p_i^{\mathrm{T}}$$

**TABLE 24.2**
**The Data for Nitrogen Removal**

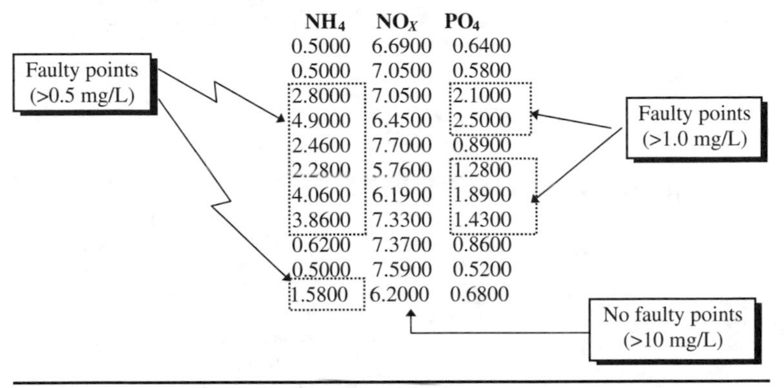

| $NH_4$ | $NO_X$ | $PO_4$ |
|---|---|---|
| 0.5000 | 6.6900 | 0.6400 |
| 0.5000 | 7.0500 | 0.5800 |
| 2.8000 | 7.0500 | 2.1000 |
| 4.9000 | 6.4500 | 2.5000 |
| 2.4600 | 7.7000 | 0.8900 |
| 2.2800 | 5.7600 | 1.2800 |
| 4.0600 | 6.1900 | 1.8900 |
| 3.8600 | 7.3300 | 1.4300 |
| 0.6200 | 7.3700 | 0.8600 |
| 0.5000 | 7.5900 | 0.5200 |
| 1.5800 | 6.2000 | 0.6800 |



The latent variables obtained essentially contain a *fingerprint* of the faulty operation under consideration, and the first two latent variables (PCs) are plotted against each other. This plot gives a visual representation of a *cluster* of data points. Depending on the data being considered, the cluster isolates faulty operational status. The term *training cluster* is used in reference to the procedure by which historical data is used to build up (or train) clusters corresponding to conditions of abnormal operation. An example of such a training cluster is shown in Figure 24.9.

The training cluster verifies the initial assumption that the faults for a particular measurement and location can be represented by a group of points in close proximity, and hence, the cluster. While this has been verified here, another more important question must be answered:

---

Would faults of *different* measurements in *different* locations produce *different* clusters?

---

The answer to this question can be both yes and no. For most measurements, the faults produce dissimilar clusters; however, some may produce clusters, which are congruent in both location and direction. Figure 24.10 is an example of discriminated clusters.

While it is apparent that the different faults are differentiated in the normal Cartesian geometry, in fault detection terms, this is only valid if these clusters are distinctly different and separate from the normal operating region for the data being considered. When the previously displayed fault clusters are combined with the corresponding NOR, the result is a monitoring chart for nitrogen removal. Figure 24.11 depicts the results.

The monitoring chart in Figure 24.11 is an example of where the faults are differentiable in space outside of the NOR. This structure is also quite effective in early fault detection, as the classification of an incoming point can be simply done by observing the location of the point, as it drifts away from the NOR. On the right-hand

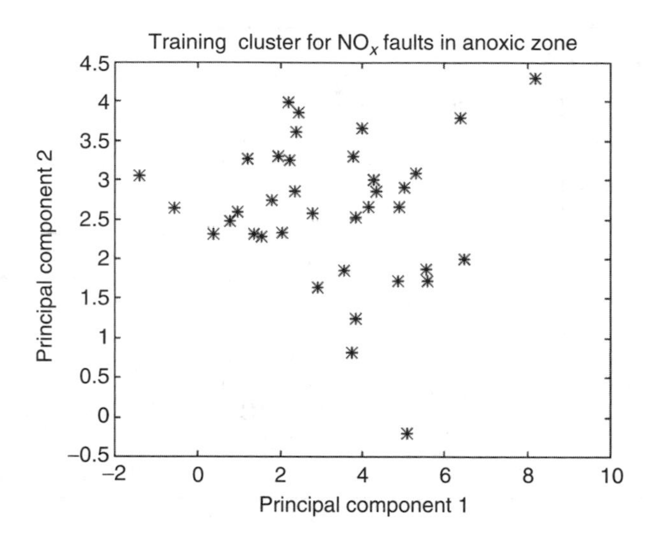


**FIGURE 24.9** Training cluster for $NO_x$ faults in anoxic zone.

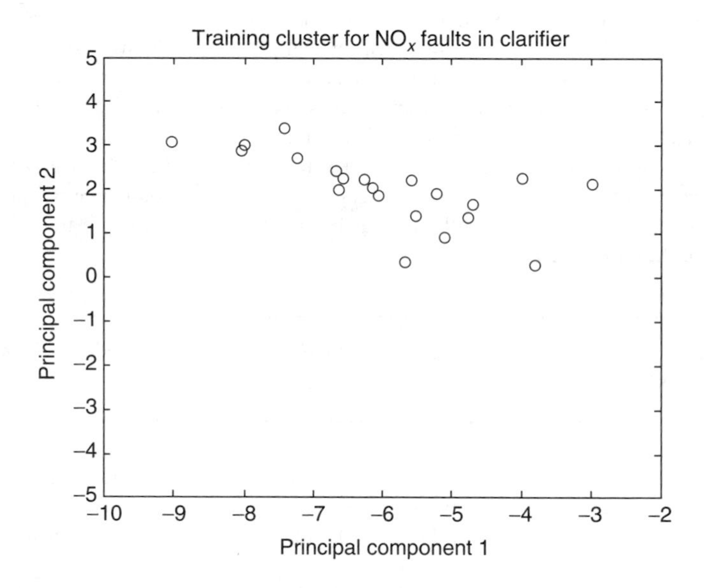


**FIGURE 24.10** Training cluster for $NO_x$ faults in clarifier.

side of the NOR in Figure 24.11, some faulty points overlap the region boundary. As the training cluster is made up of only faulty points, a modification of the normal boundary may be called for; however, this is not necessary for this example.

The development of training clusters and the NOR is the foundation for an online fault detection and classification system. While visual interpretation is the simplest method, i.e., "…incoming point is outside the NOR and belongs to a

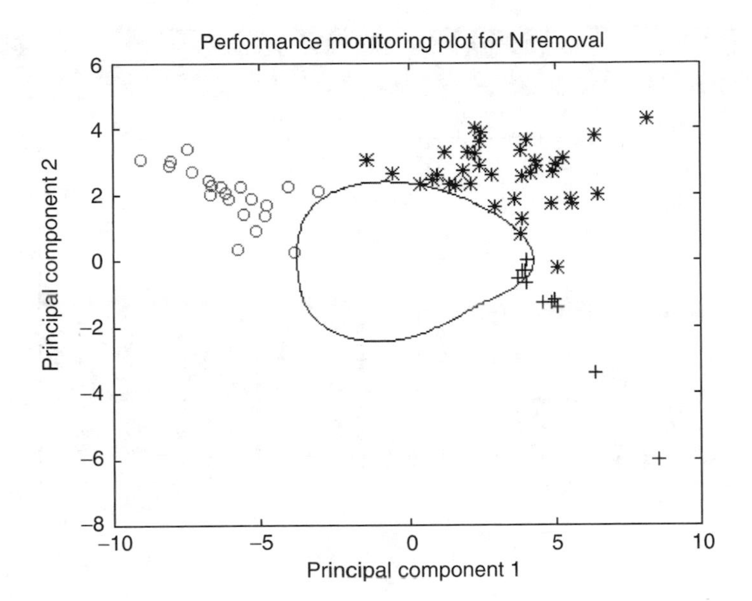


**FIGURE 24.11** Fault classification for N removal.

certain cluster…," a more robust, computer-based technique may be desirable. For more information on some of the current techniques, the reader is referred to Additional Reading following this chapter.

### Example 24.6

To illustrate an *integrated* PCA framework for process monitoring, an application within a pilot-scale plant environment running under a DCS will be discussed. A monitoring strategy using PCA has been developed and implemented for the flexible pilot-plant previously introduced in Chapter 21. Normal plant operating data is collected and used for training the PCA model. The data are first scaled to unit variance and zero mean before further processing (Figure 24.12).

Figure 24.13 illustrates the percentage in variance explained by each principal component. In this specific example, there are 25 process variables to be monitored and 13 principal components will be finally retained, capturing 70% of the total variance.

As an example, the characterization of the NOR region using the first two PCs is given in Figure 24.14. A set of normal regions has been specified for this specific application as a function of the level of confidence selected (statistical limits) using the Hotteling's statistic.

Finally, samples of the results of the online application of the strategy are provided in Figure 24.15 and Figure 24.16. The plant is being monitored using the $T^2$ statistics and an off-target condition is detected around sample 500 (Figure 24.15). This condition triggers the appearance of the contribution plot allowing the identification of the "suspected" measurements (those with a large contribution to $T^2$) associated with this specific situation.

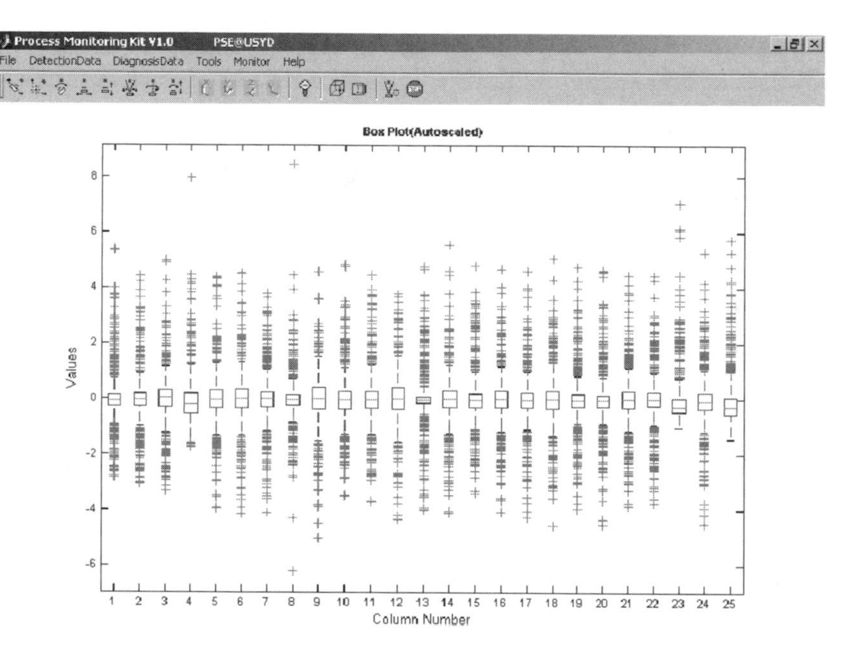


**FIGURE 24.12** Plant data auto scaled to unit variance and zero mean.

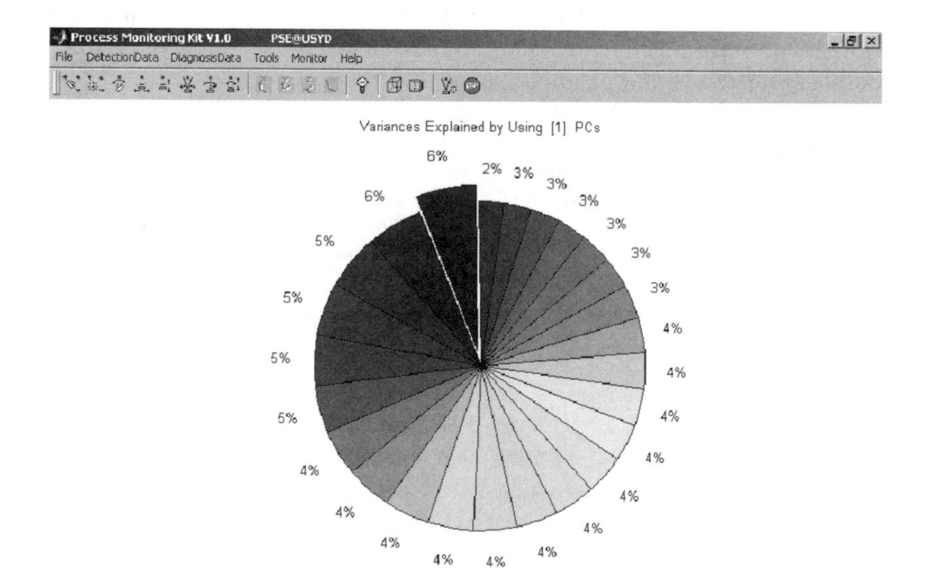


**FIGURE 24.13** Pie plot of variance explained by each PC.

Figure 25.16 illustrates a condition under which the process is moving "out of target" due to a different type of failure. The $T^2$ statistics detect the condition, and the contribution plot indicates the suspected sensors associated with this out-of-control

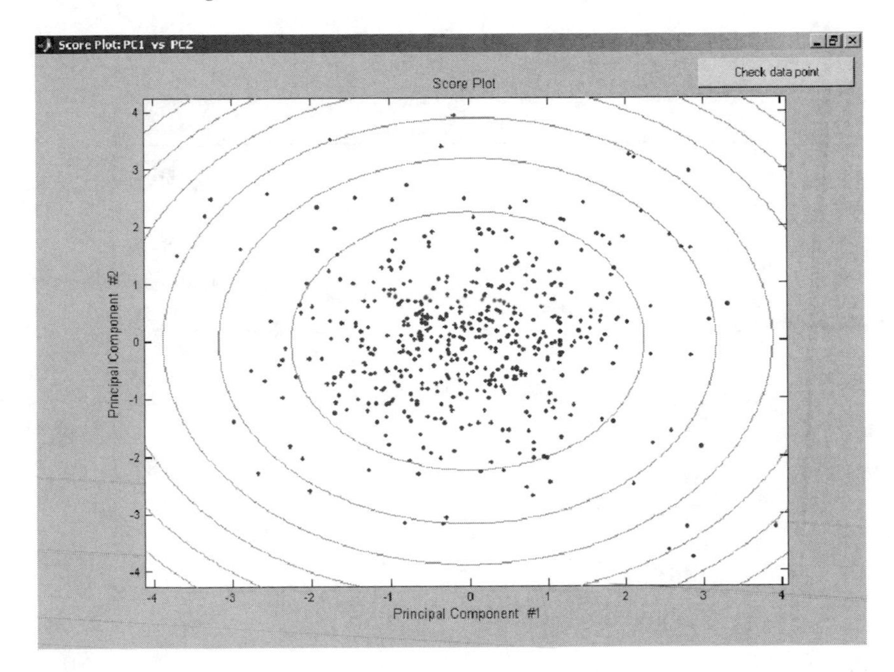


**FIGURE 24.14** NOR regions for different levels of significance are shown as contours.

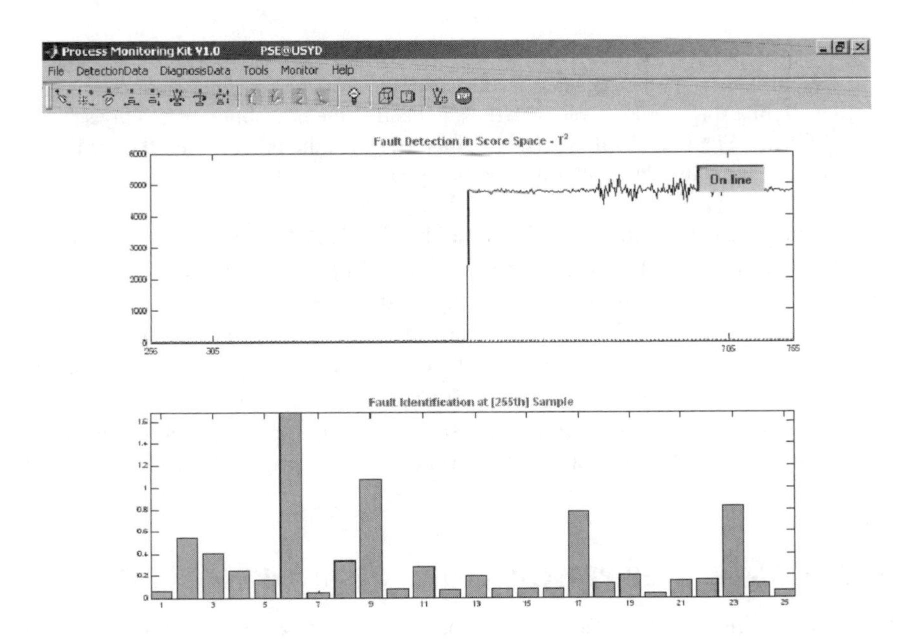


**FIGURE 24.15** Online monitoring of faulty (mean shift) condition using $T^2$ statistic and contribution plots.

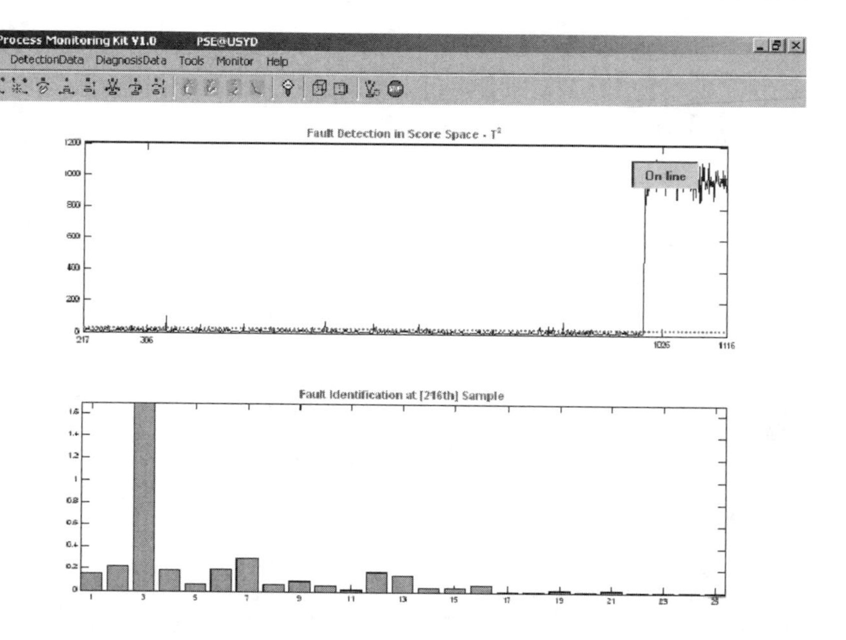


**FIGURE 24.16** Online monitoring of faulty (sensor) condition using $T^2$ statistic and contribution plots.

behavior. Compared with Figure 25.15, where several sensors are reported to have a large influence on the statistics, in this case, most of the contribution appears to come from a single process variable, thus, indicating the probability of a faulty sensor. Finally, Figure 25.17 illustrates the situation when the process after the rectification of the faulty sensor returns to normal operational status.

In many instances, the fault diagnosed by the PCA-based techniques point to equipment failures, and drifts and spikes in operating conditions. The control system is chiefly designed to deal with many of these symptoms, and if the operational performance is still poor, then, one can also pinpoint to the control system as the source of the observed behavior. Thus, the process engineer also needs the appropriate tools for monitoring the performance of the control loops and make corrections to the controllers when necessary. In that respect, control performance monitoring aids and complements the process monitoring techniques discussed before and will be studied next.

## 24.5 CONTROLLER PERFORMANCE MONITORING (CPM)

Control systems are designed and tuned based on a variety of performance specifications and the performance of the process is often tested using dynamic simulations as we have seen in Chapters 11 and 12. The control system may also be modified during the commissioning phase where the control system is tested on

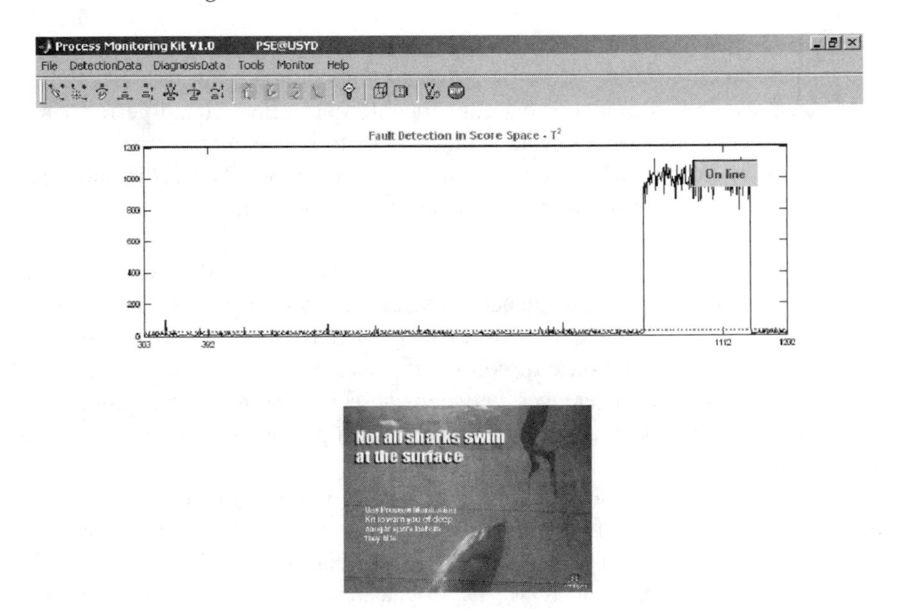


**FIGURE 24.17** Online monitoring of normal condition using $T^2$ statistic and contribution plots.

the real plant. It is, then, expected that the control system will perform according to the desired specifications. Often, however, the observed performance of the plant control system strays far from the expectations. There may be several reasons for this behavior:[9,10]

- Operating conditions may change over time
- Nonlinear effects may become significant
- Tuning parameters may turn out to be poorly chosen
- Assumed disturbance effects (magnitude and frequency) may be incorrect

Irrespective of the reasons, the control engineer is expected to detect such degradation in performance of control loops and take the necessary actions to recover the specified loop performance. This may be a daunting task as the control engineers are typically responsible for a large number of loops. Hence, an automated procedure that guides the control engineer in this effort would be invaluable. The methodologies associated with control performance monitoring (CPM) have the following goals:

- To provide information to plant personnel for determining if performance targets and response characteristics are being met
- To evaluate the performance of the controller
- To generate information to reconcile poor process performance with process fault diagnosis

It should be noted that causes of poor loop performance might not always be associated with the controller. While poor selection of control pairings and tuning parameters as well as model errors contribute to controller-related performance problems, there may also be severe changes in disturbance characteristics (oscillatory or noisy trends) and failures in sensors and actuators. Thus, it is important for the CPM to distinguish such effects and coordinate its assessment with fault diagnosis tasks.

Qin[11] categorizes control performance monitoring as either *stochastic* or *deterministic*. Stochastic performance assesses the effect of unmeasured stochastic disturbances on steady-state (routine) data, whereas deterministic performance concerns the dynamic response of controllers, such as set-point response, decay ratio, etc. Due to the differences between these two performance monitoring strategies, it may not be possible to achieve the best performance in both with the same set of controller tuning parameters.

All monitoring and assessment tools should be based on the information used by an experienced operator or control engineer to supervise the controller. Typically, this information comprises state classification, model validation (for a model-based controller such as MPC), stochastic performance monitoring and controller oscillation assessment.

### 24.5.1 State Classification

State classification is used to distinguish between two major control objectives under normal operating conditions – stochastic and deterministic control. A sound mechanism is expected to identify the transition from stochastic to deterministic state and vice versa. The difference between the setpoint and controlled variable readings is ineffective, since it has a low tolerance for fluctuations due to noise and disturbance effects. A more robust approach is to use the IAE (Chapter 12):

$$IAE = \int_{a}^{a+b} |e(t)| \, \mathrm{d}t$$

where $b$ is the length of the data window. This index is compared against a threshold $IAE_{\mathrm{lim}}$, specified by the operator. The control objective is, then, determined by the following rule:

---

If $IAE > IAE_{\mathrm{lim}}$, then, the control objective is *deterministic*, otherwise, the control objective is *stochastic*.

---

### 24.5.2 Model Validation

The validity of models in any model-based controller is always a concern. A model-based control strategy will not work effectively and safely if the discrepancy between the real process and the control model becomes significant. The residual between the predicted and measured controlled variables is constantly monitored. If the residual continuously exceeds a user-defined limit, the model is considered *poor* in describing the process at that instant. This usually suggests that a major process upset or interruption has occurred, and the control model is no longer valid. The rule can be stated as

---

If $residual > residual_{lim}$, then, the model state is *poor*, otherwise, the model state is *good*.

---

### 24.5.3 Stochastic Performance Monitoring

The most common approaches to stochastic performance monitoring start by establishing a benchmark performance that can be used to compare with the observed loop performance. Here, we shall focus on the assessment of performance relative to *minimum variance control* (MVC).[12,13] MVC is a feedback controller that achieves minimum output variance and can only be exactly achieved if the process model and the disturbance model are perfectly known. It is noted that MVC is not implemented, and is simply used as a reference (benchmark) to compare the variance of the controlled variable based on ideal MVC performance with the variance resulting from the use of the actual controller. In this way, we can judge how far the observed performance is from this ideal benchmark performance and assess the potential for making improvements in the controller.

The *variation* in the increase of the controlled output variance with respect to MVC is a sign that the control loop may not be operating as expected. On the other hand, if the variance with MVC is larger than an acceptable limit, this suggests that the operating conditions or the process may have to be modified.

The closed-loop block diagram is given in Figure 24.18. Let us consider that the process transfer functions is expressed in discrete time as

$$y(k) = g_p(q^{-1})u(k - \varphi) + v(k) \tag{24.20}$$

where $v(k)$ expresses the effect of noise and unmeasured disturbances, and $\varphi$ the delay term. Here, we are also introducing the backward shift operator, $q^{-1}$, i.e.,

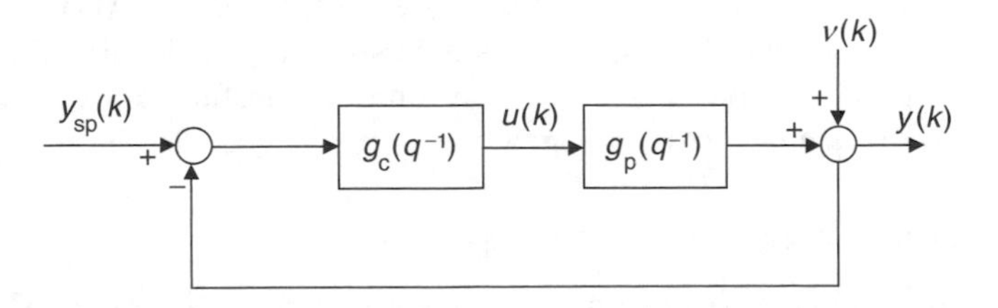


**FIGURE 24.18** Closed-loop block diagram.

$q^{-1}y(k) = y(k-1)$. This operator allows a compact description of discrete-time transfer function elements

$$y(k) = \frac{B(q^{-1})}{A(q^{-1})}u(k) = g_p(q^{-1})u(k) \tag{24.21}$$

where the polynomials $A$ and $B$ are defined as

$$A(q^{-1}) = 1 + a_1 q^{-1} + \cdots + a_{n_y} q^{-n_y}$$

$$B(q^{-1}) = 1 + b_1 q^{-1} + \cdots + a_{n_u} q^{-n_u}$$

with $n_y \geq n_u$. The closed-loop output in Eq. (24.21) can be expressed in terms of a moving average model such as

$$y(k) = [1 + \psi_1 q^{-1} + \psi_2 q^{-2} + \cdots + \psi_\varphi q^{-\varphi} + \cdots]\eta(k) \tag{24.22}$$

and the coefficients $\psi_i$ and the variance of the stochastic process $\eta(k)$can be estimated by fitting a model to the output response. The variance of the controlled variable can be calculated from the estimated coefficients and the variance as

$$\sigma_y^2 = [1 + \psi_1^2 + \psi_2^2 + \cdots + \psi_\varphi^2 + \cdots]\sigma_\eta^2 \tag{24.23}$$

If the feedback controller is an MVC, then, the $\varphi$-step ahead forecast becomes zero for terms at and beyond $\varphi$. Thus, for MVC, the controlled variable variance is given as

$$\sigma_{\mathrm{MVC}}^2 = [1 + \psi_1^2 + \psi_2^2 + \cdots + \psi_{\varphi-1}^2]\sigma_\eta^2 \tag{24.24}$$

Now, we can define a performance index, based on the calculated variances,[13]

$$PI = 1 - \frac{\sigma_{\mathrm{MVC}}^2}{\sigma_y^2} \tag{24.25}$$

The index would be zero if the performance of the implemented feedback controller matches that of minimum variance. Larger values of *PI* would indicate that the feedback controller may be far from being optimal. The calculation of PI can be carried out by solving a least-squares problem using the process data.[13]

The minimum variance establishes a lower bound for achievable loop performance and is most useful when the process time delay is very small, the process is low order and the disturbances are essentially stationary. The lack of accurate information about time delay may limit the usefulness of MVC-based indices, as well as multivariable interactions and process constraints.

### 24.5.4 Controller Oscillation Assessment

Control signal oscillation is undesirable in any application. Regardless of the controller types and control objectives, excessive oscillation could lead to unnecessary wear in final control elements and significant deterioration in controllability and operability of the process.

The objective here is to identify any ongoing, periodic oscillations, which has peak-to-peak amplitude greater than a user-defined limit. A typical control signal resembles the shape of a sine curve, monitoring peak-to-peak amplitude and calculating the period would be easy if the signal increases or decreases monotonically until the turning points are reached. Unfortunately, it is not true in reality and positive or negative *sparks* can be found during the increasing or decreasing phase of the control signal. It is, therefore, necessary to remove these sparks before correct features can be extracted from the data. Many of the mathematical signal processing techniques require data with a reasonable window length. For a kind of real-time supervisory system, this means that there will be a significant time gap between occurrence and detection of features.

As a final note, we stress the fact that most of these techniques are well established for SISO problems. Assessment of control loop performance for multivariable systems is more challenging and the reader is referred to the literature cited for further details.

## 24.6 SUMMARY

In this chapter, we have discussed the issues associated with monitoring the operational status of a process. The discussion was placed in the context of statistical process control and statistical techniques to quantify the *in-control* behavior of a process. The goal of this chapter was to introduce some statistical concepts to monitor the performance of a process over time, with emphasis on the detection of unnatural events, which may ultimately lead to degradation, or violation of the product quality specifications. The detection and further isolation of these events help allocation of a cause, which, if eliminated, results in improved reliability and control of the final product. A number of concepts were introduced, starting from the well-established control charts, moving up to projection techniques using latent variables (PCA) for process performance monitoring. The fault diagnosis and classification techniques were supplemented with controller performance monitoring techniques to identify and correct operational problems that can be attributed to the control system. While deterministic performance measures, such as decay ratio, settling time, etc., can be used to detect possible degradation in loop performance, a stochastic measure that uses the minimum variance controller as the benchmark can also offer a means to evaluate how far the tuned controller may be from the ideal MVC performance.

## REFERENCES

1. Shewhart, W., *Economic Control of Quality of Manufactured Product*, Van Nostrand, New York, 1931.
2. Robertson, T., Chen, J., Romagnoli, J.A., and Newell, B., Intelligent monitoring for quality control in a biological nutrient removal wastewater treatment pilot plant, *Proceedings of DYCOPS-5*, 1998, pp. 590–596.

3. Western Electric Company, *Statistical Quality Control Handbook*, Western Electric Company, Indianapolis, IN, 1956.
4. Kourti, T. and MacGregor, J.F., Multivariate SPC methods for process and product monitoring, *J. Quality Technol.*, 28, 409–428, 1996.
5. Jackson, E.J., *A User's Guide to Principal Components*, Wiley, New York, 1991.
6. Shaw, P.J.A., *Multivariate Statistics for the Environmental Sciences*, Oxford University Press, London, 2003.
7. McBean, E.A. and Rovers, F.A., *Statistical Procedures for Analysis of Environmental Monitoring Data & Risk Assessment*, Prentice-Hall, NJ, 1998.
8. Chen, J., Bandoni, A., and Romagnoli, J.A., Robust PCA and normal region in multivariable statistical process monitoring, *AIChE J.*, 42, 3563–3566, 1996.
9. Kozub, D.J., Controller performance monitoring and diagnosis: Experiences and challenges, In *AIChE Symposium Series 316*, Kantor, J.C., Garcia, C., and Carnahan, B. (Eds.), 94, AIChE & CACHE: New York, 1997, pp. 83–96.
10. Hoo, K.A., Piovoso, M.J., Schnelle, P.D., and Rowan, D.A., Process and controller performance monitoring: Overview with industrial applications, *Int. J. Adaptive Control Signal Process.*, 17, 635–662, 2003.
11. Qin, S.J., Control performance monitoring - A review and assessment, *Comp. Chem. Eng.*, 23, 173–186, 1998.
12. Harris, T.J., Assessment of control loop performance, *Canadian J. Chem. Eng.*, 67, 856–861, 1989.
13. Desborough, L.D. and Harris, T.J., Performance assessment measure for univariate feedback control, *Canadian J. Chem. Eng.*, 70, 1186–1197, 1992.

# Section VII Additional Reading

Among the early books on the use of control concepts in the industry, one can cite the following references:

Rhodes, T.J., *Industrial Instruments for Measurement and Control*, McGraw-Hill, New York, 1941.

Zoss, L.M. and Delahooke, B.C., *Theory and Applications of Industrial Process Control*, Delmar Publishers, Albany, 1961.

A recent text on the application of recent advances in computer and communication technologies in the petrochemical industries is by Kalani:

Kalani, G., *Industrial Process Control: Advances and Applications*, Butterworth-Heinemann, Boston, 2002.

For applications of microprocessors, computers, and distributed control systems in the process industries, the following references would also provide useful information.

Bibbero, R.J., *Microprocessors in Industrial Control*, Prentice-Hall, Englewood Cliffs, NJ, 1983.

Harrison, T.J. (Ed.), *Handbook of Industrial Control Computers*, Wiley-Interscience, New York, 1972.

Harrison, T.J. (Ed.), *Minicomputers in Industrial Control: An Introduction*, Instrument Society of America, Research Triangle Park, NC, 1980.

Skrokov, M.R. (Ed.), *Mini- and Microcomputer Control in the Chemical Process Industries: Handbook of Systems Strategies and Application Problems*, Van Nostrand Reinhold, New York, 1980.

For further references on data reconciliation, the reader is referred to the following:

Bagajewicz, M.J., *Process Plant Instrumentation: Design and Upgrade*, Technomic Publisher, Lancaster, PA, 2001.

Madron, F., *Process Plant Performance: Measurement and Data Processing for Optimization and Retrofits*, Ellis Horwood, New York, 1992.

There are many books on the application of statistical techniques for quality monitoring.

Mason, R.L. and Young, J.C., *Multivariate Statistical Process Control with Industrial Applications*, Society for Industrial and Applied Mathematics, Alexandria, VA, American Statistical Association, Philadelphia, PA, 2002.

Wang, X.Z., *Data Mining and Knowledge Discovery for Process Monitoring and Control*, Springer, New York, 1999.

For a practical application of some of the SPC concepts to a variety of industries, such as petrochemicals, chemicals, pulp and paper, food, and minerals, the reader is referred to the book by Badavas:

Badavas, P.C., *Real-Time Statistical Process Control*, Prentice-Hall, Englewood Cliffs, NJ, 1993.

Fault detection and diagnosis techniques for chemical processes have been covered extensively in a series of recent review articles by Venkatasubramanian.

Venkatasubramanian, V., Rengaswamy, R., Yin, K., and Kavuri, S.N., A review of process fault detection and diagnosis. Part I: Quantitative model-based methods, *Comp. Chem. Eng.*, 27, 293–311, 2003.

Venkatasubramanian, V., Rengaswamy, R., and Kavuri, S.N., A review of process fault detection and diagnosis. Part II: Qualitative models and search strategies, *Comp. Chem. Eng.*, 27, 313–326, 2003.

Venkatasubramanian, V., Rengaswamy, R., Kavuri, S.N., and Yin, K., A review of process fault detection and diagnosis. Part III: Process history based methods, *Comp. Chem. Eng.*, 27, 327–346, 2003.

The reader would also find the following books very useful.

Chiang, L.H., Russell, E.L., and Braatz, R.D., *Fault Detection and Diagnosis in Industrial Systems*, Springer, New York, 2001.

Himmelblau, D.M., *Fault Detection and Diagnosis in Chemical and Petrochemical Processes*, Elsevier, Amsterdam, 1978.

Russell, E.L., Chiang, L.H., and Braatz, R.D., *Data-Driven Methods for Fault Detection and Diagnosis in Chemical Processes*, Springer, New York, 2000.

Various techniques for monitoring and evaluation of the performance of controllers have been discussed in the following book:

Huang, B. and Shah, S.L., *Performance Assessment of Control Loops: Theory and Applications*, Springer, New York, 1999.

# Section VII Exercises

**VII.1.** As discussed in Chapter 20, energy integration helps to reduce utility consumption by improving the thermodynamic efficiency of the process, and for energy-intensive processes, such cost savings can be significant. However, energy integration also changes the operational characteristics of the plant. In general terms:

1. Discuss the interaction of economics, environmental impact, and energy integration
2. Discuss the impact of energy integration on the process operation and ultimately on the control system design
3. Discuss a step-by-step analysis procedure to design and analyze heat-exchanger networks to minimize the impact on the process operation and plantwide control

**VII.2.** Figure VII.1 shows an example of an HEN consisting of three hot streams and four cold streams. In this example, it is assumed that the utility cooler is using cooling water and the utility heater is using HP steam. Perform a controllability assessment and modify the proposed design if necessary.

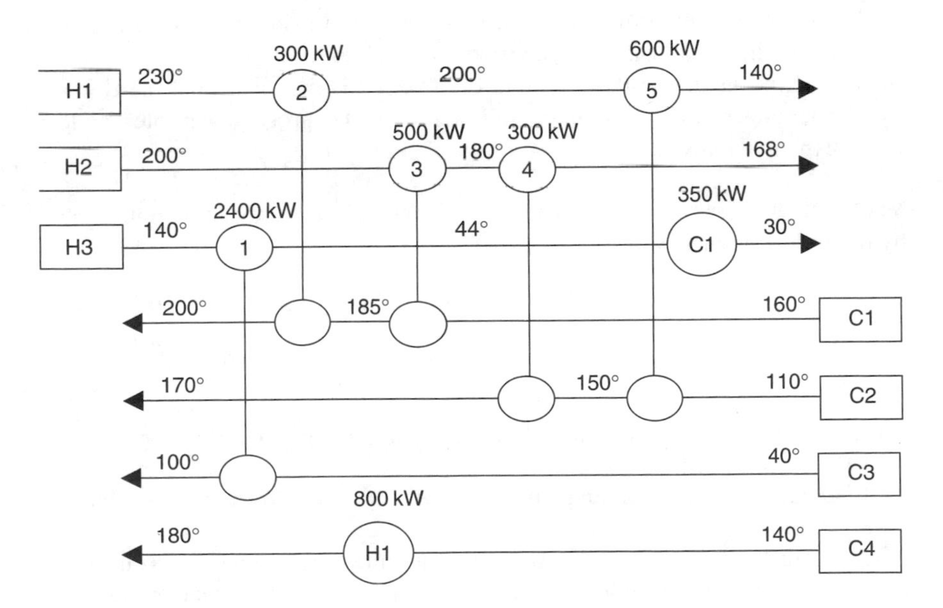


**FIGURE VII.1** Heat-exchanger network.

**VII.3.** Using the HEN obtained from the solution of Exercise VII.2, design the best possible control configuration that minimizes the interactions as well as the process network condition number.

**VII.4.** A multilayer control scheme was discussed in Chapter 22 for advanced operation and optimal control of a crystallization plant and a polymerization reactor. Consider the case of the polymerization process, which aims at controlling the particle size polydispersity index (PSPI) (a measure of the particle size distribution) and the number-average molecular weight (NAMW) by manipulating the monomer feed and temperature, assuming we have infrequent measures of the controlled variables (PSPI and NAMW):

1. Represent, in simple terms, using the block diagram configuration, the multilayer approach.
2. Describe, in simple terms, the different uses of the model within this multilayer architecture.
3. What are the consequences if the intermediate layer, MPC is eliminated from the scheme?

**VII.5.** Consider the problem given in Examples 23.6 and 23.7 of the textbook. In this case the flow rate of streams 1, 2, 5, 7, and 8 are considered to be measured:

1. Write down the balance equations and confirm the variable classification obtained in Example 23.7.
2. Assume that the flow rate of stream 3 is also measured. How will this affect the classification of the process variables?
3. Assume again that the flow rate of stream 3 is measured, but the flow rate of stream 8 is unmeasured. Classify the process variables and explain the results.

**VII.6.** The process topology of a section of an ethylene plant can be represented by the following matrix:

$$A = \begin{bmatrix} 1 & 1 & 1 & -1 & -1 & 0 & 0 & 0 & 0 \\ 0 & 0 & -1 & 1 & 0 & -1 & 0 & 0 & 0 \\ 0 & 0 & 0 & 0 & 1 & 0 & -1 & -1 & -1 \end{bmatrix}$$

and consider the total mass balance data reconciliation problem:

1. Obtain the corresponding flow diagram and write down the mass balances.
2. Assume all process stream flow rates are measured. What can you say about the classification of variables and the number of redundant equations?
3. Assume that the flow rate of stream 5 is unmeasured. Classify the process variables and find the redundant equations if any.

**VII.7.** For the problem described in Exercise VII.6, assume that the data set in Table VII.1 is given (measured values).

1. Apply the Global Test to check the consistency of a set of measurements (assume an error probability of 0.10)
2. Perform data reconciliation on the given set of data

**VII.8.** Consider the Fisher's classical data set. The data set consists of three classes with each class containing $m = 4$ measurements and $n = 50$ observations. Class 3 data were used to construct the covariance matrix of the measurements. After autoscaling and performing the eigenvalue decomposition:

$$\Lambda = \begin{bmatrix} 1.92 & 0 & 0 & 0 \\ 0 & 0.96 & 0 & 0 \\ 0 & 0 & 0.88 & 0 \\ 0 & 0 & 0 & 0.24 \end{bmatrix}$$

$$V = \begin{bmatrix} 0.64 & -0.29 & 0.052 & -0.71 \\ 0.64 & -0.23 & 0.25 & 0.69 \\ 0.34 & 0.33 & -0.88 & 0.11 \\ 0.25 & 0.87 & 0.41 & -0.09 \end{bmatrix}$$

1. Calculate the amount of variance explained when only one PC is retained
2. How many PCs should be retained for the PCA model to explain 70% of the total variance and what is the eigenvector matrix retained in this case?

**VII.9.** In Chapter 24, we have shown a way to define the ellipsoid that represents the region of normal operation. Based on the concepts of multivariate

**TABLE VII.1**
**Data for Exercise VII.7**

| Flow Rates | Measured Values | Variances |
|---|---|---|
| $F_1$ | 70.49 | 10.9 |
| $F_2$ | 7.103 | 0.2 |
| $F_3$ | 13.04 | 0.4 |
| $F_4$ | 35.38 | 2.6 |
| $F_5$ | 53.21 | 5.76 |
| $F_6$ | 23.90 | 0.9 |
| $F_7$ | 0.00 | 0.6 |
| $F_8$ | 0.0765 | 0.23 |
| $F_9$ | 54.59 | 5.8 |

statistics and the singular-value decomposition of the measurement covariance matrix:

1. Find an alternative representation of the ellipsoid based on the vector of scores
2. Illustrate the situation for the case when only two scores are retained

**VII.10.** Consider the problem described in Exercise VII.8 and assume that only the first two PCs are retained:

1. Calculate the $T^2$ statistical threshold (**Hint:** consider that the covariance matrix is estimated from the sample covariance matrix.)
2. Estimate the elliptical confidence region

# Appendix A: Linearization

## A.1 BASIC STEPS IN LINEARIZATION

Let us consider a nonlinear equation described by

$$y = (x - 3)^2 + 2 \tag{A.1}$$

The graph of this equation is depicted in Figure A.1a. Can this parabolic shape be reasonably approximated by a line in different regions of the domain of $x$? The technique that we will employ is the Taylor expansion that can be summarized as follows for the case of a one-dimensional function:

$$\begin{aligned} y &= f(x) \\ &= f(x_0) + \left.\frac{\mathrm{d}f}{\mathrm{d}x}\right|_{x=x_0}(x - x_0) + \frac{1}{2!}\left.\frac{\mathrm{d}^2 f}{\mathrm{d}x^2}\right|_{x=x_0}(x - x_0)^2 + \cdots \end{aligned} \tag{A.2}$$

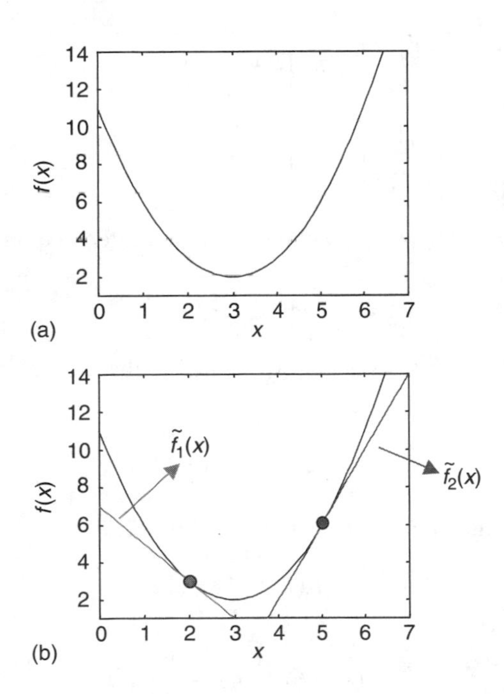


**FIGURE A.1** A nonlinear function (a) and its linear approximation (b) around two reference points (indicated as large dots).

The point of expansion $x_0$ represents the local reference point at which the approximation becomes exact. If the variation in $x - x_0$ is small, one can neglect the second- and higher-order terms in $x - x_0$, and truncate this infinite series expansion. This will result in a linear equation.

For Eq. (A.1), we note the following:

$$y = f(x) = (x - 3)^2 + 2 \tag{A.3}$$

$$\left.\frac{\mathrm{d}f(x)}{\mathrm{d}x}\right|_{x=x_0} = [2(x - 3)]_{x_0} = 2(x_0 - 3) \tag{A.4}$$

Let us consider two locations along the graph of Eq. (A.1) and develop linear equations that can describe the behavior of this function locally. These reference points are:

$$x_0 = 2, \quad y_0 = 3$$

$$x_0 = 5, \quad y_0 = 6$$

and produce the following:

$$\left.\frac{\mathrm{d}f(x)}{\mathrm{d}x}\right|_{x=2} = [2(x - 3)]_{x=2} = -2$$

$$\left.\frac{\mathrm{d}f(x)}{\mathrm{d}x}\right|_{x=5} = [2(x - 3)]_{x=5} = 4$$

The local linear equation corresponding to the first reference point then becomes

$$y = \tilde{f}_1(x) = -2x + 7 \tag{A.5}$$

One can also rephrase Eq. (A.5) to reveal the fact that $x - x_0$ is *proportional* to $y - y_0$,

$$(y - 3) = -2(x - 2)$$

The second linear function can be expressed similarly,

$$y = \tilde{f}_2(x) = 4x - 14 \tag{A.6}$$

and

$$(y - 6) = 4(x - 5)$$

Equations (A.5) and (A.6) are the linear representations of the nonlinear system given by Eq. (A.1), albeit at two different locations. Schematically this is shown in Figure A.1b.

We can make the following interrelated observations:

- *Local vs. global characteristics*: Lines display qualitatively different characteristics (slopes) at different locations along the graph. Linear approximations are truly local.
- *Region of accuracy*: As we move farther away from $x_0$, the approximation by a line becomes less and less exact (more so for $\tilde{f}_1(x)$ than for $\tilde{f}_2(x)$). Linear models have limited region of accuracy.

**Example A.1**

We shall consider the equation given below:

$$\frac{dx(t)}{dt} = x^2(t) - \frac{1}{x(t)} \tag{A.7}$$

We shall develop a linear approximate model for this nonlinear system. We use Taylor series expansion around the reference point $(x_0)$,

$$\begin{aligned}\frac{dx}{dt} &= x^2 - \frac{1}{x} \\ &\approx \left[x_0^2 - \frac{1}{x_0}\right] + \frac{d}{dx}\left[x^2 - \frac{1}{x}\right]_{x_0}(x - x_0)\end{aligned} \tag{A.8}$$

where we have neglected the expansion terms of order two and higher. Now, we proceed to define the deviation variable. The steady-state equation is given as

$$0 = x_s^2 - \frac{1}{x_s} \tag{A.9}$$

Considering that $x_0 = x_s$ and subtracting Eq. (A.9) from Eq. (A.8), we obtain

$$\frac{dx}{dt} = \left[2x_s + \frac{1}{x_s^2}\right](x - x_s) \tag{A.10}$$

The deviation variable becomes

$$\bar{x} = x - x_s$$

We finally have a linear model in terms of the deviation variable given by

$$\frac{d\bar{x}}{dt} = a\bar{x} \tag{A.11}$$

where the constant is defined as

$$a = 2x_s + \frac{1}{x_s^2}$$

## A.2 LINEARIZATION GENERALIZED

The linearization technique can be easily generalized for the case of higher dimensional state models. Let us demonstrate this with the following nonlinear ODEs:

$$\begin{aligned}
\frac{dx_1}{dt} &= f_1(x_1, x_2, \ldots, x_N, u_1, u_2, \ldots, u_M) \\
\frac{dx_2}{dt} &= f_2(x_1, x_2, \ldots, x_N, u_1, u_2, \ldots, u_M) \\
&\vdots \\
\frac{dx_N}{dt} &= f_N(x_1, x_2, \ldots, x_N, u_1, u_2, \ldots, u_M)
\end{aligned} \tag{A.12}$$

Note that we have $N$ state variables as well as $M$ input variables that are potentially time varying. We shall expand the nonlinear functions $f_i$ around the reference point $(x_{1,0}, \ldots, x_{2,0}, u_{1,0}, \ldots, u_{M,0})$ and follow the same procedure as in the previous section. This results in the following approximate linear ODEs:

$$\begin{aligned}
\frac{dx_1}{dt} &\cong f_1(x_{1,0}, \ldots, x_{2,0}, u_{1,0}, \ldots, u_{M,0}) + \sum_{i=1}^{N} K_{1i}(x_i - x_{i,0}) + \sum_{i=1}^{M} L_{1i}(u_i - u_{i,0}) \\
\frac{dx_2}{dt} &\cong f_2(x_{1,0}, \ldots, x_{2,0}, u_{1,0}, \ldots, u_{M,0}) + \sum_{i=1}^{N} K_{2i}(x_i - x_{i,0}) + \sum_{i=1}^{M} L_{2i}(u_i - u_{i,0}) \\
&\vdots \\
\frac{dx_N}{dt} &\cong f_N(x_{1,0}, \ldots, x_{2,0}, u_{1,0}, \ldots, u_{M,0}) + \sum_{i=1}^{N} K_{Ni}(x_i - x_{i,0}) + \sum_{i=1}^{M} L_{Ni}(u_i - u_{i,0})
\end{aligned} \tag{A.13}$$

where

$$K_{ij} = \left.\frac{\partial f_i}{\partial x_j}\right|_{x_{1,0}, \ldots, x_{N,0}, u_{1,0}, \ldots, u_{M,0}}, \quad L_{ij} = \left.\frac{\partial f_i}{\partial u_j}\right|_{x_{1,0}, \ldots, x_{N,0}, u_{1,0}, \ldots, u_{M,0}}$$

Note that the use of partial derivatives as the functions depend on more than one variable. Now, if $x_{1,s}, \ldots, x_{N,s}$, and $u_{1,s}, \ldots, u_{M,s}$ are the steady-state values of these variables, we have

$$\begin{aligned}
0 &= f_1(x_{1,s}, \ldots, x_{N,s}, u_{1,s}, \ldots, u_{M,s}) \\
0 &= f_2(x_{1,s}, \ldots, x_{N,s}, u_{1,s}, \ldots, u_{M,s})
\end{aligned} \tag{A.14}$$

Using the steady-state as the reference points and subtracting the steady-state equations from the dynamic equations, we have

$$
\begin{aligned}
\frac{\mathrm{d}(x_1 - x_{1,\mathrm{s}})}{\mathrm{d}t} &= \sum_{i=1}^{N} K_{1i}(x_i - x_{i,\mathrm{s}}) + \sum_{i=1}^{M} L_{1i}(u_i - u_{i,\mathrm{s}}) \\
\frac{\mathrm{d}(x_2 - x_{2,\mathrm{s}})}{\mathrm{d}t} &= \sum_{i=1}^{N} K_{2i}(x_i - x_{i,\mathrm{s}}) + \sum_{i=1}^{M} L_{2i}(u_i - u_{i,\mathrm{s}}) \\
&\vdots \\
\frac{\mathrm{d}(x_N - x_{N,\mathrm{s}})}{\mathrm{d}t} &= \sum_{i=1}^{N} K_{Ni}(x_i - x_{i,\mathrm{s}}) + \sum_{i=1}^{M} L_{Ni}(u_i - u_{i,\mathrm{s}})
\end{aligned}
\tag{A.15}
$$

With the deviation variables,

$$\bar{x}_i = x_i - x_{i,\mathrm{s}}$$

$$\bar{u}_i = u_i - u_{i,\mathrm{s}}$$

we arrive at the linear set of equations:

$$
\begin{aligned}
\frac{\mathrm{d}\bar{x}_1}{\mathrm{d}t} &= \sum_{i=1}^{N} K_{1i}\bar{x}_i + \sum_{i=1}^{M} L_{1i}\bar{u}_i \\
\frac{\mathrm{d}\bar{x}_2}{\mathrm{d}t} &= \sum_{i=1}^{N} K_{2i}\bar{x}_i + \sum_{i=1}^{M} L_{2i}\bar{u}_i \\
&\vdots \\
\frac{\mathrm{d}\bar{x}_N}{\mathrm{d}t} &= \sum_{i=1}^{N} K_{Ni}\bar{x}_i + \sum_{i=1}^{M} L_{Ni}\bar{u}_i
\end{aligned}
\tag{A.16}
$$

**Example A.2**

Consider the stirred-tank heater discussed in Example 4.3. The state equations developed form the total mass balance and the energy balance

$$A\frac{\mathrm{d}h}{\mathrm{d}t} = F_{\mathrm{in}} - F_{\mathrm{out}} \tag{A.17}$$

$$Ah\frac{\mathrm{d}T}{\mathrm{d}t} = F_{\mathrm{in}}(T_{\mathrm{in}} - T) + \frac{UA_{\mathrm{t}}(T_{\mathrm{st}} - T)}{\rho c_{\mathrm{p}}} \tag{A.18}$$

We assume that the output flow rate is a nonlinear function of the level in the tank i.e., $F_{out} = \beta\sqrt{h}$, then

$$A\frac{dh}{dt} = F_{in} - \beta\sqrt{h} \tag{A.19}$$

$$Ah\frac{dT}{dt} = F_{in}(T_{in} - T) + \frac{UA_t(T_{st} - T)}{\rho c_p} \tag{A.20}$$

We shall construct the linear model around the steady-state operating point defined by the variables involved ($h_s$, $T_s$, $F_{in,s}$, $T_{in,s}$, $T_{st,s}$). Using the results of Example 5.1, Eq. (A.19) can be written in linearized form as

$$\frac{d\bar{h}}{dt} = a\bar{F}_{in} + b\bar{h} \tag{A.21}$$

Next, we need to linearize the energy balance, which can be rewritten as

$$\frac{dT}{dt} = \frac{F_{in}T_{in}}{Ah} - \frac{F_{in}T}{Ah} + \frac{UA_tT_{st}}{A\rho c_p h} - \frac{UA_tT}{A\rho c_p h} = f(T, h, F_{in}, T_{st}, T_{in}) \tag{A.22}$$

Expanding this expression around the steady-state using Taylor series and neglecting the second- and higher-order terms, we have

$$\frac{dT}{dt} = \left[\frac{F_{in}T_{in}}{Ah} - \frac{F_{in}T}{Ah} + \frac{UA_tT_{st}}{A\rho c_p h} - \frac{UA_tT}{A\rho c_p h}\right]_{h_s,\, T_s,\, F_{in,s},\, T_{in,s},\, T_{st,s}} + c(T - T_s)$$
$$+ d(h - h_s) + e(F_{in} - F_{in,s}) + f(T_{st} - T_{st,s}) + g(T_{in} - T_{in,s}) \tag{A.23}$$

The constants are given accordingly as

$$c = \left.\frac{\partial f}{\partial T}\right|_{h_s,\, T_s,\, F_{in,s},\, T_{in,s},\, T_{st,s}} = \left[\frac{1}{Ah}\left(-F_{in} - \frac{UA_t}{\rho c_p}\right)\right]_{h_s,\, T_s,\, F_{in,s},\, T_{in,s},\, T_{st,s}}$$

$$d = \left.\frac{\partial f}{\partial h}\right|_{h_s,\, T_s,\, F_{in,s},\, T_{in,s},\, T_{st,s}}$$

$$= \left[-\frac{1}{Ah^2}(F_{in}(T_{in} - T)) + \frac{UA_t}{\rho c_p}(T_{st} - T)\right]_{h_s,\, T_s,\, F_{in,s},\, T_{in,s},\, T_{st,s}}$$

$$e = \left.\frac{\partial f}{\partial F_{in}}\right|_{h_s,\, T_s,\, F_{in,s},\, T_{in,s},\, T_{st,s}} = \left[\frac{T_{in} - T}{Ah}\right]_{h_s,\, T_s,\, F_{in,s},\, T_{in,s},\, T_{st,s}}$$

$$f = \left.\frac{\partial f}{\partial T_{st}}\right|_{h_s,\, T_s,\, F_{in,s},\, T_{in,s},\, T_{st,s}} = \left[\frac{b}{Ah}\right]_{h_s,\, T_s,\, F_{in,s},\, T_{in,s},\, T_{st,s}}$$

$$g = \left.\frac{\partial f}{\partial T_{in}}\right|_{h_s,\, T_s,\, F_{in,s},\, T_{in,s},\, T_{st,s}} = \left[\frac{F_{in}}{Ah}\right]_{h_s,\, T_s,\, F_{in,s},\, T_{in,s},\, T_{st,s}}$$

The steady-state energy balance is given by

$$\left[\frac{F_{\text{in,s}}T_{\text{in,s}}}{Ah_{\text{s}}} - \frac{F_{\text{in,s}}T_{\text{s}}}{Ah_{\text{s}}} + \frac{UA_{\text{t}}T_{\text{st,s}}}{A\rho c_{\text{p}}h_{\text{s}}} - \frac{UA_{\text{t}}T_{\text{s}}}{A\rho c_{\text{p}}h_{\text{s}}}\right] = 0 \qquad \text{(A.24)}$$

Subtracting Eq. (A.24) from Eq. (A.23) and defining deviation variables as before, we finally have

$$\frac{\mathrm{d}\overline{T}}{\mathrm{d}t} = c\overline{T} + d\overline{h} + e\overline{F}_{\text{in}} + f\overline{T}_{\text{st}} + g\overline{T}_{\text{in}} \qquad \text{(A.25)}$$

The set of linear ODEs approximating the dynamic model of the stirred-tank heater is given by

$$\frac{\mathrm{d}\overline{h}}{\mathrm{d}t} = a\overline{F}_{\text{in}} + b\overline{h} \qquad \text{(A.26)}$$

$$\frac{\mathrm{d}\overline{T}}{\mathrm{d}t} = c\overline{T} + d\overline{h} + e\overline{F}_{\text{in}} + f\overline{T}_{\text{st}} + g\overline{T}_{\text{in}} \qquad \text{(A.27)}$$

# Appendix B: Laplace Transformation

## B.1 DEFINITION

The Laplace transform is an operational method that is used for the solution of *linear* differential equations. With the use of Laplace transforms, one can replace operations, such as differentiation and integration with algebraic operations in the complex plane. Consequently, we can transform a linear differential equation into an algebraic equation in terms of a single complex variable. More importantly, one can analyze the dynamic behavior of the process without the need to solve explicitly the system of differential equations.

Let us consider the following terminology:

| | |
|---|---|
| $f(t)$ | a function of time $t$, such that $f(t) = 0$ for $t < 0$ |
| $s$ | a complex variable |
| $L$ | an operational symbol indicating the Laplace transform operation |
| $F(s)$ | Laplace transform of $f(t)$ |

A simple definition of this integral transformation can be expressed as follows:

---

The Laplace transformation $F(s)$ of a function $f(t)$ is given by the integral equation

$$L[f(t)] \equiv F(s) = \int_0^\infty f(t)e^{-st}\,dt \tag{B.1}$$

---

From the definition, we can observe that the Laplace transformation transforms a function from the time domain to the $s$-domain.

## B.2 LAPLACE TRANSFORM OF SOME TYPICAL FUNCTIONS

To understand the mechanics of the Laplace transformation, we consider the following examples.

***Constant function***: Let $f(t) = a$, where $a$ is a real constant.

$$F(s) = L[a] = \int_0^\infty ae^{-st}dt = -\frac{a}{s}[e^{-st}]_0^\infty = -\frac{a}{s}[0 - 1] = \frac{a}{s} \tag{B.2}$$

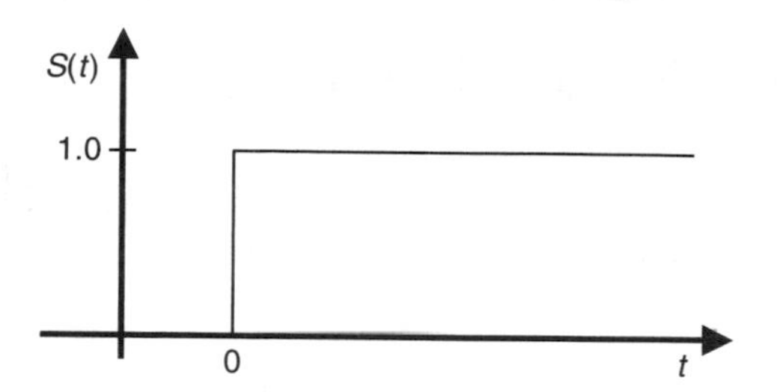


**FIGURE B.1** The unit step function.

In process control, the input variable is typically taken as a *unit step function* (Figure B.1) to analyze the dynamic response of processes. The unit step-function is defined as

$$S(t) = \begin{cases} 0, & t < 0 \\ 1, & t \geq 0 \end{cases} \tag{B.3}$$

One can show that the Laplace transformation of the unit step function is given by

$$L[S(t)] = \frac{1}{s}$$

Another typical input function is a ramp, $R(t) = at$. What would be its Laplace transform?

***Exponential function***: Let $f(t) = \mathrm{e}^{-at}$ for $t \geq 0$, then

$$\begin{aligned} F(s) = L[\mathrm{e}^{-at}] &= \int_0^\infty \mathrm{e}^{-at}\,\mathrm{e}^{-st}\,\mathrm{d}t = \int_0^\infty \mathrm{e}^{-(s+a)t}\,\mathrm{d}t = \frac{1}{s+a}[\mathrm{e}^{-(s+a)t}]_0^\infty \\ &= -\frac{1}{s+a}[0-1] = \frac{1}{s+a} \end{aligned} \tag{B.4}$$

Similarly, one can show that

$$F(s) = L[\mathrm{e}^{at}] = \frac{1}{s-a}$$

***Derivative of a variable***: Let $f(t) = \mathrm{d}^n y(t)/\mathrm{d}t^n$; then, using the definition in Eq. (B.1),

$$F(s) = L\left[\frac{\mathrm{d}^n y(t)}{\mathrm{d}t^n}\right] = s^n y(s) - s^{n-1}y(0) - s^{n-2}y^1(0) - \cdots - y^{n-1}(0) \tag{B.5}$$

In this expression, $y^n(0)$ stands for the $n$th derivative of the function evaluated at time equal to zero (initial condition).

To find the Laplace transform of a $n$th-order derivative we need $n$ initial conditions.

However, we note that for deviation variables, we have $0 = f(0) = f^1(0) = \cdots = f^{n-1}(0)$; thus, Eq. (B.5) simplifies to

$$F(s) = L\left[\frac{d^n y(t)}{dt^n}\right] = s^n y(s)$$

***Integral of a variable*:** Let $f(t) = \int y(t)dt$, then one can show that

$$F(s) = L\left[\int y(t)dt\right] = \frac{1}{s}y(s) \quad \text{(B.6)}$$

Table B.1 contains Laplace transforms of a number of common functions.

## B.3 SOME USEFUL PROPERTIES OF THE LAPLACE TRANSFORM

We have noted earlier that the Laplace operation transforms a function from the time domain to the $s$-domain. Some useful properties of this transformation now follow.

- Laplace transformation is a *linear* operation, therefore it satisfies the superposition principle

$$L[a_1 f_1(t) + a_2 f_2(t)] = a_1 L[f_1(t)] + a_2 L[f_2(t)] \quad \text{(B.7)}$$

Here $a_1$ and $a_2$ are constant parameters.

- Laplace transform of a delayed (shifted) function is

$$L[f(t - t_0)] = e^{-st_0} F(s) \quad \text{(B.8)}$$

Such functions appear frequently in process control, for instance, due to the presence of transport delays in pipes and other equipment.

- We have seen the procedure for moving from the time-domain description of a function to its $s$-domain (or Laplace domain) representation. The reverse procedure is called the Inverse transformation and is a fundamental step in the solution of differential equations. While this procedure is seldom used within the control context, we will briefly explain its principles, as it provides some insight for analysis.

**TABLE B.1**
**Laplace Transform of Some Functions**

| $f(t)$ | $F(s)$ |
|---|---|
| $\delta(t)$ | $1$ |
| $t$ | $\frac{1}{s^2}$ |
| $\frac{t^{n-1}}{(n-1)!}$ | $\frac{1}{s^n}$ |
| $\frac{t^{n-1}}{(n-1)!}e^{-at}$ | $\frac{1}{(s+a)^n}$ |
| $e^{-at}$ | $\frac{1}{(s+a)}$ |
| $\frac{e^{-bt}-e^{-at}}{a-b}$ | $\frac{1}{(s+a)(s+b)}$ |
| $\frac{(z-a)e^{-at}-(z-b)e^{-bt}}{b-a}$ | $\frac{s+z}{(s+a)(s+b)}$ |
| $te^{-at}$ | $\frac{1}{(s+a)^2}$ |
| $(1+(z-a)t)e^{-at}$ | $\frac{s+z}{(s+a)^2}$ |
| $\frac{1}{a}\sin(at)$ | $\frac{1}{s^2+a^2}$ |
| $\cos(at)$ | $\frac{s}{s^2+a^2}$ |

If $F(s)$ is the Laplace transformation of $f(t)$, then the inverse Laplace transformation is denoted as

$$L^{-1}[F(s)] = f(t)$$

## Example B.1

Solve the differential equation given by

$$5\frac{dy}{dt} + 4y = 2, \quad y(0) = 1 \tag{B.9}$$

Let us start by taking Laplace transformation of both sides:

$$L\left[5\frac{dy}{dt} + 4y\right] = L[2] \tag{B.10}$$

By using the principle of superposition and the transformation rule for the derivative, we obtain

$$5(sY(s) - y(0) + 4Y(s) = \frac{2}{s} \tag{B.11}$$

Rearranging and solving for $Y(s)$,

$$Y(s) = \frac{5s + 2}{s(5s + 4)} \tag{B.12}$$

This is the solution of the differential equation in the $s$-domain. Using inverse Laplace transformation, we can compute the solution expressed in the time domain. Consulting Laplace tables, we observe that

$$y(t) = L^{-1}\left[\frac{5s + 2}{s(5s + 4)}\right] = 0.5 + 0.5e^{-0.8t} \tag{B.13}$$

Inverse Laplace transformation rules are typically accessible through the Table of Laplace transformations found in most calculus textbooks. If the function needed is not explicitly available in such a Table, then we obtain an equivalent expression for $F(s)$ in terms of simple functions of $s$, for which the inverse of the Laplace transformation exists. This technique is known as the *partial fraction expansion.*

## B.4 PARTIAL FRACTION EXPANSION

In general terms, the procedure is as follows: Let $F(s)$ be the Laplace transformation of $f(t)$. We can decompose $F(s)$ into $r$ components, such that each component is simple enough to have its inverse transformation available:

$$F(s) = F_1(s) + F_2(s) + \cdots + F_r(s) \tag{B.14}$$

Then, the time-domain function can be obtained easily:

$$f(t) = L^{-1}[F_1(s)] + L^{-1}[F_2(s)] + \cdots + L^{-1}[F_r(s)] \tag{B.15}$$

For example, consider the Laplace-domain solution of a differential equation that yields

$$Y(s) = \frac{(s + z)}{(s + p_1)(s + p_2)} \tag{B.16}$$

where $z$, $p_1$, and $p_2$ are known real constants. This expression can be expanded into the sum of two partial fractions:

$$\frac{(s+z)}{(s+p_1)(s+p_2)} = \frac{\alpha_1}{s+p_1} + \frac{\alpha_2}{s+p_2} \tag{B.17}$$

where $\alpha_1$ and $\alpha_2$ are two yet unspecified coefficients. In general, for every partial fraction, there will be a unique set of $\alpha_i$'s that satisfy the equality. While there are several methods of calculating the unspecified constant, we will focus on the method based on the *Heaviside expansion* here.

In the Heaviside expansion, we multiply both sides of the expression by one of the denominator terms (e.g., $s + p_i$), and then set $s = -p_i$, which causes all terms except the one that contains the denominator term to vanish. Then, one can solve for $\alpha_i$ in a straightforward manner.

### Example B.2

Consider the partial fraction expansion of the expression

$$Y(s) = \frac{1}{s(s+1)(s+2)(s+3)} = \frac{\alpha_1}{s} + \frac{\alpha_2}{s+1} + \frac{\alpha_3}{s+2} + \frac{\alpha_3}{s+3} \tag{B.18}$$

Let us start by multiplying both sides by $s + 1$,

$$(s+\mathbf{1})\frac{1}{s(s+\mathbf{1})(s+2)(s+3)} = \frac{\alpha_1}{s}(s+1) + \frac{\alpha_2}{s+\mathbf{1}}(s+\mathbf{1})$$

$$+ \frac{\alpha_3}{s+2}(s+1) + \frac{\alpha_3}{s+3}(s+1)$$

Note the cancellation of the terms on both sides. Now we shall let $s = -1$, and the resulting expression will be

$$\left.\frac{1}{s(s+2)(s+3)}\right|_{s=-1} = \left.\frac{\alpha_1}{s}(s+1)\right|_{s=-1} + \alpha_2 + \left.\frac{\alpha_3}{s+2}(s+1)\right|_{s=-1} + \left.\frac{\alpha_3}{s+3}(s+1)\right|_{s=-1} \tag{B.19}$$

Observe that the coefficient associated with the denominator term $(s + 1)$ is isolated and the other terms vanish. Hence, we have

$$\frac{1}{(-1)(-1+2)(-1+3)} = -\frac{1}{2} = \alpha_2 \tag{B.20}$$

The other terms can be calculated similarly to yield $\alpha_2 = 1/6$, $\alpha_3 = 1/2$, and $\alpha_4 = -1/6$. Then, the resulting first-order terms can be transformed easily to time domain.

There are two cases where the Heaviside expansion is applied differently. The first is the case where one has repeated denominator terms (multiple roots). For instance, if the term $(s + p)$ appears in the denominator $r$ times then the expansion that incorporates that factor will look like

$$Y(s) = \frac{\alpha_1}{s+p} + \frac{\alpha_2}{(s+p)^2} + \cdots + \frac{\alpha_r}{(s+p)^r} + \cdots \tag{B.21}$$

Note that if we use the same approach as used earlier and multiply both sides by $(s + p)$ and set $s = -p$, then all terms starting with the second term become unbounded. This means that, to evaluate the unknown terms, we have to use a different procedure (note that $\alpha_r$ can be calculated normally). Let us first multiply both sides by the highest order term:

$$\begin{aligned}(s+p)^r Y(s) = {} & \frac{\alpha_1}{s+p}(s+p)^r + \frac{\alpha_2}{(s+p)^2}(s+p)^r \\ & + \cdots + \frac{\alpha_r}{(s+p)^r}(s+p)^r + \cdots\end{aligned} \tag{B.22}$$

Then, we can differentiate this expression with respect to $s$, and then set $s = -p$. By inspection of the above expression, we can guess that the next differentiation will yield $\alpha_{r-1}$, and the others will follow.

The second case involves complex factors. For example, if the denominator is expressed as

$$d(s) = s^2 + d_1 s + d_0 \quad \text{where } \frac{d_1^2}{4} < d_0, \tag{B.23}$$

then it cannot be expanded as the product of two real factors. However, it can be written as the product of two *complex* factors that involve the complex roots of the polynomial,

$$d(s) = s^2 + d_1 s + d_0 = (s + \beta + \mathrm{j}\omega)(s + \beta - \mathrm{j}\omega) \tag{B.24}$$

Hence, the expansion is obtained as

$$Y(s) = \frac{\alpha_1 + \mathrm{j}\alpha_2}{s + \beta + \mathrm{j}\omega} + \frac{\alpha_1 - \mathrm{j}\alpha_2}{s + \beta - \mathrm{j}\omega} \tag{B.25}$$

Note that the unknown coefficients are common to both terms.

**Example B.3**

Consider the following differential equation,

$$\frac{\mathrm{d}x(t)}{\mathrm{d}t} - 12x(t) = u(t), \quad x(0) = 0 \tag{B.26}$$

with a sinusoidal forcing function, $u(t) = \sin 3t$. Find the solution.

Let us start by taking the Laplace transform of both sides:

$$L\left[\frac{dx(t)}{dt} - 12x(t)\right] = L[u(t)]$$

$$sX(s) - 12X(s) = \frac{3}{s^2 + 9} \tag{B.27}$$

Solving for $X(s)$ yields

$$X(s) = \frac{3}{(s^2 + 9)(s - 12)} = \frac{3}{(s + 3j)(s - 3j)(s - 12)} \tag{B.28}$$

Let us use the expansion

$$\frac{3}{(s + 3j)(s - 3j)(s - 12)} = \frac{\alpha_1 + \alpha_2 j}{s + 3j} + \frac{\alpha_1 - \alpha_2 j}{s - 3j} + \frac{\alpha_3}{s - 12} \tag{B.29}$$

If we multiply both sides by $(s + 3j)$, then we have

$$\left.\frac{3}{(s - 3j)(s - 12)}\right|_{s=-3j} = (\alpha_1 + \alpha_2 j) + \left.\frac{\alpha_1 - \alpha_2 j}{s - 3j}(s + 3j)\right|_{s=-3j}$$

$$+ \left.\frac{\alpha_3}{s - 12}(s + 3j)\right|_{s=-3j}$$

Simplifying,

$$\alpha_1 + \alpha_2 j = \frac{3}{-18+72j} = -\frac{1}{102} - \frac{4}{102}j \tag{B.30}$$

Then, the final constant can be found by multiplying both sides by $(s - 12)$, and evaluating at $s = 12$. This gives $\alpha_1 = 1/51$. Hence, we have as the solution,

$$X(s) = \frac{-(1/102) - (4/102)j}{s + 3j} + \frac{-(1/102) + (4/102)j}{s - 3j} + \frac{(1/51)}{s - 12} \tag{B.31}$$

Using the Laplace transformation tables, the final result can be determined as

$$x(t) = -\frac{1}{51}(\cos 3t + 4 \sin 3t) + \frac{1}{51}e^{12t} \tag{B.32}$$

# Appendix C: Matrix Operations

## C.1 VECTORS AND MATRICES

An array $A$ with $n$ rows and $m$ columns is referred to as a $n \times m$ matrix and given as

$$A = \begin{bmatrix} a_{11} & a_{12} & \cdots & a_{1m} \\ a_{21} & a_{22} & \cdots & a_{2m} \\ \vdots & \vdots & \vdots & \vdots \\ a_{n1} & a_{n2} & \cdots & a_{nm} \end{bmatrix} = [a_{ij}] \tag{C.1}$$

The *transpose* of this matrix is a $m \times n$ matrix and defined as

$$A^{\mathrm{T}} = [a_{ji}] = \begin{bmatrix} a_{11} & a_{21} & \cdots & a_{m1} \\ a_{12} & a_{22} & \cdots & a_{m2} \\ \vdots & \vdots & \vdots & \vdots \\ a_{1n} & a_{2n} & \cdots & a_{mn} \end{bmatrix} = [a_{ij}]^{\mathrm{T}} \tag{C.2}$$

which basically represents the transposition of the elements of the original matrix. The addition of matrices can only be performed if the matrices have the same dimensions. The matrix addition is commutative, i.e.,

$$A + B = \begin{bmatrix} a_{11} & a_{12} & \cdots & a_{1m} \\ a_{21} & a_{22} & \cdots & a_{2m} \\ \vdots & \vdots & \vdots & \vdots \\ a_{n1} & a_{n2} & \cdots & a_{nm} \end{bmatrix} + \begin{bmatrix} b_{11} & b_{12} & \cdots & b_{1m} \\ b_{21} & b_{22} & \cdots & b_{2m} \\ \vdots & \vdots & \vdots & \vdots \\ b_{n1} & b_{n2} & \cdots & b_{nm} \end{bmatrix}$$

$$= \begin{bmatrix} a_{11} + b_{11} & a_{12} + b_{12} & \cdots & a_{1m} + b_{1m} \\ a_{21} + b_{21} & a_{22} + b_{22} & \cdots & a_{2m} + b_{2m} \\ \vdots & \vdots & \vdots & \vdots \\ a_{n1} + b_{n1} & a_{n2} + b_{n2} & \cdots & a_{nm} + b_{nm} \end{bmatrix} \tag{C.3}$$

$$= B + A$$

On the other hand, matrix multiplication requires the inner dimensions to be the same, hence, an $n \times m$ matrix $A$ can be multiplied with a $m \times p$ matrix $B$:

$$C = AB$$

$$[c_{ij}] = \left[\sum_{k=1}^{n} a_{ik}b_{kj}\right] \tag{C.4}$$

with $i = 1, 2, \ldots, n$ and $j = 1, 2, \ldots, p$ and the product matrix $C$ is $n \times p$. For example, consider the multiplication of a $2 \times 3$ matrix with a $3 \times 1$ matrix (column vector),

$$A\mathrm{v} = \begin{bmatrix} a_{11} & a_{12} & a_{13} \\ a_{21} & a_{22} & a_{23} \end{bmatrix} \begin{bmatrix} v_1 \\ v_2 \\ v_3 \end{bmatrix} = \begin{bmatrix} a_{11}v_1 + a_{12}v_2 + a_{13}v_3 \\ a_{21}v_1 + a_{22}v_2 + a_{23}v_3 \end{bmatrix} \tag{C.5}$$

The multiplication operation is not commutative, i.e., $AB \neq BA$ even if the inner dimensions agree.

### C.1.1 Special Matrices

There are matrices with special structures. The *identity* matrix, $I$, is defined as a square matrix (e.g., $n \times n$)

$$I_{4\times4} = \begin{bmatrix} 1 & 0 & 0 & 0 \\ 0 & 1 & 0 & 0 \\ 0 & 0 & 1 & 0 \\ 0 & 0 & 0 & 1 \end{bmatrix} \tag{C.6}$$

and has the property, $AI = IA = A$. The identity matrix is also a *diagonal* matrix, i.e., it has nonzero elements only along the diagonal and zeros elsewhere. Another such structurally significant matrix is a triangular matrix and one can have either an upper triangular matrix, or a lower triangular matrix, with structural zeros in the right places. Following are examples of a lower triangular and an upper triangular matrix:

$$L = \begin{bmatrix} 1 & 0 & 0 & 0 \\ 2 & 3 & 0 & 0 \\ -1 & 20 & 1 & 0 \\ 2 & 0.1 & 4 & -1 \end{bmatrix}, \quad U = \begin{bmatrix} 1 & -1 & -2 & 1 \\ 0 & 3 & 0.5 & 4 \\ 0 & 0 & 1 & 15 \\ 0 & 0 & 0 & -2 \end{bmatrix} \tag{C.7}$$

## C.2 DETERMINANT AND INVERSE

The *determinant* of a matrix is a useful operation in solving a set of linear algebraic equations. The determinant is defined for a square matrix only and can be computed using the *minors* of a matrix,

$$\det(A) = |A| = \det\begin{bmatrix} a_{11} & a_{12} & \cdots & a_{1n} \\ a_{21} & a_{22} & \cdots & a_{2n} \\ \vdots & \vdots & \ddots & \vdots \\ a_{n1} & a_{n2} & \cdots & a_{nn} \end{bmatrix}$$

$$= a_{11}\begin{bmatrix} a_{22} & a_{23} & \cdots & a_{2n} \\ a_{31} & a_{32} & \cdots & a_{3n} \\ \vdots & \vdots & \ddots & \vdots \\ a_{n1} & a_{n2} & \cdots & a_{nn} \end{bmatrix} - a_{12}\begin{bmatrix} a_{11} & a_{13} & \cdots & a_{1n} \\ a_{31} & a_{33} & \cdots & a_{3n} \\ \vdots & \vdots & \ddots & \vdots \\ a_{n1} & a_{n2} & \cdots & a_{nn} \end{bmatrix} + \cdots \tag{C.8}$$

$$\pm\, a_{1n}\begin{bmatrix} a_{21} & a_{22} & \cdots & a_{2(n-1)} \\ a_{31} & a_{32} & \cdots & a_{3(n-1)} \\ \vdots & \vdots & \ddots & \vdots \\ a_{n1} & a_{n2} & \cdots & a_{n(n-1)} \end{bmatrix}$$

One can also express the determinant as

$$\det(A) = \sum_{i=1}^{n} a_{ij}C_{ij} \tag{C.9}$$

where $j$ is any row. Here, $C_{ij}$ is called the cofactor of $a_{ij}$ and is defined as $C_{ij} \equiv (-1)^{i+j}M_{ij}$, with $M_{ij}$ being the minor of $A$ formed by eliminating $i$th row and $j$th column.

Following $3 \times 3$ matrix illustrates the computation:

$$\det(A) = \begin{vmatrix} 1 & 2 & 4 \\ 3 & 5 & 7 \\ 8 & 9 & 1 \end{vmatrix} = 1\begin{vmatrix} 5 & 7 \\ 9 & 1 \end{vmatrix} - 2\begin{vmatrix} 3 & 7 \\ 8 & 1 \end{vmatrix} + 4\begin{vmatrix} 3 & 5 \\ 8 & 9 \end{vmatrix}$$

$$= 1(5 - 63) - 2(3 - 56) + 4(27 - 40) \tag{C.10}$$

$$= -4$$

If $\det(A) = 0$, we say that the matrix is *singular*. Singularity implies that at least two columns (or rows) of the matrix are related by a constant multiplicative factor (in other words, they are linearly dependent).

Using the concept of minors, we can also define a very important property of a matrix, namely its *rank*. The rank of a matrix $A$ is the order of the highest nonvanishing minor of $A$. The rank is a measure of the number of independent rows (or columns) of $A$. For an $n \times m$ matrix, the full rank would be $r = \min(n, m)$, and the matrix is said to be *rank-deficient* if its rank is less than $r$. For a square matrix, rank deficiency also implies singularity.

The identity matrix helps us to define the concept of inverse for a square matrix,

$$AA^{-1} = A^{-1}A = I \tag{C.11}$$

The inverse of a nonsingular matrix $A$ (det($A$) $\neq 0$) can be computed as

$$A^{-1} = \frac{C^{\mathrm{T}}}{\det(A)} = \frac{\mathrm{adj}(A)}{\det(A)} \tag{C.12}$$

The inverse does not exist for singular matrices. The transpose of the matrix of cofactors $C$ is referred to as the *adjoint* of $A$. Consider the inverse of a $2 \times 2$ matrix:

$$\begin{aligned} A^{-1} &= \begin{bmatrix} 1 & 2 \\ 3 & -4 \end{bmatrix}^{-1} = \frac{\begin{bmatrix} -4 & -3 \\ -2 & 1 \end{bmatrix}^{\mathrm{T}}}{(-4-6)} \\ &= \frac{\begin{bmatrix} -4 & -2 \\ -3 & 1 \end{bmatrix}}{(-4-6)} = \begin{bmatrix} 0.4 & 0.2 \\ 0.3 & -0.1 \end{bmatrix} \end{aligned} \tag{C.13}$$

One can easily show that

$$AA^{-1} = I = \begin{bmatrix} 1 & 2 \\ 3 & -4 \end{bmatrix} \begin{bmatrix} 0.4 & 0.2 \\ 0.3 & -0.1 \end{bmatrix} = \begin{bmatrix} 1 & 0 \\ 0 & 1 \end{bmatrix} \tag{C.14}$$

## C.3 EIGENVALUE DECOMPOSITION

Eigenvalues appear in the solution of a linear system of equations and are often referred to as characteristic roots, as they correspond to the solution of the roots of a characteristic equation (Chapter 5). Consider a matrix operation on a nonzero vector,

$$A\mathrm{x} = \lambda \mathrm{x} \tag{C.14}$$

The scalar $\lambda$ is the *eigenvalue* of the square ($n \times n$) matrix $A$ corresponding to the (right) *eigenvector* of $A$. Equation (C.14) can be rearranged to yield

$$\begin{aligned} A\mathrm{x} - \lambda \mathrm{x} &= 0 \\ (A - I\lambda)\mathrm{x} &= 0 \end{aligned} \tag{C.15}$$

This holds true if and only if,

$$\det(A - I\lambda) = 0 \tag{C.16}$$

which is the *characteristic equation*. The eigenvalue decomposition of a matrix can be expressed as

$$A = M\Lambda M^{-1} \tag{C.17}$$

where

$$\Lambda = \begin{bmatrix} \lambda_1 & 0 & \cdots & 0 \\ 0 & \lambda_2 & \cdots & 0 \\ \vdots & \vdots & \ddots & \vdots \\ 0 & 0 & \cdots & \lambda_n \end{bmatrix} \tag{C.18}$$

is the eigenvalue matrix containing $n$ eigenvalues of $A$ in the diagonal. The matrix $M$ is referred to as the eigenvector matrix whose columns correspond to the eigenvector $\mathrm{x}_i$ associated with the eigenvalue $\lambda_i$:

$$M = [\mathrm{x}_1 \quad \mathrm{x}_2 \quad \cdots \quad \mathrm{x}_n] \tag{C.19}$$

Consider the $2 \times 2$ matrix,

$$A = \begin{bmatrix} 1 & 2 \\ 3 & -4 \end{bmatrix} \tag{C.20}$$

The eigenvalue decomposition of $A$ can be verified as

$$\begin{aligned} A &= \begin{bmatrix} 1 & 2 \\ 3 & -4 \end{bmatrix} = M\Lambda M^{-1} \\ &= \begin{bmatrix} 0.8944 & -0.3162 \\ 0.4472 & 0.9487 \end{bmatrix} \begin{bmatrix} 2 & 0 \\ 0 & -5 \end{bmatrix} \begin{bmatrix} 0.8944 & -0.3162 \\ 0.4472 & 0.9487 \end{bmatrix}^{-1} \end{aligned} \tag{C.21}$$

## C.4 SINGULAR-VALUE DECOMPOSITION

For matrices that are not square, the eigenvalue decomposition is not defined. For a general $n \times m$ matrix $A$ (with $n > m$), the singular-value decomposition is defined as

$$A = U\Sigma V^{\mathrm{T}} \tag{C.22}$$

where $U$ and $V$ are the left ($n \times n$) and right ($m \times m$) singular vector matrices, respectively, and they are unitary, i.e.,

$$U^{\mathrm{T}}U = V^{\mathrm{T}}V = I \tag{C.23}$$

The columns of $U$ are the left singular vectors, and the columns of $V$ are the right singular vectors. The $n \times m$ matrix $\Sigma$ is the singular-value matrix and has the form

$$\Sigma = \begin{bmatrix} \Sigma_1 & \mathbf{0} \\ \mathbf{0} & \mathbf{0} \end{bmatrix} \quad \text{(C.24)}$$

Note that the bold zeros are block zeros of appropriate dimensions, and the $r \times r$ matrix $\Sigma_1$ contains the singular values in descending order on the diagonal:

$$\Sigma_1 = \begin{bmatrix} \sigma_1 & 0 & \dots & 0 \\ 0 & \sigma_2 & \dots & 0 \\ \vdots & \vdots & \ddots & \vdots \\ 0 & 0 & \dots & \sigma_r \end{bmatrix} \quad \text{(C.25)}$$

The matrix $A$ is said to have $m$ (nonnegative) singular values (because $m < n$) and they are given as $\sigma_1 \geq \sigma_2 \geq \cdots \geq \sigma_r$ and $\sigma_{r+1} = \sigma_{r+2} = \cdots = \sigma_m = 0$. Here, $r$ is the rank of $A$. The eigenvalues and the singular values of a matrix $A$ are related by the following relationship:

$$\sigma_i(A) = \sqrt{\lambda_i(A^T A)} \quad \text{(C.26)}$$

## C.5 VECTOR AND MATRIX NORMS

The concept of norm is used to quantify the *size* or the *magnitude* of a vector or a matrix. A norm is denoted by $\|\cdot\|$, and it can be used for both vectors and matrices. A norm is expected to satisfy the following properties:

1. Norm needs to be nonnegative, i.e., $\|x\| \geq 0$,
2. Norm is positive, i.e., $\|x\| = 0$ if and only if $x = 0$,
3. Norm is homogenous, i.e., $\|ax\| = a\|x\|$ for any scalar $a$.
4. Norm satisfies the following (triangle) inequality:

$$\|x + y\| \leq \|x\| + \|y\| \quad \text{(C.27)}$$

Consider a vector,

$$x = [x_1 \quad x_2 \quad \dots \quad x_n]^T$$

The most common norm is the Euclidean norm, also known as the 2-norm:

$$\|x\|_2 \equiv \sqrt{\sum_{i=1}^{n} |x_i|^2} \quad \text{(C.28)}$$

The 2-norm gives the ordinary distance from the origin to the point **x** (note that an $n$-dimensional vector represents a point in the $n$-dimensional space), a consequence of the Pythagorean theorem. It can be generalized to the $p$-norm, given as

$$\|x\|_p \equiv \left(\sum_{i=1}^{n} |x_i|^p\right)^{1/p} \tag{C.29}$$

The ∞-norm gives the largest element magnitude in the vector,

$$\|x\|_\infty \equiv \max_i |x_i| \tag{C.30}$$

For example, for a vector x = [1  2  3  4], we have the following norms:

$$\|x\|_2 = \sqrt{1^2 + 2^2 + 3^2 + 4^2} = \sqrt{30} = 5.4772$$

$$\|x\|_\infty = 4$$

For matrices, there is an additional property,

5. A matrix norm satisfies the multiplicative property,

$$\|A \cdot B\| \leq \|A\| \cdot \|B\| \tag{C.31}$$

The Euclidean norm for a matrix (sometimes also referred to as the Frobenius norm) is analogous to the vector 2-norm,

$$\|A\|_2 \equiv \sqrt{\sum_{i,j=1}^{n,m} |a_{ij}|^2} = \sqrt{\mathrm{trace}(A^T A)} \tag{C.32}$$

where the trace (tr) operator is the sum of the diagonal elements of the matrix.

There is a special category of matrix norms called the *induced* norms, which provide information on the magnitude amplification that a signal experiences as it is acted on by a matrix. Consider a signal,

$$y = Ax$$

The gain or the amplification of the matrix $A$ is the ratio $\|y\|/\|x\|$. The induced $p$-norm is defined as

$$\|A\|_{ip} \equiv \max_{x \neq 0} \frac{\|Ax\|_p}{\|x\|_p} \tag{C.33}$$

This quantity clearly provides the amplifying power of the matrix. Other matrix norms can also be defined. An important matrix norm is the singular-value norm,

$$\|A\|_{i2} \equiv \sigma_{\max}(A) \tag{C.34}$$

where $\sigma_{\max}(A)$ is the maximum singular value of $A$.

# Appendix D: Basic Statistics

## D.1 SAMPLE MEAN AND STANDARD DEVIATION

In modeling and control of dynamic processes, we often deal with time-series data collected from the process through various measuring devices. In describing the behavior of a time series, an important characteristic is the central tendency of the data. A simple measure of this central tendency is the arithmetic average of the data samples or the *sample mean*. Considering a sequence of $n$ samples, $x_1, x_2, \ldots, x_n$ (or $x(1), x(2), \ldots, x(n)$), we can define the sample mean as

$$\bar{x} = \frac{\sum_{i=1}^{n} x_i}{n} \tag{D.1}$$

One can also define a population mean that captures the central tendency of all possible outcomes (observations) in a population. For a population with finite number of outcomes $N$, the *population mean* is similarly defined as

$$\mu = \frac{\sum_{i=1}^{N} x_i}{N} \tag{D.2}$$

The sample mean represents an estimate of the population mean.

The location of the sample mean does not adequately describe the tendencies present in a data set. The variability associated with a data set can be measured by using the *sample variance* or the *sample standard deviation*. The sample variance represents the *spread* (or the dispersion) of the data set and is defined as

$$s^2 = \frac{\sum_{i=1}^{n} (x_i - \bar{x})^2}{n - 1} \tag{D.3a}$$

The sample standard deviation is simply the square root of the sample variance, i.e., $s$. Similarly, the population variance is given by

$$\sigma^2 = \frac{\sum_{i=1}^{N} (x_i - \mu)^2}{N} \tag{D.3b}$$

and the population standard deviation would be $\sigma$. It is noted that the denominator in sample variance is $n - 1$ while for the population variance, it is $N$. In practice, the observations tend to be closer to their sample average than the population mean. If $n$ is used instead of $n - 1$, we tend to underestimate the sample variance. Thus, for consistency, we use $n - 1$ to get the sample variation.

The standard deviation has the same units as the original data, therefore it is an estimate of the spread of the data (as opposed to the variance).

## D.2 RANDOM VARIABLES AND PROBABILITY DISTRIBUTIONS

We can view each observation as being an outcome of a random experiment, thus, the variable associated with that outcome is referred to as a *random variable.* Considering that we normally have a finite number of observations, we often talk about *discrete* random variables.

The probability of an observation in a data set is closely associated with the frequency of occurrence of the value of that observation when other observations are made. In other words, the concept of probability helps us to quantify the likelihood that a specific outcome of an experiment will occur.

### Example D.1

Consider that a random variable $X$ is observed through a number of measurements and the observations are listed in Table D.1.

The probability of occurrence of a specific value of the random variable can be quantified as follows

$$P(X = x) = \frac{\text{number of occurences of } X = x}{\text{total number of observations}}$$

The results are depicted in Table D.2. Indeed, the probability observing a value of $X = 4$ is 0.25 or 25%. Note that the probabilities add up to 1.

The *probability distribution* of a random variable $X$ is a description of the set of possible values of the variable along with the probability of each possible

**TABLE D.1**
**The Observations of a Random Variable**

| Observations | Variable Value |
|---|---|
| 1 | 5 |
| 2 | 4 |
| 3 | 5 |
| 4 | 6 |
| 5 | 3 |
| 6 | 4 |
| 7 | 5 |
| 8 | 5 |

**TABLE D.2**
**The Probability of Occurrence of the Variable**

| Variable Value $X$ | Probability |
|---|---|
| 5 | 0.5 |
| 4 | 0.250 |
| 6 | 0.125 |
| 3 | 0.125 |

outcome. This is a very useful summary of the data set, offering the opportunity to model the behavior of the data set using well-defined functions.

We can define a *probability mass function* (pmf), $f_X(x)$, as the set of possible values of the random variable $X$, i.e., $P(X = x)$, in the interval [0, 1]. Furthermore, the *cumulative distribution function* (cdf) can be defined as

$$F_X(x) = P(X \leq x) = \sum_{x_i \leq x} f(x_i) \tag{D.4}$$

This definition helps us offer another version of the mean or expected value of a random variable as

$$\mu_X = E(X) = \sum_x xf(x) \tag{D.5}$$

Accordingly, the variance can be expressed as

$$\sigma_X^2 = E(X - \mu_X)^2 = \sum_x (x - \mu_x)^2 f_X(x) \tag{D.6}$$

The standard deviation would again be the positive square root of $\sigma_X^2$, i.e., $\sigma_X$. These definitions that are based on weighted averages differ from the sample mean and the variance.

## D.3 GAUSSIAN (NORMAL) DISTRIBUTION

So far we have considered discrete random variables. If the range of a random variable $X$ contains an interval of real numbers, then $X$ is considered a *continuous* random variable. Use of continuous random variables, despite the fact that measurements may be finite, facilitates the definition of models if the resolution of the sample scale is high.

Now, the probability distributions can be expressed as a continuous function (indeed, an area) through the pdf. For instance, the probability of a variable in an interval would be given as

$$P(x_1 \leq X \leq x_2) = \int_{x_1}^{x_2} f_x(\xi)\, d\xi \tag{D.7}$$

The most common distribution for modeling random experiments is the Gaussian, or the normal distribution. Many observed variables in nature are in fact the sum of many independent random effects, and are very well approximated by a Gaussian random variable. The *Gaussian* distribution plays a fundamental role in probability theory and statistics. One reason is that the sum of independent observations tends to this distribution rather quickly in practice. With broad conditions this can be proved mathematically, and the result is called the *Central Limit Theorem.*

---

*The central limit theorem* states that the observed measurements that arise from the cumulative effect of many small but unidentified independent sources should closely resemble the normal distribution.

---

A random variable $X$ with a pdf,

$$f_X(x;\mu,\sigma) = \frac{1}{\sqrt{2\pi}\sigma} e^{-(x-\mu)^2/2\sigma^2} \quad \text{(D.8)}$$

has a *normal distribution* with the expected value (mean) of $\mu$ and the variance of $\sigma^2$. We will denote the normal distribution as $N(\mu,\sigma)$. Figure D.1a illustrates normal pdfs for various realizations of the model parameters.

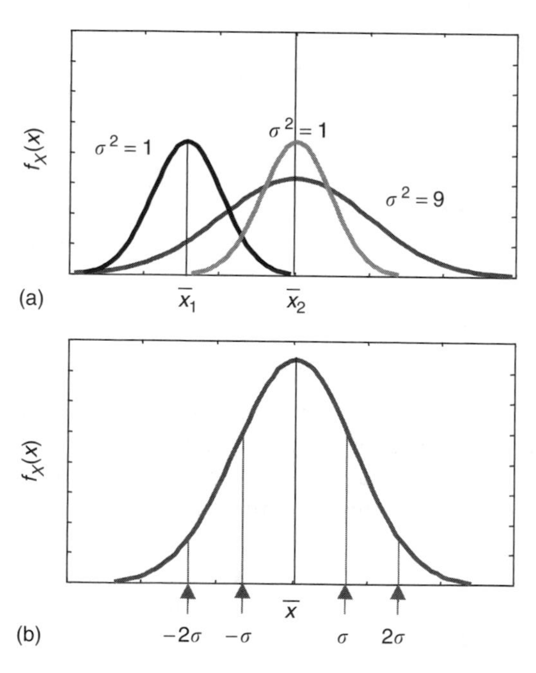


**FIGURE D.1** Normal distribution plots. (a) different means and variances, (b) location of 1- and 2-sigma deviations.

From Figure D.1b, we can also observe that if the data set follows a normal distribution, then 68.27% of the values lie within one standard deviation (SD) of the mean (on either side). This is a very useful result and can be extended to larger spreads as well:

$$P(\mu - 2\sigma < X < \mu + 2\sigma) = 0.9545$$
$$P(\mu - 3\sigma < X < \mu + 3\sigma) = 0.9973 \qquad \text{(D.9)}$$

## D.4 CORRELATIONS AND COVARIANCE

When we deal with two or more random variables, it would be useful to characterize how they vary together. An effective measure of this relationship is the *covariance*. Let us start by defining a *joint* pdf, first. For two random variables $X_1$ and $X_2$, the joint pdf, denoted as $f_{X_1X_2}(x_1, x_2)$, is used to determine the probability that $(X_1, X_2)$ take values in a region $R$ by the double integral of $f_{X_1X_2}(x_1, x_2)$ over the region $R$, with the property,

$$\int_{-\infty}^{\infty}\int_{-\infty}^{\infty} f_{X_1X_2}(x_1, x_2)\, \mathrm{d}x_1\, \mathrm{d}x_2 = 1 \qquad \text{(D.10)}$$

The *covariance* measures the linear association between variables, and it can be defined as

$$S_{X_1X_2} = E[(X_1 - \bar{x}_1)(X_2 - \bar{x}_2)] = \sum_R (x_1 - \bar{x}_1)(x_2 - \bar{x}_2) f_{X_1X_2}(x_1, x_2) \qquad \text{(D.11)}$$

Another measure of relationship between two variables is the *correlation* and is defined as

$$\rho_{X_1X_2} = \frac{S_{X_1X_2}}{S_{X_1}S_{X_2}} \propto [-1,1] \qquad \text{(D.12)}$$

If two variables have a nonzero correlation, they are said to be correlated.

If we consider that we have a data matrix $\mathbf{X}$, then the sample covariance matrix $S$ for an $n$-dimensional problem is defined as follows,

$$S = \text{cov}(\mathbf{X}) = N^2(\mathbf{X}^{\mathrm{T}}\mathbf{X})$$

and

$$S = \begin{bmatrix} s_1^2 & s_{12} & \cdots & s_{1n} \\ s_{12} & s_2^2 & \cdots & s_{2n} \\ \vdots & \vdots & \ddots & \vdots \\ s_{1n} & s_{2n} & \cdots & s_n^2 \end{bmatrix} \qquad \text{(D.13)}$$

Each element of this matrix is given by

$$s_{ij} = \frac{n\sum x_{ik}x_{jk} - \sum x_{ik}\sum x_{jk}}{[n(n-1)]} \tag{D.14}$$

where $x_{ik}$ is the $(i, k)$ element of the data matrix $\mathbf{X}$, $s_i^2$ the population variance of the $i$th variable and $s_{ij}$ the covariance between the $i$th and the $j$th variables. We note that if the variance is used instead of the sample variance in Eq. (D.13), one obtains the covariance matrix denoted by $\Sigma$.

## D.5 $\chi^2$ AND $F$ DISTRIBUTIONS

Let $X_1, X_2, \ldots, X_n$ be normally and independently distributed random variables with zero mean ($\mu = 0$) and unit variance ($\sigma^2 = 1$). Then, the random variable

$$X = X_1^2 + X_2^2 + \cdots + X_n^2 \tag{D.15}$$

has the pdf

$$f(x) = \frac{1}{2^{n/2}\Gamma(n/2)} x^{(n/2)-1} e^{-x/2} \tag{D.16}$$

where $x > 0$ and the $\Gamma$ function is defined as

$$\Gamma(r) = \int_0^\infty x^{r-1} e^{-x}\, dx \quad \text{for } r > 0$$

The random variable $X$ is said to follow the $\chi^2$ distribution with $n$ degrees of freedom, which is denoted by $\chi_n^2$.

The $F$ distribution is a very important distribution in statistics and the random variable $F$ is defined to be the ratio of two independent $\chi^2$ random variables, where each variable is divided by its number of degrees of freedom:

$$F = \frac{X/n}{Y/m} \tag{D.17}$$

The random variable $F$ has the pdf,

$$f(x) = \frac{\Gamma((n+m)/2)(n/m)^{n/2} x^{(n/2)-1}}{\Gamma(n/2)\Gamma(m/2)((n/m)x + 1)^{(n+m)/2}} \tag{D.18}$$

where $0 < x < \infty$. Thus, the random variable $F$ is said to follow the $F$ distribution with $n$ and $m$ degrees of freedom and is denoted by $F_{n,m}$.

The percentage points of the $\chi^2$ distribution and the $F$ distribution as a function of their degrees of freedom can be readily found in standard statistics textbooks.[1]

**TABLE D.3**
**$F$ Distribution Table for $F_{n,m}(\alpha = 0.10)$**

| m/n | 1 | 2 | 3 | 4 | 5 | 6 | 7 | 8 | 9 | 10 | ∞ |
|---|---|---|---|---|---|---|---|---|---|---|---|
| 1 | 39.863 | 49.50 | 53.593 | 55.832 | 57.24 | 58.204 | 58.906 | 59.44 | 5986 | 60.195 | 63.328 |
| 2 | 8.526 | 9.000 | 9.162 | 9.243 | 9.293 | 9.326 | 9.349 | 9.367 | 9.381 | 9.392 | 9.491 |
| 3 | 5.538 | 5.462 | 5.391 | 5.285 | 5.266 | 5.252 | 5.240 | 5.230 | 5.216 | 5.200 | 5.134 |
| 4 | 4.545 | 4.325 | 4.191 | 4.107 | 4.051 | 4.010 | 3.979 | 3.955 | 3.936 | 3.920 | 3.761 |
| 5 | 4.060 | 3.78 | 3.620 | 3.520 | 3.453 | 3.405 | 3.368 | 3.339 | 3.316 | 3.297 | 3.105 |
| 6 | 3.776 | 3.463 | 3.289 | 3.181 | 3.108 | 3.055 | 3.015 | 2.983 | 2.958 | 2.937 | 2.722 |
| 7 | 3.589 | 3.257 | 3.074 | 2.961 | 2.883 | 2.827 | 2.785 | 2.752 | 2.725 | 2.703 | 2.471 |
| 8 | 3.458 | 3.113 | 2.924 | 2.806 | 2.727 | 2.668 | 2.624 | 2.589 | 2.561 | 2.538 | 2.293 |
| 9 | 3.360 | 3.007 | 2.813 | 2.693 | 2.611 | 2.551 | 2.505 | 2.469 | 2.440 | 2.416 | 2.159 |
| 10 | 3.285 | 2.924 | 2.728 | 2.605 | 2.522 | 2.461 | 2.414 | 2.377 | 2.347 | 2.323 | 2.055 |
| ∞ | 2.706 | 2.303 | 2.084 | 1.945 | 1.847 | 1.774 | 1.717 | 1.670 | 1.632 | 1.599 | 1.000 |

**TABLE D.4**
**$\chi^2$ Distribution Table**

| $n\backslash\alpha$ | 0.995 | 0.990 | 0.950 | 0.90 | 0.10 | 0.050 | 0.005 |
|---|---|---|---|---|---|---|---|
| 1 | 0.00004 | 0.00016 | 0.0039 | 0.0158 | 2.7055 | 3.8415 | 7.879 |
| 2 | 0.01003 | 0.0201 | 0.103 | 0.211 | 4.605 | 5.991 | 10.597 |
| 3 | 0.072 | 0.115 | 0.352 | 0.584 | 6.251 | 7.815 | 12.383 |
| 4 | 0.207 | 0.297 | 0.711 | 1.064 | 7.779 | 9.488 | 14.860 |
| 5 | 0.412 | 0.544 | 1.145 | 1.610 | 9.236 | 11.071 | 16.750 |
| 6 | 0.676 | 0.872 | 1.635 | 2.204 | 10.645 | 12.592 | 18.548 |
| 7 | 0.989 | 1.239 | 2.167 | 2.833 | 12.017 | 14.067 | 20.278 |
| 8 | 1.344 | 1.646 | 2.733 | 3.490 | 13.362 | 15.507 | 21.955 |
| 9 | 1.735 | 2.088 | 3.325 | 4.168 | 14.684 | 16.919 | 23.589 |
| 10 | 2.156 | 2.558 | 3.940 | 4.865 | 15.987 | 18.307 | 25.188 |

A set of $F$ distribution and $\chi^2$ distribution tables is summarized in Tables D.3 and D.4, respectively. These distributions have found a critical role in hypotheses testing about the variance of two independent normally distributed populations.

## REFERENCES


1. Montgomery, D.C. and Runger, G.C., *Applied Statistics and Probability for Engineers*, Wiley, New York, 1994.

# Index

## D

# G



# H

## I



## J



## K



## L

## M

# P

## Q



## R

# S

## T

## U



## V



## W



## X



## Z